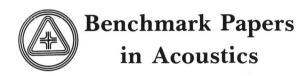

Benchmark Papers
in Acoustics

Series Editor: R. Bruce Lindsay
Brown University

Published Volumes and Volumes in Preparation

Benchmark Papers
in Acoustics / 5

———— A *BENCHMARK* ® Books Series ————

MUSICAL ACOUSTICS, PART I
Violin Family Components

Edited by
CARLEEN M. HUTCHINS
Catgut Acoustical Society, Inc.

Published by
Dowden, Hutchinson
& Ross, Inc.
523 Sarah Street, Stroudsburg, Pa. 18360

Distributed by ACADEMIC PRESS

LIBRARY OF CONGRESS CATALOGING IN PUBLICATION DATA

Hutchins, Carleen Maley, comp.
   Musical acoustics.

   (Benchmark papers in acoustics, v. 5)
   CONTENTS: pt. 1. Violin family components.
   1. Music--Acoustics and physics. 2. Violin. I. Ti-
tle.
ML3817.H89   787'.1'01   74-11377
ISBN 0-471-42540-0

Distributed by
ACADEMIC PRESS
A Subsidiary of Harcourt Brace Jovanovich, Publishers

This book is gratefully dedicated to
Morton A. Hutchins
John C. Schelleng
Harriett M. Bartlett
and
the memory of
Frederick A. Saunders
who have helped to make my share
of the work possible.

Editor's Acknowledgments

I should like to thank all those who have made this volume possible:

First, the authors of the papers included—as well as those not included—whose deep interest in the acoustical nature of the violin can be reflected only briefly in such a selection.

Second, the editors and publishers of the various journals who have freely given permission for use of the papers.

And last but not least the members of the Catgut Acoustical Society who have supported me through two years of work in preparation with technical advice, editorial and organizational criticism, secretarial assistance, free copying, understanding, and a sense of humor: John C. Schelleng, Robert E. Fryxell, Lothar Cremer, Virginia Apgar, Harriett M. Bartlett, Martha Taylor, Tania Proskouriakoff, H. Burritt Miller, Morton A. Hutchins, William A. Hutchins, Donald A. Fletcher, Elizabeth McGilvray, and James J. Kalled.

Acknowledgments
and Permissions

ACKNOWLEDGMENTS

THE INDIAN ASSOCIATION FOR THE CULTIVATION OF SCIENCE
Indian Association for the Cultivation of Science Bulletin
  On the Mechanical Theory of the Vibrations of Bowed Strings and of Musical Instruments of the
    Violin Family, with Experimental Verification of the Results: Part I
Proceedings of the Indian Association for the Cultivation of Science
  Experiments with Mechanically-Played Violins

INSTITUTE OF INDUSTRIAL SCIENCE, UNIVERSITY OF TOKYO—*Report of the Institute of
    Industrial Science, University of Tokyo*
  On the Study of Violin and Its Making

PERMISSIONS

The following papers have been reprinted with the permission of the authors and the copyright holders.

ACADÉMIE DES SCIENCES—*Comptes Rendus des Séances de l'Académie des Sciences*
  Sur le chevalet du violoncelle

ACOUSTICAL SOCIETY OF AMERICA—*Journal of the Acoustical Society of America*
  Acoustical Effects of Violin Varnish
  The Bowed String and the Player
  The Mechanical Action of Violins
  Regarding the Sound Quality of Violins and a Scientific Basis for Violin Construction
  The Violin as a Circuit
  The Wolf in the Cello

AKADEMIE DER WISSENSCHAFTEN IN GÖTTINGEN—*Nachrichten der Akademie der Wissenschaften in
    Göttingen: II. Mathematisch–Physikalsiche Klasse*
  Die Geige aus der Sicht des Physikers

AMERICAN STRING TEACHERS ASSOCIATION—*American String Teacher*
  The Hazards of Weather on the Violin

JOHANN AMBROSIUS BARTH, LEIPZIG—*Annalen der Physik*
  Über die innere Reibung und die Strahlungsdämpfung von Geigen

CATGUT ACOUSTICAL SOCIETY—*Catgut Acoustical Society Newsletter*
  The Action of the Soundpost
  The Influence of "Bow Pressure" on the Movement of a Bowed String: Part I
  The Influence of "Bow Pressure" on the Movement of a Bowed String: Part II
  On the Characteristics of Resonance Spruce Wood
  Übertragungseigenschaften des Streichinstrumentenstegs

W. H. FREEMAN AND COMPANY FOR SCIENTIFIC AMERICAN, INC.—*Scientific American*
 The Physics of Violins

S. HIRZEL VERLAG, STUTTGART—*Acustica*
 Contribution à l'étude des cordes du violoncelle
 Sur les phénomènes transitoires des cordes vibrantes

MARTINUS NIJHOFF B.V.—*Physica*
 The Vibrations of the Violin Bridge

THE PHYSICAL SOCIETY OF JAPAN—*Journal of the Physical Society of Japan*
 The Vibrational Properties of Wood I
 The Vibrational Properties of Wood II

SOCIETA ITALIANA DI FISICA—*Il Nuovo Cimento*
 Misura dell'attrito interno e delle constanti elastiche del legno

SPRINGER-VERLAG Berlin, Heidelberg, New York
 Holz als Roh- und Werkstoff
 Die Bestimmung der Schubmoduln des Fichtenholzes
 Zeitschrift für Physik
 Über die innere Reibung von Holz

TAYLOR & FRANCIS LTD.—*Philosophical Magazine*
 On the "Wolf-Note" in Bowed String Instruments

Series Editor's Preface

The Benchmark Papers in Acoustics constitute a series of volumes that make available to the reader in carefully organized form important papers in all branches of acoustics. The literature of acoustics is vast in extent and much of it, particularly the earlier part, is inaccessible to the average acoustical scientist and engineer. These volumes aim to provide a practical introduction to this literature, since each volume offers an expert's selection of the seminal papers in a given branch of the subject — that is, those papers which have significantly influenced the development of that branch in a certain direction and introduced concepts and methods that possess basic utility in modern acoustics as a whole. Each volume provides a convenient and economical summary of results as well as a foundation for further study for both the person familiar with the field and the person who wishes to become acquainted with it.

Each volume has been organized and edited by an authority in the area to which it pertains. In each volume an editorial introduction is provided that summarizes

these volumes will serve as a working library of the most important technical literature in acoustics of value to students and research workers.

The present volume, *Musical Acoustics, Part I: Violin Family Components,* has been edited by Carleen M. Hutchins, Permanent Secretary of the Catgut Acoustical Society, Montclair, New Jersey. Through its well-chosen articles it provides a unique survey of the acoustical properties of the violin and related string instruments. It constitutes the first of two Benchmark volumes to be devoted to the violin family. The author, long a student and maker of violins, is an internationally recognized authority on this type of musical instrument. The nature of the book is discussed in greater detail in the editor's Introduction.

R. Bruce Lindsay

Parts of the Violin

English	German	French	Italian
back, bottom	Boden	fond	fondo
bassbar	Bassbalken	chaîne	catena
block	Eckklotz	tasseau	tassello
bow	Bogen	archet	archetto
bridge	Steg	chevalet	ponticello
f-hole	f-Loch	ouie	foro armonico (ff)
fingerboard	Griffbrett	touche	tastiera
lining	Reifchen	contre-éclisse	controfascia
maple	Ahorn	érable	acero
neck	Hals	manche	manico
purfling	Adern	filetage	filetti
rib	Zarge	éclisse	fasce
scroll	Schnecke	volute	riccio
soundpost	Stimme	âme	anima
spruce	Fichte	sapin	abete
string	Saite	corde	corda
table, top	Decke	table	piano armonico
violin	Geige	violon	violino

Contents

II. THE BOWED STRING

III. THE BRIDGE

IV. THE SOUNDPOST

V. WOOD FOR THE VIOLIN FAMILY

VI. VARNISH

Contents by Author

Introduction

Technical research has revealed a great deal about the acoustical behavior of the violin and other stringed instruments, but much remains obscure. Even the most sensitive and sophisticated measuring equipment and techniques available have not identified beyond question all the complicated vibrational characteristics that are built into the finest concert violins. It is now possible, however, to measure and explain many differences between mediocre and excellent violins, and to use known resonance characteristics to produce instruments of high musical quality. But the reasons for the differences, real and alleged, between "excellent" and "superb" instruments remain open to question.

The violin as a musical instrument must be studied in context in order to be comprehended fully. Not only the instrument itself, but the player, the style of music being played, the acoustics of the environment, and the musical taste and background of the listeners are involved in the final evaluation. Understandably, therefore, violin makers and musicians who have traditionally coped with all these variables intuitively have been sceptical of quantitative measurements that take into account only one or two factors without regard for others. Analysis alone is not enough.

To progress beyond the present stage in stringed instrument research, there must be an underlying foundation of information and interchange of ideas that includes the total relevant situation so that specifically focused research efforts can be related to a consistent framework. This will involve creative thinking and coordination in disciplines as widely separated as psychoacoustics, musical composition and performance, materials research, and vibration analysis. Herein lies an unprecedented opportunity to apply modern systems theory to an age-old device that is one of the unique products of human ingenuity and a pleasurable sensory experience.

Historical Development of the Violin

In its early development during the sixteenth and seventeenth centuries, the violin was used along with vocal music to carry certain parts (as were members of the viol family). For a time it was used by traveling or street musicians and was considered a vulgar instrument. Gradually it came into favor and surpassed the viols, particularly in the church and concert music of the day. With its development into the instrument we know today by such early makers as Gasparo da Salo (1542–1609) and the Amati family (mid-sixteenth century), the violin came to be a used as a solo instrument and in combination with other members of the violin family. The character of the music, the styles of playing, and the uses to which the violin was put in the Baroque and Renaissance periods were vastly different from the musical practices and concepts of today.

In studying the acoustics of the early concert halls of Europe, Fritz Winckel (1907–   ) (1) has suggested that certain instruments may have been developed to suit the size and acoustical characteristics of the places in which they were to be played. Apparently the highly arched, sweet-toned violins of Jacobus Stainer (1621–1683) and the Amati were preferred in the seventeenth century to the flatter and more powerful instruments of Antonio Stradivari and Giuseppe Guarneri of the early eighteenth century.

Another factor that must be taken into account was the effect on the violins of the gradual rise in frequency of the musical A from around 425 Hz in the early eighteenth century to an A of 440 Hz (and slightly higher), which is in use today. This rise apparently created a demand for more power from the violins, for we find that practically all those made before 1800 have had their necks lengthened and the angle of neck to body increased. In doing this the violin makers skillfully grafted the original scroll-and-peg box onto the longer neck, and joined the neck into the upper block, instead of fastening it with nails as Stradivari and other makers did originally.

The effect of these changes—increased frequency and string length, and a more acute angle of strings at the bridge—was to increase the load on the violin top sufficiently so as to require a heavier bass bar (see the diagram in Paper 1, p. 14). This set of changes poses the interesting question as to whether the violins considered "best" today were those most prized when they were made. There is no question, however, that it is the more powerful violins with flatter overall plate archings, such as those made by Stradivari and Guarneri, that are in greatest demand today to meet the needs of larger concert halls and to compete with the greatly enhanced power of modern wind and percussion instruments. At the same time there is a great concern that the much-prized older instruments are beginning to give out under the demands of modern usage and the ravages of time.

Choosing an Instrument

When correlated with objective tests, comments by string players assisting in the evaluation of experimental instruments have shown that the individual's method and training, as well as the use to which he proposes to put his instrument, create demands that are not best served by instruments of a single standard type. Instruments satisfactory in a large hall differ from those that are satisfactory in a small room. An aspiring soloist needs an instrument that will respond to his lightest bow stroke, that can be played fortissimo with a sense of reserve power, that will not "break" under a variety of bowing techniques, and that will respond when changing from one note to another with clarity and ease throughout its entire range. The tone must "project" and carry well to the back seats of a large hall, yet be sweet and clear in a small room or chamber music hall.

A less ambitious violinist wishing to play chamber music and an occasional solo concert or sonata recital will usually settle for clarity and mellowness of tone, yet will need easy overall response to his method of playing. And, perhaps more importantly, with practice on his chosen instrument he should be able to feel an integral relationship between it and his own conditioned responses.

For the well-trained violinist wishing to play for pure enjoyment, sharing in chamber music sessions with friends, or following the listings of players in the *Directory of the Amateur Chamber Music Players* (2) throughout the world, there is a veritable jungle of instruments from which to choose. From among the violins that the connoisseurs readily admit are not the greatest in the world but are very good, the amateur needs to discover one that will satisfy his special needs. Will it blend with the other instruments in his quartet? Will it respond to his method of playing? Will it be satisfying for practice in a small hotel room and yet be suitable for recitals in the local church and school? Will it change drastically with temperature and humidity, so that it is virtually unplayable at certain extremes? Is its age and condition such that it will be useful for some time to come, or is it nearly worn out and in need of expensive repairs?

For relatively inexpensive student instruments, the tone and playing qualities are largely a matter of chance. Each year, for the last several years, well over 40,000 student violins have been imported to the USA from the German and Japanese violin-making industries. These violins are made to standard dimensions and nicely varnished, but, as one importer remarked, thus far practically no quality control for tone is involved. According to Yankovskii (3), the Russian violin-making industry has developed a method for control of tone quality in the automated mass production of student violins. At the present time, although the technology is available, there is no such development in the USA.

Problems of the Violin Maker

The violin maker has perforce to become as much a judge of people and of playing techniques as of instruments, as he tries to evaluate a musician's needs when playing, his musical tastes and training, and the scope of his playing in order to decide which type of violin will suit a particular musician. In many instances the final adjustments of bridge, soundpost, and force from the strings are so critical that a slight change in any one will drastically alter the response of the entire instrument.

A violin maker knows from long experience that what will work well for one violin or violinist will not necessarily do for another; he has built up a set of intuitive standards and responses from his successes as well as from his failures. He knows, for example, that a prospective customer trying a series of instruments will find all of them different from the one he has been playing. Indeed it has become customary for dealers in fine instruments to let a musician borrow a violin, viola, or cello for at least a month so that he can begin to adjust to the new responses needed and really sense the potential of the new instrument.

One famous connoisseur tells the story of a player who owned an unresponsive, weak-toned cello and came looking for a better one. Leaving his own in the shop, he went home with a cello that had been used extensively by a fine concert artist, although it did not have a famous name. He did this reluctantly, insisting that his own sounded much better. At the end of the month of steady playing on the borrowed cello he came back to the shop and compared it with his own, which in no way had been adjusted or changed. The borrowed cello was so much easier to play that he accused the dealer of having altered his own for the worse in order to sell him a new one. In reality it was his own set of responses that had changed, not those of the instrument. Many other examples could be given of the extreme importance of the conditioned responses of a player to a given instrument, whether in playing, teaching, judging, testing, or making.

Evaluating an Instrument

Often an audience or jury of experts is asked to evaluate the tone and playing qualities of a series of violins. This can be a useful method of evaluation, but it has its dangers. To get a fair evaluation of each instrument, more than one player—usually two or three—are needed to demonstrate the musical qualities of a group of instruments, and neither the judges nor the players should know which instrument is being played. It is important to vary the order so that each violin appears at least once in first, second, third, and fourth places. Careful selection of the musical pieces should be made to ensure a wide variety of musical expression and demands on each instrument. In addition, the players themselves should be of varied background and experience. If one player, for example, plays a Stradivarius violin and is used to the type of playing needed to make a Stradivarius speak, another should be expert on a violin of somewhat different playing qualities, such as a Gagliano or a Guarnerius. Then perhaps a third should be a reasonably good violinist but considerably less expert than the other two.

The earliest known use of a jury was recorded by Savart (4) in the evaluation of his trapezoidal violin. Herman Meinel (see Paper 3) used juries of connoisseurs in his evaluation of violins. He avoided vibrato and material of musical interst as distractions from analytical listening; instead, he used scales, broken thirds, arpeggios, and the like. He considered the results significant.

During the thirty years in which Frederick A. Saunders experimented with many aspects of the violin, he conducted many listening tests that he felt had a certain validity, especially in comparison with the more objective tests he was also using. One variable that he found important was that, when a series of violins was heard in the same piece of music with the same player, it was the instrument in second place that tended to get the most votes (5). His challenge for the psychologists and physiologists is only now coming into focus through the work of Boomsliter and Creel (6). Other listener variables that inevitably enter into any jury evaluation of violin tone are fatigue, musical background, room acoustics, and the weather.

A simple way to obtain a relative evaluation of four or five instruments is to hand them one at a time to a violiinist and ask for comments. A good player can usually pick the one he likes with considerable consistency. He will often describe them in a variety of terms that are meaningful to him but difficult or impossible to correlate with objective measurements, since his comments are apt to be colored by factors other than those inherent in the instrument in hand. Is there a similarity in tone and playing qualities to his own violin? Are room acoustics and weather affecting certain instruments more than others? Is the musician himself under stress from physical or emotional problems? Nevertheless, serious flaws can be uncovered in this way.

The Underlying Technical Simplicities of the Violin

In essence, the violin is a set of strings mounted on a wooden box that contains an almost enclosed air space. Its usefulness as a musical instrument depends on the interaction of vibrations in strings, wood, and air. For a delineation of violin structure and names of parts, see the diagram in Paper 1, page 14.

The *strings*, always circular in cross section, ideally would have complete flexibility and uniformity from end to end. The termination at each end, at the nut and the bridge, is sufficiently rigid so that for practical purposes it may be said that the rates of string vibration are determined wholly by their tension, weight, and length—without consideration of motion at the ends due to compliance. In other words, Father Mersenne's (1588–1648) seventeenth-century rules apply with exactness: string frequency is directly proportional to the square root of tension, and inversely proportional to length and square root of mass.

The unique action of the flexible string, the readiness with which beneath the bow it sustains a periodic wave rich in harmonics, is the most characteristic feature of the violin. Richness in harmonics occurs primarily because of two features: (1) speed of travel along the string is almost the same for vibrations of high harmonic frequency as for the lowest, and (2) transitions between "sticking and slipping" are practically discontinuous. The rather simplified explanation of the bowed string—namely, that friction between bow and string pulls the latter as a spring

5

sideways until friction can no longer hold, so that the string snaps backward until it is able to take a fresh hold, and so on again and again—is less than a half-truth. To understand its real action we must look to the wonderful adaptability of the flexible string itself (see p. 143). One component that is apt to be ignored is the rosin on the bow. The friction that controls bowing is not that between the horsehair of the bow and the gut (or other material) of the string but that between the rosin and rosin. The string does not perform properly until it has acquired a layer of rosin from the bow.

A vibrating string itself produces almost no audible sounds, because its diameter, on the order of 1/1,000 of the wavelength, is entirely negligible for the purpose of radiation. Vibrating areas having dimensions more nearly comparable with wavelength are required, hence the need for an additional sound-radiating device. In the violin, and other stringed instruments that are small compared with the longest waves they must project, this sound-radiating device takes the form of a box.

Obviously the design of the violin's *acoustical radiator* cannot imitate the sounding board of the piano because of its necessarily small size. The single sounding board of the piano is effective because its dimensions exceed the length of the air waves radiated throughout much of its musical scale. The upper and lower sides radiate with near independence, so that the radiations from the front and back sides of the soundboard do not cancel each other, even though the two surfaces of the single board move according to an identical pattern and in exactly the same degree. When a single sounding board is smaller than the wavelengths of the sounds coming from it, radiations from the front and back sides of the board tend to cancel each other as observed from a distance. One important function of the box of the violin is to prevent this canceling effect.

In the violin the motions of the top and back plates are different,so that the air displaced by the box is not constant in volume. In its lowest octave this radiation approximates that of an expanding and contracting sphere undergoing the same pulsations in volume, the "simple source" of acoustic theory. This would be true even if the back plate of the violin were stationary, which it is not. In the upper octaves where the wavelengths of sound produced are smaller than the dimensions of the box and no volume change necessarily occurs, "doublet" radiation becomes important.

This volume change of the violin body, so important to the sounds in its lower octaves, is made possible by the asymmetrical interior arrangement of the soundpost and bassbar. When the bow is pulled across the string parallel to the top of the violin, a rocking motion of the bridge is set up so that the force which the two feet of the bridge exert downward tends to be a push–pull action with one foot down and the other up, in exactly opposite phase. If the box of the violin had complete bilateral symmetry, there would be no volume change, for the left foot of the bridge would completely undo the effect of the right ("right foot" always meaning "right" as the violin player sees it). As it is, however, the *soundpost* is inserted beneath the right foot of the bridge between the top and the back so that the top tends to become immobilized at this point. Thus the soundpost functions somewhat like the fulcrum of a rocking lever, stabilizing the motion of the right foot of the bridge, while enhancing the motion of the left foot. This causes important volume changes in the violin body to occur in step with the left foot of the bridge (see Paper 19). This creation of asymmetry in the violin is the chief function of the soundpost, which the French call the "soul" (*l'âme*) of the violin. Its position, the way it is shaped and fitted into place, as well as the wood from which the post is made, can be highly critical to the performance of an instrument.

The *bassbar,* running lengthwise of the top approximately under the string of lowest tuning, tends to keep the vibrations of the upper and lower areas of the top plate in step with the left foot of the bridge. It is glued in such a way as to lend static strength to the thin wood of the top plate, which, together with the soundpost, must support a downward force from the strings of about 16 to 20 pounds in the violin. Its shaping is one of the most critical tasks in the proper thinning or "tuning" of the top plate. The *bridge* serves to transmit vibrations from the strings to the body, or corpus, of the instrument. Its correct proportions and adjustment are matters of great importance.

The *f-holes,* on either side of the bridge, have two chief acoustical functions: (1) to reduce the stiffness of the floor on which the bridge stands, and (2) to form a Helmholtz resonator. The rocking motion of the bridge, in addition to being affected by soundpost and bassbar, is conditioned by the stiffness of the wood between the f-holes, particularly the width and thickness of the wood between the two upper ends of the f-hole openings. The adjustment of this action might be thought of as providing a suitable output "circuit" for the bridge. Although the subject seems not to have been investigated quantitatively, such measurements as have been made verify the expectation that the inner edge of the left f-hole moves with much less restraint than the outer edge.

The second function of the f-holes is to strengthen the sound in the lowest octave of the violin. They are not simply openings to "let the sound out of the box" generally over the frequency range. Together with the walls of the box, the f-holes, or ports, form a Helmholtz resonator (7). Some unsung inventor discovered the usefulness of holes in the body of stringed instruments long before the invention of the somewhat similar modern "reflex" bass of loudspeakers. The air in and about the ports swings in and out against the compression and rarefaction of the air within, thus providing the mass and stiffness necessary to a simple harmonic vibrator. Within a considerable range about the frequency of resonance, radiation due to motion of the air in the f-holes reinforces radiation arising directly at the outside surface of the box. According to simple theory, this enhancement becomes zero at one half-octave below resonance. The frequency of the resonator depends on the volume of air enclosed and the equivalent area of the f-hole openings.

The *top plate* of the violin is almost universally made of spruce, a wood light in weight and lively of vibration, with relatively low internal friction. The *back plate* is usually of maple, although other woods are used, such as pear, sycamore, and beech. Since the grain of the wood of both top and back plates runs lengthwise of the instrument, the elongated violin shape is in harmony with the nature of the wood

7

employed. The wood has a far greater modulus of elasticity (is much stiffer) along the grain than across it, so that in a crude sense bending waves travel across the plate and along it in about the same amount of time. The top and back plates of the violin are arched in both directions. This *arch* serves not only to help support force from the strings, in wood that is in places less than 2 millimeters thick, but also to create characteristic differences in the tone quality of the instrument. Musicians and violin makers are well aware of the differences in timbre between the highly arched violins of Stainer and those of Stradivari and Guarneri in which the arches are much flatter. A great deal has been written on the effect of arching shapes on tone quality, but definitive studies have yet to be made.

The Literature

Between Félix Savart's (1791–1841) hand-operated cogwheel machine (8) for measuring frequency and the modern computer, there is a long history of ingenious experiments and apparatus for probing the secrets of the violin. There has been no steady development of the field but rather islands of activity — first with Savart in the early nineteenth century and then later with Hermann von Helmholtz (1821–1894) and Lord Rayleigh (1842–1919) (9). It was not until the early twentieth century, beginning with the work of C. V. Raman (Papers 7, 8, and 9), that a consistent stream of work began to emerge, largely stimulated by developments in electronic, acoustical, and optical equipment sensitive enough to measure what goes on in the violin.

That contemporary investigators have not yet been able to isolate unequivocally all the essential elements characteristic of an excellent violin is at once a challenge to them and a tribute to the early masters who developed the instrument by purely empirical methods.

The papers included in this volume, as well as those in a companion volume, *Musical Acoustics, Part II: Violin Family Functions* (hereafter referred to as *"Musical Acoustics, Part II"*) represent a selection from the literature of research on the violin family that seems characteristic and interesting, not only from an historical point of view to indicate the variety of equipment and methods used, but also to bring into focus important findings for the modern investigator as he attacks an old problem with new tools. Publication of all the papers that the editor considers valuable is out of the question because of the limitations of space. Selection is necessarily in part a matter of personal judgment; it is nonetheless with regret that many papers have had to be omitted.

References

1. Personal communication, Technical University Berlin.
2. Amateur Chamber Music Players, Inc., Vienna, Va.

*3. Yankovskii, B. A. "Methods for the Objective Appraisal of Violin Tone Quality," *Soviet Phys. Acoust.*, **11**(3), (Jan.–Mar. 1966).

*4. Savart, Félix. *Mémoire sur la construction des instruments à cordes et à archet*, Deterville, Paris, 1819.

*5. Saunders, F. A., "Studies of the Instruments of the Curtis String Quartet," *Overtones* (April 1940), The Curtis Institute of Music, Philadelphia, Pa.

*6. Boomsliter, P. C., and W. Creel. "Research Potentials in Auditory Characteristics of Violin Tone," *J. Acoust. Soc. Amer.*, **51**(6), Pt. 2 (1972).

7. Helmholtz, Hermann von. "Sensations of Tone," selections from an English translation of *Die Lehre von den Tunernpfindungen als physiologische Grundlage für die Theorie der Musik*, 3rd ed., 1–6, 36–49, 49–56 (1895), in *Acoustics: Historical and Philosophical Development*, R. Bruce Lindsay, ed., Dowden, Hutchinson & Ross, Inc., Stroudsburg, Pa., 1973, pp. 324–350.

8. Savart, Félix. "On the Sensitivity of the Ear," translated from Poggendorf's *Ann. Physik. Chemie*, **20**, 290 (1830), in *Acoustics: Historical and Philosophical Development*, R. Bruce Lindsay, ed., Dowden, Hutchinson & Ross, Inc., Stroudsburg, Pa., 1973, pp. 203–209.

9. Strutt, John William (Lord Rayleigh). "On the Theory of Resonance," selection from *Scientific Papers of Lord Rayleigh* (Dover Publications, Inc.), Vol. I, 33–55 (1961), in *Acoustics: Historical and Philosophical Development*, R. Bruce Lindsay, ed., Dowden, Hutchinson & Ross, Inc., Stroudsburg, Pa., 1973, pp. 376–398.

---

*These four papers appear in *Musical Acoustics, Part II*.

I
General Papers in Violin Acoustics

The six papers included in this section have been selected to give an introduction to the work and thinking in violin acoustics, as well as an overview of the research in various parts of the world. Paper 1, "The Physics of Violins," by Carleen M. Hutchins, sketches briefly the history and development of the violin and provides a reference diagram of its various parts with some discussion of the way in which they function. The second paper, "The Mechanical Action of Violins," represents the early work of Frederick A. Saunders in the Cruft Acoustics Laboratory of Harvard University when Leo Beranek and Harry H. Hall were among his assistants. Saunders' other major papers (see p. 27) are included in *Musical Acoustics, Part II*. Paper 3, "Regarding the Sound Quality of Violins and a Scientific Basis for Violin Construction," is in essence a summary of the work of Hermann Meinel during nearly 20 years of investigations on the violin in Germany. Another two of his papers appear in *Musical Acoustics, Part II* (see page 46).

The fourth paper, "On the Study of Violin and Its Making," by Hideo Itokawa and Chihiro Kumagai, indicates some of the research done in Japan following the work of F. A. Saunders and others. Paper 5, "The Violin as a Circuit," also shows the influence of F. A. Saunders, since its author, John C. Schelleng, worked closely with him for several years. This article represents the first application of electrical circuit theory to the function of the violin as a whole.

Finally, Paper 6, "Die Geige aus der Sicht des Physikers," by Lothar Cremer, provides a summary of the investigations on the violin by the research group under the direction of Cremer at the Technical University of Berlin using the advanced technology and mathematics of the 1960s and 1970s.

11

VIOL FAMILY, a somewhat earlier development than the violin family that eventually supplanted it, encompassed a large number of instruments of varying shapes and sizes. This drawing from the *Syntagmatis Musici* of Michael Praetorius (1619) shows three examples of the viola da gamba (*1, 2, 3*), a viola bastarda (*4*) and a viola da braccio (*5*). They have not all been drawn to the same scale.

The Physics of Violins

Modern acoustics is making it possible to account for the exquisite performance of the violins made by the Italian masters. The results promise a further evolution in the instruments of the violin family

by Carleen Maley Hutchins

During the Renaissance there grew up in Italy two new families of musical instruments, both of them stemming from primitive stringed instruments of the Middle Ages such as the rebec and lute. The earlier of the two groups to emerge was the viols; the later, following about a century afterward, was the violins. The violin was not an outgrowth of the viol but a somewhat later development from similar sources, and the two were lively competitors for a long time. Composers wrote distinctive music for each kind of instrument, and each had its virtuoso performers. Eventually the violin family, having a richer and more powerful sound, supplanted the older group—except for the largest and lowest-pitched instrument, which survives as the bass viol.

As this story unfolds it will be clear that viols still have more than mere historical interest. For the moment I shall describe the viols briefly. They came in a variety of shapes and sizes [see illustration on opposite page], most of them having a flat back unlike the beautifully arched back plate of the violins. They had five, six or more strings, more slackly tuned than violin strings and supported on a flatter bridge. Often their finger boards were crossed by gut frets resembling the metal ridges on the finger board of a guitar. Their wooden sounding boards were lighter and more flexible than those of the violin family.

Exactly who invented the violin is not clear. It may have been Andrea Amati, who in any case founded the great Cremona school of violinmakers. Amati died around 1580; within 150 years or so his descendants and their pupils, particularly Antonio Stradivari and Giuseppe Guarneri, had brought the art of violinmaking to such an extraordinarily high level that it is only now that one dares

to dream of equaling or surpassing it. These early masters must have had an open mind toward the little that was known in their time about the physics of sound. Their successors deserve credit for having lovingly preserved an art, but certainly not for advancing a science. In effect they have formed a cult that has been plagued with more peculiar notions and pseudo science than even medicine.

Today the well-developed science of acoustics is applicable to the understanding and making of violins. For the past 30 years or so a handful of interested physicists, chemists, musicians and some people who, like me, began as amateurs have been applying it. In fact, we have organized ourselves informally as the Catgut Acoustical Society. Much of what

has been learned is still empirical, but it is nonetheless interesting and valuable. In this article I shall try to touch on at least the high spots of our studies.

In essence a violin—as well as its larger, deeper-voiced relatives, the viola and the cello (properly the violoncello)—is a set of strings mounted on a wooden box containing an almost closed air space. Some energy from the vibrations induced by drawing a bow across the strings (precious little energy, it turns out) is communicated to the box and the air space, in which are set up corresponding vibrations. These in turn set the air between the instrument and the listener into vibration; in other words, they produce the sound waves that reach his ears. That is the main story. The sound of a violin, putting aside the acoustics of

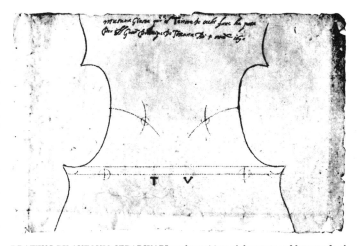

DRAWING BY ANTONIO STRADIVARI marks positions of the upper and lower ends of "f-holes" in a tenor viola. At top he has written: "Exact measurements for the sound holes of the tenor made expressly for the Grand Prince of Tuscany, the 4th day of October, 1690."

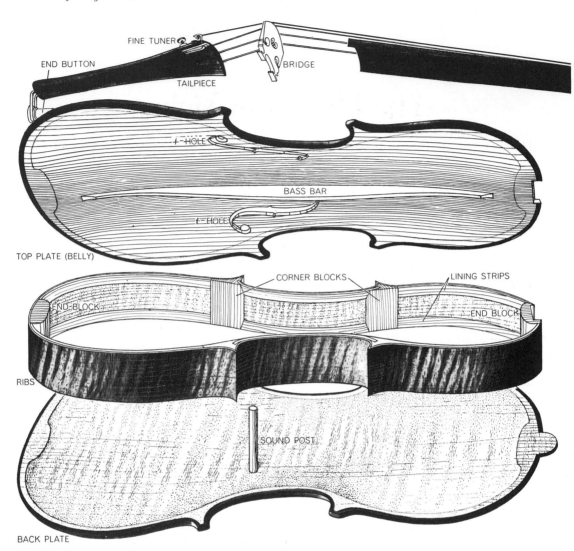

FINE TUNER

END BUTTON

TAILPIECE

BRIDGE

f-HOLE

BASS BAR

f-HOLE

TOP PLATE (BELLY)

CORNER BLOCKS

LINING STRIPS

END BLOCK

END BLOCK

RIBS

SOUND POST

BACK PLATE

ANATOMY OF VIOLIN INSTRUMENTS is essentially the same for the violin, viola and cello. The exploded view of the viola on these two pages shows the top plate, ribs, back plate and devices for stringing at left, the neck, scroll and finger board at top right. Immediately below, a section of the top plate illustrates the bilateral symmetry of the grain of the spruce wood. At bottom

the room in which it is played and the skill of the player, depends on the transfer of vibration from string to sounding box to air.

The Basic Violin

Before getting into this apparently innocent problem, which turns out to be a veritable jungle of unknowns, it is worthwhile to pause for a moment to examine the instrument itself. Violin strings are usually made of metal, pig gut or gut wound with fine silver or aluminum wire. The sounding box consists of a front plate and a back plate, both arched slightly outward to form broad bell-like shapes, and the supporting ribs, or sides. The back plate is carved with chisel, plane and scraper, traditionally from a block of curly maple seasoned for at least 10 years and not kiln-dried. (Pear or sycamore wood are sometimes used.) It can be a single piece or two pieces carefully joined. In thickness the back plate varies from about six millimeters in the center to almost two millimeters just inside the edges (from 1/4 inch to 5/64 inch). The sides are pieces of matching curly maple, thinned down to a millimeter all over, bent into shape and glued to spruce or willow blocks set in the corners and at the forward and rear ends of the plates.

The top plate, usually spruce, is split lengthwise from a log and then joined so that the wood of the outside of the tree is in the center of the top, making the grain bilaterally symmetrical. In thickness the top plate ranges from two to three millimeters, and a pair of beautifully shaped "f-holes" are cut into each

4

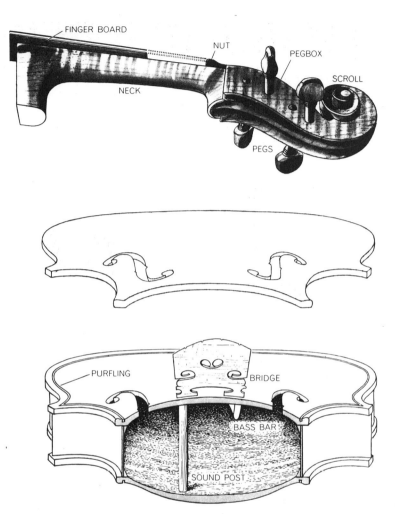

FINGER BOARD

NUT

PEGBOX

SCROLL

NECK

PEGS

PURFLING

BRIDGE

BASS BAR

SOUND POST

right a cross section through the middle of the instrument shows the relative positions of the bridge, bass bar and sound post. The purfling consists of three very narrow strips of wood that are set in a shallow groove around the edge of both the top and back plates.

open question. In a few years we hope to have some answers.

I might anticipate a bit here to mention a point that illustrates the subtlety of some of the problems in understanding the violin. Does the purfling serve any purpose other than decoration? It happens that the wood of the plates underneath the purfling is extremely thin. After years of playing, the glue that holds the purfling strips in their grooves begins to crack, in effect creating a vibrating plate with very thin edges. Frederick A. Saunders, professor emeritus of physics at Harvard University, has suggested that this may be a factor in the improved tone of an instrument that has been played for a long time.

The combined tension of the four strings of a properly tuned violin comes to around 50 pounds. As a result about 20 pounds is directed straight down through the bridge and against the delicate eggshell-like sounding box. To distribute the load and help the top plate withstand the downward component of string tension the viol makers glued to it a strip of wood running lengthwise down the middle. Whether by accident or by a stroke of genius, one of the earlier violinmakers moved the bar to one side so that one foot of the bridge rested above it. The strip, made of spruce, is now called the bass bar, since it is under the foot of the bridge on the side of the string of lowest tuning.

To support the other foot of the bridge there is placed approximately underneath it a vertical post, also made of spruce, called the sound post. It is carefully fitted and held in place by friction between the front and back plates. The acoustical function of the sound post has been a matter of debate for many years. The tone of a violin can be so greatly altered by small changes in the position, tightness and wood quality of the sound post that the French call it the soul (*l'âme*) of the instrument. Removing it altogether makes the violin sound rather like a guitar.

Although the modes of vibration of the plates exhibit great diversity throughout the frequency range, the bridge must always have some rocking motion to receive power from the string. In the important lower half of the range the sound post and the adjacent foot of the bridge have relatively little motion, thus providing in a sense a fulcrum that serves to transfer maximum travel to the bridge foot standing over the bass bar.

This too is getting a little ahead of the story, which begins when a bow is drawn across one or more of the strings of a violin. Vibrating strings have been stud-

side of the plate. All around the outside of each plate, near the edge, is cut a shallow groove in which is inlaid the "purfling," consisting usually of two strips of black-dyed pearwood and a strip of white poplar.

Other materials may be mentioned: curly maple for the neck, ebony for the finger board, rosewood or ebony for the tuning pegs and tailpiece, hard maple for the bridge. The outside of the instrument is treated with filler and varnished. Filler, varnish and glue all contribute to the over-all characteristics of the violin,

but there is no definite evidence to show that 300 years ago any of them was superior to the materials available now. In fact, the Catgut Acoustical Society is working to discover new substances that may be even more effective than the old. But it is a slow, painstaking search.

These were the general specifications for the Cremona fiddles and, with minor variations, for all good instruments since. Whether there is a mysteriously unique virtue in any of the woods or finishes, or whether some other types might not do as well for various purposes, is an

5

15

32.703 41.203 55.000 65.406 73.416 97.999 130.813 146.832 195.998 220.000 261.626 293.665 329.628

C_1 E A C_2 D G C_3 D G A C_4 D E

EADG

ADGC

CGDA

LARGE BASS

SMALL BASS

NEW CELLO

NEW VIOLIN FAMILY will consist of eight instruments, twice the number of the present violin family. Of these the large bass and the treble violin are yet to be constructed; the small bass and the tenor and soprano violins are newly designed instruments; the new cello and vertical viola are rescaled instruments; the violin is the only member of the original family to be included in the new

6

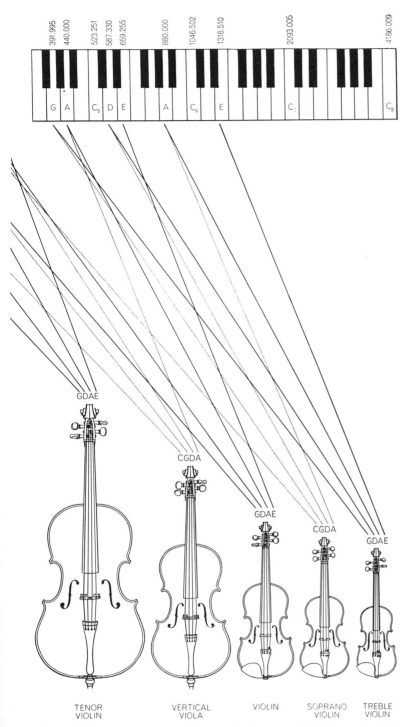

391.995 440.000 523.251 587.330 659.255 880.000 1046.502 1318.510 2093.005 4186.009

G A C₅ D E A C₆ E C₇ C₈

GDAE
CGDA
GDAE
CGDA
GDAE

TENOR VIOLIN
VERTICAL VIOLA
VIOLIN
SOPRANO VIOLIN
TREBLE VIOLIN

group unchanged (*see illustration on next page*). The notes to which the four strings of each instrument are tuned can be read from the piano keyboard and the associated colored lines indicating the strings. The numbers at top show the frequencies in cycles per second.

ied since the time of Pythagoras. An early 19th-century French physicist, Félix Savart, showed that the bowed string has a multitude of harmonics; then the great German physicist Hermann von Helmholtz elucidated the types of vibration that distinguish the bowed string from the plucked string. In our own century the Indian physicist Sir C. V. Raman made an exhaustive investigation of the vibration of bowed violin and cello strings. It would be fair to say that the reaction of a string to a bow is quite thoroughly understood.

In spite of the vigorous vibration of the moving string, the sound from the string alone would be all but inaudible. It has too little surface area to set an appreciable amount of air in motion. Trying to make music with an unamplified string would be like trying to fan oneself with a toothpick. What happens is that some portion of the energy supplied by the player to the bow—perhaps 5 to 10 per cent—is communicated to the wooden body of the violin through the complex motions of the bridge. (Of all the energy that the player feeds into the violin, 1 or 2 per cent emerges as sound. The rest goes off as heat.) The vibrations of the bowed string at any instant include dozens of energetic harmonics with amplitudes falling off as frequency increases. Each of the frequencies present shakes the wooden box—"forces" it to vibrate—at its particular rate. Obviously the amplitude of vibration depends on the strength, or amplitude, of the forcing vibration.

The Resonant Box

If this were all there was to it, matters would be simple, and all tones would be amplified equally. But the wooden structure itself has scores of frequencies at which it tends to vibrate naturally. The coincidence of such a frequency of resonance in the wood with the frequency of a string harmonic will result in an enhanced transfer of energy from string to box and a correspondingly greater amplification of that particular tone. Therefore the actual response of a violin to the playing of various notes is an enormously complex affair, but a good violinist must unconsciously and automatically deal with it and compensate for it every time he plays.

The scientific violinmaker is interested in all of these wood resonances, but he usually finds the resonance of lowest frequency an adequate guide during construction. This is called the main wood resonance. He is also interested in the lowest natural frequency (here only one

7

seems to have any measurable importance) of the enclosed air space, called the main air resonance. Tests show that a good violin usually has its main wood resonance within a whole note of 440 cycles per second: the note A, to which the second highest string of the instrument is tuned.

Some instruments have a "wolf note," almost always at the frequency of the main wood resonance. When this note is played on any string, the tone warbles unsteadily, often breaking by a whole octave somewhat as the voice of an adolescent boy does. The wolf note occurs when the string and the wood form a pair of mechanically coupled circuits; a beating action occurs because energy is cyclically shuttled back and forth between them. Violas and cellos are notori-

ously subject to wolf-note trouble. Even some of the finest cellos have bad wolf notes. It is possible to ameliorate this difficulty in a variety of ways, for example by tuning a length of string between the bridge and the tailpiece to the actual frequency of the wolf note. This absorbs enough energy to control the wolf. So far, however, the ideal method of control has not been found.

Inside the box of the violin is the air chamber, or resonating cavity, which communicates directly to the outside by means of the f-holes in the top of the instrument. As I have said, the enclosed air has so far been found to add measurable resonance to a range of tones surrounding one note on each instrument. The pitch of this main air resonance, or air tone, can be approximately located

by blowing across the f-hole, as one might blow across the top of an empty bottle. (When one f-hole is covered lightly, this pitch is lowered.)

The frequency of the air tone is controlled by the volume of air enclosed by the box of the instrument and the combined area of its f-hole openings. The larger the air volume, the lower the frequency; the larger the f-hole area, the higher the frequency. These two variables can be calculated roughly. I have found that to raise the resonance of the enclosed air a whole tone requires approximately a 20 per cent reduction in air volume or a 59 per cent increase in f-hole area. Anyone looking at the handsomely shaped f-holes of a violin can appreciate that it is not practical to try to raise the frequency of the air

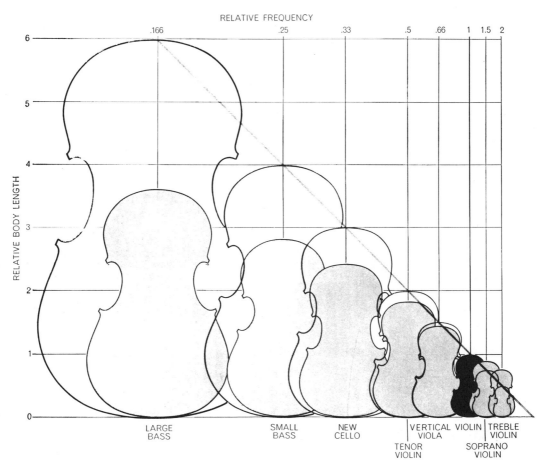

RELATIVE FREQUENCY

THEORETICAL AND ACTUAL SIZES of new instruments are compared. According to theory an instrument of a given relative frequency would have to be of the size represented by the colored outlines. In practice the size is that represented by the light gray areas. The relative frequency of an instrument (e.g., treble violin) is obtained by dividing the frequency of one of its strings (e.g., A) by the frequency of the corresponding string on the violin: 880 divided by 440 equals 2. The conventional viola, cello and double bass (*not shown in illustration*) have relative body lengths (using the violin as unity) of 1.17, 2.13 and 3.09 respectively.

8

18

resonance even a semitone by changing the size of the holes.

A number of workers—particularly Saunders, the late Hermann Backhaus of the Technische Hochschule in Karlsruhe, Hermann Meinel of Berlin, Gioacchino Pasqualini of Rome and E. Rohloff of the University of Greifswald—have developed methods of studying the resonances of violins. One of the most useful is the "loudness curve" originated by Saunders. It is also called the curve of total intensity, because it shows at each measured frequency the combined strengths of all the harmonics.

Loudness Curves

In making a loudness curve the violin is bowed normally, but without vibrato, at semitone intervals over its entire range to produce the loudest tone possible at each note. A General Radio sound-level meter, such as is used to measure levels of applause on television shows, records the loudness of each tone. It often comes as a shock to a musician to discover that his instrument is much louder at certain notes than at others. Try as he will he cannot possibly make them all register at an equally high level on the sound-level meter.

At the right are displayed loudness curves of a few violins, good and poor. In the good one the main wood resonance and the main air resonance fall approximately seven semitones, or a musical fifth, apart. (A fifth is the interval from "do" to "sol" on the diatonic scale. The frequency of each note is in the ratio of three to two for the note below it.) In some poor instruments the main wood and air resonances may be as much as 12 semitones, or an octave, apart (frequency ratio, two to one), giving two areas of strong resonance with a wide range of weak response between. The curve for one poor instrument shows only one area of strong resonance—the air resonance—with the wood contributing virtually nothing in the way of resonant reinforcement. A $5 violin with a curve almost as bad was used by Saunders for some time as his "standard" of badness. When I took the wretched thing apart and balanced it for good tone production, it showed an overall increase of loudness and an even spacing of peaks. At this point it was named Pygmalion. When it was played behind a screen in alternation with an excellent Cremona violin, the two were voted equal in tone by a college music department audience. In fairness it should be added that the skilled musician playing behind the screen was never in any doubt

as to which was the superior instrument.

An octave below the main wood resonance there is almost always another strong peak of loudness that we label "wood prime." It can be called a subharmonic. It is well known in acoustics that if one harmonic of a complex tone is strengthened, the ear will hear an increase in loudness of the note as a whole with a slight change in quality but no change in pitch. By this process the wood peak is strengthened by the tone of the main wood resonance an octave above. The subharmonic of the main wood reso-

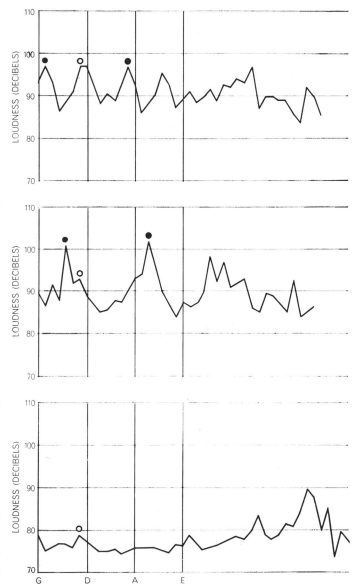

VIOLIN "LOUDNESS CURVES" compare maximum sound levels produced at semitone intervals by a good 1713 Stradivarius (*top*), a poor 250-year-old violin of doubtful origin (*middle*) and a poorer, somewhat older instrument credited to P. Guarneri (*bottom*). Only the first shows desirable spacing and strength of wood (*black dot*), "wood prime" (*gray dot*) and air resonances (*open circle*). Letters at bottom indicate tuning of open strings.

9

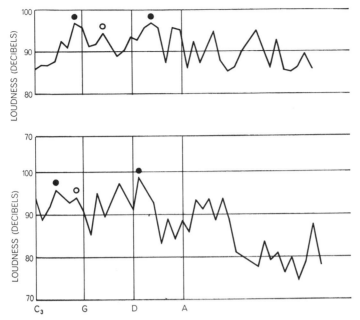

VIOLA LOUDNESS CURVES compare the responses of a conventional (*top*) and a vertical viola (*bottom*), both made by the author. The convention of dots and colored lines used in this illustration and that below is the same as in the illustration on preceding page.

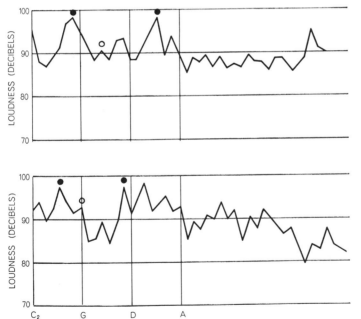

CELLO LOUDNESS CURVES compare the responses of a conventional (*top*) and a new cello (*bottom*), both made by the author. Note that the main wood and air resonances of the new viola and cello are near the two open middle strings. The loudness curves on this and preceding page are based on tests performed by Frederick A. Saunders of Harvard University.

nance benefits the lower tones of the violin, viola and cello.

The subharmonic of the main air resonance does not show on curves of conventional instruments because it falls below the bottom notes of the instruments. Spacing the main wood and air resonances about a half-octave apart spreads these peaks so that the air-tone peak falls nicely in the middle of the octave between the wood resonance and its subharmonic. In hundreds of tests of violins, violas and cellos this arrangement of wood and air resonances emerges as one of the characteristics of the good instruments.

Experimental Instruments

I have built a series of experimental violins and violas to test the effect of moving the frequencies of the main resonances up or down the scale. In a pair of violas of similar pattern with identical f-holes, one was made with sides half an inch high to decrease the air volume; the other had sides two inches high, giving a large air volume. Normally the sides of a viola are about 1¼ inches high, and the air tone is found to be in the range from B to B flat (233 cycles per second) on the G string. In the viola with the smaller air volume the air tone, as expected, moved up the scale to D sharp (300 cycles per second). In the one with the larger air volume the air tone moved downscale near A (220 cycles per second).

In both of these altered violas the normally strong tones of the B to B flat on the G string were missing, because the air resonance was no longer there to reinforce them. Musicians playing the instruments discovered interesting features. Neither one was suitable for playing the two-viola quintets composed by Mozart. The composer had written so well for the normally strong tones of the viola that the oustanding parts lacked their full expressive qualities when the experimental instruments were used. The strong resonance of the air tone was not where musicians expected to find strength, nor where Mozart had counted on it.

The most interesting feature of the two violas was that the thin, shallow instrument had a full, rich tone and a particularly strong, low C string, where the normal viola is notably weak. This was because the air tone had been shifted upscale enough so that its subharmonic came into useful range near 150 cycles per second on the low C string.

The thick viola with the two-inch ribs,

10

on the other hand, had a thin tone, and the lower range of its C string was weak, partly because the air tone had been moved from its normal position. Many musicians playing the two violas in alternation have remarked with astonishment at the full, rich tone of the thin one with the small air volume. Ribs half an inch high are structurally not very practical, but application of the principles involved has made possible the construction of good small violas.

In studying the resonances of violins I have discovered that in the best violins the main wood and air resonances invariably fall within a semitone or two of the frequency of the two open middle strings, the wood resonance corresponding to the higher-tuned string. When the early violinmakers hit on this arrangement, the muses must have been smiling. It is, quite simply, the way in which most good violins have been made ever since.

This is not true of the viola and cello. In these instruments as they are now built the wood and air resonances fall three to four semitones higher with respect to the frequencies of the open middle strings than they do in the violin. The reason is simple enough: the viola and cello are built smaller than optimum size to make them a convenient playing size. As a result the resonances are too far above the lower notes of the instruments, and these suffer in strength and quality. I shall have more to say about this matter later.

Tap Tones

At the moment I should like to consider a different problem. Assuming one knows what the violin should be like when it is finished—where its resonances should fall and so on—how are these aims achieved in the process of construction? How does one make it come out the way one wants it? In addition to careful workmanship and accurate measurements the traditional method of the violinmaker has been to listen to the "tap tones" of the front and back plates.

In the final thinning and graduating of the top and back plates of a violin, the maker traditionally holds the plate near one end in his thumb and forefinger, taps it at various points with a knuckle and listens carefully to determine the pitch of the sounds he hears. These sounds are called the tap tones of the plates. The ability to judge the proper relation of tap tones of the free top and back plates in this manner is an important part of the art of violinmaking. With the ear alone it is extremely difficult to make out the frequencies of these tones, particularly in the case of the top plate, where the complicated structure with f-holes and bass bar creates at least two, and sometimes as many as five, strong natural resonances below 600 cycles per second.

Saunders and I, together with Alvin Hopping of Lake Hopatcong, N.J., have developed a method that makes it possible to determine the tap-tone frequencies in a free plate with considerable

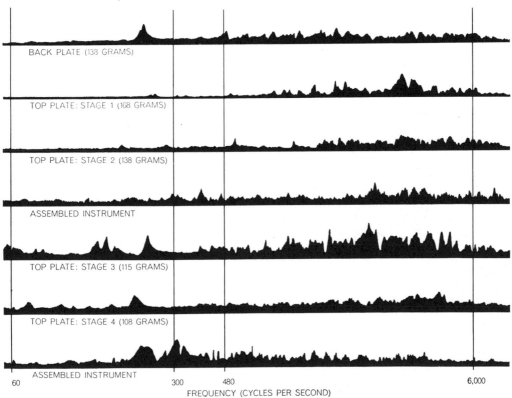

FREQUENCY-RESPONSE CURVES of top and back plates of a viola and of the assembled instrument at various stages are depicted. Although the tests run from 20 to 20,000 cycles per second, most of the response to the magnetic driver used to vibrate the wood falls in the range of 60 to roughly 10,000 cycles per second. The four frequencies indicated here are those at which checks were made to ensure that the recording film was synchronized with the audio-generator. Height of peaks represents amplitude of response.

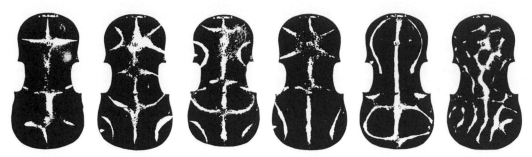

CHLADNI PATTERNS discussed in the text were made on a brass plate by Saunders at frequencies of 260, 340, 435, 520, 780 and 1,600 cycles per second. The plate is supported horizontally by a bolt at the center of its upper half; bottom end rests on a padded block.

accuracy. First we cut a flat brass plate in the shape of the violin plate. We dust it with powder and bow it at various points around the edge to set up different modes of vibration. Where the plate vibrates, the powder is bounced away, piling up along the nodal lines where there is no vibration. From these "Chladni" patterns on the brass [*see illustration above*] we are able to predict where a principal nodal point in the frequency test range will fall on the mid-line of a real violin plate. Since clamping on a nodal point does not affect the vibration pattern, we then clamp the violin plate at this point and set it into vibration at its exact center

by means of a magnetic driver, activated by an audio-frequency generator, that can be varied from 20 to 20,000 cycles per second. The response of the wood plate to the input signal, which has variable frequency but constant amplitude, is picked up by a microphone and fed to an oscilloscope or a sound-level meter. The amplitude and frequency of the points of greatest response can be recorded manually. Better still, a "photostrip" can be made by pulling a film across the oscilloscope face at a speed synchronized with the sweep of the audio-frequency generator.

Once we had established the testing procedure we could address ourselves to

a question that had been worried for several hundred years and that had been answered in a number of different ways: What sounds should the top and back plates of an instrument produce before they are joined?

In 1840 Savart reported that "a top of spruce and a back of maple tuned alike produced an instrument with a bad, weak tone." He took the plates off a number of Stradivarius and Guarnerius violins (imagine!) and tested them, finding that the tap tones varied "between C sharp 3 and D3 (in the octave above middle C) for the top, and D3 and D sharp 3 for the back, always one tone or one semitone difference, the

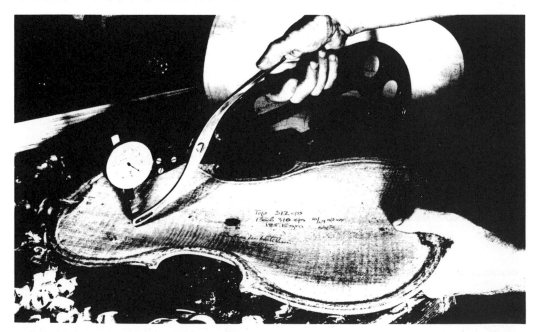

MEASURING PLATE THICKNESS makes it possible to determine where thinning can be done while maintaining a fairly uniform pattern of thickness. The measuring device, or caliper, consists of a dial gauge attached to one arm of an extended metal U.

12

back being higher than the top." Some violinmakers have held that the back should be a tone lower than the top; others, that the plates should be tuned to the same frequency.

My own findings are as follows: In the range of 120 to 600 cycles per second there may be one, two and possibly three peaks in the back plate and perhaps two or three more than that in the front. When the peaks of the front plate alternate with those of the back and the adjacent peaks are within about a semitone of one another, I get a good instrument. When the peaks coincide or are more than a tone apart, I get a bad one. Moreover, an average of the frequencies of the tap tones from front and back turns out to be just about seven semitones below the main wood frequency of the finished violin.

These conclusions are drawn from more than 400 photostrips of top and back plates of 35 instruments in the process of construction. After the plates were tested the instruments were assembled and then judged for tone quality by three criteria: (1) loudness test, (2) photostrips of the completed instrument and (3) actual playing by professionals. Then one plate, usually the top one, was removed and thinned, tested again and the instrument assembled for reappraisal. The back plate was thinned only when the top plate became so thin that it could no longer support string tension with safety. The entire thinning and testing process was sometimes repeated four or more times until each violin, viola or cello was judged to be good. So far I have spent six years on the program.

With our tap-tone test it is possible to follow the position of the main wood vibrations as they drift to lower frequencies when the wood is thinned and becomes more flexible. With a little practice one learns how to remove a few grams of wood from certain areas with a scraper or small plane and to estimate that the plate peaks (strong natural resonances) will move downscale, say 10 cycles per second. In some cases such a shift can make the difference between a good and a poor instrument.

As a kind of acid test of the theory I made a cello with the plate peaks matching; this is of course exactly wrong. During the next two years I gave the cello to several different cellists to play. All of them admired the workmanship and tried to be complimentary about the tone and playing qualities. The more forthright of them said that the tone was harsh and gritty in spots and weak in others and that the instru-

MAGNETIC DRIVER used in frequency-response tests is placed at the exact center of a top plate. The wires leading from the driver are connected to an audio-frequency generator that activates the driver over a range of frequencies from 20 to 20,000 cycles per second.

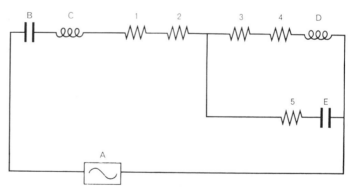

SIMPLIFIED ELECTRICAL CIRCUIT shows the nature of the two main resonances discussed in text. The current from a constant-amplitude alternating-current generator (*A*) is analogous to the force applied by a given string to the bridge; this force is proportional to the string tension and amplitude of string vibration. The first capacitor (*B*) is analogous to a stiffness associated with the elasticity and dimensions of the wood; the first inductance coil (*C*) is analogous to a mass moving with the velocity of the bridge-string contact and having a kinetic energy equal to that in the wood. In instruments of the violin family the stiffness and mass of the wood largely determine its over-all response and the frequency of the main wood resonance. The main air resonance is determined largely by opposition of the air to compression when the f-holes are closed (*E*) and the mass of the air near the open f-holes (*D*). The five resistors in the circuit represent mechanical or acoustical resistances.

13

ment was particularly hard to play softly.

Finally I took the plates off, tested them again and removed about 10 grams of wood from the edges of the top plate so that the peaks of the top alternated with those of the back. In this condition Mischa Schneider played the cello in a concert by the Budapest String Quartet and pronounced it to be *magnifico*.

The greatest difficulty with the tap-tone test on a finished instrument is that both the top and back plates must be off at the same time so that they can be tested under the same conditions and without the complication of drift in the measuring equipment. The removal of both plates is a touchy operation even for an expert. With the help of several co-operative violinmakers, however, we have been able to test the plates of a few good old violins. More such tests are needed for definitive comparisons.

New Violins

It has been hundreds of years since the violin won its battle with the viols. The victory was not an unmitigated blessing. The variability of the shape of the viols, and particularly their flat back plate without a complicated set of resonances, meant that the instruments could be built in a variety of sizes that easily covered the entire range of pitch represented by the piano keyboard. On the other hand, the violin family leaves substantial gaps in coverage and, as has already been pointed out, its two deeper-voiced members do not have optimum musical characteristics.

Buried in private collections and museums there is a neglected but rich repertoire of polyphonic string music from the Renaissance period written for viols. Their characteristically thin and nasal, but uniform and distinctive, timbre blended well with the clavichord and cembalo, which were played at the courts of Renaissance nobles. Many of these gentlemen kept a chest of viols, usually consisting of six instruments, two each of the treble, alto and tenor sizes.

For contemporary performance of the viol repertoire, however, the old instruments are unsuitable. They do not have the variety of timbre that the violin has taught the modern ear to expect, and they do not have nearly the power to satisfy the requirements of a concert hall of even moderate size. On the other hand, the present family of violin, viola and cello have too much inequality in timbre and too great gaps in pitch to play the music as it was written.

The need for new instruments of the violin type has been considered by musicians and violinmakers for many years. The present work of developing the new instruments was initiated when Henry Brant, the composer-in-residence at Bennington College in Vermont, came to us with the problem. Brant felt that modern musicians, faced with the need to find an expressive language appropriate to the present day, are ever seeking to extend the powers of the bowed string instruments. The violin family remains the composer's most eloquent and expressive vehicle among all the instruments so far devised in Western music, but its members have been essentially unchanged for 200 years. More and more the need is being recognized for a gamut of graduated instruments of the violin type, with each member well enough developed to meet the test of solo as well as ensemble playing.

Changing the classical dimensions of the violin to create instruments of varied sizes and tunings has been tried many times without success. Now the necessary knowledge is at hand. I have indicated that the variables in the design of the violin are close to optimum. The object is to keep the two main resonances on the two open middle strings in spite of changes in size and tuning. It can readily be appreciated that it is no mean task to arrive at the correct proportions among physical variables—size, thickness and stiffness of wood, tightness of stringing and so on—that will produce the desired result in the resonance. For me it has meant years of literal cut and try, but with the help of scaling theory I am now close to having a set of empirical rules for making a genuinely complete family of instruments of the violin type. In doing this I have drawn heavily on the knowledge gained by other violinmakers who have tackled the same problem but without success because they did not have the benefit of modern acoustical physics. I have already built revised versions of the viola and cello, enlarging them somewhat to bring the resonances down to the frequencies of the open middle strings. As a result my viola has to have a peg at the bottom, like a cello, and is played between the knees. In addition I have added two new instruments to the family (one replaces the bass). This past January the six scaled members of the violin family were tried out at an informal concert, which a number of professional musicians found interesting and challenging as well as aesthetically pleasing. The smallest and the largest of the new instruments have not yet been finished and are giving the most trouble. Although scaling theory

tells us what to do, we are up against the limits of available materials and the human physique. For the smallest instrument, which is tuned an octave above the violin, material of sufficient tensile strength for strings is the major problem. Few materials have the strength to vibrate within the requisite range of frequencies and still provide strings long enough to allow the player to finger consecutive semitones. In the largest instrument the designer faces the mechanical problem of making it possible for the musician to bow and finger simultaneously.

Other violinmakers have experimented with instrument size. In the 19th century Jean Baptiste Vuillaume introduced a new model of the viola with an exceptionally large air volume, constructed on the scientific principles of Savart. He also developed a huge double bass, known as the *octobasse*, that was tuned by means of levers. Fred Dautrich of Torrington, Conn., spent much of his time during the 1920's and 1930's working on a graded series of instruments of the violin type that he called the vilonia, the vilon and the vilono. I have been fortunate enough to obtain a set of these. They are of such excellent workmanship and proportions that it has been possible to modify them slightly by applying scaling theory and adapt them to our present series of instruments.

In the past few years J. C. Schelleng, formerly of the Bell Telephone Laboratories, has been studying the violin as a circuit, one of the standard techniques of acoustics in which the various mechanically vibrating parts are treated in a manner analogous to the elements of an electrical circuit. Although the violin is exceedingly complicated, it possesses many simplicities not usually recognized. These, along with the fundamental physics of the instrument, permit the definition of "circuit elements" and lead to relations difficult to obtain empirically. This circuit concept is already being of great help in perfecting the new instruments, defining such problems as string tension, the mass of the box and the stiffness of the plates.

To sum up, I believe that, without ignoring the precious heritage of centuries, the violinmaker should become more conscious of the science of his instrument, and that the acoustical physicist should see that here is a real challenge to his discipline. We really ought to learn how to make consistently better instruments than the old masters did. If that challenge cannot be fulfilled, we should at the very least find out the reasons for our limitations.

14

24

The Author

CARLEEN MALEY HUTCHINS has been designing and constructing violas and other stringed instruments of the violin family for the past 15 years. Her first step toward a career as a luthier came in 1942, when, as she describes it, "I bought an inexpensive weak-toned viola because my musical friends complained that the trumpet I had played was too loud in chamber music, as well as out of tune with the strings—and besides they needed a viola." The viola was unsatisfactory and Mrs. Hutchins turned for help to her uncle, William Harvey Fletcher (not the well-known acoustical physicist Harvey Fletcher), who had made many violins himself. Fletcher declined to try his hand at making a viola and instead directed his niece to a dealer who could supply her with the requisite books, blueprints and wood. In 1947 Mrs. Hutchins took a leave of absence from the Brearley School in New York, where she had taught science since 1937, to have her first child. Encouraged by her husband, she also embarked on the task of making her first viola, a job that took two years. Mrs. Hutchins has produced 55 instruments, selling some of them to help pay for further research. A graduate of Cornell University, where she studied entomology, Mrs. Hutchins retired from teaching in 1949. For the past 12 years she has collaborated with Frederick A. Saunders of Harvard University in the study of the acoustics of the instruments of the violin family. At present she is continuing this work under her second Guggenheim Fellowship.

Bibliography

THE MECHANICAL ACTION OF INSTRUMENTS OF THE VIOLIN FAMILY. F. A. Saunders in *The Journal of the Acoustical Society of America*, Vol. 17, No. 3, pages 169–186; January, 1946.

THE MECHANICAL ACTION OF VIOLINS. F. A. Saunders in *The Journal of the Acoustical Society of America*, Vol. 9, No. 2, pages 81–98; October, 1937.

.MISURA DELL'ATTRITO INTERNO E DELLE COSTANTI ELASTICHE DEL LEGNO. I. Barducci and G. Pasqualini in *Nuovo Cimento*, Vol. 5, No. 5, pages 416–446; October 1, 1948.

REGARDING THE SOUND QUALITY OF VIOLINS AND A SCIENTIFIC BASIS FOR VIOLIN CONSTRUCTION. H. Meinel in *The Journal of the Acoustical Society of America*, Vol. 29, No. 7, pages 817–822; July, 1957.

SUBHARMONICS AND PLATE TAP TONES IN VIOLIN ACOUSTICS. Carleen M. Hutchins, Alvin S. Hopping and Frederick A. Saunders in *The Journal of the Acoustical Society of America*, Vol. 32, No. 11, pages 1443–1449; November, 1960.

Editor's Comments on Paper 2

2 Saunders: *The Mechanical Action of Violins*

Professor Frederick A. Saunders (1875–1963) first became interested in violin acoustics in 1933 when Henry Shaw, then treasurer of General Radio Company, asked him if there was any way in which the tone qualities of a good modern cello could be measured and compared with those of a fine, expensive, eighteenth-century cello. A violin–viola player himself, Saunders thought the answer would be relatively easy to find. Thirty years later he said, "There is still no measurable physical difference, that we can think of, which clearly distinguishes the fine early violins from the fine modern ones."

The first part of Saunders' violin research was done at the Cruft Acoustics Laboratory in the Physics Department of Harvard University with the help of several graduate students and many fine string players, of whom the most prominent was Jascha Heifetz. The second part was carried on after his retirement in a small laboratory at Mount Holyoke College in South Hadley, Massachusetts. It was during this second period that C. M. Hutchins, J. C. Schelleng, and R. E. Fryxell became associated with the work, jokingly calling themselves the Catgut Acoustical Society (see p. 455).

Published at intervals of about 7 years, his four major papers (listed below) demonstrate the wide range of Saunders' study of the violin. He carried on an extensive correspondence with Meinel, Pasqualini, Backhaus, Raman, and others, collecting many of the reprints included in this volume. He lectured on violin acoustics and had instruments evaluated by both auditors and players. One study of this nature in which William Moennig and Son took part is included in *Musical Acoustics, Part II*, "Studies of the Instruments of the Curtis String Quartet," *Overtones* (April 1940).

Born in London, Ontario, in 1875, Saunders attended the University of Toronto and in 1899 obtained his doctorate in physics at Johns Hopkins. He taught physics at Haverford, Syracuse, Princeton, and Vassar before accepting a professorship in the Department of Physics at Harvard University in 1919. He is best known to physicists for his work in spectroscopy on the Russell–Saunders coupling and to several generations of students for his textbook *A Survey of Physics* (Holt, Rinehart and Winston, Inc., New York, 1943). He was a Fellow of the National Academy of Sciences, and a past President and Fellow of the Acoustical Society of America. A biography of F. A. Saunders and his family background can be found in *William Saunders and His Five Sons*, by Elsie M. Pomeroy (The Ryerson Press, Toronto, 1956).

Major Publications on the Violin by F. A. Saunders

In *Musical Acoustics, Part I*

Saunders, F. A. "The Mechanical Action of Violins," *J. Acoust. Soc. Amer.*, **9**, 81–98 (Oct. 1937).

In *Musical Acoustics, Part II*

Saunders, F. A. "The Mechanical Action of Instruments of the Violin Family," *J. Acoust. Soc. Amer.*, **17**(3), 169–186 (Jan. 1946).

Saunders, F. A. "Recent Work on Violins," *J. Acoust. Soc. Amer.*, **25**(3), 491–498 (May 1953).

Hutchins, C. M., A. S. Hopping, and F. A. Saunders. "Subharmonics and Plate Tap Tones in Violin Acoustics," *J. Acoust. Soc. Amer.*, **32**(11), 1443–1449 (Nov. 1960).

2

Copyright © 1937 by the Acoustical Society of America

Reprinted from *J. Acoust. Soc. Amer.*, **9**, 81–98 (Oct. 1937)

The Mechanical Action of Violins

F. A. SAUNDERS
Harvard University, Cambridge, Massachusetts
(Received June 17, 1937)

INTRODUCTION

VIOLINS have aroused the interest of people of all sorts for several centuries. The beauty of their form, of the wood itself, and of the varnish pleases the eye, their tones charm the ear, and the range of their expressiveness appeals to the souls of artists. The coldly calculating scientific man has paid less attention to them as vibrating systems than they deserve. The strings, the bridge, the two "plates" (top and bottom), and the sides of the violin form a complex system, closely coupled together, in which each part is affected by the others, all with different natural habits of vibration, damping, etc., of their own, but strictly bound by their close relationship and thus forced to give up their individuality for the good of the group.

About a century ago F. Savart in Paris made many experiments on their action, seeking among other things to find in the vibrations of the plates by themselves the secret of the success of the completed instrument. Sir William Huggins, the astronomer, investigated the action of the sound-post which connects the top and back at a point near one foot of the bridge, without, however, always taking into account both the mass and the elasticity of this member. In more recent times Raman[1] made the first measurements on the mechanical work required to make a violin "speak" properly at different frequencies. A long series of experiments have studied the action of the strings themselves, without always emphasizing the fact that what we hear comes not from the strings but from the body of the violin. H. Backhaus[2] in Germany has made many tests showing (among other things) something about the way in which the plates in the completed instrument subdivide into a complicated vibrating pattern when a particular tone is being produced. Most recently H. Meinel[3] has carried out exhaustive experiments with modern apparatus of high quality, the results of which will be referred to at appropriate places in this article.

With the generous support of Mr. Henry S. Shaw of Boston, who appreciates both the musical and scientific sides of this problem, a group in the Harvard physics department has been engaged for some years in the study of several of the purely physical aspects of the

---

[1] C. V. Raman, Phil. Mag. **39**, 535 (1920).
[2] H. Backhaus, Zeits. f. Physik, **62**, 143 (1930) and **72**, 218 (1931).
[3] H. Meinel, Akust. Zeits. **2**, 22 and 62 (1937).

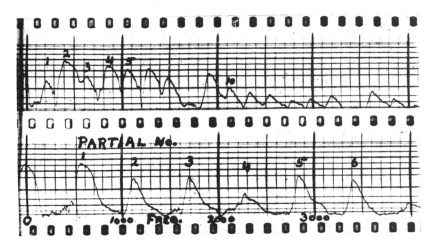

FIG. 1.

violin. The experiments of Raman have been extended, several of those now described by Meinel have been independently carried out, and in certain other ways modern technique has been applied to violin behavior. The aim has been to discover the characteristics of the best instruments and the reasons why poor ones fail. This aim is not yet achieved, but a report on the present state of the work may be worth giving.

It should be emphasized that only quite recently have the results of the development of electron tubes made it possible to measure the intensity of a sound easily and accurately, and to analyze a tone with equal facility into its component harmonics, measuring the frequency and the strength of each. Variations of tone quality under many influences can now be examined in a time expressible in minutes while formerly it would have taken days.

THE ANALYZER

The most important instrument which we have used is the harmonic analyzer constructed by H. H. Hall[4] who has given an account of its design and action. A high quality moving coil microphone, whose response is uniform over a wider range of frequency than is here needed, picks up the sound from near the violin, and the output of this microphone passes to the analyzer. If the tone is sustained uniformly for a time which has

[4] H. H. Hall, J. Acous. Soc. Am. **7**, 102 (1935).

usually been set at 3.8 seconds, the analyzer delivers a strip of photographic paper upon which is recorded a complete analysis of this tone. This consists of a scale of frequency (vertical lines) running from 0 to 10,000 cycles per second ("c.p.s."), and a scale of horizontal lines giving the logarithm of the sound intensity, which is related to the loudness of the sound, though not proportional to it. A third trace appears on the photograph which is a record of the motion of the luminous spot of a cathode-ray oscillograph. This spot remains at the bottom of the strip at any frequency when no sound of that frequency is audible, but rises to a peak at each frequency that is present as a partial tone in the complex tone received by the microphone. Each peak shows the intensity of a separate component in the sound, with the exception that there is a disturbance in the region of zero frequency which is not connected with the sound. The height of each peak is read off in decibels (db) from the record. Thus Fig. 1 shows a part of the analyses of two sample tones obtained from a Stradivarius violin (our No. 20), one, A on the G string (which we shall call "A on G"), showing intensities of its partial tones of

23·40·25·34·30·31·24·10·26
·20·17·16·11·12·15·5·17·12·7

and the other D on A giving the values

36·24·25·17·25·23·10·10·9·6·10·3.

29

The frequencies, read from the 10th partial, were 212 and 572 c.p.s., respectively. The frequency scale is intended for identification only, not for precise measurement. One obtains records showing sometimes over 20 partials in the lower notes, falling to 3 or 4 at the top of the scale.

THE FREQUENCIES STUDIED

The fundamental tones ranged from open G (here called G_0), whose frequency is 193 on our scale (based on $A = 435$), to the second octave E on the E string (called E'' on E) with a frequency of 2607 c.p.s. Each semitone was studied on the G string from G_0 to G'; the D and A strings were treated similarly; the E string was carried up two octaves. Thus in all in each analysis 64 notes were played, but the upper notes on each of the three lower strings were duplicated on the string above, so that actually only 46 separate (fundamental) frequencies were used. The harmonics of the upper notes, however, were studied up to a frequency of 10,000 c.p.s., or to the D♯ which lies nearly two octaves above the highest musical note in the normal playing range (E''). The photographs necessary for such a study of one violin can easily be taken, repeating each one once, in the course of an hour. Not counting the time required for the photographic operations (on 125 ft. of paper), three hours more yield most of the information which the analyses furnish for our purposes. During this time one tabulates the strength of the partial tones (perhaps 1400 values) and plots "a response curve" for the instrument, which discloses its main mechanical peculiarities.

These analyses do not give the relative phases of the partial tones. It seems by now to be well established by the work of many investigators that these phases are of no importance in determining the musical quality of complex tones as long as the sounds produced are not excessively loud. This is corroborated by the fact that though many of the phases change in passing from *down* to *up* bowing, the ear detects no change of quality.

MACHINE BOWING

In the first experiments a constant and accurately controllable method of bowing was neces-

sary. A revolving group of celluloid disks was used, as described later. Some analyses were taken with this bow and compared with hand bowing analyses. The individual results differed somewhat, due to causes which we shall consider in a moment. It was later found that when the results were plotted so as to yield a "response curve" (soon to be described) this curve was the same by either method of bowing. Subsequently hand bowing was used, as it is far more convenient, and musicians accept the results obtained by means of it with less prejudice than they would if the celluloid-disk bow had been used. They cannot help feeling that their cherished instrument will be misrepresented if treated so unconventionally.

HAND BOWING

The variations in the quality of a tone (i.e., in the relative strengths of its partial tones) pro-

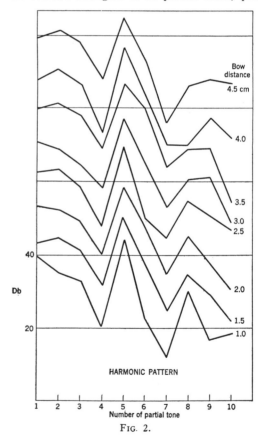

FIG. 2.

duced by hand bowing have been traced to variations in bow pressure, bow speed, and the distance of the bow from the bridge (here called bow distance). Of these, changes in bow pressure alter the tone quality very little, changes in speed (alone) somewhat more, and changes in bow distance probably most of all. Even so, the variations are not very great compared with changes in quality in passing from note to note on the instrument. To illustrate the change in quality when bow speed and pressure are held approximately steady and bow distance is varied, the observations plotted in Fig. 2 were taken. This shows sound intensity vertically and the number of each partial tone horizontally. To avoid confusion the resulting "harmonic pattern" for the note is redrawn at a different height for each bow distance. The first 10 partials only were plotted here, though more were measured. The note played was the open A on a cheap and strident violin, and the tone was very rich in harmonics. The figure shows on the whole the same harmonic pattern, i.e., weak partials 4 and 7, and strong 5 and 8, but the details vary somewhat. Such variations in tone quality are repeatable, and the analyses detect them, but they are not easily heard. If a midposition for the bow is maintained (nearer the bridge for the highest notes) changes in tone quality are reduced to a quite negligible amount, especially as the bow pressure and speed are also held approximately constant.

It is interesting to consider the reasons why the tone quality might be expected to change with the bow distance. If a string, excited by a bow, were giving the fundamental tone only, one would expect a rather weak response if the bow were applied at its middle point, because the excursions of the string could nowhere else be so great as they are under the bow. If the bow were moved nearer to one end the vibration of the middle would increase greatly, and this increase would continue until the bow came very close to the end, when the vibration could no longer be produced at all. Thus the strength of the fundamental tone would vary in a simple way with the bow distance. In like manner, if the string were vibrating in eight segments, giving its 8th partial, the bow would be expected to excite it weakly if it were applied at a point in the

middle of a segment (e.g. at a bow distance one-sixteenth of the length), but more tone of this partial would be produced if the bow were moved nearer to one of the nodes (but not too near).

With due precautions, successive analyses of the same note were found to agree surprisingly well. Table I shows some representative pairs of analyses of different notes given by a good modern German violin. The numbers are the intensities of the successive partial tones as read off the analyses, and are seen to be in close agreement until the high harmonics are reached. The variations in the harmonic pattern are exhibited in Table I. The notes C on A and C♯ on E have their main strength in the first partials (the fundamental tones); the others do not, but vary among themselves. The same note E played on two strings (G and D) gives very much the same pattern, characterized in this case by strong partials 4, 7, 8, and 9, and weak 3 and 10.

The high harmonics are the most difficult to keep steady by hand bowing. A momentary wavering of the bow during the time in which one particular harmonic is being recorded on the analyzer is enough to cut down its intensity greatly, especially when it is weak. For several reasons we have been led to use in most of our studies only a few of the stronger partial tones. By taking many observations of these one smooths out accidental variations.

In bowing for analysis care was always taken not to introduce any tremolo, i.e., the finger was held still. The finger was pressed very firmly on

TABLE I. *Sample analyses.*

Note	Intensities of Partials in db			
B♭ on G	34·38·30·32·34	40·25·31·25·25	26·26·28·19·12	16·17·17·19
	32·38·26·32·35	36·26·31·24·30	23·20·23· 5·10	10·17·15·16
E on G	43·36·25·43·25	29·31·29·31·11	21·15·10	
	43·36·25·45·23	30·31·28·30· 8	22·24·10	
E on D	43·34·20·44·30	20·32·25·26·12	20·20·17·12	
	41·31·20·44·30	26·36·26·28· 8	15·15·20·10	
C on A	47·32·35·20·27	37·18·12·17·10	11·17·20·10	
	47·31·33·22·26	38·21·20·13·10	8·19·20· 8	
F on E	36·44·42·39·35	10·25·15·22·20	22·12·10	
	39·44·44·41·30	14·22·22·19·23	23·10·15	
A on E	40·39·33·33·32	27·18·12·17·10		
	40·37·33·33·35	28·23·17·19· 5		
C on E	33·46·45·23·31	25·22·22· 7		
	35·47·43·25·30	31·22·18·10		
C♯ on E	47·31·23·35·17	17·21· 5		
	45·31·23·36·12	15·22·15		

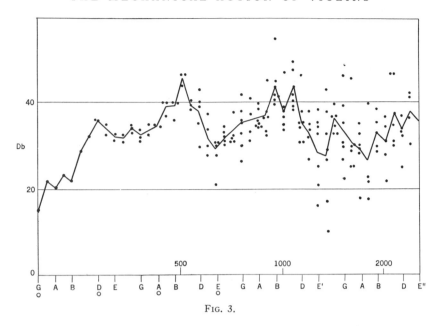

FIG. 3.

the string, as a loose pressure reduces the upper partials somewhat. The strings which were at the moment not in use were always damped (by idle fingers) to prevent resonance effects in them. The loudness of the notes used would be classed musically as about mezzopiano.

THE SOUND STAGE

All analyses were made in a special room to avoid effects due to sound reflection from the walls. In this room the walls, ceiling, and floor are covered with sound-absorbing material of about 50 percent absorbing power, but one corner of the room contains a "sound stage" built according to the ideas of Bedell.[5] For a distance of 8 ft. out from this corner there extends a covering (over two walls, ceiling, and floor) designed to prevent the reflection of sound entirely. This consists of 4 inches of rock wool, over which are stretched 10 or 12 layers of thin cotton sheeting with spacings which average about 1 inch. The open part of the stage is partly shielded from the rest of the room by heavy curtains. Tests were made of this stage by Dr. S. A. Buckingham which showed that if the source is placed as near as possible to the corner, and

[5] E. H. Bedell, J. Acous. Soc. Am. 8, 118 (1936).

the microphone within a meter of it, in the middle of the stage, no effects large enough to be of importance in this investigation will occur which are due to sound reflected from the walls. We have thus almost duplicated ideal conditions, such as one might have out-of-doors, far from any reflecting surfaces.

THE EMISSION PATTERN OF THE VIOLIN

Several observers (but especially Meinel) have shown that the sound emitted by a violin varies in quantity and in quality with the direction from the instrument. When any note is produced, and the ear is in a definite situation with respect to the violin, certain partial tones will be louder than at other angles, and others weaker. The violin thus possesses what might be called an emission pattern, which varies with the frequency, and is most marked at high frequencies. The quality of a violin thus varies with the position of the hearer, and the "true" quality cannot be obtained without going through the labor of measuring all notes at all possible angles. This average value if obtained would not correspond to what one hears with ears (almost) in one direction only. Moreover, in an ordinary room with reflecting walls, we hear the instrument (in a sense) from several different directions at once. In this case also the three-dimensional interference pattern of sound which is set up in the room itself distorts the quality, and would do so even if the violin radiated equally in all directions.

These considerations make it seem unwise to spend too much time in securing high accuracy in the measurement of tone quality. The best method is to take such measurements as would show what vibrations the instrument itself is capable of executing with the greatest ease, and draw conclusions about the quality from averages of many observations. We have accordingly chosen to measure the sound emitted in the direction which was most convenient to fix, namely perpendicular to the middle of the top of the violin. The microphone was held in this line and at a distance of 75 cm from the top surface. Meinel has used a direction 45° away from the vertical, such as might be taken by a listener standing on the player's right when the violin is held at 45° to the horizontal. Some investigators have referred to this direction as the one in which the violin radiates the greatest amount of sound. Meinel's data show that this is true for certain frequencies, but not in general. Meinel also found that the back of a violin radiates sound of high frequency feebly, a result of considerable importance to the violin maker, and consistent with the fact that some old violins were decorated by carving on the back, that is, had thick and irregular areas there which would have diminished the usefulness of the back if it had been very important. In any case, whether the back is a good radiator or not, it is awkward to reach it with measuring instruments while the violin is being played. Our choice of

direction seemed on the whole to be a good compromise.

It is not hard to see why a violin radiates sound differently in different directions. The old experiments of Chladni on plates sprinkled with sand, and the newer ones of Backhaus on violins, show that the surfaces break up into vibrating areas which move in opposite phases, and thus partially neutralize each other in pairs as far as their radiating power goes. If these areas were mapped for pure-tone excitation, they might prove to be simpler in form and more interpretable; but this would probably not be a very profitable research, considering how complicated the plates of a violin are with variations of thickness, curvatures, constraints, and added masses.

RESPONSE CURVES

One may best study the analyses by a graphical method. A horizontal scale of semitones is laid off covering the necessary range. Beginning with the average of the two analyses of the lowest note, G_0, one places a point vertically over G_0 at a height corresponding to the intensity for this (first) partial tone. At the G one octave above, one plots the intensity of the second partial, at the D above that, the third, and so on, as far as one pleases. It is unnecessary to go beyond the fourth partial, as one has data enough without, and also one begins to meet frequencies among the partials which are not exactly equal to those of the notes on the scale of equal temperament. Notes not in agreement in frequency could not be expected to give the same results, especially if they happened to lie in a region where the response of the violin was changing rapidly. Proceeding next to the G♯ on G one plots four points for its first four partials, and so on, note by note, up the scale. When we reach the first D we have D on G and D_0 to plot, and therefore two points per semitone. At G′ on G we have three, and farther up we have as many as seven points; at the top of the scale this number dwindles down to one again. The multiple values at each frequency do not coincide, though they often come very close. It is hardly to be expected that they would. A note like G on E, for instance, will occur as the first partial of G on E and the fourth of G_0. It would be

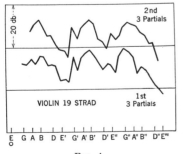

FIG. 4.

FIG. 5.

unreasonable to expect an equal amount of energy of this frequency to be emitted in these two cases. The E string fundamental tones contribute values above the average, and on the whole the fourth partials lie below. The accuracy of the average cannot be expected to be high; it is low enough to justify our taking arithmetic means of logarithmic values. The absolute heights of the peaks thus disclosed are much less important than their positions on the frequency scale.

Figure 3 shows a portion of such a graph. One can see in this figure that all three values at C on A lie well above any value a semitone away, while at C on E five out of six points lie below any of the points a semitone below. Thus the existence of peaks at C on A, B on E, and C$\sharp$ on E is shown without doubt. In spite of the usual loudness of the open E(E_0) all the points at this frequency are low in this case, a marked peculiarity of this particular violin (violin 9, a cheap one). Joining the averages for each note on this figure discloses a very definite "response curve" with peaks corresponding to the natural resonance frequencies of the violin, and hollows where the sound emission is weak. The situations of these peaks and hollows are quite definite; the errors already pointed out affect their height and depth only, but in case of well-marked peaks and hollows no possible error could reduce the peaks to nothing, or fill up the hollows.

Two methods might be used to illustrate the accuracy with which the resonant peaks appear. The first is to plot a response curve from observations on the first three partials only, and another from the partials 4, 5, and 6. These constitute totally independent observations. They do not overlap throughout the whole range of the violin, but the accuracy with which they agree in the region where they do overlap is well shown in Fig. 4, where only this range is plotted. The peaks agree within a semitone in position, and approximately (in most cases) in height.

The second method of testing the accuracy of the response curves is to take a second complete analysis of the same instrument. This has been done with satisfactory results (see, for instance, the two lower curves of Fig. 11, below 3500 c.p.s.).

Similar response curves were taken by Meinel (called frequency curves by him) in quarter-tone intervals. They show the same constancy, though with his smaller steps he was able to show sharper peaks, especially at high frequencies, than we can. He implies, however, that these have little musical significance since we do not play with quarter-tone subdivisions. Another reason why such sharp peaks may not be important is that most players introduce a tremolo effect by rocking the finger tip on the string. The resulting small changes of pitch tend to inhibit sharp

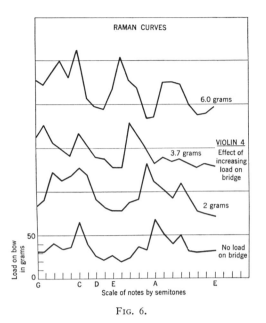

FIG. 6.

resonances, but have less influence on wider peaks.

RAMAN CURVES

Before going on to consider the response curves of different types of violins it will be well to mention other ways of obtaining them. We have repeated the experiments of Raman with this end in view. For this purpose an automatic bow was constructed consisting of several (usually six or eight) disks of thin (0.125 mm) celluloid 8.5 cm diameter, clamped together on a shaft so that their edges rub on the string as the shaft revolves, and cover a length on the string equivalent to that touched by the hairs of a bow in ordinary playing (about 6 mm). The celluloid disks were shaped under pressure between hot spherical forms so as to make them into parts of spherical shells of radius about 15 cm. This gave their edges more flexibility. These edges were heavily resined, and then produced a tone which was uniform when the edges were all lined up to touch the string properly. Even a musical ear could hardly distinguish the resulting tone from that produced in the usual way.

This "bow" (Fig. 5) was rotated by a silent variable speed motor, and the horizontal bow shaft was so supported that it swung about a

perpendicular (horizontal) axis and could be counterpoised so as to stand in mid-air, or to press on the string with a known pressure. With this device, the bow speed, distance, and pressure were all fixed and measurable.

One of Raman's experiments consisted of measuring the least weight on the bow necessary to make the violin "speak" properly. Until one tries the experiment one has no idea how definite this weight is. It is almost always determinable within an accuracy of 10 percent, and at fairly high bow speeds the precision is still greater. The weight necessary to produce a proper tone may vary from 10 grams to over 60, so that these variations are never in danger of being obscured by the experimental error. The necessary weights become smaller at higher frequencies, and depend on the bow speed also. A high bow speed leads to the most accurate results. Above the low notes on the E string the observations become difficult and uncertain, so that the range of the study of the responses of a violin by this method is limited.

Typical "Raman curves" for a violin are given in Fig. 6, which shows a family of curves, the first with no mute on the bridge, the others with mutes of increasing weight. Raman has noted the effect of a mute in lowering the natural periods of vibration of the wood of the violin (i.e., shifting them to the left in this figure), while the peak usually found near D_0 is unaffected. This peak has been known since the days of Savart to be due to the vibrations of the

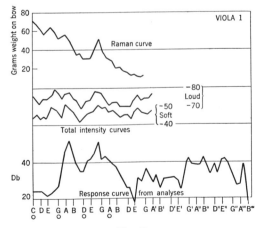

FIG. 7.

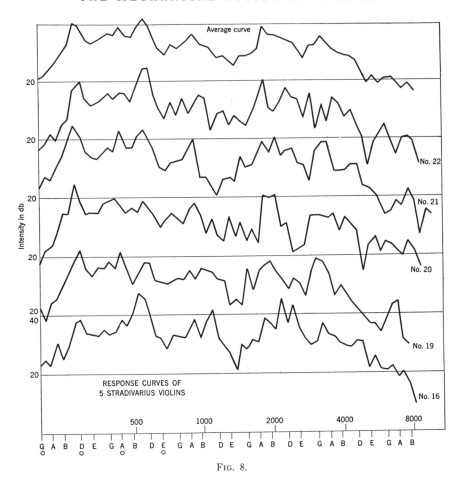

FIG. 8.

air inside the body. It is altered only by changes in the volume of the air, the area of the *f*-holes, or the nature of the gas inside. In the figure we see the air-peak constant at C on G, while the main body peak at A_0 shifts down to F on D with a total load of 6 grams. In general the peaks become more pronounced as the load is increased. Conversely, lightening the bridge shifts them slightly toward higher pitches.

A pretty demonstration is furnished by filling the inside of a violin with a stream of carbon dioxide from a small rubber tube, keeping the instrument horizontal. As the gas is heavy, it displaces the air, and the rich, full tone due to the air resonance can be traced as it goes down a musical interval of a minor third during the entrance of the gas. The air-peak can be observed

on a Raman curve for such a gas-filled violin, and is found to be shifted as expected, while the body peaks are, of course, quite unaffected.

WOLF-NOTES

Every peak shown by a Raman curve is the haunt of a wolf-note, which appears when the load on the bow is insufficient to "kill" it. This is in complete agreement with the explanation of these notes by Raman. If energy enough is supplied one can ride over these peaks without ever causing a wolf-note, but otherwise one can produce and observe them very well, especially with an automatic bow. There is, of course, no escape from wolf-notes with any instruments which show peaks (and all do), but they appear

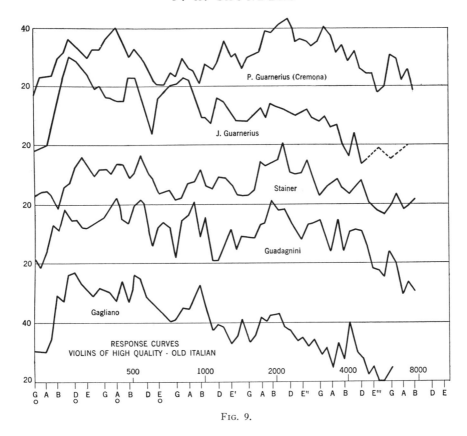

FIG. 9.

only when one attempts to play softly. The wolf-note due to the air-resonant peak is sometimes much less prominent than usual, especially in violins whose wood is relatively thick.

TOTAL INTENSITY CURVES

The existence of such an instrument as the noise meter makes it possible to measure accurately on a decibel scale the intensity of any sound over a very wide range of frequencies. A violin tone presented to the microphone of this instrument causes it to respond simultaneously to all the partial tones present, and yield us the total sound intensity produced. If most of this intensity is concentrated in the fundamental tone, which is usually the case on the peaks of the response curve, these peaks can readily be obtained. One merely has to play each note on the instrument as *loud* as possible, without allowing the tone to "break," and find the

places of maximum intensity with the meter. The peaks may also be obtained fairly well by bowing as *gently* as possible still producing a good tone. Minor peaks may, however, appear on the total intensity curves when a particular harmonic coincides with a major peak of the response curve. Fig. 7 shows at the bottom a response curve obtained from analyses, above this two total intensity curves, soft and loud, and at the top a Raman curve. The instrument used in this case was a viola. The Raman and the total intensity curves indicate a peak at F or F♯ on the C string which does not appear on the response curve. It is due to the second partial, an octave higher, which coincides with the main body peak of this instrument. The resemblances in all these curves are very marked. The noise meter method is very quick; one can usually find the main peaks of an instrument in five minutes. The upper frequencies, beyond the playing range (E″ on E for the violin) are inaccessible, and the minor peaks difficult to find by this method.

RESPONSE CURVES FOR STRADIVARIUS VIOLINS

Through the kindness of their generous owners we have been permitted to make tests of many high quality violins obtainable in the neighborhood of Boston. Five of these were made by A. Stradivarius in the years from 1684 to 1721, the last three (Nos. 16, 19, 20) having his characteristic flat model, the other two (Nos. 21 and 22) being higher, and somewhat like an Amati. The response curves of these five instruments are given in Fig. 8, and at the top is an average curve of all five. Their air-peaks lie at C♯ (on G) or D₀. Nos. 16 and 22 have pronounced body peaks at C to C♯ (on A), the others minor ones in this region. No. 16 is unique in giving marked response at every C♯ within the playing range of the instrument and relatively low response at every F.

The average curve might be expected to show the real characteristics of the product of Stradivarius, but this expectation is not justifiable for several reasons. These instruments are not in their original condition as they left his hands. Not only have they suffered damage from cracks, etc., but their bass bars have been replaced by stronger ones to counterbalance an increase in string tension due to the rise of musical pitch (nearly a minor third) in two centuries. No physicist would expect a tuning fork to act as it did before, if it had cracked and been welded together again, or if part of it had been replaced by a piece of heavier metal. Moreover the number of instruments here considered is far too small to make the average significant, and two of them are of a different shape from the other three. We plan, of course, to study many more instruments as occasion offers and time permits.

In spite of these difficulties the average curve deserves some attention. We note that many minor peaks present in the individual curves have been smoothed out in the average, and are therefore unimportant. We are left with five major peaks; the lowest in pitch is the air peak, the others body peaks, of which the highest lies above the normal playing range (E″) of the

violin. These peaks are spaced 12, 9, 12, and 10 semitones apart, this separation averaging less than an octave, but being unexpectedly even. The response is high below E₀, except for the fundamentals of the low G string notes. It is low on the E string between 1000 and 1700 c.p.s., high from 1700 to 2200, and again high around 2500. The low value of the actual tone at E₀ is made still more conspicuous by the weakness of its second partial (E′ on E), but partly compensated by the strength of the upper partials. There is not much evidence here of any characteristic region, or "formant," of specially high response.

The tone quality as disclosed by this curve is shown to have innumerable variations. One may pick it out for any note by comparing the strength of the fundamental with that of the second partial (12 semitones on), the third (7 more), the fourth (5 more), the fifth (4 more), etc. Thus G₀ has a very weak fundamental, strong partials 2 and 3, medium 4, 5, and 6, weak 7 and 8, etc. But C♯ on G has extremely strong partials 1 and 2, medium 3, 4, and 6, weak 5, etc. B♭ on A is characterized by weak partials 3 and 6 and very strong partial 4. It would be hard to find two harmonic patterns on this instrument which are alike. Perhaps this continual change of tone quality from note to note constitutes one of the charms of a violin, but it makes it impossible to state any one grouping of strength, or pattern, among the partial tones which truly represents the "fine old Italian tone" that one hears so much about. An even more shocking result is that it is possible to duplicate the tone quality of this average Stradivarius *at certain notes* by the harmonic pattern given by a $5.00 violin that we use as our "standard of badness." The same would be

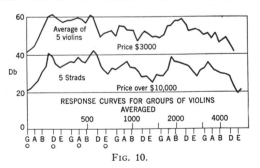

FIG. 10.

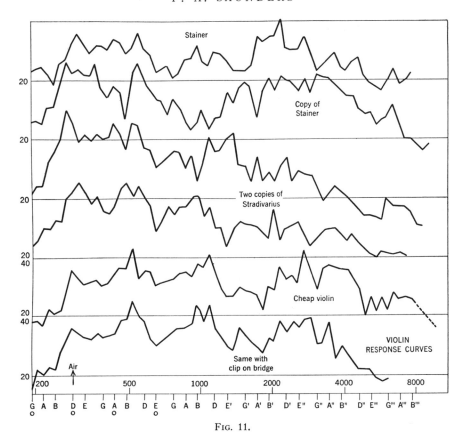

FIG. 11.

as true, of course, of an individual Stradivarius as it is of the average. If one tries to identify the Italian tone by taking the average harmonic pattern of *all* the notes of the violin one reaches an almost uniform distribution of strength in the low partial tones, fading away slowly as we pass to the highest ones. Would a uniform or only slowly changing strength of the partial tones produce a good violin tone, or not? Certainly it would not be like the tone of any existing violin, though it would probably be very nearly the tone we receive from the violin section of a large orchestra when all the players are in unison. It would not be difficult to make an instrument in which the agitation of the string was transmitted to the moving system of a high quality loud speaker. This would then yield on the whole the same tone quality as an orchestra of a large number of different violins sounding together. Perhaps such an instrument would be voted to be monotonous and uninteresting, but the experiment ought to be tried.

OTHER ITALIAN INSTRUMENTS

Five other good instruments of Italian origin have been tested. Their response curves are given in Fig. 9. The Stainer and Peter Guarnerius are each in excellent condition and are built on a high model. The Guadagnini is a famous instrument, once played by Spohr. The J. Guarnerius is considered to have a genuine top; the rest of the instrument is probably not his, though it is old, and Italian. The response curves of the P. Guarnerius and the Stainer are very similar, showing low response in the frequency region 600–1700 c.p.s. The J. Guarnerius and the Gagliano are somewhat similar, and both a little weak in the high frequencies. The Guadagnini is strong throughout, except for a few narrow "valleys."

The average curve of these five instruments is plotted in Fig. 10 against the average of the five Stradivarius instruments. The similarity of these two curves is so great that it is difficult to see why there should be so great a difference in price between these two groups. One might in fact prefer the average curve of the cheaper instruments as it does not drop quite so low in the region 1100 to 1900 c.p.s. On this account one can say that the average harmonic pattern given by these instruments would be more uniform than that of the more expensive ones, though any one violin by itself gives the usual wide variety of tone qualities.

Violins of Lower Price

In Fig. 11 there are five more response curves, and at the top the curve for the Stainer already given is repeated. Next below this comes the curve for a good copy of a Stainer. This instrument and the two Stradivarius copies below were very carefully made by Professor Koch of Dresden, as exact copies of particular instruments which he had studied. They are rated by various intelligent people as excellent modern instruments, producing a tone very nearly like old Italian violins.

The Stainer copy shows a relative weakness in the frequency region 600–1200 c.p.s. while the real Stainer's weak region stretches from 600 to 1800. The copy shows a sharp deep valley at 500 c.p.s. (B on A). Though its air-resonant peak is not in the same place, a great many of its body peaks coincide in position with those of the Stainer, a circumstance probably due to the similarity of form and workmanship.

The Stradivarius copies show a weakness in the high frequencies (above 2000 c.p.s.), compared to the Stainer, which tends to make their tone less strident and more agreeable.

Below these come the curves of our "standard of badness." The second one of these, the lowest curve in the figure, was obtained with a paper clip weighing 0.74 gram fastened to the bridge directly under the E string. The two curves show marked similarity (as they should) up to 3000 c.p.s. This instrument radiates on the whole less sound than the better ones in the region 200–1000 c.p.s. (though the Stainer is not much

better in this region). It is, however, quite strong in the range 2000 to 4500 c.p.s., or farther. This gives the tone a shrill and unpleasant quality, the worst of which (due to frequencies higher than 4000 c.p.s.) is eliminated by the inertia of the clip, which seems to filter out these high vibrations from the string, and prevent their affecting the body and thus becoming audible. There remains in this cheap instrument the deficiency in the strength of the low frequencies, a fault which cannot be cured without a more serious operation. It will be interesting to make studies of other modern, carefully made instruments to see how their average curve compares with those of the best violins. This we have not yet had time to do.

Aluminum Violins

Through the courtesy of Professor J. E. Maddy of the University of Michigan, we have been allowed to test three instruments made of aluminum under his direction on a good Stradivarius model. The plates are welded to the sides. The analyses of tones produced by these instruments indicate a lack of response in the lowest octave of the instrument similar to that found by Meinel in a violin whose wood is too thick. The explanation is probably that the metal cannot be made light enough, and still have the necessary elasticity and mechanical strength. The density of the metal is some five times as great as that of spruce wood, and this handicap cannot easily be overcome. Above about 500 c.p.s., however, these instruments respond extremely well. They have a good tone in this

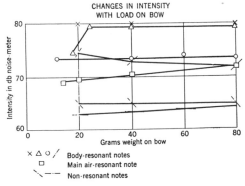

FIG. 12.

range of frequency, and show the highest efficiency (on open strings) that we have measured (see below).

ARE OTHER SOUNDS PRODUCED BESIDE THE PARTIAL TONES?

"Scratch" noises are doubtless produced in violins by the rubbing of the bow over the string, but they are usually weak. They would show on our analyses in the same way as a hiss does, or any other aperiodic disturbance; that is, the whole length of the record would show a continuous agitation, probably stronger in the high frequency end, as hissing sounds usually are. While disturbances do occasionally show, there is no trouble in getting records free from them, that is, not containing any effect as great as 1 db on our intensity numbers. No effect smaller than that could be seen, nor probably heard. Even if much scratch sound were present its volume ought not to be very different from one instrument to another, and it would thus not be a factor in comparing the quality of two violins.

It was thought worth while to test whether or not sounds were present whose frequency was above 10,000 c.p.s., the upper limit of our analyses. This was done indirectly by means of a sensitive flame which "ducked" when sound over 5000 c.p.s. reached it and was sensitive up to 15,000 c.p.s. Using our worst violin in the low part of its range, there was no doubt that the tone was very bad, but the sensitive flame remained undisturbed. The analyses of these notes showed nothing above 5000 c.p.s., and the flame gave the same answer. As higher notes were played, whenever the analyses indicated some important partials over 5000 c.p.s. the flame responded to them. We infer that the unpleasant quality of the lower notes on this instrument is not due to sounds above 10,000 c.p.s., and therefore that they are not likely to be present in the higher notes either. They probably occur as upper harmonics of the very highest notes, but they are very weak. This is shown by the strength of the lower harmonics of these notes which come within our range. Thus if high C' on E is played, its frequency is about 2060, and its successive harmonics are 4120,

6180, 8240, 10300, 12360, etc. Of these we could not record the two last mentioned; but if the one at 8240 is too weak to show on the record (as was often the case) it is safe to infer that the higher ones are negligible.

VARIATIONS OF EMITTED SOUND INTENSITY

A series of tests has been made by means of a General Radio Company noise meter showing how the quantity of sound emitted by a violin at a given pitch in a given direction varies when the bow distance, the bow pressure, or the bow speed is altered. R. B. Abbott[6] has already recorded a few measurements of this sort indicating that when everything else is held constant the total sound emitted does *not* increase with increase of bow pressure. We have tested this with a number of violins and for notes which corresponded to air-resonant and body-resonant peaks, as well as hollows in the response curve. The results vary slightly but not in any systematic way. Fig. 12 shows a few of these observations, in which we used the automatic bow mentioned above. The load on the bow varied from 16 to 80 grams. In only one case was the total intensity of the sound much altered. This was when the load was changed from 18 to 24 grams on one body-resonant note, and the intensity rose from 74.5 to 79.5 db. Subsequent increase in the load to 80 grams made no further change. This initial rise was due to the fact that at 18 grams load the wolf-note had not been "killed" and the tone was not steady. Similar effects could have been produced by too small a load on other resonant notes. In all the other cases tested the changes with bow pressure were very small. On one body-resonant note the intensity fell from 74.5 to 72 db as the load increased from 20 to 80 grams. On one nonresonant note an increase of 1 db was observed in the same range. On one air-resonant note an increase of 2.5 db was recorded, but observations on other instruments showed that their air-resonant notes did not gain in loudness with increase of bow pressure. We conclude therefore (with Abbott) that on the whole no change in loudness of any systematic sort occurs when the

[6] R. B. Abbott, J. Acous. Soc. Am. **7**, 111 (1935).

bow pressure is increased, if the bow distance and speed are kept constant.

Changes in intensity with *bow distance* are well known. When the bow pressure and speed are kept constant but the bow distance is lessened, the intensity increases at a rate of about 2 db per cm shift, the energy in the low partials increasing faster than in the high (determined by a set of analyses). Increase of *bow speed* causes a more rapid rise of intensity. In the range of speed from 10 to 60 cm per second the increase in intensity is not quite linear, being more rapid at the lower speeds. The average increase is between 3 and 4 db for each increase of speed of 10 cm per second.

THE RANGE OF INTENSITY OF THE VIOLIN

The measurements of total intensity mentioned above included a set in which the player made as much sound as possible, and another which was as soft as possible, in each case consistent with producing a good tone. The difference in intensity between these two curves might be called the expression range of the instrument. The usual range is about 30 db at any ordinary distance from the instrument, varying between 25 and 35 db. It varies very little among good instruments. At the peaks of the response curves the loudest tone is much louder, but so also is the faintest tone, because one cannot produce a good soft tone on account of the wolf-note there. It frequently happens (though not always) that the range is least where the actual intensity is greatest. The accuracy with which the range can be measured is unsatisfactory, as it is the difference between two numbers, each of which may vary by 2 db or so with the player's skill.

As examples, a Stainer violin gave a range variation on the G string (over an octave) of from 26 to 30 db, on the D string 26 to 34, on the A string 27 to 32, on the E string 25 to 35 in the first octave and 22 to 35 in the second. A P. Guarnerius (Cremona) violin gave similarly 23 to 28 on the G, 24 to 29 on the D, 23 to 30 on the A, and 22 to 30 and 19 to 29 in the two E string octaves. An excellent Stradivarius (No. 19) gave 23 to 29 on the G string; further

measurements of the range on this instrument were not taken. A $5.00 violin gave 27 to 31 on the same string. Considering the inaccuracy of the measurements one can conclude only that the range has an average value of about 25 to 30 and is approximately constant over the whole instrument. There is (from incomplete observations) no indication that the best instruments have a materially greater range than cheaper ones.

FRICTION BETWEEN BOW AND STRINGS

The fact that increased pressure of the bow on the strings produces on the average no increase in loudness of the tone is a very curious one. From a physical standpoint we see that more work must be done at higher pressures, and this must be converted into other forms of energy; if not into sound, then probably into heat in the string and bow. The question naturally arose whether the force of friction varied in this case in the conventional manner, i.e., directly with the force pressing the bow against the string. If these two have the usual relation, their ratio is the coefficient of friction, and this is a constant quantity. Experiments were made to test this relation.

A violin was suspended on a sort of bifilar pendulum support so that the frictional force could drag the instrument horizontally. A light spring with variable tension could be adjusted to balance the frictional force, and keep the violin at rest. The stretch of the spring measured this force. The measurements showed that the coefficient of friction was constant (value 0.206) over a range of bow pressure from 0 to 80 grams. The bow used was the automatic one with celluloid disks; the value might possibly differ with a hair bow.

One concludes therefore that the extra energy lost at high bow pressures must go into heating the string. The motion of the string under these conditions should be investigated. Throughout the range of weight used the tone produced remained good (a Gagliano violin was used), but beyond 80 grams weight the tone always "breaks down" at the speed used and the string does not vibrate properly.

A Supposed Low Frequency Resonance in Violins

It has been stated that a violin ought to have a strong natural frequency of vibration at a frequency just below the lowest tone (193 c.p.s.) of the instrument. Physical principles show that such a resonant point would greatly increase the response of the instrument on the lower notes of the G string, where it is weak. The fact that the fundamental tone is weak for these notes, as shown by the response curves, indicates the absence of such a natural resonance, or at least a lack of effect due to it. A direct search was made for it by the Raman method, tuning an ordinary G string down to C below, i.e., to a frequency of 129 c.p.s. The resulting curves showed no trace of a peak on three instruments tested in the range from 129 to 193 c.p.s., within which it has been stated to lie.

A test made by tapping a violin with a finger showed the existence of a natural frequency of the instrument in this region but this was easily found not to be due to anything in the body, but to be the fundamental vibration of the tailpiece to which the upper ends of the four strings are fastened. The effect of this vibration is probably negligible, but it might be helpful if its pitch were raised to coincide with one of the lower notes on the G string. This will be tried.

The Mechanical Efficiency of Violins

In the past several measurements have been made of the total amount of sound energy emitted by a violin compared to the amount of mechanical work that must be done in order to produce the tone. This ratio is the efficiency. We have made no measurements of the efficiency, but have measured the work put into several instruments with the apparatus used for the Raman curves, and also the sound intensity as recorded by a microphone at a standard distance. These data yield numbers proportional to the efficiency, and useful for comparisons.

Since the sound intensity does not increase with bow pressure (if speed and distance are kept constant) the efficiency must vary inversely as the bow pressure used. The lightest bow pressures gave efficiencies which varied at different notes by a factor of 2 or 3 on one instru-

ment, and did not differ by as much as 10 percent in the case of two instruments whose prices were in the ratio of 600 : 1. The efficiency has a maximum value at the frequencies of the resonant peaks, especially high at the air-resonant point. It is evident from these measurements that when a player feels that a good instrument sings when he barely touches it, he is not acting as a good measuring machine. In fact the attempt to use efficiencies as a means of distinguishing good from bad instruments seems doomed to failure.

Variations in the Action of a Violin

Moisture swells wood which is not protected from it, adds to its weight, and alters its elasticity. Each of these changes must affect the natural frequencies of the wood of a violin. These effects should be investigated by means of a properly planned set of experiments. Some of the changes observed by players to occur in a little-used instrument when steady playing is resumed may be due to absorption of moisture by the instrument from the player. It is a well-known fact that the human body loses moisture to the air rapidly, so that it could be acquired by a dry violin by mere proximity to a player even without actual contact. Some of the observed moisture effects may be due to the gut strings; metal-wound strings with a gut core are not immune to changes in moisture. As long, however, as violins are left with their inner surfaces in a condition ready to part with or to absorb moisture, we shall always have trouble. It would appear to be an obvious improvement to so treat the wood of the instrument that no moisture could be absorbed by it. It is stated that this experiment is being tried in Germany with good results. Modern technical methods involving thorough drying at low pressure and high temperature and impregnation with some moisture-repellent substance can be applied without difficulty, and the experiment seems to be a very promising one, unless the vibration of the wood is interfered with by such treatment.

The tension of the strings produces a large stress, especially in the top of the violin. The component of this force tending to drive the bridge into the top has been known to produce a caving-in after the lapse of a long time, indicating a wearing out of the wood itself under these extreme conditions. Changes in the behavior of an instrument due to the forces involved in high or low tuning are easily demonstrated. Some of our measurements have shown a shift of peaks in the Raman curves due to relaxing one or more strings, and a regular progression of peaks toward higher frequencies with increase of tension when the pitches of all the strings were raised. The increase in the standard of pitch which has occurred since the time of Stradivarius has undoubtedly strained old instruments injuriously. New and stiffer bass bars have been put in such violins to strengthen them, thereby altering both the mass and elasticity of the tops, and increasing their departure from symmetry. Such changes might be expected to have a considerable effect on the tone quality.

One sometimes meets the statement that the parts of a violin need to get used to one another's presence and learn to cooperate before the instrument can work properly. This might be called the psychology of wood, but it certainly cannot be said to be good physics. One may feel sure that a physical (or chemical?) change must be responsible for any real effects of this sort.

Elastic fatigue is an effect well known in solids, and wood is probably not free from it. If this occurs in violins it is likely to be fatal in time, as rest is no cure for it. In the absence of careful physical measurements one can hardly say whether a certain amount of this type of elastic change is good or bad. Perhaps it is part of the process of "breaking-in" a new violin, but if so one would expect that a change noticeable in a few months of playing would progress and ruin the instrument in a few years.

A more attractive explanation of the changes occurring early in the life of a violin is that the glue may crack partly open, or change in its own properties in such a way as to give the plates of the violin more freedom than they had at first. The purfling in particular offers a possibility. This consists of a little ditch, as it were, dug a third (or even half) of the way through the wood of the violin near the edges, and subsequently filled up with thin wooden strips and glue. Most of the old instruments we have

seen had a microscopic crack in the glue in the purfling which made the main area of the plates of the violin much more free for vibration than they can have been before. It seems (in the absence of a better hypothesis) reasonable to suppose that the rawness of a new violin is in part at least due to the fact that its plates are too tightly bound to the sides, and that a crack in the purfling is desirable, and actually occurs during the process of "ageing" a new instrument. Experiments on this point will be undertaken.

Do the Best Violins Vibrate Long After the Excitation Ceases?

It has been stated that plucking a string is a sufficient test of an instrument; the best instruments ringing on like a bell, while the poor ones stop at once. We propose to test this by a proper experiment, but in the meantime a few remarks may be made. One puts energy into a violin by plucking (or bowing) its string. That energy is lost in internal friction (viscosity) in the string or the wood, or it is radiated out as sound energy. If a violin sounds for an unusually long time it is certainly losing energy by radiating it, or we should hear nothing. It must therefore lose less than normal in internal friction. If violins differ in this respect it must be either that the wood itself (including the varnish) must vary in internal viscosity, or the glue at the edges may offer greater or less chances for frictional energy losses. It is not impossible that a freshly glued instrument may be bad in this respect until the glue has partly cracked free. The internal viscosity of the wood leads us to consider the varnish.

The Varnish

Much has been written on this subject. From a standpoint of cold physics the varnish may alter the mass of the wood, its elasticity, and its viscosity. The mass change is certainly small. If the varnish penetrates the wood it may drive out water imbedded there. It has been observed that a violin may actually weigh less after varnishing than before.

Whether the varnish does or does not penetrate the wood is a matter of very great importance which needs investigation. Fluorescence under ultraviolet light seems to furnish a promising test, and some experimenters have begun with it. The elasticity of the wood is doubtless changed by the varnish, and probably the wood is stiffened by it, the more, the deeper the penetration. If so, the laws of physics tell us that the effects of added mass and added stiffness tend to compensate each other, as they act oppositely on the vibration rate.

The effect of viscosity needs investigation. Meinel shows an interesting pair of curves for a new violin taken before and after varnishing. There is a very small change, but several of the response peaks are slightly diminished in height by the varnish, practically none being shifted. This indicates that the mass and elasticity effects due to the varnish have canceled each other, and the varnish has slightly increased the viscosity. An unvarnished violin should, therefore, ring out for a longer time, after being plucked, than a varnished one. Most violin makers seem to agree that the varnish improves the tone. Hence, if one may generalize from Meinel's experiment, an increase in viscosity is desirable, and the best instruments will ring for a *less* time than poorer ones. More experiments are, however, badly needed.

CRITERIA FOR THE BEST VIOLINS

The quality of the "best" violins is a matter of taste. A biting tone is preferred by some, a softer tone by others. Large violins were once thought to be better; now we prefer the tone of smaller ones. Many violinists are as much affected by the action of their instruments under their hands as by the sounds they produce. Price is to a certain extent a matter of fashion. The tone quality ought to be the most important criterion, and the most desirable response curve would probably be one that is practically uniform over most of the range of the instrument. This ideal is unattainable, but the best instruments may well turn out to be those that approximate most closely to it. Further experiments will be carried out to test this hypothesis.

This research would have been impossible without the enthusiastic cooperation of many people. We are indebted to Dr. H. H. Hall for the analyzer and for much time spent in its use; to Dr. F. V. Hunt for many consultations; to R. B. Watson for skillful help in the recent use of the analyzer; to Miss H. Waldstein for much assistance in observations; to D. W. Mann for the design of the Raman apparatus; and to many friends for their kindness in letting us test their instruments.

Editor's Comments on Paper 3

3 Meinel: *Regarding the Sound Quality of Violins and a Scientific Basis for Violin Construction*

Before obtaining his schooling in acoustics, Dr. Hermann F. Meinel (1904–   ) was a thoroughly trained Master Violin Maker, one of many generations in a family whose interests were traditionally centered in musical instruments in Germany, near Markneukirchen and Klingenthal. Recognizing the narrow limits of constructing violins on an empirical basis, he entered the University of Leipzig, and eventually obtained his doctorate in acoustics under the guidance of Professor H. Backhaus at Griefswald. Meinel's interests and abilities placed him in an excellent position to relate the fast-expanding knowledge of acoustics to a study of the violin, and in 1934 he wrote an essay on violin acoustics that won an award from the Prussian Academy of Sciences. The list of Meinel's publications (see below) shows that his research on the violin has never ceased.

Reproduced in these volumes are three papers, chosen from Meinel's publications, which indicate the general scope of his work on stringed instruments. Probably the best known and most widely cited is from *Elek. Nachri. Tech.* of 1937, translated by Roger Kenneth Walter, which documents the great importance of wood thickness, as compared with other factors in violin making, with respect to vibrational modes, volume, and timbre of sound produced. In other papers he has studied the properties of outstanding violins, the methods of evaluating tone quality and performance, the effects of varnish, arching of the plates, and properties of wood. The article from *J. Acoust. Soc. Amer.* presented in this volume forms a good introduction to much of his work. Two other papers appear in *Musical Acoustics, Part II*.

On page 53 of the article that follows, Meinel discusses the possibility of improving the behavior of a particular violin in a given frequency range by removing wood from a specific area. Meinel remarks that improvement does not always result but depends on the physical state of the violin in question. This point deserves greater emphasis than it has received, since it highlights a basic problem in violin research. A small change that will markedly improve one instrument may adversely affect another. If the configurations of modes of vibration and relative stiffness and damping throughout the wood were the same from instrument to instrument, standard corrections might be prescribed. These factors, vary widely however, due to variations in the wood and methods of construction, precluding any such standardization at the present time. Application of pertubation theory to the study of vibration modes in the violin may provide some helpful answers.

Selected Publications by Hermann F. Meinel

*"Über Frequenzkurven von Geigen," *Akust. Z.*, **2** (Jan. 1937), S. Hirzel Verlag, Leipzig.
"Über Frequenzkurven von Geigen," *Akust. Z.*, **2** (Mar. 1937), S. Hirzel Verlag, Leipzig.
*"Über die Beziehungen zwischen Holzdicke, Schwingungsform, Körperamplitude und Klang eines Geigenkörpers," *Elek. Nachr. Tech.*, **14**(4) (1937), Verlagsbuchhandlung Julius Springer, Berlin.

―――――――――

*Included in *Musical Acoustics, Part II*.

"Über einige neuere physikalische Untersuchungen an Geigen," *Forsch. Fortschr.*, **14**(14) (May 1938), Academie-Verlag GmbH, Berlin.

"Zur schalltechnischen Prüfung der klanglichen Qualität von Geigen," *Z. Tech. Physik*, **19**(10) (1938), Johann Ambrosius Barth Verlag, Leipzig.

"Akustische Eigenschaften klanglich hervorragender Geigen," *Z. Tech. Physik*, **19**(11) (1938), Johann Ambrosius Barth Verlag, Leipzig.

"Akustische Eigenschaften klanglich hervorragender Geigen," *Akust. Z.*, **4**(2) (1939), S. Hirzel Verlag, Leipzig.

"Akustische Eigenschaften von Geigen verschiedener Klangqualität," *Akust. Z.*, **5**(3) (1940), S. Hirzel Verlag, Leipzig.

"Über Frequenzkurven von Geigen," *Akust. Z.*, **5**(5) (1940), S. Hirzel Verlag, Leipzig.

"Akustische Eigenschaften von Geigen," *Physik. Blätter* (1948), Physik-Verlag GmbH, Mosbach/Baden.

"Zur Stimmung der Musikinstrumente," *Acustica*, **4**(1) (1954), S. Hirzel Verlag, Zürich.

†"Regarding the Sound Quality of Violins and a Scientific Basis for Violin Construction," *J. Acoust. Soc. Amer.*, **29**(7), 817–822 (July 1957).

---

†Included in this volume as Paper 3.

3

Reprinted from *J. Acoust. Soc. Amer.*, **29**(7), 817–822 (1957)

Regarding the Sound Quality of Violins and a Scientific Basis for Violin Construction*

H. Meinel

Forschungsinstitut für Musikinstrumentenbau des Deutschen Amtes für Material -und Warenprüfung, Berlin, Germany

(Received October 15, 1956)

Response curves, important both for steady sound and transients, give a far-reaching insight into the objective characteristics of violins: good ones exhibit large amplitudes at low frequencies and small ones at high frequencies, a broad minimum near about 1500 cps, and larger amplitudes between about 2000 and 3000 cps. The musical subjective significance of these physical properties is mentioned briefly. In general, the sound pressure radiated from a violin follows the inverse-distance law, being independent of frequency. The influence of wood thickness is very important, that of the varnish is comparably small. Pine has a greater damping at high frequencies than at low frequencies. This seems to be a good acoustical reason for making important parts of stringed instruments of pine. Sapwood is better than heartwood. Similarly, some kinds of varnish produce more damping at high frequencies than at low frequencies. If a resonance curve is to be imitated in detail, it is necessary to change carefully the wood thickness of certain parts of the violin body. The applicability of present-day scientific knowledge to the construction of violins is here outlined.

INTRODUCTION

THE physical behavior of violins is reviewed as background for a discussion of technical details that influence the quality of the violin sound. More specifically, we first consider response curves of violins, the law of radiation, and the transient response. Then we look at the influence of kind of wood, varnish, and wood thickness. Finally, we summarize present status of violin making on a scientific basis.

A. REGARDING THE SOUND QUALITY OF VIOLINS

A comparison of old Italian violins with present-day ones is an obvious way to start research on tone quality. For example, response curves have been measured for such comparison. F. A. Saunders and his cooperators in the United States deserve praise in this connection as does also G. Pasqualini in Italy. Similarly, H. Backhaus and collaborators in Germany are to be named.

Response Curves

Whereas some have used electrical excitation in recording response curves, we generally prefer to bow the violin in an automatic way. To do so is indeed more troublesome, because it requires sound analyses in small frequency intervals. It does, however, supply curves under rather natural conditions, and that is what the present-day investigation sometimes needs very urgently. Furthermore, these curves are easily reproducible.

If the comparatively small sound pressures at high frequencies vary in their production or measurement by even 10%, the appearance of the resonance curve is changed very slightly, but the difference is readily audible. This surprisingly strong subjective effect arises from the fact that these variations usually concern at least several (with low-frequency sounds,

many) partials in the high-frequency region, where the ear is very sensitive.

Examples of response curves we obtained[1] for each string are represented in Fig. 1, where the frequency is plotted on a logarithmic scale horizontally and the sound pressure measured at a distance of about one meter is plotted vertically on a linear scale. These curves have been obtained from the best Stradivari which I ever had the opportunity to test. This was the concert instrument of Joseph Joachim and Karl Klingler, a violin of fascinating, fine tone quality. Such a group of curves exhibits all the sound spectra generally possible under the conditions of bowing used. In regard to the superb violin timbre the following peculiarities are significant.[2]

(1) Large amplitudes at low frequencies in the response curve mean large amplitudes for the low harmonics of the sounds. Subjectively, this means that the sounds are agreeably sonorous and that they "carry" well.

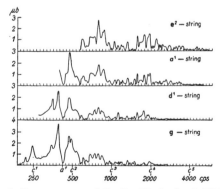

FIG. 1. Response curves of Joachim-Klingler Strad (1715), a violin of fascinating, fine tone quality.

* Presented at the Second International Congress on Acoustics, held in conjunction with the Fifty-First Meeting of the Acoustical Society of America, Cambridge, Massachusetts, June 17–23, 1956.

[1] H. Meinel, Akust. Z. 2, 22–33 (1937); 2, 62–71 (1937); 5, 283–300 (1940).
[2] H. Meinel, Akust. Z. 4, 89–112 (1939).

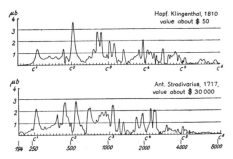

FIG. 2. Response curves of a Strad (a violin of very good tone quality) and a Hopf, Klingenthal (a violin of mediocre tone quality).

(2) Small amplitudes at high frequencies (above about 3000 cps) give the sound a harmonious softness and a fine, pure response; see also the results of F. A. Saunders[3] and G. Pasqualini.[4]

(3) Small amplitudes near 1500 cps prevent a very nasal character.[2] Such a condition is likewise very favorable to the tone quality of other instruments.[5,6]

(4) If the region from 2000 up to 3000 cps is stressed the sound acquires a very agreeable, pithy, and dull brightness. Less good violins do not exhibit these signs of quality to the same degree as shown in Fig. 1.

For the present we are working mostly with simpler curves obtained in a manner which is detailed in the literature,[7] such as those shown in Fig. 2 for Stradivari and Hopf. Again we found that the Stradivari exhibits smaller amplitudes at high frequencies and larger amplitudes at low frequencies. There is an impressive minimum near 1500 cps and then a rise in the region from 2000 cps to about 3000 cps. Therefore, these curves also demonstrate the most essential characteristics of a violin of good tone quality.

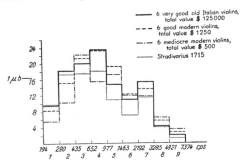

FIG. 3. Average response curves of groups of violins of different tone quality.

[3] F. A. Saunders, J. Franklin Inst. **229**, 1–20 (1940).
[4] G. Pasqualini, Ist. naz. elettroacust. O. M. Corbino (I.N.E.A.C.) **20**, 1–48 (1939); **44**, 1–19 (1943).
[5] E. Thienhaus, Akust. Z. **6**, 34–45 (1941).
[6] W. Lottermoser, Physik. Bl. **4**, 103–109 (1948).
[7] H. Meinel, Akust. Z. **2**, 22–33 (1937).

We have still further simplified by forming average values over intervals of a fifth, for groups of violins of equal worth: that is to say, a group of distinguished old Italian violins, a group of good modern violins, and finally a group of mediocre present-day ones. As Fig. 3 shows, again the best violins have the smallest amplitudes at high frequencies, the largest amplitudes at low frequencies, small amplitudes in the vicinity of about 1500 cps, and a stress of the frequency range from about 2000 up to about 3000 cps. Thus the groups also exhibit the differences which have been stated for single violins. That is why we consider them as typical. For comparison, the curve of the Joachim-Klingler Stradivari has also been drawn in.

Curves for some bad violins[8] are shown in Fig. 4 in comparison with the average curve of the best old Italian master-violins. Here we see that bad violins exhibit large deviations from the mean curve of old Italian violins. Therefore, we consider the mean curve

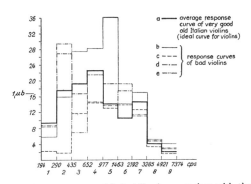

FIG. 4. Response curves of bad violins in comparison with the average response curve of very good old Italian violins.

to be the ideal one for violins; we concede, of course, that the individually different, artistic taste allows deviations. From the physical point of view, however, these deviations are relatively small, especially at high frequencies; see also Fig. 3.

The differences mentioned above refer to beauty of sound, response, carrying power, and quantity of sound which are important properties. If we want to obtain a still more comprehensive picture of a violin's tonal qualities, it is necessary—in addition to the response curves shown in Fig. 1—to record for each string, to make measurements of the dynamic range (the difference between largest and smallest amplitudes), and of the efficiency.[9] In performing all these tasks, several standard violins, some of which have been in my personal possession for more than 20 years, have been faithful helpers. All of them represent instruments physically tested and artistically judged.

[8] See particulars in Akust. Z. **5**, 124–129 (1940).
[9] F. A. Saunders, J. Acoust. Soc. Am. **9**, 96 (1937).

49

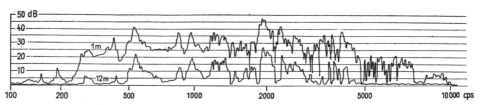

FIG. 5. Violin response curves at distances of 1 and 12 meters. Note that the difference is very near to 21.5 db as required by the inverse-distance law.

Law of Radiation

The response curves shown in Figs. 1–4 were recorded at a distance of about one meter. We now consider the question whether at a larger distance there are any other changes aside from the effect of sound divergence. In winter, when there was deep snow, the surface of which was very loose, we excited violins' electrically to radiate from an open window into the outside without any reflection. The recording microphone was taken as far as 12 meters distance into the open air, without damaging the surface of the snow. At different distances we recorded response curves. These demonstrate that in the absence of reflection the sound pressure varies inversely as the distance and that this relationship is practically independent of frequency. See for instance Fig. 5 which shows the curves obtained at distances of 1 and 12 meters, for which the theoretical difference is 21.5 db. On the whole the evidence indicates that the sound is finally spreading spherically, in spite of the selective radiation at high frequencies.[7] There results no evidence for a preferred radiation at special nodal lines.

If one performs such tests in an unsuitable room or commits any other mistake (such as measurement very near to the violin, 50 cm or less) the $1/r$-law is apparently not satisfied (see Fig. 6). In such a case there is a larger frequency dependence for the decrease of amplitudes and a considerable influence of the various nodal lines of the violin body on the spreading of sound. Also we see that within a distance as small as 12 meters (Fig. 5) the high-frequency sound of a violin is no longer significant. Should the curves not be recorded in the open air, but in a crowded concert hall, the high frequency amplitudes would be even more suppressed due to their stronger absorption.[10]

According to experience, the piercing sound of a violin being played in a concert hall dies away within a short distance. This observation agrees with our measurements, according to which a good violin—a violin with good carrying power—has the high point in its sound spectrum at low and medium frequencies. By contrast, the sound of a violin that has large amplitudes at high frequencies is markedly reduced by the absorption in the concert hall. The effect is even more noticeable subjectively, because the ear is particu-

larly sensitive to sound in frequency region concerned. Such violins do not possess good carrying power.

Many a violinist, playing in an orchestra, wants to hear distinctly the sound of his own violin and eventually to take the lead with his instrument, too. He is therefore fond of a somewhat more penetrating sound. That may be warranted, however, without having anything to do with the longing for fine tone quality. On such concertmaster violins really small amplitudes at high frequencies are not to be found. For this reason I always entrust—if possible—a group of experts with the selection of high-quality violins before testing the instruments. Thus I obtain reasonably clear results.

Transient Response

In order to clarify questions relative to response I but recently showed a prominent concertmaster a very good present-day violin with small amplitudes in high frequencies, in accordance with observations noted above. He felt obliged to say that this violin sounds beautifully and responds more easily than his own, somewhat piercingly sounding Italian violin. Perhaps one may explain this result on the basis of Raman's[11] well-proved theory that the force, required for obtaining the partial vibration of a string, is inversely proportional to the ordinal number. Applied to the transient response, it means that in setting the bow upon the string just at the beginning of the bow pressure, it is the high partial vibrations of the string and thus the high harmonics of the resulting airborne sound that are being produced first, and the lower ones later. If

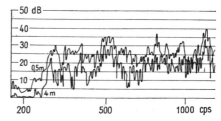

FIG. 6. Violin response curves at distances of 0.5 meter (too near the violin) and 4 meters. The curves differ by much less than the 18 db that would be required by the inverse-distance law.

---

[10] H. O. Kneser, Akust. Z. **5**, 256–257 (1940).

[11] C. V. Raman, Indian Assoc. Sci. Bull. **15**, 62 (1918).

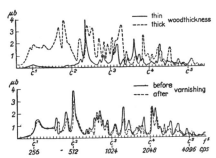

FIG. 7. The influence of wood-thickness is relatively great; the influence of varnish is comparatively small.

the high harmonics have small amplitudes, they are scarcely audible in the transient. The sound that appears without this preliminary high-frequency noise articulates better. This difference of response between violins possessing strong and weak high-frequency amplitudes is audible, especially distinctly by the player himself.

On this occasion we may further refer to the work of J.-G. Helmbold,[12] who in another direction reveals distinctly the relations between the response curve and the nature of transients. Inasmuch as the damping of a violin with its influence on the duration of the transients expresses itself in the resonance curves, one may say that the kind of transients is also contained in the response curve recorded from the stationary sound. Thus if one duplicates the response curve of distinguished violins he likewise duplicates the transient response.

B. THE RESPONSE CURVE AS A TECHNICAL TOOL

We now consider how the response curve can help practical violin making. The influence of wood-thickness changes[7] is relatively great (see Fig. 7). Though not to be disregarded, the influence of varnish is comparatively small. Further influences of the construction have likewise been tested.[7,13,14] Now all these features which change the response curves need to be investigated exactly, in order to make it possible to construct violins exhibiting the desired characteristics.

Kind of Wood

For the most comprehensive researches in respect to the important influence of wood on the damping of violins we are indebted to E. Rohloff.[15] According to his findings, the logarithmic decrement for bending vibration—such as in stringed instruments—from 10 up to 10 000 cps is independent of the frequency.

However, that did not seem right to us. Pasqualini,[16] who also carried out comprehensive investigations, took no especial interest in violin wood and therefore did not study its behavior at high frequencies. E. Ptacnik found a rise in damping of pine with increasing frequency, but according to E. Skudrzyk,[17] he distinguished between tone-pine on the one hand and timber-pine on the other, and the age of the wood. Thus we had to start from the beginning.

Wood such as used in making musical instruments is being tested in this research which is still in progress: pine (spruce), maple, pear, cherry, oak, elm, mahogany, rosewood, ebony, and other kinds.

Between 100 cps and 5000 cps we study rods, about 400 mm long, 20 mm wide, and several millimeters thick. Sticks of half the width yield nearly the same results, so any influence of radiation damping, even in high frequencies, could not be stated from our investigations.† The test method which—as far as we know—has not yet been employed previously, is to fix the rod at one end and to excite electrically the free end at various frequencies. Hanging up the rods at two vibration nodes did not prove to be satisfactory, particularly at high frequencies. The damping is determined by means of the resonance curves, recorded capacitively. Figure 8 shows some results.

The damping of the various kinds of wood used for making musical instruments is different: maple, oak, elm, rosewood, mahogany, and other sorts show a smaller increase of the damping with the frequency than does the typical violin-resonance wood, pine. We find that from 100 cps up to 5000 cps the damping of pine is doubled, a change almost twice as great as for the other kinds of wood. The significance of these results must be judged according to the acoustical demands on the various instruments. As to violins, the requirements are: small sound amplitudes at high frequencies, large amplitudes at low frequencies. Owing to its greater damping at high frequency, pine distinctly complies with the requirements better than do other kinds of wood. It seems this peculiar damping property is an acoustical reason for making the important belly of stringed instruments of pine.

By means of this method one can also demonstrate differences of damping in pine samples which only differ as to their position inside of a piece of tone-wood. Sapwood is better than the heartwood, because it has a distinctly stronger damping at high frequencies. This is a significant result; and there the far-reaching, empirical experience of the violin maker is obvious. The bellies of the stringed and fretted instruments are joined in such a way that the sapwood is placed at the center and heartwood at the edge.

[12] J.-G. Helmbold, Akust. Z. 2, 256–261 (1937).
[13] G. Pasqualini, Ist. naz. elettroacust. O. M. Corbino 22, 622–639 (1940).
[14] F. A. Saunders, J. Acoust. Soc. Am. 25, 491–498 (1953).
[15] E. Rohloff, Z. Physik 117, 64–66 (1940); see also, F. Krüger and E. Rohloff, Z. Physik 110, 58–68 (1938); and E. Rohloff and W. Lawrynowicz, Z. tech. Phys. 5, 110–111 (1941).

[16] J. Barducci and G. Pasqualini, Ist. naz. elettroacust. O. M. Corbino 87, 1–32 (1948).
[17] E. Skudrzyk, Acustica 4, 249–253 (1954).
† It seems, however, that the latest still running investigations made by vacuum show an influence of radiation damping with the highest frequencies here investigated.

51

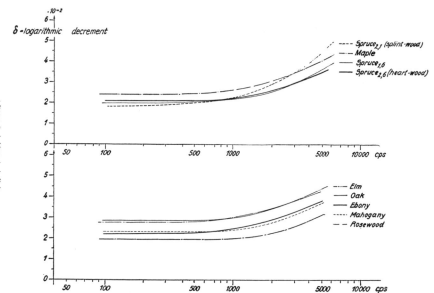

FIG. 8. Logarithmic decrement of different kinds of wood. According to present measurements, the damping of pine (spruce) exhibits a greater increase with the frequency than does the damping of other kinds of wood.

The experiments show, incidentally, that rosewood has small damping at both low and high frequencies, so that it is particularly suitable for fretted instruments that do not produce very much sound. This result of measurement is likewise in accord with many decades of empirical experience.

Influence of Varnish

The same method of measuring damping, by means of vibrating wooden rods, is also used for our (still running) varnish inquiries. From Fig. 9 in comparison with Fig. 8 we see that the varnishing increases the damping. Furthermore, we see that a hard varnish increases the damping at high frequencies less than does a soft varnish.[18]

These measurements of the influence on damping of wood and varnish confirm acoustical results[7.2] formerly obtained by recording in quite another way the response curves of violins. Thus a procedure lies before us for investigating kind of wood and of varnish

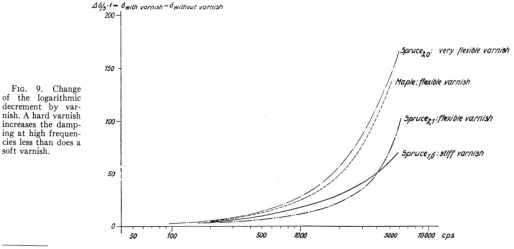

FIG. 9. Change of the logarithmic decrement by varnish. A hard varnish increases the damping at high frequencies less than does a soft varnish.

[18] This varnish was produced by Dr. Ing. Karl Letters, Köln-Lindenthal, Sielsdorfer Str. 9. During the investigations the soft varnishes were not quite dry.

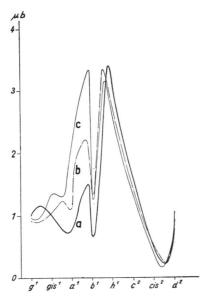

FIG. 10. Influence on the response curve of certain changes in the wood thickness at the left middle bow of the back. Condition *c* for most wood removed.

with reference to its acoustical adaptability to violin making.

Influence of Wood Thickness

In connection with wood thickness, I want to point out particularly a result of F. A. Saunders.[14] He discovered that the loudness can be distinctly increased by diminishing the scoop at the corners. This disclosure appears to me to be of great importance. By means of just such discrete changes in wood thickness, restricted to certain parts of the violin body, we are able to copy small significant properties. Having already formerly started experiments[19] of this kind, we are now continuing them again. Thus, for instance, a resonance range near 500 cps (just above a^2), being originally too weak, is reinforced by gradually removing wood from the left middle bow of the back. (See Fig. 10; the wood was thickest for condition *a*, thinnest for condition *c*.) The same effect can also be obtained by diminishing the wood thickness at certain other parts of the violin body. Probably the effect is influenced by the position of the nodal lines. The effect does not always occur; obviously it depends on the physical state of the violin at hand.

―――――――
[19] H. Meinel, Elek. Nachr. Tech. **14**, 119–134 (1937).

Unduly thinning the wood spoils the result. An elastic aftereffect seems to exist: there may be continual change in an acoustical effect for several days after the change of wood thickness. This may be caused by adjustments in the internal stress set up by removal of the wood, which take place after a few days. Sometimes the desired effect is really obtained, but unwished for effects appear at the same time in other frequency ranges. Everyone who surveys the importance of these few statements will, I think, understand that we are now facing the real problem of making a violin which has the desired acoustical effects. We likewise recognize the extent and difficulty of the work still before us.

C. STATUS OF SCIENTIFIC VIOLIN MAKING

Even before we succeed in mastering the response curves completely, partial results may be obtained, that enable us to recognize whether or not the way taken will be the right one. Figure 11 shows the curve

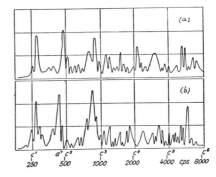

FIG. 11. Response curves of a Strad (above, a) and a modern violin (H. Meinel, below, b) measured by Dr. Karl Steiner, Tübingen, 1950. The approach of the modern violin to the Italian model is good, but the result is not yet entirely satisfying.

of a Stradivari and that of a violin made about 1950 according to the scientific-technical experience of that time. These curves were recorded by K. Steiner at the University of Tübingen, Württemberg. The violin in question is not a consciously intended imitation of the Stradivari, but the striking similarity of the three most important maxima and of the mean amplitudes at high frequencies show that one can arrive at a useful approach to Italian models by use of general principles now available. Although we are standing at the beginning, in view of what is still to be done, there is no doubt that research on violins will solve the most important problems in a measurable time.

Editor's Comments on Paper 4

4 Itokawa and Kumagai: *On the Study of Violin and Its Making*

The research of Dr. Hideo Itokawa (1913–) has ranged widely from aircraft design to brain-wave acoustics (in which he received his doctorate), medicine, oceanography, computer science, and space technology. A professor at the University of Tokyo from 1948 to 1967, he was Visiting Professor at the University of Chicago in 1953. In recent years he has established the Systems Research Institute of Tokyo, the RAND (Research and Development) Company Ltd., and the Associates of Ocean Techniques Development. At present he is Director of the Japan Computer School and a member of the International Space Research Academy.

His work in violin acoustics, represented by the paper included here, was done at the time of his studies of monitoring techniques for levels of anaesthesia in surgical patients. During his trips to the USA he visited Saunders several times to discuss problems of violin research. Of particular interest are Itokawa's rigorous test methods and his experimental findings on many aspects of violin acoustics.

Mr. Yohei Nagai (1938–), who has kindly translated this paper from Japanese to English, is himself an acoustical engineer currently with the Nippon Gakki Company Ltd., in Hamamatsu, Japan. He is a graduate of the University of Tokyo and from 1969 to 1971 studied in France as a French Government Scholarship Engineer, after which he continued his acoustical studies at the Technical University of Berlin, under the direction of Lothar Cremer. Some of his work is reported in Paper 6 by Cremer. His current research covers the acoustics of musical instruments — especially the winds, piano, and electronic organ.

4

On the Study of Violin and Its Making

HIDEO ITOKAWA and CHIHIRO KUMAGAI

*This article was translated expressly for this
Benchmark volume by Yohei Nagai, Nippon Gakki
Company Ltd., Hamamatsu, Japan, from* Report
of the Institute of Industrial Science,
University of Tokyo, **3**(*1*), *5–19 (1952)**

Introductory Remarks

A sound source usually has three necessary elements — first, the vibrational system, second, the communicative part, which generally includes resonant and filtering systems, and, third, the radiation.

In the case of the violin, the string takes the role of the first part, the bridge works as the second part, transmitting and filtering, and the third part consists of the top and back plates as well as the air enclosed in the violin body, which constitutes a Helmholtz resonator with the f-holes.

Thus we can consider a violin as one type of sound-generating apparatus to be investigated as follows:

1. Frequency characteristics.
2. Tone quality (spectrum).
3. Transient characteristics.
4. Efficiency.
5. Directivity.

6. Frequency Characteristic

In 1914, Raman measured the minimum forces applied to the bow in order to make a sound from the violin over the whole region of tones (i.e., frequencies). The curve thus obtained was named a "Raman curve," which is shown in Figure 3; it represents a type of frequency characteristic for the violin body. The first problem

---

*This translation begins with Section 6, page 5, in the original.

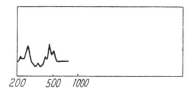

Figure 3. Raman curve.

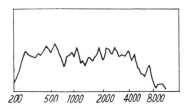

Figure 4. Response curve obtained by Saunders by means of the early method.

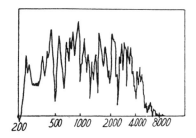

Figure 5. Response curve obtained by Saunders by means of the later method.

that arises in the measurement of a frequency characteristic is the method of exciting the violin body. To obtain the desired sound, Raman fixed a box horizontally and ran the violin appropriately along a track installed parallel to the bow. On the other hand, Saunders made a performer play the violin at each semitone over the whole spectrum of tones. Then, by superposing together all the results and taking the average, he obtained his frequency characteristic. An example of the results he obtained is shown in Figure 4. Saunders then developed another method. That is, a metal wire was inserted in place of the G string. This wire was excited with a sine wave electromagnetically and the frequency characteristic obtained. The curve, however, was very complicated, as seen in Figure 5. To simplify it, the whole spectrum was divided into six regions and the average taken for each region. The resulting six average responses were used as a measure of the frequency characteristic. Table 2 shows the results thus obtained; the numerals given are the values in decibels for respective levels. Subsequently, Saunders made similar measurements for a number of violins by both methods. Kawai (8) applied a sound wave to the violin body, to excite it.

Table 2

Frequency	196~ 349	349~ 784	784~ 1568	1568~ 3136	3136~ 4186	4186~ 6272
9 violon, averages, old method	37.0	42.4	38.0	38.6	37.5	27.3
Same by new method	13.7	23.3	25.3	26.3	23.8	11.6
Difference due to method	23.3	19.1	12.7	12.3	13.7	15.7
7 Strads, new method only	14.1	23.7	25.7	25.9	23.6	10.9
5 Strads, old method only, but corrected to new	14.6	24.4	24.0	25.4	23.1	11.7
10 Strads, both method	14.2	23.8	24.7	25.7	23.9	10.7

The experiment in this connection by the present authors, however, was made as follows. A vibrator was attached on the side of the bridge and a sound wave transmitted to the violin body (Figure 6). The wave thus vibrated the top plate, soundpost, back plate, body as a whole, and the air inside. The sound emitted was measured by means of the sound pressure at the position just above the horizontal violin, 7 cm from the bridge, and the desired frequency characteristic was obtained. The characteristic of the amplifier used was completely flat, as seen in Figure 7. The microphone then was of the velocity type and had a calibration curve as shown in Figure 8. In the violin excitation, the frequency was varied continuously with a beat

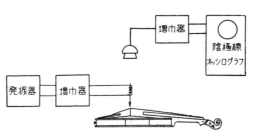

Figure 6. Frequency characteristic measuring device: (1) oscillator, (2) amplifier, (3) amplifier, (4) cathode-ray oscillograph.

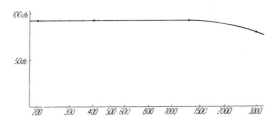

Figure 7. Characteristic of the amplifier (for voltage).

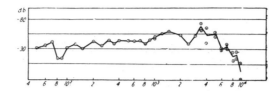

Figure 8. Characteristic of the microphone.

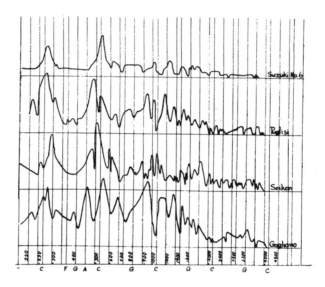

Figure 9. Characteristics of some violins.

frequency oscillator, and peaks and troughs obtained in ·the measurement. The results are shown in part in Figure 9. In the figure, the violin Seiken, trial 1, was produced completely by Kumagai, one of the present authors, including wood selection, design, treatment, working, construction, varnishing, and adjustment. Figure 10 shows the measuring apparatus used, and Figure 11 the violins studied.

The frequency characteristics obtained are irregular, with peaks and troughs appearing one after another. In actual performance with a violin, the tone produced continuously changes amplitude. This is also the case with Stradivarius violins observed by Saunders, as shown in Table 2. In the Stradivarius in general, the frequency characteristic is more or less flat, and the peaks are lower in the middle region. A violin with a frequency characteristic of quite irregular peaks is generally inferior; and an abnormal peak in the high frequency region is characteristic of a cheap violin. The peak of lowest frequency in a characteristic curve is caused by the Helmholtz resonator of air in the violin body. This result is in good agreement with the frequency value calculated by the approximation of f-holes as ellipses. The peak following the first peak corresponds to a fundamental vibration in the front plate; this will be discussed later.

Figure 10. Apparatus measuring a frequency characteristic of the violin.

Figure 11. Three violins measured: trial Seiken 1, Italian Puglisi, and Suzuki 6, from left to right.

When a relatively large-sized microphone such as the velocity type used here is placed close to the sound radiation from the violin body, as described previously, a stationary wave occurs between them, thereby causing an error in the measurement. If then the distance from a speaker as a sound source to the microphone is varied by moving the latter, a peak appears at the point of an integral multiple of the half-wavelength. Therefore, the distance was taken at 7 cm in the experiment. Nevertheless, some error was caused at frequencies over 2,500 cps.

The amplitudes of vibrations in the violin body and their distribution were then measured directly. In the measurement, barium titanate porcelain was used for the pickup, and the receiving end was contacted to the violin body. The voltage induced in the porcelain was amplified and read in the Braun tube. The lever ratio of the pickup was made sufficiently large so as to allow vibration of the violin body; its construction is shown in Figure 12.

The vibration amplitudes in the violin body and the frequency characteristic were measured as shown in Figure 13.

When the frequency corresponds with one of the fundamental vibrations in the front plate, the amplitude of this fundamental mode becomes extremely large, thus giving a peak in the amplitude of the plate. On the front plate there appears then a distinct nodal line with the phases on each side of the nodal line oposite to each other; thus the sound radiation is canceled out, reducing the output considerably. Therefore, the frequency characteristic of amplitude for the violin plate is expected to be much different from the frequency characteristic of sound radiation. However, the measurements of frequency characteristics with the microphone at various distances directly above the violin body showed no differences in the peaks, except for those due to a cavity resonance and to a stationary wave. Therefore, it appears that the frequency characteristic for the maximum amplitude may be used directly as a characteristic of the violin body. In the present experiment, however, for convenience in measurement the frequency characteristic curve was taken by placing a microphone in front of the violin body, at a distance of 7 cm (Figure 14).

The peak at lowest frequency is due to a cavity resonance of the violin body. In this case, the amplitudes of the plates are small, but the sound radiation from the f-holes is very large. The positions of this and the second peak are very important in violin design; they determine the magnitudes of sounds in the low-frequency region. The compass of a violin is four octaves, from about 196 to 3,000 cps (Figure 15). It is generally considered that sound is difficult to produce especially in the low-frequency region (the D- and G-strings). The dimensions of a violin are nearly fixed because of the problem of playing technique. Therefore, the peak of lowest frequency is even up at 300 cps, which is far above the lowest G_2 tone; therefore, the tones on the G-string cannot be much amplified. In designing the violin, however, the resonance frequency can be brought to a suitable position by the choice of area for f-holes and the arching of the plates.

Figure 12. Pickup.

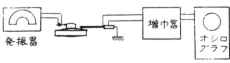

Figure 13. Apparatus measuring a vibration amplitude in the violin body: (1) oscillator, (2) amplifier, (3) oscillograph.

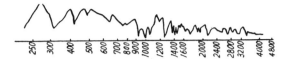

Figure 14. Frequency characteristic of the violin body.

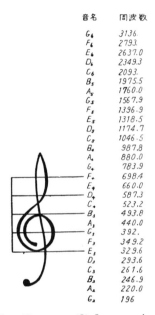

音名	周波数
G_6	3136.
F_6	2793.
E_6	2637.0
D_6	2349.3
C_6	2093.
B_5	1975.5
A_5	1760.0
G_5	1567.9
F_5	1396.9
E_5	1318.5
D_5	1174.7
C_5	1046.5
B_4	987.8
A_4	880.0
G_4	783.9
F_4	698.4
E_4	660.0
D_4	587.3
C_4	523.2
B_3	493.8
A_3	440.0
G_3	392.
F_3	349.2
E_3	329.6
D_3	293.6
C_3	261.6
B_2	246.9
A_2	220.0
G_2	196

Figure 15. Compass of the violin: (1) tones, (2) frequencies.

The second peak corresponds to a fundamental vibration of the front plate; the form of this vibration is most simple. The frequency of this second peak is mainly related to the mass and rigidity of the front plate.

The intensity is then the amplitude of resonance, which is therefore inversely proportional to the amplitude reduction constant of the plate. This constant then is the vibration loss of plate divided by the mass, so the amplitude in resonance is inversely proportional to the vibration loss and is proportional to the mass.

The sharpness in the peak then can be considered as Q in the resonance circuit. Q is inversely proportional to the amplitude reduction constant, and so the sharpness of the peak is inversely proportional to the constant.

In an actual violin, however, it is desirable for the frequency characteristic to be uniform as far as possible. This is because, if there is a salient peak, in producing a sound of fundamental tone or overtone in its vicinity, the balance with other tones is considerably damaged.

By choosing appropriately the vibration loss of plate material and the effective mass in the vibrational portion, the amplitude reduction constant is thus determined, and hence the form of the peak. In actuality, however, it is difficult to know

61

the effective mass of the vibrational portion in the violin performance. The usual practice is therefore to draw a response curve experimentally by the cut-and-try procedure.

7. Calculation of the Cavity-Resonance Frequency

A calculation will be made to confirm the observation that the peak of lowest frequency in the resonance curve is due to the Helmholtz resonator, consisting of air in the violin body, and the two f-holes. The violin body with two f-holes can be considered as the Helmholtz resonator shown in Figure 16. In the figure, air volume V corresponds to the inner volume of the violin body and opening A and A_2 to the two f-holes.

Figure 16. The violin body can be considered as a type of Helmholtz resonator.

The increase in pressure when the air volume V in the cavity has been compressed by dV adiabatically is given as

$$
\begin{aligned}
p &= -\rho c^2 \left(\frac{dV}{V} \right) \\
&= -\rho c^2 (dV_1 + dV_2)/V \\
&= -\rho c^2 (s_1 \xi_1 + s_2 \xi_2)
\end{aligned} \tag{1}
$$

where

$\quad p$: excess pressure
$\quad \rho$: density of air
$\quad c$: velocity of sound
dV_1, dV_2: changes in volume due to the motions of air mass in the openings A_1 and A_2, respectively
$\quad \xi_1, \xi_2$: displacements of air masses in the openings A_1 and A_2, respectively.
$\quad s_1, s_2$: equivalent cross-sectional areas of the two openings

Therefore, the equations of motion for the air mass in A_1 and A_2 are

$$
\begin{aligned}
M_1 \ddot{\xi}_1 &= -\rho c^2 (s_1 \xi_1 + s_2 \xi_2) s_1 / V \\
M_2 \ddot{\xi}_2 &= -\rho c^2 (s_1 \xi_1 + s_2 \xi_2) s_2 / V
\end{aligned}
$$

where M_1 and M_2 are the masses of air in A_1 and A_2, respectively.

62

For an actual violin, $s_1 = s_2 = s$, and $M_1 = M_2 = M$. Then, by substituting

$$\xi_1 + \xi_2 = X$$

the equation is reduced to the ordinary differential equation

$$M\ddot{X} = -\rho c^2 \cdot 2s^2 X/V$$

Substituting,

$$\frac{\rho s^2}{M} = C_0$$

$$\ddot{X} = -2C_0 C^2 \frac{X}{V}$$

Hence, the proper frequency is obtained as

$$f = \frac{C}{2\pi} \sqrt{\frac{2C_0}{V}}$$

Here C_0 is the conductivity for f-holes. It is difficult, however, to obtain the value for the actual f-holes. Therefore, an approximation by the ellipses as in Figure 17 is made. The conductivity for the elliptical openings has been already calculated by Rayleigh (14).

$$C_0 = 2 \sqrt{\frac{S}{\pi}} \cdot \frac{\pi}{2\sqrt{\cos \phi} \cdot F(\sin \phi)}$$

Here
s = cross-sectional area of the elliptical openings = $\pi a^2 \sqrt{1 - e^2}$
$e = \sin \phi$ = difference ratio of the ellipse centers

$$F(\sin \phi) = F(e) = \int_0^{\pi/2} \frac{d\theta}{\sqrt{1 - e^2 \cos^2 \theta}}$$

By actual measurement,
$s = 5 \text{ cm}^2$
$a = 4 \text{ cm}$
$e = \sqrt{1 - \left(\frac{s_1}{\pi a^2}\right)^2} = 0.994, \phi = 84°$

Substituting $F(e) = 3.8$,
$C_0 = 3.31$

Figure 17. Approximation of two f-holes by the ellipses.

By actual measurement,

$$V = 1.84 \times 10^3 \text{ cm}^3$$
$$C = 3.4 \ \times 10^4 \text{ cm/s}$$

Substituting in the preceding formula for f,

$$f = 325 \text{ cps}$$

The value thus obtained is in approximate agreement with 300 cps at the lowest frequency in Figure 9 obtained experimentally.

8.   Calculation of a Fundamental Frequency for the Front Plate

The following calculation was made to confirm the observation that the peak of lowest response next to the air peak, obtained experimentally in the response curve, corresponds to a fundamental frequency of the front plate, and further to clarify the effects of factors related to this vibration.

The front plate is approximated by a plate with flat surface, symmetrical in the longitudinal direction. The proper frequency is calculated by Itokawa's procedure (15). The twelve points are taken as in Figure 18. That is,

(1) $\theta_1 = \quad 0 \qquad r_1 = 0.28939 \qquad w = 0, \quad \dfrac{\delta w}{\delta r} = \ 0$

(2) $\theta_2 = \ \dfrac{2}{3}\pi \qquad r_2 = 0.51298 \qquad w = 0,$

(3) $\theta_3 = \ \dfrac{4}{3}\pi \qquad r_3 = 0.9157 \qquad w = 0,$

(4) $\theta_4 = \ \dfrac{\pi}{2} \qquad r_4 = 1.0000 \qquad w = 0, \quad \dfrac{\delta w}{\delta r} = \ 0$

64

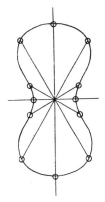

Figure 18. To calculate a frequency in the front plate theoretically, the boundary conditions are given at the positions marked with 0.

For the flexure w at each point, the preceding boundary conditions are also assumed. That is, at all points in (1), (2), (3), and (4), the flexure is taken to be zero; and for the boundary points $\theta = 0$ and $\theta = \pi/2$, the fixed condition, the slope = 0, is used. The equation of vibration for the plate in spherical coordinates can be written as

$$\rho h \ \frac{\delta^2 w}{\delta t^2} \ + \ \frac{E h^3}{12(1 - \mu^2)} \nabla^4 w = 0$$

where ρ is the specific gravity of the plate, h its thickness, E Young's modulus, and μ Poisson's ratio.

For the stationary vibration, with

$$w = \phi(r, \theta)^{e^i w^t}$$

$$\left(\frac{\delta^2}{\delta r^2} \ + \ \frac{1}{r} \cdot \frac{\delta}{\delta r} \ + \ \frac{1}{r^2} \frac{\delta^2}{\delta \theta^2} \right) \phi - k^4 \phi = 0$$

where

$$w = \sqrt{\frac{E h^2}{12 \rho (1 - \mu^2)}} \cdot k^2$$

Substituting the boundary conditions in the following general solution for ϕ,

$$\phi = \sum \left[A_n J_n(kr) + B_n I_n(kr) \right] \left[C_n \cos n\theta + D_n \sin n\theta \right]$$

w is obtained as an eigenwert. For the preceding values, by numerical calculation, the following was then obtained:

$$k = 4.89$$

Substituting the measured values $E = 1.05 \times 10^5$ kg/cm²

$$p = 0.4 \times \frac{1}{980} \text{ g-s}^2/\text{cm}^4$$
$$\mu = 0.3$$

and from the actual length on the longer side of 35.5 cm,

$$k = \frac{4.89}{35.5/2} = \frac{4.89}{17.7} = 0.278$$

Hence,

$$f = \frac{1}{2\pi} \sqrt{\frac{E}{21\rho (1 - \mu^2)}} \cdot h \cdot k^2 = 688 \text{ cps}$$

Compared to this calculated value, the experimental value is $f = 550$ cps, which is about 20 percent smaller. The cause for this seems to be that in actuality $\theta_1 = 0$ and $\theta_4 = \pi/2$, and thus the fixed boundary conditions do not hold; since they are close to the supporting conditions, the boundary conditions tend to lower the frequency.

The preceding formula for f is sufficient to give information on the effects of factors related to the proper vibration of the front plate. The following are thus revealed:

1. The frequency is proportional to the thickness of the plate.
2. The frequency is proportional to the square root of the ratio between the modulus of elasticity and the specific gravity.
3. The value of k is also determined by the flat surface of the plate. This can also be shown approximately by calculation.

9. Effect of the Position of the Soundpost on the Frequency Characteristic

It is generally believed that the characteristic of a violin is delicately changed by the position of the soundpost. Therefore, the frequency characteristic was measured for a trial violin by varying the position of the soundpost as in Figure 19. The results are shown in Figure 20. As seen in the figure, the heights of peaks in the high-frequency region are increased as the position of the soundpost is moved closer to the bridge; and the sound in the high-frequency region is thus improved.

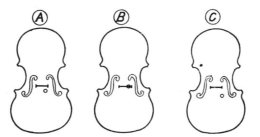

Figure 19. Three different positions of the soundpost, as indicated by 0.

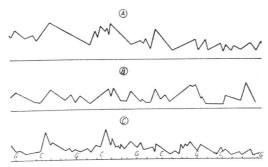

Figure 20. Frequency characteristics in the violin body for the three different positions of the soundpost.

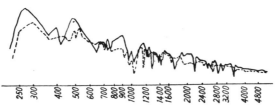

Figure 21. Change in frequency characteristic for the violin body by adjustment of the position of the soundpost is evident.

Then the impedance in the back plate is added to the effective mass in vibration, as described in section 6. This is applied to the impedance in the front plate through the soundpost between the two plates, almost directly below the bridge. The position of the nodal line is thus changed by the position of the soundpost. The impedance is of course changed by the fact of contact between the soundpost and back plate.

It is pointed out by a number of workers that the sound radiation from the back plate is very small (13). According to the authors' results, the amplitude in the back plate is far smaller than in the front plate.

An examination of Figure 20 shows how the frequency characteristic is changed by the additive impedance from the back plate and soundpost. Therefore, a desired characteristic in the low-frequency region can be obtained by suitable adjustment of these members. Figure 21 shows a uniform characteristic attained by such a procedure.

10. Vibration Transmission Characteristic in the Bridge

Bridges with shapes as shown on the left side of Figure 22 are used currently; the material is maple. Saunders reported in his paper that the frequency characteristic in the violin body can be adjusted by means of the width of the bridge at its shoulder; and squeaking sounds in the high-frequency region can be prevented by using a paper clip just below the E-string. The vibration of strings is transmitted to the violin body mostly through the bridge, and so it is evident that the shape of the bridge plays an important role. To examine this role the vibration transmission characteristic of the bridge may be utilized. For this purpose, barium titanate porcelain was inserted between the bridge and violin body, and the voltage induced in it was measured. Alternatively, in place of the violin body, an iron surface plate of extremely large impedance was used; the measuring procedure was similar to the

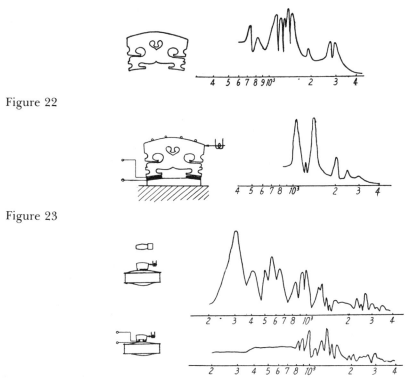

Figure 22

Figure 23

Figure 24

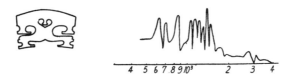

Figure 25

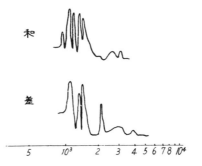

Figure 26. Addition of the vibrations at the bridge two legs, and the difference between: (1) addition and (2) difference.

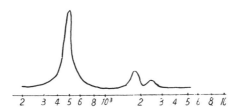

Figure 27. Characteristic of the bridge with a damper.

former case. The results thus obtained were taken as vibration transmission characteristic of the bridge (Figure 22).

The results of measurements are shown in Figures 23, 24, and 25; Figure 23 is for the iron surface plate and Figure 24 for the violin body. In both figures, the frequency characteristic for sound radiation from the violin body is also given.

As seen in Figure 25, by increasing the cutting in the bridge, the peak heights in the high-frequency region are reduced; the peaks are now in a lower-frequency region, and the region with peaks is increased.

In the frequency characteristic of sound radiation from the violin body, the peak heights are low especially between 2,000 and 4,000 cps. Because of this the harmonics of sound in the violin performance are changed, and thus also the tone color.

The directions of vibrations in the bridge in violin performance can be examined by use of a differential transformer. The voltages induced at both the legs of

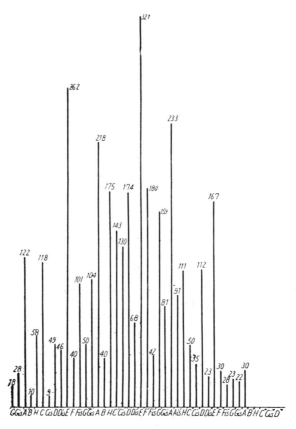

Figure 28a. Kreutzer Sonata, Beethoven.

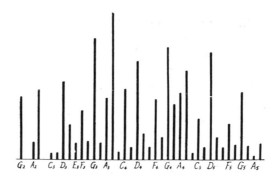

Figure 28b. Concerto No. 1 in G Minor, Max Bruch.

70

the bridge were measured, the values added, and the differences between obtained. In this way, a relation in phases could be obtained.

The results are shown in Figure 26; the characteristics are similar for both the added and the subtracted values. It thus appears that there occur vibrations of a phase and its antiphase simultaneously at both the legs.

The characteristic curve was taken with a sound damper attached to the bridge. The peak is then single, and also very sharp; it is far down in a low-frequency region. The effect of a damper is thus exhibited. The response in the violin body is now in the low-frequency region; but there exist few harmonics (Figure 27).

11. Tones Appearing Most Frequently in Violin Performance

The tones in a violin range from G_2 to E_6, as shown in Figure 13. If the intensities of respective tones differ fron one tone to another as described already, not only the tone colors, but the sounds themselves should also differ. In violin design it is nearly impossible for every tone to have its best intensity and color; therefore, steps must be taken to satisfy those conditions for the more important tones.

The frequencies of each tone appearing in the performance of a violin were examined by the authors as follows. All the violin pieces broadcast from NHK in

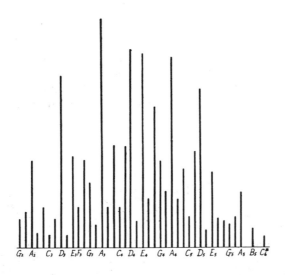

Figure 28c. Violin Concerto in D-major, Tchaikovsky.

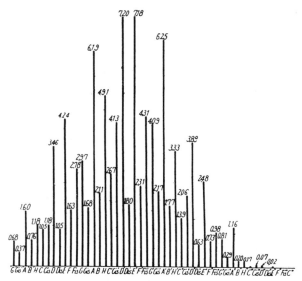

Figure 29. Occurrences of respective tones for the violin music pieces broadcast in 1948.

1948 were collected. For each piece of music, the tones were sorted, and then grouped in 1, ½, . . ., for the crotchet, quaver, . . ., respectively. The performance times for respective tones were then computed in seconds, according to the slow and fast in a metronome. The frequencies of each tone appearing in a piece of music thus obtained are shown in Figure 28a for Beethoven's Kreutzer Sonata, in part b for Max Bruch's Concerto No. 1 in G minor, and in part c for Tchaikovsky's Concerto in D-major. Figure 29 shows the total results for the musical pieces broadcast in 1948, which appears to provide general information on the frequencies of appearance of the tones for a violin. As seen in the figure, D_4 and E_4 are the highest in occurrence, followed by A_3 and A_4. Therefore, these tones may be important for violins. It is thus necessary to pay special attention to the four tones, in both tone volume and quality.

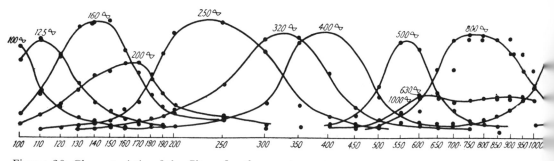

Figure 30. Characteristic of the filters for the spectrometer.

12. Measurement of the Tone Quality, and the Results

To examine tone quality, instead of the mechanical vibration already described, a human being played the violin with a bow. The tones produced were received in a microphone and then amplified; the outputs in respective bands were obtained by means of twenty filters (Figure 31) with characteristics as shown in Figure 30. The performers were Mr. Takuyo, a professional, and Mr. Kumagai, an amateur. The tones used were A_4 on the E-string, A_3 on the open A-string, D_4 on the A-string, E_4 on the open E-string, and B_4 on the E-string. A comparison was also made with vibrato and without. In addition, a viola was played, producing the same tones. Several violins were used, including Suzuki, Puglisi, trial 1, and Seiken, and also French and German violins. Figure 32 shows the zone wave filters and spectrograph used. The results obtained are shown in part in Figure 33. In all cases, the fundamental tone is unexpectedly small; sound energy is mostly in partial tones in the higher-energy region. This can be explained by the phenomenon of a combination tone (difference tone), as also described by Fletcher (16). The tone A_3 of 440 cps, for example, is made up of 440, 880, and 1320 cps. Of this series, even if 440 does not exist, the difference tone $1320 - 880 = 440$ by the combination of 880 and 1320 occurs, giving out the same pitch of sound; the tone is thus explained by the nonlinearity characteristic of the ears.

In examining the sound spectra in Figure 33 with the concept just described, the following idea occurs; that is, two methods are available for producing any given pitch. One is by the fundamental tone, and the other is mainly by the difference tone. To produce the tone A, it may be either by simple tone 440 or by the combined 880 and 1320. In the former case, it is the generating tone itself, that is, A. Which of the two tones sounds better to the human being?

According to the authors' experiment, although not conclusive, in a superior violin the difference tones are generally small, and reliance is made on the generating tones. One feature of such violins is that the tone quality is not changed by distance or because of some obstacle; this is an important property to the violin in performance. If these observations are correct, the following are possible. In a violin

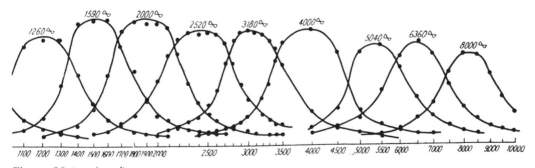

Figure 30 *(continued)*.

73

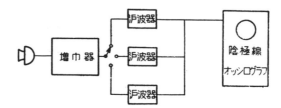

Figure 31. Apparatus for analyzing the frequencies: (1) amplifier, (2) filter, (3) filter, (4) filter, (5) cathode-ray oscillograph.

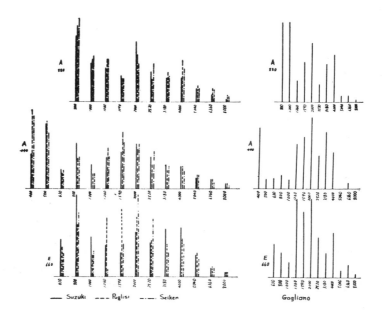

Figure 32. Zone wave filters and spectrograph used.

Figure 33. Sound spectrum of violins.

dependent mainly on the difference tones, the sound is more susceptible to damping from a distance and at the same time diffraction is minimized, since the frequencies of the carrying waves for the difference tones are high. Furthermore, owing to the high frequencies of the carrying waves, the sound is somewhat offensive to the ear. To prove this, however, further experiments are necessary in the future.

13. Varnish and Damping Constant

The problem of varnishing a violin is considered very important both by performers and violin dealers. Some people even go to the extreme of stating that the superiority of a violin is determined only by the varnish applied. Apart from the artistic value of a color itself, the varnishing is believed to have an influence that is physical in meaning, that is, moisture prevention, an increase in the elastic strength of the plate, an increase in the specific gravity, and the damping constant. There is however a question as to whether a violin is so influenced by the varnish as is pointed out generally. In this respect, an experiment was made by the authors with the apparatus shown in Figure 34. Sample specimens were prepared from the material (spruce) left over in making the trial Seiken. To these, the following were applied as varnish, respectively: (1) lacquer varnish, (2) ester gum solution, (3) short oil varnish with ester gum, and (4) short oil varnish with phenol resin. The damping constants before and after the varnishing were measured, and the difference in logarithmic damping constants was taken. The procedure of the experiment was similar to Akiyama's (7). That is, a sample specimen of 30 by 2 by 0.5 cm was suspended with two strings, and the measurement was made by means of a bending vibration. At each end of the specimen tiny iron sheets were attached; one was excited with the electromagnetic coil of a telephone receiver, and the vibration was picked up at the other iron sheet similarly. The suspension was at the positions of nodes of the bending vibration in the sample specimen, which was excited at the fundamental frequency of bending vibration. Upon turning off the switch, the damping curve is registered in the electromagnetic oscillograph via the receiving coil, and the photo of the oscillogram obtained. From this the logarithmic damping constant is calculated. According to the results obtained, it is 1.38×10^{-2} before varnishing, which increased by 20 percent to 1.67×10^{-2} when varnished on one side. The proper frequency is considered not to have been much changed due to

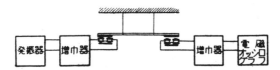

Figure 34. Apparatus for measuring the damping constant: (1) oscillator, (2) amplifier, (3) amplifier, (4) electromagnetic, oscillograph.

cancellation of the increase in mass of 5 to 10 percent with the change in the modulus of elasticity; this is in agreement with the results of Meinel et al. The increase in damping constant reduces the sharpness (Q) at the resonance position, and so the top of the peaks in the frequency characteristic curve may be somewhat flattened. In actuality, the comparison in frequency characteristic before and after the varnishing for Seiken attests to this fact. In this way, the uniformity in frequency characteristic required of a superior violin is improved; on the other hand, due to the large value of damping constant, there also are drawbacks, such as in the ease of violin performance and the decrease in tone volume. The problem is then a compromise between good and bad effects; any other influence due to varnishing is not considered at the present stage. Generally, a violin is varnished only on the outside, with the inside intact. From the standpoint of moisture prevention, the varnishing may have to be done on the inside. In this case, however, the damping constant will be considerably increased, thereby affecting the tone volume.

14. Extension in the Tones

The extension in tone of a sounding body is generally related to the transient characteristic; for the acoustic (hearing) sense, it has a large influence on the tone color. If the extension of tone is small, that is, the damping large, the tone tends to be less vivacious. This also applies to the violin. With extension, the tones become more enriched and refined. What affects this is, of course, the vibration loss in wood and strings due to internal friction; the smaller the logarithmic damping constant in the wood, the larger becomes the extension of its tones. As described already, the amplitude reduction constant is expressed as $2\Delta = b/m$, with mass m in the denominator. Therefore, the extension of tones can be adjusted by means of the mass in the vibrating portion. By the use of a stiffener attached to the violin as described later, for example, the extension of tones could be largely increased. Furthermore, as also described previously, the impedance added to the front plate can be varied by the position of the soundpost. By adjusting this position, therefore, the extension of tones from the four strings can be appropriately produced. In the present experiment, for convenience, the time from the start of the vibration of a string until the sound therefrom has disappeared completely for the hearer was measured with a stopwatch. By this method, the values obtained are always constant. According to the results obtained by the authors, the time is about 3 s on average for the E-string, 4 s for the A-string, 8 s for the D-string, and 9 s for the G-string. The vibration loss in wood increases generally by applying varnish to it, which thus reduces the extension of tones in the violin. According to the experiment made, if a rapid-cure varnish is applied thinly, it does not much penetrate into the wood; therefore, the extension of tones is hardly affected.

76

15. Experiment on Drying in a Vacuum Chamber

According to the measurements by Saunders, an old violin is generally lighter in weight than a new violin. The average weight for seven Stradivari is 283 g, and for another six old violins 374 g. The average for thirty new violins is 410 g, about 7 percent heavier than that for the Stradivari. The weight of the trial violin made by the authors was 420 g. The lightness of a violin may be a desirable factor, since it is then easier to vibrate the material. The lighter weight of old violins may be caused by their aging. In this connection the authors examined whether the properties of a new material would be changed by forced drying in a vacuum chamber.

Vacuum drying (18, 19) is a modern means of drying wood developed recently; it is now attracting fair interest as a quick, practical method. Usually, of course, it is applied for drying green wood. In the present experiment, however, the spruce wood naturally dried for use in violins was used; we examined whether further drying could be obtained by the vacuum method.

The drying furnace used is shown in Figure 35. A stick of spruce wood, 20 by 2 by 0.5 cm, was supported in a porcelain cylinder, as shown in Figure 36. A nichrome wire was wound around the cylinder; the voltage was controlled with a slide resistance (Figure 36), and the temperature with a thermocouple.

The drying conditions used were as follows:

Temperature (°C)	Pressure (mmHg)	Time (h)
80	200	5
80	100	7
80	50	5
80	10	7

The results were

	Damping constant	Mass
Before drying	1.84×10^{-2}	8.05 g.
After drying	2.03×10^{-2}	7.24 g.

When the sample specimen treated was kept in the atmosphere at a relative humidity of about 70 percent with a temperature of 30°C for one week, it returned to the measured values of the state before drying. On examination, the reduction in mass and the increase in damping constant immediately after the drying were nearly the same, at about 10 percent each. It thus appears that the increase in damping constant was not caused by the change in internal friction, but by the apparent change due to the reduction in mass. According to available figures, the equilibrium

77

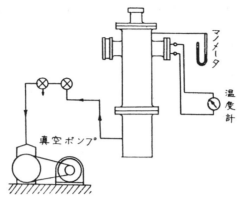

Figure 35. Vacuum-drying apparatus used: (1) manometer, (2) thermocouple, (3) vacuum pump.

Figure 36. Heater used in vacuum drying: (1) sample specimen.

moisture content in wood is 13 percent for a temperature of 30°C and relative humidity of 70 percent. It can thus be assumed that the sample specimen was brought nearly to an absolutely dry state by the vacuum-drying method.

16. Forms of Vibrations in the Front Plate

Peaks in the response curve for the violin body correspond mainly to the normal vibrations in the front plate. To improve the response curve, therefore, these normal vibrations in the front plate must be changed. In the present experiment, the forms of normal vibrations were measured for the violin plate.

The measurement procedure is indicated in Figure 37. As already described, when the excitation frequency applied to the violin body corresponds to one of the

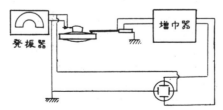

Figure 37. Apparatus for measuring the forms of vibrations in the violin body: (1) oscillator, (2) amplifier.

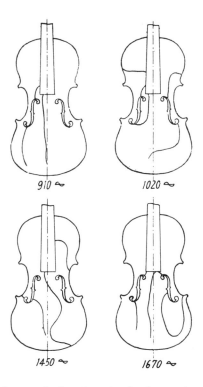

Figure 38. Example of the forms of vibrations in the front plate.

normal vibrations in the front plate, this normal vibration only is enhanced; therefore, the vibration in the front plate has a distinct nodal line. The frequency of this normal vibration can be determined from the peak in the response curve. The location of the nodal line is then determined as follows. Using the pickup described in section 6, the outputs of the pickup and the outputs of the oscillator are simultaneously exhibited on a Braun tube as ordinates and abscissas, respectively; a Lissajous figure is obtained. The shape of this figure is generally elliptical. But the direction of its inclination is changed at the position of a node, from which, therefore, the nodal line can be determined.

The measured results for one violin are shown in Figure 38. Figure 39 shows the normal vibrations when the excitation is in a direction parallel to the front plate. Two nodal lines appear perpendicular to this direction: one passes nearly through the position of the bassbar, and the other through the position of the soundpost.

The forms of vibrations thus observed are totally common to other violins. Therefore, they are not related to the distribution in the front plate of either mass or stiffness, but may be determined by the method of excitation applied and the construction of a violin. (If stiffness in the front plate is taken as the combined bassbar and soundpost, the forms of vibrations are of course determined by this stiffness.)

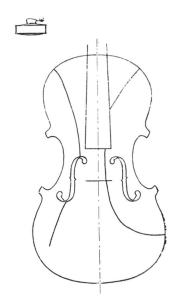

Figure 39

Measurements in performance with a bow, as in an actual case, were not made by the authors. However, the results of such measurements by Kessler at the Massachusetts Institute of Technology (20) will be cited. Generally, when the violin is played with a bow, many forms of vibrations in the violin body are induced by the harmonics of vibrations in the strings. In this case, if the frequency of a fundamental vibration in the string corresponds to that of a fundamental vibration in the front

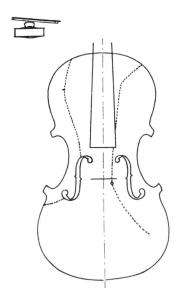

Figure 40

80

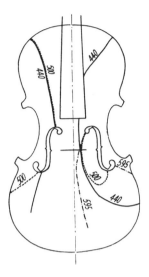

Figure 41

plate, the amplitude of this fundamental vibration in the front plate becomes extremely large; therefore, the form of this vibration can be measured. The result of such measurement is shown in Figure 40. The forms of vibrations shown also appear in a viola and cello. This result is in good agreement with that obtained with sine-wave excitation by the authors.

In further experiments it was shown that, for the few harmonics immediately above the case just mentioned, the nodal lines also pass through nearly the same positions (Figure 41). For the harmonics of higher orders, the forms of vibrations in the front plate become complicated; the positions of peaks are also close together.

In the equation of vibration given in section 8, ph gives the mass per unit area, and $Eh^3/[12(1-\mu)^2]$ the stiffness. Therefore, if the material is fixed, the mass is proportional to the thickness and the stiffness to the cube of the thickness.

It is thus seen that the resonance frequencies in the plate are proportional to its thickness.

As will be described in the following section, the rules for the forms of vibrations in the front plate thus revealed play an important role in artificially changing the frequency characteristic of a violin.

17. Improvement of the Characteristic of a Violin

As described already, in the low-frequency region the amplitude of the plate is large, but the intervals between peaks are widely separated. In the tones on the G-string, the fundamental tone is weak. It is therefore desirable to bring the frequency response as low as possible. One such method for bringing the resonance frequencies down is to increase the bulge of the violin body. Or, alternatively, the area of f-holes may be reduced, which, however, lowers the intensity of sound radiation. By the factor indicated, the spectrum in the low-frequency region is determined. But it is therefore related to the positions of harmonics for the tones and also to tone color, so care must be taken. It is then also desirable to lower the resonance frequencies in the front plate as far as possible. The resonance frequency in the plate (15) is generally given as

$$f = \frac{h}{2\pi} \sqrt{\frac{E}{12\rho(1 - \mu^2)}} \cdot k^2$$

Therefore, for the given material, the thickness of the plate may be reduced. By this method the sharpness in peaks can also be reduced.

For plates of like shapes, the ratios of frequencies between harmonics are generally the same. By reducing the thickness of the plate, therefore, the higher harmonics are brought down; and overall the region of frequencies becomes small. The consequence is then a reduction in tone volume in the high-frequency region; besides, the tone color becomes less vivacious. "The tone becomes confined if the plate is thin," is often said; this is due to this phenomenon. To cope with this problem, the whole region concerned will be divided into the lower and higher regions and an attempt made to bring up the response in the latter region. Substituting the following in the resonance frequency formula,

$$\frac{Eh^3}{12(1 - \mu^2)} = s \qquad \rho h = m$$

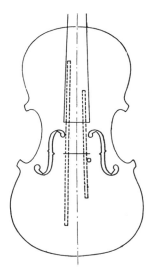

Figure 42

then

$$f = \frac{1}{2\pi} \sqrt{\frac{s}{m}} \cdot h^2$$

Consequently, if stiffness only is increased for the modes of vibrations in the high-frequency region, the desired purpose can be achieved.

One method of doing this is to utilize the fact that nodal lines in the low-frequency region always pass through the same positions, as described already. By attaching stiffness at the positions along the nodal lines, therefore, hardly any effects of stiffness and mass are produced in the low-frequency region; but in the high-frequency region, where the forms of vibrations are now changed, the effect of stiffness is produced (Figure 42).

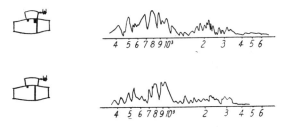

Figure 43. Effect of the stiffeners.

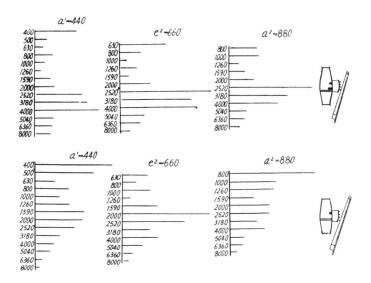

Figure 44

The procedure described was tested for the trial violin. The response curve was changed as shown in Figure 43. Measuring the forms of vibrations showed that several in the low-frequency region were not changed at all. But in the higher region, frequencies moved up; in the response of the violin body, the sound radiation of higher tone was increased. And in the sound spectrum when played with a bow, the harmonics were increased, as shown in Figure 44. The tones reaching the ears became more vivacious, and the tone volume on the E-string was increased. Furthermore, the extension in tones was also increased greatly.

The compass of a violin is usually up to about 3,000 cps. Therefore, if the harmonics produced are included, the region of frequencies is extended considerably. The tone color, however, is impaired by too large a volume of harmonics. A suitable compromise is then necessary, which may be obtained by adjustment of the position of the bridge.

18. Conclusion

Various aspects of a violin have been studied both experimentally and theoretically, including its construction, strength, manufacture, varnishing, frequency characteristic, tone quality, and form of vibration. Furthermore, to confirm the results of studies, a violin was constructed in trial.

In the course of the present study on the violin itself, voices from music circles have called for a study of the bow. This is now in progress, and the results will be published in the future.

The authors are deeply indebted to the following people for their aid in the present work: Hirotsugu Shinozaki, Sukekazu Hattori, Sadatada Maki (NHK), and performers Takuyo, Hisataka, Masako Suwa, and Chieko Sakurai. Thanks are also due to the late Professor Masuno and others for varnishes, Yasuo Kanazawa for the computation of frequencies in the front plate, and Takeo Shiga for the making of barium-titanate porcelain.

The studies described were made in Itokawa Laboratory, The Institute of Industrial Science, during the 5 years from 1947 to 1952 with research funds from the Ministry of Education for fiscal 1950 and 1951.

References

1. *Encyclopedia of Violin*
2. Raman, C. V., *Phil. Mag.*, **39**, 535 (1920).
3. Backhaus, H., *Z. Physik*, **62**, 143 (1930).
4. Saunders, F. A., *J. Acoust. Soc. Amer.*, **9**, 81 (1937).
5. Saunders, F. A., *J. Franklin Inst.*, **229**, 1 (1940).
6. Saunders, F. A., *J. Acoust. Soc. Amer.*, **17**, 3 (1946).
7. Akiyama, *Bull. Sci. Eng. Res. Lab.*, **1**(3), 38 (1947).

8. Kawai, H., *Shindo (Vibration)*, **1**, 1 (1947).
9. Fukada, E., *J. Phys. Soc. Japan*, **5**, 5 (1950).
10. ———. *J. Acoust. Soc. Japan*, **7**, 2 (1951).
11. Kohashi, Y., *J. Acoust. Soc. Japan*, **8**, 1 (1952).
12. Itokawa and Kumagai, *J. Acoust. Soc. Japan*, **7**, 1 (1951).
13. Meinel, H., *Akust. Z.*, **2**, 22, 62 (1937).
14. Rayleigh, *Theory of Sound, Vol. II*, §306, p. 178.
15. Itokawa, H., "A General Solution on the Vibrations in Flat Boards," *Oyorikigaku (Applied Mechanics)*, **2**, 10 (1949).
16. Fletcher, H., *Speech and Hearing*, p. 245.
17. Ishibashi, M., "Paints and Varnish."
18. Matsumoto, B., "Methods of Drying Woods."
19. Wood Technological Association of Japan, *Tech. Note No. 26*.
20. Kessler, J. A., *J. Acoust. Soc. Amer.*, **19**, 5 (1947).

Editor's Comments on Paper 5

5   Schelleng: *The Violin as a Circuit*

Mr. John C. Schelleng (1892 –   ) has been actively engaged in violin research since his retirement in 1957 as Director of Radio Research in the Bell Telephone Laboratories at Holmdel, New Jersey. His pioneering application of circuit-theory concepts to the violin stems from a combination of his engineering skills and his lifelong interest in the cello and quartet playing. As one of the small group associated with F. A. Saunders' violin research in the 1960s, Schelleng is a founding member of the Catgut Acoustical Society, and, since Saunders' death, the recognized leader of its technical activities. Four of his major papers to date are included in this volume — Papers 5, 12, 19, and 27.

Schelleng is responsible for many of the comments, abstracts, and translations in this Benchmark effort. Without his help in many ways, this work could not have been brought to fruition.

Selected Publications by John C. Schelleng

"The Violin as a Circuit," *J. Acoust. Soc. Amer.*, **35**(3), 326 – 338 (1963).
"The Acoustical Effects of Violin Varnish," *J. Acoust. Soc. Amer.*, **44**(5), 1175 – 1183 (1968).
"The Bowed String and the Player," *J. Acoust. Soc. Amer.*, **53**(1), 26 – 41 (1973).
"The Physics of the Bowed String," *Sci. Amer.* (Jan. 1974).

In the *Catgut Acoustical Society Newsletter*

"On Damping of Wood Vibrations and the Effect of Shear; Also Some Thoughts About the Violin Arch," No. 2, 1964.
"On the Physical Effects of Violin Varnish," No. 4, 1965.
"Power Relations in the Violin Family" (abstract), No. 5, 1966.
"On the Physical Effects of Violin Varnish: Part II. Experimental Results," No. 7, 1967.
"On the Physical Effects of Violin Varnish: Part III. Estimation of Acoustical Effects," No. 8, 1967.
"Letter to Editor About Teaching Advantages in Using Tenor and Soprano Violins," No. 8, 1967.
"On Vibrational Patterns in Fiddle Plates," No. 9, 1968.
"On Polarity of Resonance," No. 10, 1968.
"The Sound of Strings," No. 10, 1968.
"Requirements for Sounding Board Material," No. 11, 1969.
"The Bowed String" (abstract), No. 11, 1969.
"Negative Resistance and the Bowed String," No. 12, 1969.
"Pressure on the Bowed String," No. 13, 1970.
"Wood Translucence and Violin Making," No. 13, 1970.
"The Action of the Soundpost," No. 16, 1971.

5

Copyright © 1963 by the Acoustical Society of America

Reprinted from *J. Acoust. Soc. Amer.*, **35**(3), 326–338 (1963)

The Violin as a Circuit

JOHN C. SCHELLENG

310 Bendermere Avenue, Asbury Park, New Jersey
(Received 6 August 1962)

The paper applies elementary circuit ideas to bowed-string instruments and their component parts. Parameters are defined and calculations based on simple circuit diagrams for the main resonance and the air resonance; curves describe their combined performance. The relative importance of circuit resistances—wood loss, radiation, and wall-surface loss—is discussed. Wall-surface loss is an important component of air decrement. No material improvement is to be expected from change in decrement or enclosure volume.

A theory for the wolfnote is given in terms of the beating of two equally forced oscillations, together with a criterion for its occurrence and a method for its elimination.

The paper analyzes principles of dimensional scaling between members of the violin family and shows why the cello and viola are more susceptible to wolftone than the violin.

A study of impedance requirements in wood shows that flexural similarity depends on the parameter c/ρ (compressional velocity over density); high values are in general favorable in the top plate. In the violin, cross-grain losses probably exceed those along the grain.

INTRODUCTION

THOUGH the use of circuit concepts is a standard practice in acoustics, in the specific field of the bowed-string instrument they have hardly been emphasized to the degree which their usefulness justifies. The violin family presents many unsolvable problems; its shape and the peculiarities of its materials were certainly not selected with regard to convenience in analysis. This, however, only emphasizes the need for understanding the simplicities that do exist and may even condone a certain amount of oversimplification. It is, therefore, with no thought of novelty that this paper applies elementary circuit ideas to bowed-string instruments, but rather with the belief that something can be gained through representation by circuit concepts and diagrams even though some of the results are only roughly quantitative. These relations lead naturally to such related topics as the relations between different instruments of the family from point of view of dimensional scaling and the physical requirements of the wood.

LIST OF SYMBOLS

A	equivalent piston area
E	Young's modulus
F	force
H	thickness
K	characteristic impedance of string, $(T\mu)^{\frac{1}{2}}$
L	maximum safe load
M	mass
P_0	barometric pressure, 10^6 dyn/cm²
Q	quality factor of a resonance, π/δ
R	resistance
S	stiffness
$\bar{S}$	area of surface of cavity
T	string tension
U	volume velocity
V	volume of enclosure
W	potential energy per unit area
Z	impedance (Appendix II)
a,b	subscripts for air, body; dimensions of a rectangular plate
c	speed of compressional waves; in air, $c \doteq 3.45 \times 10^4$ cm/sec
d	diameter of port
f	frequency; subscript denotes resonance
i	$(-1)^{\frac{1}{2}}$
l	length of string or plate
l_1	length of string from bow to bridge
l_2	$l-l_1$
l_{bs}	length of string having frequency that of body resonance
m	mass loading on bridge
r	subscript for radiation; also radius of curvature of a plate

s	amplitude of vibration
t	time
u	particle velocity
Ω	$f/f_b = \omega/\omega_b$; $\Omega_s = f_s/f_b$
Δ	frequency of separation of fundamental pair (Appendix II)
α	
α_w	acoustical absorption coefficient
γ	ratio specific heats
δ	log decrement to base e, $= \pi/Q = 2.30\delta_{10}$
ϵ	low-note scaling ratio (Sec. VII)
η	resonance scaling ratio (Sec. VII)
λ	air wavelength
λ_s	wavelength on string
μ	mass of string per cm
ρ	density; for air $\rho_0 = 1.2 \times 10^{-3}$ gm/cc
σ	length (breadth) scaling ratio (Sec. VII)
ω	$2\pi f$; a subscript denotes resonance

I. GENERAL CHARACTERISTICS

In contrast with most of the wide-band radiation systems of today, such as the horns of acoustics and microwaves and "hi-fi" loudspeaking systems, the 17th-century creators of the violin of necessity accomplished their "broadbanding" by distributing through its wide frequency range many relatively narrow resonances, rather than using one or two bands of nearly aperiodic response. The frequencies desired extend from about 200 to 6000 cps, a span of five octaves. In the upper octaves the wood provides body resonances in number sufficient to give a quasiuniformity of response. In the lower octaves this series comes to an end, and the lowest or next-to-lowest resonance—in the vicinity of 460 cps—is commonly called the "main body resonance" because of its pronounced effects. Without reinforcement below this point, response would fall off at a rate of 12 dB per octave or more. Air resonance similar to that in loudspeakers is employed to sustain the volume for the better part of another octave, that is, resonance of the air chamber breathing through the f holes.

It is common knowledge that even a fine violin has strong and weak regions in its frequency range, but the effect is by no means as extreme as the measured characteristics suggest. F. A. Saunders and co-workers[1] point out that the subjective feeling of uniform strength, which a good violin evokes, depends markedly on the well-known effect in which the ear credits the fundamental with an increase of volume actually brought about by a strengthened harmonic. This subjective reduction of depressions occurring in an objective response curve contributes not only to uniformity of loudness, but gives subtle and interesting differences in tone color.

In Fig. 1(a) is shown a violin with names of various parts. Figure 1(b) represents its circuits in terms of electrical symbols in accord with standard conventions of

acoustics for the direct or impedance type of analogy. Letters on the circuit correspond to those in Fig. 1(a).

II. THE CIRCUIT IN BRIEF

The circuit begins at point B where the bow rubs across the string giving rise to a negative resistance. In the simplest Helmholtz mode[2,3] for the bowed string, the string at every moment comprises two straight sections either side of a discontinuity which shuttles from end to end of the string around a narrow lenticular path. Ideally the string clings to the bow except for brief recovery periods in which short negative pulses occur. From the circuit point of view, the important result is that the bow-string contact B is a constant-velocity generator. The weight applied to the string by the hand provides a condition necessary for vibration, but is unimportant in its effect on amplitude and velocity.

It is inherent in the Helmholtz concept that capture and release of the string are timed by the shuttling discontinuity that provides the trigger by which the pulse is regenerated. Since sharpness of discontinuity depends on properly phased harmonics more than fundamental, it is necessary to bear them in mind in any question concerning frequency produced. One naturally expects the occurrence of a frequency for which reactance seen by the bow at the fundamental is zero; if this were so, however, the frequency near the main resonance (to be discussed) would depart intolerably from the natural frequency of the string. Thanks to harmonics, whose impedance is independent of the main resonance, this effect is small.

The motion of the contact is in series with the two parts into which it divides the string. Except for high harmonics, Sec. AB is essentially positive mechanical

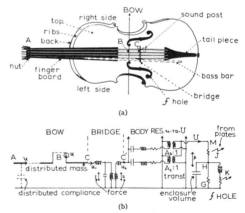

(a)

(b)

FIG. 1. Violin and an equivalent circuit. A, B, C, and K appear on both diagrams.

[1] C. M. Hutchins, A. S. Hopping, and F. A. Saunders, J. Acoust. Soc. Am. **32**, 1443–1449 (1960).

[2] H. von Helmholtz, *On the Sensations of Tone* (Dover Publications, Inc., New York, 1954), pp. 80–88 and 384–387.

[3] C. V. Raman, Indian Assoc. Cultivation of Science, Bull. No. **16**, 11 (1918).

reactance, and Sec. BC, except as modified by bridge impedance, is negative; together they form a series resonant circuit. For our purposes, the string may be treated as a lossless transmission line.

Bridge C is the transducer that accepts power from the string and transfers it to the body, which in turn excites the air within and surrounding it. Since the motion of the bridge in its own plane may be regarded essentially as that of a rigid body, it acts in the lower octaves primarily as a transformer.[4] Presumably its compliance is important to normal transmission at higher frequencies, though Minnaert and Vlam consider that its main function is to permit yielding to extraneous (e.g., torsional) motions.

For present purposes, it is simplest to regard the sound post as an important part of the body. It shares with the ribs the function of connecting the back to the source of vibration, and is necessary for strength; it is extremely important as a means for providing the dissymmetry needed for effective radiation, and plays a crucial role in determining the frequency and geometric form of the natural modes of the box.

The use of an enclosure—a box with vibrating walls—is an ancient device in instrument making. Even though the f holes of a violin were narrowed to the point of eliminating them as emitters of sound, the enclosure would still be essential since it is the variations of its volume which give the character of a simple source, these changes in volume being a difference effect between oppositely phased parts of its surface.

The bridge stands with one foot near the soundpost and the other over the bassbar. With no losses and no radiation, the bridge would see the top plate with its many vibrational modes as a complicated reactive circuit.[5] In terms of Foster's reactance theorem, there would be a series of frequencies with zero reactance, each separated from its neighbor by a frequency of infinite reactance. Losses in material and by radiation modify the reactance curve and add a curve of finite resistance. The curve, however, remains a very bumpy one. For each frequency the motional response of the body to the force exerted by the bridge is the summation of responses of the various modes. The low resistance of a resonant mode tends to "short-circuit" the others. The body thus acts like a number of series resonant circuits in parallel, as shown in Fig. 1(b), but even at resonance resistance must be several times the characteristic impedance of the string.

It does not follow that acoustical peaks must be associated with points of lowest impedance. There is at least one important exception—the air mode that interposes an impedance maximum tending to restrict bridge motion. This part of the circuit [HG in Fig. 1(b)], being described in terms of volume velocity, is shown connected with the mechanical circuits by a mechanical-to-acoustical transformer at T, with transformer ratio A to 1. Since A is different for each mode, separate transformers are shown; all these feed into the same circuit MHG.[6]

III. THE MAIN BODY RESONANCE

Meinel[7] has traced the nodal lines appearing on top and back of one good violin for the seven lowest modes of vibration. (See also reference 8.) The surface is divided with more or less clearness into many small areas at the higher frequencies, but a certain simplicity marks the lower three octaves. It is often necessary in tracing a nodal line to follow it along a plate to the edge, to cross the ribs directly or peripherally, thence along the other plate, and so forth. For each of the following resonances, 366 cps (the lowest), 690, 977, and 1380, but not at its "main" resonance at 488, the entire body was divided into only two areas separated by one endless nodal line. At the "main" resonance there were three areas. The number of areas equals the number of lines plus one. He and others have found a definite tendency for large areas to be oriented lengthwise so as to include the bass bar.

In the low octaves the restraining effect of the sound post near the right foot of the bridge leads to a greater motion at the left than at the right foot. Bridge bassbar and top plate thus rock about a nodal point near the post.

Instrument makers have commonly located the main resonance about 15 semitones above the lowest tone in the violin, 17 in the viola and cello.

The following method has been used to determine equivalent series stiffness and mass, S and M. The frequency of resonance was measured with different small loads m clamped to the bridge with mass centered at the string notch. The wolftone, if there is one, can be used as relative indicator of resonance. Since $\omega = [S/(M+m)]^{\frac{1}{2}}$, it follows by differentiation that at $m=0$:

$$M = -\tfrac{1}{2}f/(df/dm) \quad \text{and} \quad S = -2\pi^2[f^3/(df/dm)]. \quad (1)$$

In this manner, data from Raman[9] and Saunders[10] have been used to calculate M and S for one cello and one violin:

	f(cps)	S(dyn/cm)	M(g)	$(SM)^{\frac{1}{2}}$, cgs ohms
Cello	176	1.13×10^8	92.	1.02×10^5
Violin	500	1.76×10^8	17.8	0.56×10^5.

It is interesting that the stiffnesses for these two instruments of widely different size are not very different. Such relations will be examined in Sec. VII on scaling.

This resonant circuit has been characterized by S and M, but any two of the four quantities above can be used. Violin makers, in particular those who use electronic techniques,[1] explicitly consider one of these the

---

[4] M. Minnaert and C. C. Vlam, Physica 4, 361–372 (1937).
[5] The mass of the bridge added to the body leads to natural modes somewhat different from those of the body alone.

[6] Strictly speaking, the radiations from plates and f holes should be shown as from a single simple source.
[7] H. Meinel, Elekt. Nach. Tech. 4, 119–134 (1937).
[8] F. Eggers, Acustica 9, 453–465 (1959).
[9] C. V. Raman, Phil. Mag. 32, 391–395 (1916).
[10] F. A. Saunders, J. Acoust. Soc. Am. 25, 491–498 (1953).

location of the resonance in the range of the instrument. A second is not used except as it is implicit in rules for dimensioning and in particular in the selection of wood.

The foregoing estimation of M and S is useful in dealing with a narrow band about resonance, as later in connection with the wolfnote. The stiffness measured is predominately that of the body, strings being accountable for less than 10%. The air circuit (see following section) also contributes significantly to both mass and stiffness.

The third circuit constant is resistance. It can be obtained by dividing $(SM)^{\frac{1}{2}}$ by Q, the quality factor of the resonance. Q may be found from measurements of logarithmic decrement δ, such as those by Saunders[11] ($\delta_{10} = \pi/2.30\,Q$). His values of δ_{10} range between 0.062 and 0.14. Clearly the measurement should be made for the particular instrument studied.

For the principal mode, Meinel maps the amplitude of motion over the two plates of a violin.[7] This provides data for estimating the equivalent simple source and resulting radiation. Net change in volume is equated to volume displaced by piston area A conceived as moving with the bridge-string contact; in this way one can estimate series radiation resistance referred to the same point for which equivalent mass and stiffness have been measured. The motion at the string is greater than that at the left foot of the bridge by a factor of about 1.5. From Meinel's data:

Amplitude left foot	28 μ
Amplitude bridge slot (lever ratio 1.5)	42 μ
Average amplitude over plates	7 μ
Area, two plates (ribs neglected)	1000 sq cm
Since $42A = 7 \times 1000$, $A =$	170 sq cm.

Considering the body as a small source, radiation resistance $\pi \rho_0 f^2 A^2/c$ turns out at 470 cps in this example to be 700 cgs mechanical ohms. This is a component of the total resistance to bridge motion, viz., $(SM)^{\frac{1}{2}}/Q$. With the value previously mentioned for a particular violin as the numerator, and as denominator Saunders' smallest value of Q, the total resistance is 5900 cgs ohms. Radiation resistance is therefore 12% of the total. Lacking measurements on the same instrument, we have chosen data so as not to overestimate radiation. Usually it will be considerably higher than 12%. Radiation efficiency, however, is not to be confused with over-all efficiency, which is much lower owing to inefficiency of conversion at the bow.

Similar study at other body resonances might prove interesting.

IV. AIR RESONANCE

The plate motions that produce the simple source of the previous section cause changes in the opposite sense in the air within the body. The equivalent piston area is the same. If frequency is very low, the air passes through the f holes without change of pressure; if high, the air is trapped and compression occurs; the cavity

with its ports in this oversimplified concept is thus a parallel-tuned circuit [HK in Fig. (1b)],[8] and has an obvious similarity to reflex-bass enclosures in loudspeakers. (To avoid anachronism the comparison should be reversed.) Of the many modes possible, only the lowest in frequency is important. Similarity to a Helmholtz resonator is obvious.

However, with as peculiar a shape as that of the f hole and with nonrigid walls, one hardly expects to use Rayleigh's $\omega = c(d/V)^{\frac{1}{2}}$, even though he showed[12a] that with elliptical ports of small eccentricity the same frequency occurs as with a circular port of equal area. As a matter of fact, the expression yields rough estimates with f holes, the area of one being considered with half the volume.[13] The frequency thus calculated will be a semitone or so too low.

Along with the measurements of logarithmic decrements of body resonance, Saunders[11] measured decrements associated with the decay of transient oscillations at air resonance for many violins, old and new. The total decrement comprises components from several causes: useful radiation, surface absorption, viscosity in the air, and wall motion (loss in the wood). It would be useful to know their relative importance.

Radiation decrement. It can readily be shown that the logarithmic decrement to base 10 caused by radiation is

$$\delta_r = (2\pi^3 V/\lambda^3)/2.30 = 27V/\lambda^3, \qquad (2a)$$

which applies only to the lowest mode, for which volume is very small compared with a cubic wavelength. Wall motion is assumed blocked.

Wall-surface loss. This is the component ascribed to the surface regarded as stationary. Absorption coefficients derived from architectural acoustics may be used. By contrast, the violin is much smaller than a wavelength at resonance, and pressure and phase are substantially the same throughout the volume. For small absorption coefficients α_w, the logarithmic decrement to base 10 of a large enclosure (all boundaries of same material) is[14a] $\delta_w = 0.543cS\alpha_w/Vf$. It can be shown that equal energy densities produce mean squares of sound pressure on the walls of violin and room which are in the ratio 1/2 to 1; decrements are in the same ratio.[15]

[11] F. A. Saunders, J. Acoust. Soc. Am. 17, 169–186 (1945).

[12] Lord Rayleigh, *Theory of Sound* (Dover Publications, Inc., New York, 1945): (a) Article 306, p. 179; (b) Article 225; (c) Article 214, Eq. (2).

[13] If the cavity were bisected by a thin longitudinal wall, there would be no effect on frequency of air resonance. Calculation indicates that frequency is lowered about one-fifth tone by the air mass added by plate thickness.

[14] L. L. Beranek, *Acoustics* (McGraw-Hill Book Company, Inc., 1954): (a) p. 305; (b) p. 300.

[15] The reason becomes apparent on consideration of any oblique mode in a rectangular space in comparison with the (0,0,0) mode, an approximation to which can be realized by adding a port, as in the Helmholtz resonator. In the zero mode, p^2 averaged along an edge exceeds that for an oblique mode by one factor of 2, over the walls by two factors of 2, and over the volume by three such factors, since the average value of cosine squared is $\frac{1}{2}$. Decrement is proportional to p^2 averaged over the walls (rate of dissipation), divided by p^2 averaged over the volume (energy storage). This adds a factor of 4/8, or $\frac{1}{2}$, to the ratio of the decrement for the zero order divided by that for the oblique mode.

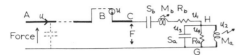

FIG. 2. Simplification of Fig. 1(b), showing air resonance and one body resonance.

Hence wall-loss logarithmic decrement in the violin is

$$\delta_w = 0.027 c \bar{S} \alpha_w / V f = 0.027 \bar{S} \alpha_w \lambda / V. \qquad (2b)$$

Viscosity in the air. It is assumed that with ports as wide as the *f* holes in the standard violin, this is a negligible effect.

Loss within the wood. In the simplified circuit of Fig. 2, a transient in HG would suffer loss from resistance R_b unless wall motion were blocked during measurement. Calculation suggests the effect to be appreciable but possibly not large. However, it seems desirable to exclude from the definition of this decrement effects of important circuit elements such as R_b, which are shown explicitly. One would include effects of modes, if any, that are not directly excited by the bridge.

Discussion. The losses contributing most to the decrement of the air circuit as thus defined seem, therefore, to be wall-surface loss and radiation. For a given frequency a change in volume affects these components oppositely, one varying inversely and the other directly. Account also must be taken of the fact that $\bar{S}$ may be a function of volume. A question often raised concerns the best height of the ribs, dimensions of plates being determined by other considerations. In Fig. 3 are plotted computed values of these components and their sum. Account was taken of the fact that $\bar{S}$ includes the area of the ribs. Resonance frequency (wavelength) is held constant by changing width of *f* holes or by use of additional ports. The α_w used is 0.04, the value for wood floors on solid foundation,[14b] architecturally a small value.

For assumed values of parameters, the two components are equal for a rib height of 5 cm. Their sum has a minimum between 3 and 4 cm. This may be compared with the rib height used by Stradivarius, 3.0 cm. It is true that the minimum is not a sharp one, and that its calculated value will shift somewhat with better data. Nevertheless, the agreement seems significant, as does the consistency with Saunders' data. His median value of 0.115 contains a significant amount resulting from wall motion. The value shown by Fig. 3 is 0.105. It would, of course, be a mistake to lose sight of the considerable spread from instrument to instrument—from 0.09 to 0.14—and the fact that the computation depends on choice of particular values of α_w and f.

Other air modes. Because air resonance is important in the lower register, it is sometimes supposed that the many natural modes of the cavity must play an important part in the upper ranges. It is argued that the volume is large enough to support dozens of modes

within the range, and that these must be helpful. The experimental evidence, however, is that they are of little or no value. Saunders[10] reported that "there appeared to be upper resonances near 1300, 2600, and 3660" in a certain violin, but his general conclusion was that the output from the *f* holes is unimportant except near the lowest resonance.

To be effective a natural mode has to satisfy two requirements: it must be energetically excited by the walls, and it must be "impedance-matched" to the *f* holes. The lowest mode satisfies both by design. It is excited because the wall motion provides the net changes in volume[7,8,16] required to induce "zero order" pressure changes. Secondly, its very existence depends upon there being an air flow through the *f* holes, that is, *upon radiation.* By contrast consider the other modes: their geometric structures, except by accident, are unrelated to those of the wood, and there is no obvious basis for expecting power transfer to the air within. Moreover, such excitation as may fortuitously occur will not necessarily cause the breathing through the *f* holes needed for radiation; the holes are apt to be too small or too large or in the wrong place.

V. COMBINATION OF AIR AND BODY MODES

Air and main body resonances are not isolated means, but in good instruments are matched for best total effect to insure a strong lower register. In studying circuit behavior, they need to be considered together. The problem, which is complicated, will here be limited to simple conditions.

To this end, consider the circuit of Fig. 2 in which one resonant body mode is assumed to predominate, its series circuit being in series with the shunt resonant circuit representing the air circuit, and its impedance being sufficiently great to control wall velocity.[17] Here the acoustical circuit HG has been transformed into its mechanical equivalent in terms of linear velocity at C, the top of the bridge. Assuming the configuration of the body mode to be independent of frequency, radiation resistance of the plates (without contribution from the *f* holes) is proportional to frequency squared.

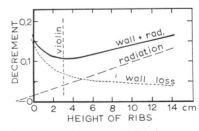

FIG. 3. Two main components of air decrement.

[16] H. Backhaus, Z. Physik **62**, 143 (1918); **72**, 218 (1931).
[17] A more refined calculation would forego the last assumption. The impedance measurements that Eggers[8] made on a cello are of interest in this connection.

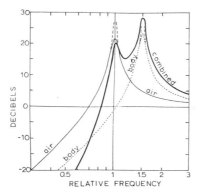

Fig. 4. Relative sound pressure, air resonance, and one body resonance. In the violin other peaks occur at the right.

Following the method of Appendix I, relative performance is calculated for the following typical condition:

Air resonance at relative frequency	1.0
Q_{air}(mean of Saunders' values)	= 12.
Body resonance at relative frequency	1.5
Q_{body} (Saunders' mean)	= 14.5.

The dotted curve of Fig. 4 depicts radiated response without air resonance, a condition that might be realized by using very large volume or by closing the f holes; the violin now has the advantage of enclosure, but not of air resonance. If resonance is now permitted and adjusted to an interval of a fifth (frequency ratio of 3/2) below body resonance, the light solid line gives the improvement in decibels. Finally, the combination is shown by adding the two, giving the heavy solid line.

The "air" curve has several features to notice. Most prominent, of course, is the resonance peak. At lower frequencies the air advantage falls off until, at relative frequency 0.7, half an octave, it becomes zero. Within practical bounds, a difference of Q(or δ) would not affect this conclusion, though it could change response radically within 10% of resonance. In violas air resonance is not infrequently placed eleven semitones (frequency ratio 15/8) above the lowest fundamental, which consequently suffers an "air-resonance disadvantage" of 8 dB, though the second and third harmonics can be in a very strong position. The curve is somewhat too optimistic at its high-frequency end because of oversimplification in circuit representation. It seems safe to say, however, that there is some gain over an entire octave and 9 dB or more over half an octave.

The curve for body resonance alone, instead of falling to very low values at high frequencies, approaches a horizontal asymptote, a point of considerable significance. This behavior, which holds for all modes above resonance and which has its counterpart in direct-radiation loudspeakers, is the result of two opposing

tendencies. The conformation of a mode and, therefore, its equivalent mass are taken to be independent of frequency; hence, the velocities set up by a given applied force vary *inversely* with frequency. But the sound pressure radiated by a given velocity is *directly* proportional to frequency and thus annuls the effect of the decline of velocity. The way in which the various modes will combine depends upon the phase relations of their simple sources; it is plausible to suppose, however, that the lower modes thus provide a more or less level table land on which the higher ones erect their peaks, and that this is an important contributor to violin tone.

Another result of Appendix I is an expression for the ratio of sound pressures produced at body and air resonances:

$$p_{body}/p_{air} = (f_b/f_a)^2(Q_b/Q_a). \qquad (3)$$

This agrees reasonably well with ratios found from single-frequency curves by Saunders (see reference 11, p. 173).

How important are the Q's? In Fig. 4 the effect of an increase in either Q beyond usual values produces an elevation of level within a very narrow band at resonance, an effect as apt to be harmful tonally as helpful. There seems little to seek in a Q higher than that which gives a 3-dB bandwidth of one tone—a Q of 8 or 9—and there may be something to avoid in bandwidths that can be straddled by adjacent semitones. The importance of resonances, in other words, is to provide broad foothills rather than sharp peaks.

What is the best volume to be used? It is significant that in approximate Eqs. (1), (2), and (3) of Appendix I, volume does not appear as an explicit term affecting radiation. It is true that resonant impedance of the shunt circuit increases as volume decreases, suggesting an advantage because more power can be abstracted from the constant velocity source. But this is opposed by reduction of radiation decrement and, therefore, of radiation efficiency, leaving a change of only second-order importance.

There is another consideration, however. In good instruments the "piston area" A (see Sec. III) is made as large as possible, so large in fact that the impedance offered to the motion of the top plate by air resonance is by no means negligible. The sign of its reactance component at a frequency somewhat below resonance is positive, so as to tend to cancel the negative reactance of the body. If the volume is now made too small, this cancellation can be complete, so that the impedance into which the string works is a relatively low resistance capable of interfering with the normal operation of the bow. Something of this sort seems to be the reason for the airtone "wolfnotes."[18]

---

[18] The behavior and elimination of airtone wolfnotes have been studied by F. A. Saunders (private communication). Impedance measurements by Eggers,[8] Fig. 18, are of interest in this connection.

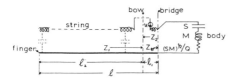

FIG. 5. Elements of wolfnote circuit.

VI. THE WOLFNOTE

The most troublesome wolfnote, however—a cyclic stuttering response to the bow on the heavier strings, particularly in cellos—can occur when the fundamental is within a half-tone or less of the main resonance; it may occur in otherwise fine instruments. Its behavior immediately suggests beating and coupled circuits. The best explanation has been one published by C. V. Raman 45 years ago.[3,9] Having much in common with his theory, the present one, which is stated in the language of circuits, differs in one important respect.

Through most of the frequency range the impedance presented by the bridge is high compared with the characteristic impedance of the strings, perhaps ten times or much more. Trouble may ensue at resonance when this ratio is well below ten, and the Q is in the range found by Saunders.

In this study it is the impedance presented to the bow that is the most informative. Calculations must take account of the distributed nature of the mass and compliance of the string, hence requiring standard methods of computation for transmission lines, as indicated in Fig. 5 and Appendix II. Resistance in the string itself is neglected.[19] For generality, equations are written in terms of dimensionless ratios: impedance relative to K, the Q of the bridge circuit, Ω and Ω_s, and the fractional part of the string length between bow and bridge. Since wolfnote is not sensitive to the latter, two parameters Q and $K/(S_bM_b)^{\frac{1}{2}}$ remain as the fundamental data prescribing circuit behavior when impedance seen by the bow is considered as a function of frequency.

In the normal situation, the string presents to the bow an unambiguous impedance; it is that of a simple series tuned circuit having positive resistance low enough to be matched by the negative of the bow-string contact, and reactance that passes through zero at the frequency of operation as shown for the fundamental by the broken line in Fig. 6(b). Though still not a simple problem, in view of the complicated impedance pattern in which the various components must seek a compromise frequency, the situation contains no obviously tempting invitation to misbehavior. By contrast, consider the impedance pattern at the main resonance as shown in Fig. 6, in which for a given string length (e.g., $\Omega_s = 1.028$) there is not one but three frequencies at which reactance is zero. Steady oscillations can conceivably occur at any

[19] H. Backhaus, Z. Physik **18**, 98 (1937).

of these frequencies.[20] If the negative resistance is insufficient to cope with the high resistance at the point of tuning, it may, nevertheless, be adequate at the two outside points. That is to say, if bow pressure is insufficient for the normal vibration, it still may be enough for the outside pair because of their lower resistance. Let these frequencies be $(\Omega' + \Delta/2)$ and $(\Omega' - \Delta/2)$. If amplitudes are equal, total string velocity will be proportional to

$$\cos 2\pi(\Omega' + \Delta/2) + \cos 2\pi(\Omega' - \Delta/2)$$
$$= 2 \cos 2\pi\Omega' \cos \pi\Delta t, \quad (4)$$

the familiar expression for a beat. Speaking in terms of "instantaneous" frequency instead of Fourier components, this is a wave of average frequency Ω' pulsating at frequency Δ. Frequency is always Ω', but there is a phase reversal in passing through zero amplitude.

Raman[9] criticized G. W. White's suggestion[21] that it is a beating process on the ground that one of the frequencies must be that of a free oscillation that will soon decay. I believe the conclusion incorrect that there is no beating. The two oscillations suggested here are equally forced.

These two oscillations (referred to as the "fundamen-

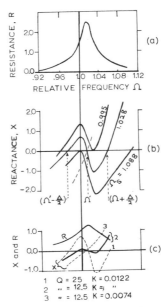

FIG. 6. Impedance seen by bow ($Z_1 + Z_2$ of Fig. 5); (a) and (b) for a bad wolfnote; in (c), 1, 2, and 3 show consecutively improved conditions.

[20] Even though the negative sign of reactance slope raises a question of stability, with the highly nonlinear behavior of the bow-string contact, the stabilizing effect of harmonics on a string free of phase distortion will in borderline cases probably permit oscillations in spite of the slope.

[21] G. W. White, Proc. Cambridge Phil. Soc. **18**, 85–88 (1915).

tal pair") are not the only sinusoidal components that move through the whole beat cycle substantially unchanged. The same is true of the even harmonics of Ω'. Raman showed that the octave becomes prominent at the beat minimum. Curiously enough, this prominence is not because the even terms have grown, but because the odd terms have subsided; the sawtooth at its maximum amplitude contains even terms of about the same amplitude and phase as at the minimum. This is brought out clearly and simply by a graphical separation of odd and even terms in the wave shown in Fig. 7(a).[22] The maximum displacement of the even components shown in Fig. 7(b) is about half that in Fig. 7(a), and the slopes of the long sections are the same, matching the same bow velocity. Addition of Figs. 7(b) and 7(c) shows how the intermediate discontinuities have been cancelled by the odd terms of Fig. 7(c). Removal of the odds will, therefore, bring the "octave" back.

An important difference between a linear and a nonlinear generator needs mention. In the former it is possible for oscillation to occur at one of the fundamental pair alone. The principle of superposition gives them complete independence. But with bowing, the recurrence rate depends by virtue of the necessary harmonics on the string length and the string's simple phase characteristic. Neither of the pair can exist without the other because its frequency is so different from a recurrence rate possible on the string. On the other hand, by cooperation they can produce an instantaneous frequency acceptable to the string, equal to half the trigger rate of Fig. 7(b). Like Siamese twins, they can exist as a pair, but not otherwise.

If this theory is correct, it should be possible from experimental evidence to show that an epoch of maxima has undergone a phase shift of π relative to the preceding one. This can, in fact, be seen in Raman's oscillograms (Plate I of reference 3), which show simultaneous motion of bridge and string, that of the string being suggested in Fig. 8. At the epoch where the amplitude of the bridge motion is growing most rapidly, the string, as Raman indicates, has a clean sawtooth displacement. At this moment bridge amplitude is matched to bow velocity and pressure. The amplitude of bridge motion continues its growth for a time, but the sawtooth shows signs of deterioration in the form of new discontinuities midway between those of the series just considered.

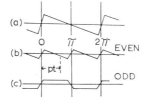

FIG. 7. Separation of even and odd terms of idealized string displacement.

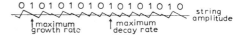

FIG. 8. String displacement through one wolfnote beat.

This new series is destined to be the sole series of discontinuities during the next period of bridge–motion growth. To see this in the original oscillograms requires close examination, preferably with a magnifying glass,[23] but it is readily followed in Fig. 8. If we place a zero adjacent to each clearcut discontinuity at the left and a 1 midway to the next later discontinuity, and continue this alternate naming through the octave period to the next clearcut stage, we find that the discontinuities of the latter are named 1, not 0. All three transitions shown in the original give the same result. This indicates a phase shift of 180° in instantaneous frequency in passing through the minimum, and this, of course, is what a beat requires.

The bad situation shown in Figs. 6(a) and (b) is greatly improved in Fig. 6(c) by reducing Q from 25 to 12.5. Though the S shape remains, the reactance curve is nearly flat in the region of interest, the fundamental pair closer together, and their resistances just slightly less than at midband. It is doubtful that a wolf could occur. Curve (3) shows the further advantage of reducing characteristic impedance of the string through reduction of weight. The side frequencies have now disappeared. That light strings' help is, of course, well known.

The rate at which beating occurs is consistent with the difference in frequency within the fundamental pair. Applying Fig. 6(c) to the cello, the indicated rates for conditions (1) and (2) are, respectively, 16 and 8 per sec. The delays of bridge maximum with respect to string maximum are calculated as 0.36 and 0.17 of a beat cycle for the same respective conditions, agreeing with Raman's[9] 0.25, which would have been found here had Q been taken at the more typical value of 17.5.

Finally it should be re-emphasized that conditions at harmonic frequencies may have some connection with wolfnote, and that (1) the distance of the strings from the nodal line about which the bridge swings and (2) the angle of bow motion certainly do have an effect. One experimental condition was to bow the cello C string (the lowest) *underneath* rather than over the strings, with hand held as high as possible without the bow touching the wood. On this most "wolfish" of strings, the wolftone, as expected, disappeared.

Wolftone criterion. It is desirable to show graphically the relations prevailing under different conditions of susceptibility to wolftone. To this end Fig. 9 provides dimensionless coordinates on which can be exhibited essential parameters applying to all instruments of the violin family, viz., $K/(S_b M_b)^{\frac{1}{2}}$ and logarithmic decre-

---

[22] Odd components of $f(pt) = [f(pt) - f(pt+\pi)]/2$; even components $= [f(pt) + f(pt+\pi)]/2$.

[23] The reproductions in a communication to *Nature* are not clear enough.

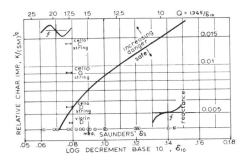

FIG. 9. Wolfnote criterion.

ment. The heavy line shows the locus of conditions for which the midpoint of the S curve in Fig. 6 is horizontal, i.e., on the verge of having three intersections; it is the boundary between safe and questionable conditions.

On the basis of data previously cited for one cello and Saunders' median δ, points are plotted for the three strings that are used to produce the frequency of main resonance. For the cello D string, the point falls close to the safe area as one expects from experience. For the heavier G string, the point is definitely away from the "safe" region, also in agreement with experience; being much used in this position, this string causes the cellist the most annoyance. In going to the C string, one would expect and one finds that it may be very difficult indeed. Fortunately this note is played on this string relatively infrequently. The point representing a violin indicates relative freedom from wolftone.

Eliminating the wolftone. The instrumentalist is rarely interested in an explanation. What he wants is to have the curse removed. The idea occurred to the author and to others[8] to insert a narrow-band suppressing circuit in series with the bridge motion. This is most conveniently done by attaching to one of the strings between bridge and tailpiece a mass of a few grams chosen to tune the string end to the wolfnote. On the cello tuning can be done by ear by tapping the load with the eraser of a pencil used endwise. Without tuning there is no guide to adjustment. A second precaution, not always necessary but to be recommended, is use of a suitably lossy or nonringing substance such as the rubber of a large pencil eraser. Calculation indicates a desirable Q to be 10 or less, so as to give a wider band than the main resonance. The less obstinate the wolf, the farther the load may be placed from the bridge. Hutchins finds molding clay very convenient, since it has desirable loss and its mass is easy to adjust.

Avoidance of wolftone by means external to the body may give more freedom to the maker in other respects, such as ability to use wood of high c/ρ (see Sec. VIII).

VII. DIMENSIONAL SCALING

The process of tuning a violin includes more than adjusting string tension; the luthier must first "tune"

its pattern of wood resonances with respect to those of string and air. If he is designing an instrument to occupy a new frequency range, all of these resonances are changed with respect to those of instruments that have become conventional to him. If he wishes to change size without changing frequency range, either to accommodate smaller hands or to allow larger ones to work to better advantage, an understanding of the principles governing modified dimensional scaling will be helpful.

Complete scaling is theoretically no problem; all that is needed is thoroughgoing change of all dimensions in proportion to change in air wavelength, and use of identical materials. (Actually the difficulties of even one-to-one scaling—that is, copying—are considerable, partly because of unavailability of identical materials; this ever-present problem is not considered here.) In general, complete scaling is not practicable because of the impossible demands that it places on strength or size of the player, or for other reasons. Failure to understand scaling leads to errors in construction; for example, in making a viola[24] luthiers have sometimes chosen a plate thickness in proportion to that of a violin; however, this is too much.

Before undertaking to scale, it is necessary to decide on basic aims and the compromises one is willing to accept. Discussion here will be limited to conventional instruments; for concreteness we shall rather arbitrarily regard the violin as the most suitable starting point, and ask what this leads one to expect about the viola and cello and their strings. There are three basic ratios to consider:

(1) The lowest frequency of the "new" instrument is to equal that of the violin divided by ϵ. That is, air wavelength is multiplied by that factor. For the viola $\epsilon=1.5$, for cello 3.0.

(2) The instrument being thought of as lying on its back, its shape as seen from above, is similar to that of the violin, and all horizontal dimensions are multiplied by σ. For a 16½-in. viola $\sigma=1.17$, for a typical cello 2.1.

(3) The pattern of body resonances is to remain the same logarithmically, and wavelength of the main resonance is to be multiplied by η. Usual practice places η at 1.33 for the viola and 2.66 for the cello.

These factors are defined so as to be greater than unity in going to viola or cello. Plate shapes not being strictly the same, σ involves a small compromise.

In a completely scaled instrument, the air resonance takes care of itself. Incompleteness will be evident in width and thickness of ribs and width of f holes. It seems desirable to scale length of f holes. Ribs can be scaled according to principles of air resonance (Sec. IV).

A reservation with respect to the arch needs mention. The rigidity of a plate depends on (1) its stiffness in flexure and (2) its two-dimensional curvature. With a flat plate only the first enters. (The effect on resonances

[24] I am indebted to C. M. Hutchins for information on this and other matters.

is somewhat analogous to that in a piano string where two forms of stiffness, tension, and rigidity occur.) While our analysis in strictness, therefore, applies only to a flat violin (thickness graded, however), it is believed to give numbers of significance even considering the arch.[25]

For a flat plate of arbitrary shape and graded thickness, the basic relation is that the frequency of a flexural mode is directly proportional to the scale of thickness and inversely to the square of a horizontal dimension. We assume that the main effect of the ribs, aside from providing enclosure and coupling between plates, is that of mass loading at the edges, which are by no means immobile.[7,8] The idea that, in scaling, total mass of ribs should remain proportional to mass of plates is suggested by the fact that in the cello they are proportionately thinner than in the violin, their height being proportionately greater.[24]

Thickness. Subscripts 1 and 2 refer respectively to the model and the "new" instrument. H is thickness at some reference point. We now have

$$f_1 = kH_1/l^2 \quad \text{and} \quad f_2 = kH_2/l^2\sigma^2, \tag{5}$$

l being length, and since $f_1/f_2 = \eta = H_1\sigma^2/H_2$,

$$H_2/H_1 = \sigma^2/\eta. \tag{6}$$

Stiffness. Consider design in two steps—first to an intermediate instrument completely scaled by factor σ and indicated by primed symbols. A general relation for stiffness S of dimensionally similar shapes is

$$S'/S_1 = \sigma. \tag{7}$$

Altering intermediate to new, change of thickness affects stiffness in proportion to the cube, or

$$S_2/S' = (H_2/H')^3 = (H_2/\sigma H_1)^3; \tag{8}$$

and from (6) and (7),

$$S_2/S_1 = \sigma^4/\eta^3, \tag{9}$$

the ratio of stiffness entering into corresponding resonances.

Mass. Corresponding masses are proportional to total mass of the plate:

$$M_2/M_1 = \sigma^2 \cdot \sigma^2/\eta = \sigma^4/\eta. \tag{10}$$

---

[25] The tendency for deformations to be predominately inextensional (potential energy purely flexural) is discussed in Rayleigh's *Theory of Sound*, Article 235 b. The fact that the arch does have an effect in acoustical behavior has recently suggested that similarity of behavior should result if in scaling we maintain the same ratio between flexural and extensional potential energies. An elementary analysis indicates that this is obtained simply by scaling altitude of arch according to σ^2/η, the same factor used in Eq. (6) in scaling thickness; that is, by holding constant the ratio, arch/thickness, rather than the more naturally used ratio, arch/horizontal dimension. Since arch is a means of adjusting prominence of high frequencies with respect to lows, and since low-voiced instruments do not necessarily require the same balance as high-voiced, exact scaling of the arch may not be the best.

Body impedance.

$$(S_2M_2)^{\frac{1}{2}}/(S_1M_1)^{\frac{1}{2}} = \sigma^4/\eta^2. \tag{11}$$

Ultimate strength. Ratio of ultimate strength of forms completely scaled is

$$Y'/Y_1 = \sigma^2, \tag{12}$$

and if as in the flexural strength of isotropic materials,

$$Y_2/Y' = (H_2/H')^2 = \sigma^2/\eta^2, \tag{13}$$

$$Y_2/Y_1 = \sigma^4/\eta^2. \tag{14}$$

This is the same as for body impedance, Eq. (11).

Strings. General opinion, borne out by the need of old instruments for stronger bass bars, is that string tension T with a factor of safety is set with relation to ultimate strength of the body.[24] Therefore, from Eq. (14)

$$T_2/T_1 = \sigma^4/\eta^2. \tag{15}$$

It follows from Mersenne's law for strings that

$$\mu_2/\mu_1 = \sigma^2\epsilon^2/\eta^2. \tag{16}$$

From (16) and (15) ratio of characteristic impedance is

$$K_2/K_1 = \sigma^3\epsilon/\eta^2. \tag{17}$$

"Wolf ratio." Finally taking ratio of Eqs. (17) and (11), we have

$$[K_2/(S_2M_2)^{\frac{1}{2}}]/[K_1/(S_1M_1)^{\frac{1}{2}}] = \epsilon/\sigma, \tag{18}$$

expressing the relative impedance positions in Fig. 8.

In order to check these relations experimentally, instruments compared must embody the uniformities of construction assumed. In individual instruments this is difficult assurance to gain. However, certain general relations may be tested. Thus, Eq. (6) indicates why in violas and cellos thickness is less than for complete scaling since $\eta > \sigma$. Again according to Eq. (15), one expects cello-string tension to be 2.7 times as great as in the violin ($\sigma = 2.1$, $\eta = 2.7$); for two sets of strings compared, the ratio was 2.4.

Equation (18) is particularly interesting since it explains why wolftone trouble increases in going from violin to viola to cello; the reason is simply that the sizes of the instruments have not been increased in proportional to the air wavelength, and maximum safe tension has been insisted upon in the strings. The chronic susceptibility of the cello to this trouble is the price paid for the convenience of a small instrument, small compared with one completely scaled.

Since *with scaling* A is proportional to σ^2, and f at body resonance to $1/\eta$, radiation resistance that depends on A^2f^2 is proportional to σ^4/η^2. This is identical with the expression that Eq. (11) gives for body impedance. It follows that a body-resonance radiation decrement, which is the ratio of these quantities, is invariant. If it is true that body losses depend primarily on the wood, we may conclude that total decrement is also invariant in the scaling process, since Rohloff finds the decrement of wood to be independent of frequency.[26] Within these limits, this justifies the name "wolf ratio" used with Eq. (18). In absence of scaling, we require measurement of both decrement and $K/(SM)^{\frac{1}{2}}$ for the application of Fig. 9 to individual instruments.

VIII. ON THE REQUIREMENTS OF WOOD

Violin makers have always attached great importance to the selection of wood, not only as to species but also the characteristics of the particular piece to be used. Acousticians have measured elastic properties, density, and damping coefficients, and the more scientifically minded makers are trying to take advantage of such procedures. It is important to relate these measurements to the luthier's problem in as simple a manner as possible.

An important question is: When are two pieces of wood acoustically equivalent? Should one try to match both the elastic modulus and the density?

Along and across the grain elastic properties are different velocities of compressional waves appearing to be in the order of three or four to one. We assume a fixed ratio and consider the elastic behavior determined by a single modulus E, density, and scale of thickness. As in the previous section, flat plates of graded thickness are assumed.

Flexural similarity and c/ρ. Having given a reference plate (subscript zero), let it be required to duplicate its acoustical behavior in a plate of different material (subscript 1). Consider any point and a line through it in the plane of the plate, the line being part of the linear wavefront of a flexural wave. Such a wave is characterized by a torque per unit length lying along the front and traveling along with it. This torque is directly related to the potential energy of the medium, which is momentarily located in the stiffness at the point under consideration. The kinetic energy is similarly associated with the transverse velocity of the mass per unit area (rotational energy being negligible in the thin plates of the violin). Flexural behavior is similar in the two plates when their stiffnesses per unit length and densities per unit area are equal. Stiffness per unit length is proportional to E and to the cube of thickness (as in the analogous static problem of the beam). Hence,

$$E_0 H_0{}^3 = E_1 H_1{}^3. \tag{19}$$

Equal densities per unit area require that $H_0\rho_0 = H_1\rho_1$, or

$$H_1/H_0 = \rho_0/\rho_1. \tag{20}$$

With $(E/\rho)^{\frac{1}{2}} = c$, it follows that

$$c_1/\rho_1 = c_0/\rho_0. \tag{21}$$

This means that *even if we are not able to duplicate c and ρ separately, reactive behavior remains the same if*

[26] E. Rohloff, Ann. Physik, No. 5, 38, 177–198 (1940).

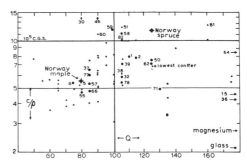

Fig. 10. c/ρ vs Q for various species of wood (based on Barducci and Pasqualini).

their ratio remains invariant. However, ratio of thickness must change as required by Eq. (20).

c/ρ *and circuit parameters.* The ways in which c, ρ, and c/ρ enter into circuit relations may now be indicated. Consider a violin plate of standard shape (width vs length) and assume it distorted in the pattern of a tap-tone vibration. A stiffness and a mass will be involved whose changes with certain variables are indicated by the following proportions:

Stiffness $S \propto EH^3l^{-2}$

Mass $M \propto \rho Hl^2$

Tap-tone frequency $f_b \propto (S/M)^{\frac{1}{2}} \propto cHl^{-2}$

Thickness $H \propto (1/c)f_bl^2$ (22)

Impedance $(SM)^{\frac{1}{2}} \propto (E\rho)^{\frac{1}{2}}.H^2 \propto (\rho/c)f_b{}^2l^4$ (23)

Mass $M \propto (\rho/c)f_bl^4$ (24)

Stiffness $S \propto (\rho/c)f_b{}^3l^4$. (25)

Since f_b and l are determined by considerations unrelated to the wood, plate impedance mass and stiffness depend on the wood only through the parameter c/ρ, thickness on c.

Comparison of species. Barducci and Pasqualini[27] measured c and ρ for 85 species of wood. In Fig. 10 their data have been adapted to display the violin-wood parameter c/ρ as a function of its Q. Most of their species are plotted, but to avoid confusion only those represented by four or more specimens and indicated by black circles are numbered. (Numbers give the order in their table.)

A fact immediately evident is that *Picea excelsa* (the spruce which in Europe has traditionally been used in top plates) is high, whereas *Acer platanoides* (the maple used in backs) is low. The former has few neighbors to be candidates for substitution, the latter several. For reasons to be mentioned, horizontal separation of points is difficult to interpret. Comments largely drawn from Howard's *Timbers of the World* (MacMillan and

[27] I. Barducci and G. Pasqualini, Nuovo cimento 5, 416–446 (1948).

Company Ltd., London, 1951) are listed below for some neighbors of *Picea excelsa*.

#58	*Populus alba*	Close, hard, tough texture. Sounds promising.
#60	*Populus nigra*	Same characteristics as #58.
#59	*Populus canadensis*	
#51	*Pinus cembra*	Knots prevalent, otherwise promising.
#81	*Thuja plicata*	A cedar. Perhaps good, except for splits and shakes.
#82	*Tilia europaea*	A linden. Sounds good.

The craftsman may rule some of these out for nonacoustical reasons.

Relation of top and back. Measurements by Meinel (Fig. 16 of reference 7; also reference 8) show that at main resonance (and presumably generally) the back contributes materially to the equivalent simple source of the violin. Though the top is the more important as radiator, the back can by no means be neglected as a contributor; and the proper matching of one to the other seems essential to insure a strong, simple source over the frequency range. The conjecture, therefore, appears justified that the relation of impedances of the plates [Eq. (23)] needs to be maintained by keeping the values of c/ρ in the same approximate ratio of 2 to 1, which practical experience has led to, and which Fig. 11 shows for *Picea excelas* and *Acer platanoides*.

Other values of c/ρ.[28] On the assumption of some such balance between top and back, what is to be expected when, instead of using the customary spruce, the c/ρ of the top of a new violin is made lower, as it sometimes is, by using woods lower in Fig. 10? Equation (23) indicates that its body impedance will be increased. The oscillating force that a string is able to deliver to the bridge for a given amplitude of string motion is proportional to tension. Hence with unchanged strings the velocity produced in the radiating surfaces and hence the sound pressure will be correspondingly lower. The extent to which this disadvantage can be overcome depends on the willingness of the performer to use heavier strings and the ability of the structure to withstand the greater load. These considerations make it obvious why maple would be a very bad choice for the top plate. The 2-to-1 impedance increase that was mentioned in the previous paragraph would cause a 6-dB loss unless possibly cancelled by doubling string tension, and it is not obvious what material having a still lower c/ρ would be suitable for the back. The weight of the box would be almost doubled.

Oppositely, higher ratios make more power available and deserve careful trial. However, the problem is not simple. The same lowering of impedance which adds to

acoustical output increases vulnerability to wolfnote. Though probably acceptable in a violin, this might be serious in a cello unless a wolf eliminator is used. Strength of wood is another consideration. When string tension is proportioned to strength of structure, it appears that a different wood parameter, $\phi/(c\rho)$, measures relatively the upper limit of sound pressure produced. Here ϕ is the strength function of the wood (e.g., bending, shear, tension across grain, etc.), in which the structure is most vulnerable as used in the violin.

Damping requirements of wood. In measurements of elastic characteristics of wood, the usual emphasis on properties along the grain has led to a preponderance of values quoted[27] between 60 and 130, whereas the Q of the principal body resonance of the assembled instrument[11] ranges from 10 to 20. It has, of course, been appreciated that cross-grain Q's are the lower by a factor near 4 and must certainly depress the resultant. The difficulties of the problem have precluded an estimate of what resultant Q one should expect from a plate.

There is one qualitative consideration worth mentioning; namely, the relative narrowness of the instrument and its vibrational patterns in comparison with length, and the resulting tendency to emphasize the effect of cross-grain constants, both as regards to potential energy of vibration and energy loss.

Consider a flat, rectangular plate of wood "supported" (hinged) along its edges and vibrating in its lowest flexural mode.[12b] We may suppose that if ratio of length a to width b equals that of length to average width of the body of the violin, the energy relations will be somewhat comparable. At the center of the plate, the principle curvatures lie parallel and perpendicular to the grain, and, if we ignore Poisson's ratio, the potential energy of deformation[12c] per unit area is $W \propto EH^3(r_a^{-2}+r_b^{-2})$ for an isotropic material, r being radius of curvature. For wood (anisotropic), with $c=(E/\rho)^{\frac{1}{2}}$ and, for a given s, $r_a \propto a^2$, and $r_b \propto b^2$,

$$W \propto s^2H^3\rho(c_a^2a^{-4}+c_b^2b^{-4}). \qquad (26)$$

Here the first term corresponds to the long dimension a, the second to the short one b. It is immediately evident that although c_b in wood is much smaller than c_a, the inverse fourth power of width can be a powerful influence in emphasizing cross-grain energy if width is, in effect, much smaller than length.

Similarly, it follows that rate of energy loss at the center is

$$dW/dt \propto s^2H^3\rho(c_a^2a^{-4}Q_a^{-1}+c_b^2b^{-4}Q_b^{-1}). \qquad (27)$$

The cross-grain term is now further emphasized relative to the other by the inverse of its smaller Q. We should, therefore, expect the Q of a violin (corrected for radiation resistance) to lie nearer to that indicated by cross-grain wood samples than to that obtained with samples cut along the grain; that is, nearer to some compromise value between 30 for the spruce plate and 20 for the maple, than to a compromise between 125 and 80.

---

[28] Gleb Anfiloff, Znatie-Sila (February 1961). According to a summary by G. Pasqualini, the importance of c/ρ was emphasized by Mr. Anfiloff.

The effect of the arch may be to lower the Q still further.

It may be, therefore, that simple wood loss and radiation are enough to account for the low Q's of violins. The very fact that these instruments are built-up structures may accentuate unfavorable strains not as yet sufficiently studied, perhaps such as shear along the grain. Other mechanisms for increasing losses, of course, deserve consideration, such as pre-stressing of bassbar and top-plate, as proposed by Rohloff.[26]

ACKNOWLEDGMENTS

It is a pleasure to express my indebtedness to Professor F. A. Saunders for the benefit of his long and fruitful experience in this field, and to Mrs. Carleen M. Hutchins for her insight into the problems of scientific violin making.

APPENDIX I

Refer to Fig. 2, right half, beginning at C where force F is applied by string. Subject to approximation that impedance of air circuit is negligible compared with reactance of body,

$$|u_1|^2 = F^2/\{s_b M_b[1/Q_b^2 + (\omega/\omega_b - \omega_b/\omega)]\}. \quad (A1)$$

Current-producing radiation is $(u_1 - u_2)$ (external surfaces and f holes) and equals $(u_3 + u_4)$. Its radiation resistance is that of a simple source:

$$R_r = k\omega^2, \quad (A2)$$

where $k = \rho_0 A^2/4\pi c$. Since impedance of $S_a = -iS_a/\omega$, $R_a = S_a Q_a/\omega_a$, and impedance of $M_a = iS_o\omega/\omega_a^2$, it follows

that

$$|u_1 - u_2|^2/|u_1|^2 = [1/Q_a^2 + (\omega/\omega_a)^2]/ \\ [1/Q_a^2 + (\omega/\omega_a - \omega_a/\omega)^2]. \quad (A3)$$

This is a function of V only insofar as Q_a depends on it implicitly (see Fig. 3) or above approximation unacceptable.

With f holes closed, power radiated is $(1) \times (2)$.

With f holes open, power radiated is $(1) \times (2) \times (3)$.

Except near resonance, relative variation of radiated power with frequency is

$$[1 - (\omega_b/\omega)^2]^{-2} \cdot [1 - (\omega_a/\omega)^2]^{-2}. \quad (A4)$$

By use of Eqs. (1)–(3), the ratio of sound pressures produced at the two resonances assumes a simple form:

$$p_b/p_a = (f_b/f_a)^2 \cdot (Q_b/Q_a). \quad (A5)$$

APPENDIX II

Refer to Fig. 5 and take $R = (SM)^{\frac{1}{2}}/Q$, $\omega_b = (S/M)^{\frac{1}{2}}$,
$\bar{Z}_b = Z_b/(SM)^{\frac{1}{2}} = 1/Q + i(\Omega - 1/\Omega)$.

The line over other impedances indicates similar normalizing to $(SM)^{\frac{1}{2}}$.

$$\alpha_1 = 4\pi l_1/l_{bs} \quad \text{and} \quad \alpha_2 = 4\pi l_2/l_{bs}.$$

With this nomenclature, calculations were made as follows:

$$\bar{Z}_1/\bar{K} = Z_1/K = \frac{\bar{Z}_b/\bar{K} + i\tan\alpha_1\Omega}{1 + i(\bar{Z}_b/\bar{K})\tan\alpha_1\Omega}$$

$$\bar{Z}_2/\bar{K} = Z_2/K = -i\cot\alpha_2\Omega.$$

Erratum: The Violin as a Circuit
[J. Acoust. Soc. Am. 35, 326–338 (1963)]

JOHN C. SCHELLENG

301 Bendermere Avenue, Asbury Park, New Jersey

AT several points in the paper, use was made of measurements made by F. A. Saunders on logarithmic decrements of violins. I have discovered that, whereas his results are in terms of the Naperian base, I erroneously interpreted them as to the base 10. This does not affect the theoretical developments of the paper, but it does affect some numerical comparisons covering power losses, as follows: It can no longer be said that the calculation for air decrement agrees with his measurement unless we use a considerably lower absorption coefficient than 0.04, the value for wood floor on solid foundation borrowed from architectural acoustics. A lower value, however, is credible. The radiation efficiency at the principal body resonance should be 28% or more, rather than 12% or more. In Fig. 9, the Saunders comparison data should be moved to the left by a factor of 2.3. This means that curves such as those in Fig. 6 will in typical cases be more S-shaped than those in Fig. 9. On page 337 under "Damping Requirements of Wood," the range of Q's for the assembled instrument should be 20 to 50 instead of 10 to 20.

Editor's Comments on Paper 6

6 **Cremer:** *Die Geige aus der Sicht des Physikers*
English translation: *The Violin from the Point of View of the Physicist* (abstract only)

Professor Lothar Cremer (1905 –   ) received his training at the Technical University of Berlin, obtaining his doctorate in 1933. After one year at AEG-Telefunken and another at the Heinrich-Hertz-Institute of Berlin in the department of Erwin Meyer, he spent 10 years as chief engineer at the Institute of Mechanics at the Technical University of Berlin. After the war he did consulting in acoustics in Munich and since then has been a full professor at the Technical University of Berlin and Director of the Institute of Technical Acoustics, as well as of the acoustical laboratories of the Heinrich-Hertz-Institute. He is a Fellow of the Acoustical Society of America and Member of the Academy of Sciences of Göttingen.

In addition to about one hundred papers, Professor Cremer has published three books: *Die wissenschaftlichen Grundlagen der Raumakustik* (Vol. 1, Zürich, 1949; Vol. 3, Leipzig, 1950; Vol. 2, Stuttgart, 1961), S. Hirzel Verlag; *Körperschall* (with Manfred Heckl), Berlin-New York, Springer-Verlag, 1967; *Vorlesungen über technische Akustik*, Berlin, Springer-Verlag, 1971.

The work of his research group at the Technical University of Berlin has been productive in many areas of acoustics, including the acoustics of stringed instruments, of which an overview is given in the paper presented in this section.

Selected Publications by Lothar Cremer

Cremer, L., and H. Lazarus. "Der Einfluss des Bogendruckes am Anstreichen einer Saite," 6th Intern. Cong. Acoust., Tokyo (Aug. 21 – 28, 1968).

*Reinicke, W., and L. Cremer. "Application of Holographic Interferometry to Vibrations of the Bodies of String Instruments," *J. Acoust. Soc. Amer.*, **48**(4, pt. 2), 988 – 992 (Oct. 1970).

†Cremer, L. "Die Geige aus der Sicht des Physikers," *Nachr. Akad. Wiss. Göttingen; II. Math. Physik. K.*, **12** (Jan. 1971).

‡Cremer, L. "The Influence of "Bow-Pressure" on the Measurement of a Bowed String": Part I, *Catgut Acoust. Soc. Newsletter*, No. 18, 1972, pp. 13 – 19; Part II, *Catgut Acoust. Soc. Newsletter*, No. 19, 1973, pp. 21 – 25.

Cremer, L., and F. Lehringer. "Zur Abstrahlung geschlossener Körperoberflächen," *Acustica*, **29**(1), 1973.

Cremer, L. "Der Einfluss des 'Bogendrucks' auf die selbsterregten Schwingungen der gestrichenen Saite," Acustica, **30**(3), 119–136 (1974).

---

*Included in *Musical Acoustics, Part II*.
†Included in this volume as Paper 6.
‡Included in this volume as Papers 13a and 13b.

6

Reprinted from *Nachr. Akad. Wiss. Göttingen: II. Math. Physik. Kl.*, **12**, 223–259 (Jan. 1971)

Die Geige aus der Sicht des Physikers[*]

Von *Lothar Cremer*, Berlin-Charlottenburg

Öffentlicher Vortrag, gehalten vor der Göttinger Akademie der Wissenschaften am
11. 6. 1971

Aus der Sicht des Physikers ist jedes Streichinstrument ein Schallsender, bei dem zunächst eine zwischen zwei Auflagern gespannte Saite infolge der Gleichbewegung eines Reibungskräfte übertragenden Bogens selbsterregte Schwingungen ausführt, die viele Obertöne enthalten. Da die Saite sehr dünn ist, auch im Vergleich zu den Luftwellenlängen der höchsten Obertöne, strahlt sie selbst kaum Schall ab. Die erzeugten Schwingungen werden erst dadurch hörbar, daß das eine Auflager, der sogenannte Steg, zugleich einen Übertrager von Kräften und Bewegungen zu einem festen elastischen Instrumentenkörper darstellt, bei dem jedenfalls zwei begrenzende biegesteife Platten, genannt Boden und Decke, einigermaßen vergleichbar mit den Wellenlängen des Hörschalls sind, weshalb dieser Körperschall zur Abstrahlung von Luftschall führt.

Es geht in diesem Vortrag nur darum, wie weit der Physiker die vier angedeuteten Vorgänge, Selbsterregung einer Saite, Übertragung am Steg, Schwingungen des Instrumentenkörpers und Abstrahlung dieser Schwingung, erklären, eventuell sogar quantitativ behandeln kann.

Nun werden sicher viele von Ihnen darüber erstaunt sein, daß eine Spielart, die seit mehr als 2000 Jahren gepflegt wird, und daß Instrumentenkörper, deren beste Exemplare vor mehr als 200 Jahren gebaut wurden, noch nicht restlos erforscht sein sollten, zumal es sich hier doch nur um lauter Probleme der klassischen Mechanik und Akustik handelt, die spätestens mit Helmholtz ihren Abschluß fand. Gewiß — neue Effekte sind hier nicht zu erwarten. Und doch ist es so, daß wir, je mehr uns die moderne Physik, insbesondere die Elektronik, die Möglichkeit der Messung dieser subtilen Kräfte und Bewegungen an die Hand gibt, um so mehr erkennen müssen, daß das Streichinstrumentenspiel mehr physikalisch ungelöste als gelöste Probleme enthält.

Ich muß Sie dabei bitten, nicht enttäuscht zu sein, daß der Physiker heute noch nichts, zumindest nichts durch entsprechende hörpsychologische Versuche Erhärtetes, darüber aussagen kann, warum sich die Preise der Streichinstrumente um vier Zehnerpotenzen unterscheiden können, noch warum es so viel besser klingt, wenn zum Beispiel Menuhin eine Geige anstreicht, als wenn unsereins das tut.

---

[*]*Editor's Note:* An abstract in English follows this article.

Beginnen wir mit dem Problem der Schwingungserzeugung durch Anstreichen. Kein geringerer als Hermann von Helmholtz [1] hat sich bereits,
zunächst experimentell und dann theoretisch, damit befaßt. Er konnte diese
Bewegungen nur aus den Lissajous-Figuren erschließen, die er in einem auf
einen markierten Saitenpunkt gerichteten Mikroskop sah, das in Saiten-
richtung vibrierte. Wir können diese Bewegung im Besitz der elektronischen
Verstärkung heute viel einfacher dadurch ermitteln, daß wir eine metallische,
also leitende Saite, am Beobachtungspunkt in ein Magnetfeld bringen (Bild 1).

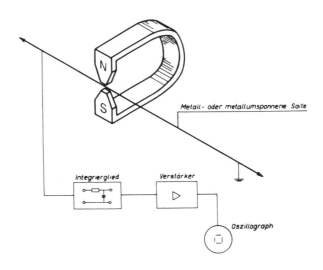

Bild 1: Elektrodynamische Abtastung der Saitenbewegung
(nach H. A. Müller)

Sie wirkt dadurch wie ein elektrodynamisches Mikrofon, indem in der Saite
eine der Geschwindigkeit proportionale Spannung induziert wird, die wiederum elektrisch integriert werden kann, so daß man am Oszillographen den
Zeitverlauf des Ausschlags sieht [2]. Der Anstrich der Saite erfolgt dabei
am besten durch ein umlaufendes Band, wie es u. a. schon von Meinel [3]
benutzt wurde. Legt man die Saite zusätzlich horizontal, so läßt sich auch die
zur Saite senkrechte Kraft, die der Geiger etwas unkorrekt als Bogendruck
bezeichnet, regulieren. Bild 2 zeigt die schon von Helmholtz angegebenen
Zickzacklinien, deren Auf- und Abstiegszeiten sich verhalten wie die Abstände
des Beobachtungspunktes von den Saitenenden. Dieser Verlauf leuchtet insbesondere am Anstreichort ein. Der langsam steigende Teil entspricht dabei
der Mitnahme der Saite durch den Bogen, der schnell abfallende dem Zurück-
gleiten. Faßt man die Bewegung aller Saitenpunkte zusammen, so ergibt sich
eine parabolische Hüllkurve, innerhalb der ein Dreieck mit konstantem
Peripheriewinkel umläuft (Bild 3 oben).

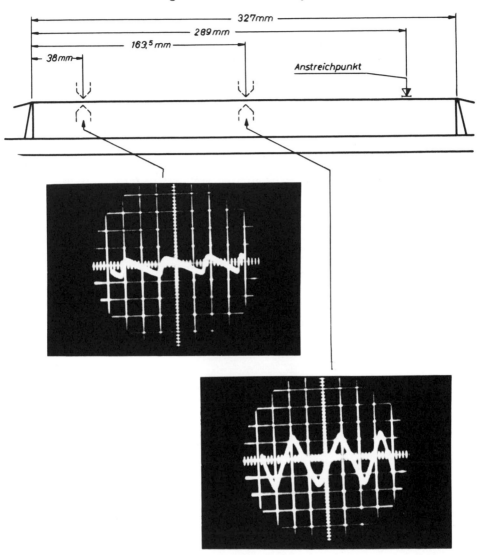

Bild 2: Oszillographische Aufzeichnung des Ausschlages an zwei verschiedenen Orten
(nach H. A. Müller)

Helmholtz konnte zeigen, daß diese Bewegung, die wir mit seinem Namen kennzeichnen wollen, sich als freie Schwingung der Saite ergibt, wenn der erwähnte zeitliche Zickzackverlauf auch nur an einer Stelle, zum Beispiel der Anstrichstelle, vorgegeben wird. Es sei erwähnt, daß schon Helmholtz auch andere, kompliziertere Bewegungen beobachtete. Da sie aber einen unschönen Klang ergeben, also nicht den normalen Geigenklang, seien sie hier außer acht gelassen. Aber auch diese komplizierteren Bewegungen lassen sich als freie Schwingungen beschreiben.

Es sieht hiernach so aus, als ob die Verluste je Periode klein sind gegen die schwingenden kinetischen und potentiellen Energien.

Dem aber widerspricht die Tatsache, daß eine bei aufgelegtem Bogen gezupfte Saite, entgegen einer freien Saite, sofort abgebremst wird. Die Gleitreibungskräfte sind also keineswegs vernachlässigbar. Die durch sie bedingten Verluste sind sicher sehr viel größer als die interessierenden Energieübertragungen am Steg, aber auch als andere Verluste, die in der Saite oder an den Einspannstellen entstehen.

Trotzdem handelt es sich um eine sogenannte entdämpfte Eigenschwingung, weil der größte Teil der beim Gleiten entstehenden Verluste in der Haftphase

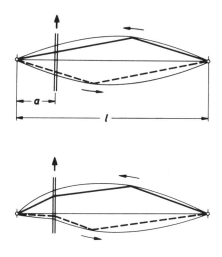

Bild 3: Helmholtzbewegung

oben: ohne konstante Reibungskraft   unten: mit konstanter Reibungskraft

durch eine gleich große, mittlere Kraft wieder zugeführt wird, und weil für die Schwingungen außerhalb der Anstrichstelle nur die kleinen Differenzen gegenüber dieser mittleren Kraft interessieren.

Diese konstante mittlere Kraft bewirkt eine dreieckförmige Auslenkung, die der Helmholtzbewegung superponiert werden kann (siehe Bild 3 unten).

Jeder Geiger und namentlich jeder Cellist kann diese zusätzliche statische Saitenverlagerung beim Wechsel der Strichrichtung deutlich beobachten.

Bestimmt man aus den nach beiden Seiten unterschiedlichen Ausschlägen die statische Verschiebung und die Schwingungsausschläge, so kann man hieraus das Verhältnis von Verlustenergie je Periode zu schwingender Energie errechnen. E. J. Völker [4] hat hierfür bei freilich großem Bogendruck Werte bis zu 0,4 erhalten.

Für die überlagerte freie Bewegung ergibt die Helmholtzsche Näherung im Einklang mit der Messung, daß der maximale Ausschlag η_{max} — und damit auch die vom Steg übertragene Kraft — wie H. A. Müller [2] experimentell bestätigte, linear mit der Bogengeschwindigkeit v_B wächst (Bild 4a) und daß er umgekehrt proportional dem Abstand der Anstreichstelle a vom Steg ist:

$$\eta_{max} \sim \frac{v_B}{a}.$$

Gerade von der letzten Abhängigkeit macht jeder Spieler Gebrauch, indem er bei forte näher am Steg, bei piano näher am Griffbrett anstreicht.

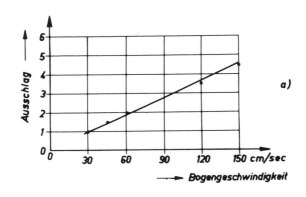

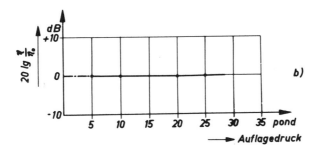

Bild 4: Saitenausschlag in Abhängigkeit
a) der Bogengeschwindigkeit   b) des „Bogendrucks"
(nach H. A. Müller)

Aber er weiß auch, daß er in Stegnähe einen höheren Bogendruck braucht, und daß dieser auch am gleichen Ort den Eindruck der Lautstärke des Klanges bei gleicher Bogengeschwindigkeit ändert. Dabei ist der Maximalausschlag, wie H. A. Müller [2] ebenfalls experimentell belegen konnte, erstaunlich unabhängig vom Bogendruck (Bild 4b). Das steht mit dem Eindruck wachsender Lautheit nicht im Widerspruch. Dieser Eindruck hängt mit einem Anteil an

Obertönen zusammen, die andererseits nur wenig zum Ausschlag beitragen.
(Etwas ganz Ähnliches stellt das Zuschalten der sogenannten Mixturen, d. h.
von Obertonregistern bei einer Orgel, dar. Sie ergeben eine wesentliche Erhöhung der empfundenen Lautheit, obschon sich der resultierende Schalldruck
objektiv nur wenig erhöht.)

Eine befriedigende Theorie der gestrichenen Saite verlangt also die Berücksichtigung des Einflusses des Bogendrucks, der in der Helmholtzschen Beschreibung durch freie Schwingungen nicht enthalten ist.

Eine solche Erweiterung der Theorie wurde — vor etwa 50 Jahren — von
C. V. Raman [5] versucht, allerdings nur dadurch, daß er die Helmholtzsche
Kinematik der Saitenbewegung, die sprungförmige Geschwindigkeitsänderun-
gen ergibt (Bild 5), als gegeben ansah, und daß er für die zu jeder dynamischen

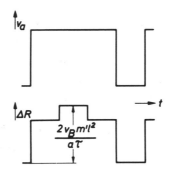

Bild 5: Schnelle (oben) und Reibungskraft (unten) am Anstreichort $a = \frac{1}{4}l$
(nach C. V. Raman)

Theorie gehörigen Verluste die einfachste Annahme machte, nämlich die, daß
alle Teilschwingungen gleich gedämpft sind, d. h. einzeln erregt gleich schnell
abklingen.

Denkt man sich die Verluste am Steg konzentriert, so hieße das, daß dort
das Verhältnis von transversaler Kraft zu transversaler Geschwindigkeit, also
die sogenannte Stegeingangsimpedanz, einen frequenzunabhängigen Widerstand darstellen würde.

Bild 5 zeigt einen von Raman errechneten Verlauf der zusätzlichen Reibungskraft ΔR für den einfachen Fall $a/l = \frac{1}{4}$. Daß dabei neben den Sprüngen
zu den Zeiten des Gleit- und Mitnahmebeginns noch zwei andere auftreten,
hängt damit zusammen, daß während der Mitnahme Sprungwellen zwischen
Steg und Bogen hin- und herlaufen. Es sieht hiernach so aus, als ob die höchste
Reibungskraft R_{max} während der Mitnahme erreicht wird. Das ist aber physikalisch nicht der Fall. Während der Mitnahme ist die Reibung eine Haftreibung, die nur einen Höchstwert R_{Hmax} annehmen kann. Erst wenn er überschritten wird, tritt Gleiten ein, und die Mitnahme verlangt, daß er wieder
erreicht wird. Wir haben also das Ramansche Bild durch zwei unendlich

schmale senkrechte Linien über Gleit- und Mitnahme-Zeitpunkt zu ergänzen, die noch etwas über den Höchstwert in der Mitte der Mitnahmezeit ragen.

Trotz seiner willkürlichen Annahmen erhielt Raman ein mit der Spieltechnik in Einklang befindliches Resultat. Zur Erzeugung der Helmholtzbewegung war ein Mindestwert der Änderung der Reibungskraft und damit des Bogendrucks erforderlich, der wiederum der Bogengeschwindigkeit proportional und dem Abstand vom Steg umgekehrt proportional war. Bild 6 zeigt die von Lazarus [6] an einem ausbalancierten umlaufenden Band an einem Monochord gemessenen Bogendruckkräfte in Abhängigkeit vom Anstreichort,

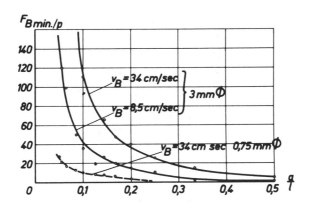

Bild 6: Gemessener min. „Bogendruck" in Abhängigkeit vom Anstreichort
(nach H. Lazarus)

die beide Abhängigkeiten, zumindest der Tendenz nach, bestätigen, die also offenbar von der Ramanschen Annahme über die Verluste unabhängig sind. Man wird dieses Anwachsen des benötigten „Bogendrucks" gegen den Steg hin allgemein aus dem bereits erwähnten Anwachsen der Ausschläge mit Verkleinerung des Steg-Bogen-Abstands erklären können. Zu größeren Ausschlägen gehören größere Verluste, und ihre Deckung verlangt größere Reibungskräfte.

Mißt man die Abklingzeiten der Saitenschwingungen — bei entsprechender Filterung im Empfänger — in Abhängigkeit von der Frequenz, so zeigt sich bereits bei starren Stegen — dem Monochord also — (siehe Bild 7, das Messungen von W. Reinicke [7] entnommen ist), eine erhebliche Abnahme mit der Frequenz. Noch stärker ist sie — im interessierenden Frequenzbereich — bei der auf den Steg eines Instrumentes aufgespannten Saite.

Trotzdem sind die „Nachhallzeiten" der großen Streichinstrumente, Cello und Kontrabaß, bei ihren tiefsten Grundtönen oft länger als die des Raumes, in welchem sie gespielt werden [8].

Diese größere Dämpfung der höheren Teiltöne bedeutet nun, daß die Knicke der Helmholtzbewegung in Wirklichkeit in sehr abgerundeter Form auf den Bogen zulaufen, was übrigens schon wegen der Biegesteife der Saite angenommen werden muß. Diese Abrundungen werden nun beim Passieren der Anstreichstelle um so mehr „geschärft", je größer der Bogendruck und somit die maximale Haftreibung ist [6], indem es von ihr abhängt, wie weit die Mitnahmezeit verlängert, die Gleitzeit verkürzt wird.

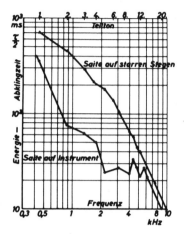

Bild 7: Abklingzeiten von gezupften Saiten
(nach W. Reinicke)

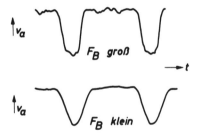

Bild 8: Oszillogramm der Schnelle am Anstrichort
oben: großer „Bogendruck"   unten: kleiner „Bogendruck"
(nach H. Lazarus)

Bild 8 zeigt zwei Oszillogramme der Schnelle bei großem und kleinem Bogendruck. In keinem Falle gibt es die idealen Sprünge. Sie sind aber bei großem Bogendruck besser angenähert. Wer mit der Analyse eines Zeitverlaufes in Teiltöne vertraut ist, übersieht, daß der obere Verlauf offenbar der obertonreichere ist.

Nun fällt bei diesem auf, daß die Schnelle auch während der Mitnahme Schwankungen aufweist, wo man eigentlich den konstanten Wert der Bogengeschwindigkeit erwartet.

Diese Erwartung setzt allerdings voraus, daß die Bogenhaare nicht etwa in der Richtung ihrer Spannung ihrerseits unter den Reibungskräften nachgeben. Um dies zu kontrollieren, wurde von Ottmer [9] mit der in Bild 9 gezeigten Apparatur die Kraft gemessen, die nötig ist, um die Bogenhaare mit

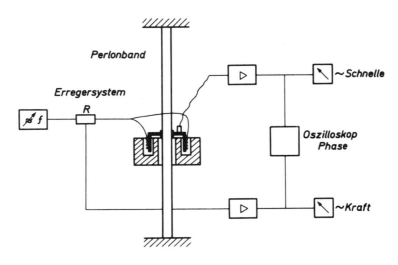

Bild 9: Messung der longitudinalen Bogenimpedanz
(nach D. Ottmer)

bestimmter Schnelle longitudinal zu bewegen und festgestellt, daß sie — ungeachtet ihrer komplizierten Frequenzabhängigkeit — immer viel größer ist, als die Kraft, die nötig ist, der Saite die gleiche transversale Schnelle zu erteilen. In der Sprache des Akustikers heißt das, daß die longitudinale Bogenimpedanz immer groß ist gegen die transversale Saitenimpedanz. Alle bestehenden theoretischen Behandlungen nahmen das zwar an, aber der Nachweis der Richtigkeit dieser Annahme mußte einmal erbracht werden.

Die im Oszillogramm (Bild 8) auftretenden Schwankungen sind dagegen ohne weiteres daraus erklärbar, daß der Bogen exzentrisch an der Saite angreift, das Oszillogramm aber seine Schwerpunktsbewegung wiedergibt, und daß die Saite zusätzliche Torsionsbewegungen ausführt. Ihr Vorhandensein wurde von Tippe [10] dadurch erwiesen, daß er auf der Saite ein leichtes unter 45° zur transversalen Bewegung geneigtes Blättchen anbrachte (siehe Bild 10), das bei Torsionsbewegung einen auf eine Fotozelle fallenden Lichtstrom steuerte. Dabei wurden die (in Bild 10 unten über der transversalen

Saitenbewegung gezeigten) Torsionsbewegungen dadurch besonders hervor-
gehoben, daß die Saitenlänge auf eine Resonanz mit einer Torsionseigen-
schwingung abgestimmt war. Man erkennt deutlich, daß die Schwankungen

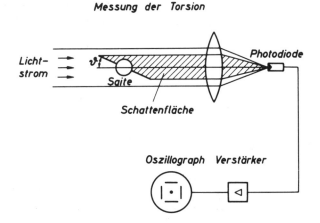

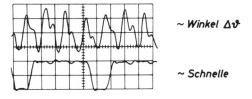

angestrichen   bei $\frac{x_a}{e} = \frac{1}{4}$

gemessen   bei $\frac{x_m}{e} = \frac{3}{4}$

Bild 10: Messung der Torsion der Saite
(nach W. Tippe)

der transversalen Schnelle die Periodizität der Torsionsschwingungen aufweisen. — Inwieweit solche Torsionsresonanzen auch am Instrument auftreten, oder ob die Torsionswellen über Steg und Sattel reflexionsfrei hinweg-
laufen und dahinter absorbiert werden, ist noch nicht geklärt. Wenn das letzte
der Fall ist, könnte aus den Torsionswellen auf den Zeitverlauf der anders kaum
meßbaren Reibungskräfte zwischen Saite und Bogen geschlossen werden.

Sicher ist, daß infolge der Torsion und infolge der Biegesteife eine im Mitnahmebereich auf die Anstrichstelle zulaufende Transversalwelle nicht, wie

man zunächst annahm, vollständig reflektiert wird. Dies ergeben sowohl Rechnungen wie Messungen, die von Eisenberg [11] an einer 30 m langen, 3 mm dicken Perlonsaite durchgeführt wurden. Ein bei der in Bild 11 gezeigten Anordnung links erzeugter Tonimpuls passiert die Meßstelle 1 vor dem Auftreten auf den Bogen. Dort entstehen je ein reflektierter und ein durchgelassener Torsionswellen- und Transversalwellen-Impuls, die an den Stellen 1 und 2 registriert werden. Die so gemessenen Werte erweisen sich als hinreichend mit der Rechnung übereinstimmend.

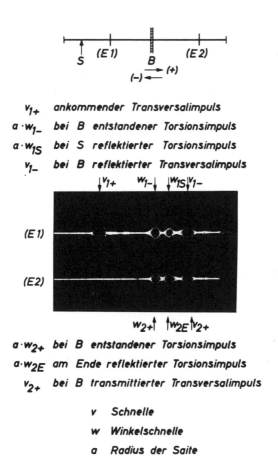

v_{1+} ankommender Transversalimpuls

$a \cdot w_{1-}$ bei B entstandener Torsionsimpuls

$a \cdot w_{1S}$ bei S reflektierter Torsionsimpuls

v_{1-} bei B reflektierter Transversalimpuls

$a \cdot w_{2+}$ bei B entstandener Torsionsimpuls

$a \cdot w_{2E}$ am Ende reflektierter Torsionsimpuls

v_{2+} bei B transmittierter Transversalimpuls

v Schnelle

w Winkelschnelle

a Radius der Saite

Bild 11: Reflektierte und durchgelassene Wellen an einem aufgelegten Bogen
(nach P. Eisenberg)

Dem Musiker mag das Studium einer 30 m langen, 3 mm dicken Personsaite als etwas weit entfernt von den interessierenden Instrumenten erscheinen. Der Physiker aber muß die Daten so variieren, daß das, was auch beim Instrument zu erwarten ist, optimal erfaßbar wird.

Zu solchen „Seitentälern" gehören auch Untersuchungen an Modellen, wie sie die analoge Umwandlung mechanischer Systeme in elektrische bietet. Wir hatten ursprünglich versucht, das ganze Problem ins Elektrische zu übersetzen, also auch die Saite durch einen Kettenleiter aus Spulen und Kondensatoren, mußten dies aber aus der bekannten Erfahrung aufgeben, daß elektrische Systeme infolge der Ohmschen Widerstände viel gedämpfter sind als mechanische. Dagegen gelang es Heinrich [12], die zwischen Haft- und Gleitreibung schwankenden Reibungskräfte mit Hilfe von Leistungstransistoren durch ein elektrisches Antriebssystem mit ähnlichen Eigenschaften zu ersetzen, zumindest, was die „fallende Kennlinie" der Gleitreibung mit wachsender Relativgeschwindigkeit anbelangt, wie sie in Bild 12 (nach neueren Messungen von Lazarus [13]) gezeigt ist. Die zur Relativgeschwindigkeit Null gehörige Haftreibung, die sich auf einem senkrechten Ast bewegt, konnte freilich bei der analogen Anordnung nur durch eine steile Gerade ersetzt werden (siehe Bild 13). Offenbar hängt damit die in den Versuchen von Heinrich zutage getretene Einsattlung der transversalen Schnelle im Mitnahmebereich zusammen. Was aber die Versuche brachten, war der als Stromverlauf in Erscheinung tretende Kraftverlauf, der deutlich zeigt, daß der der maximalen

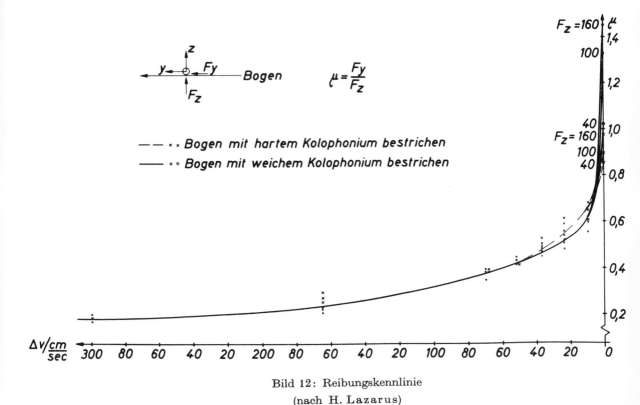

Bild 12: Reibungskennlinie
(nach H. Lazarus)

Haftreibung entsprechende Gipfel sowohl zu Beginn wie zu Ende der Gleit-
periode kurzzeitig überschritten wird. Auch konnte die Beeinflussung des
Gleit- und Mitnahmeeinsatzes in Abhängigkeit von diesem Höchstwert und
von systematisch veränderten Stegimpedanzen gezeigt werden.

Die richtige Erfassung der Stegimpedanz stellt übrigens die größte Schwierig-
keit dar, die der Vorausberechnung des stationären Zustandes entgegen-

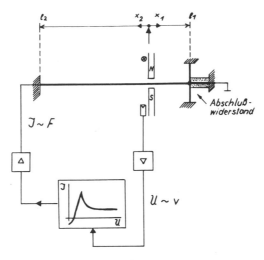

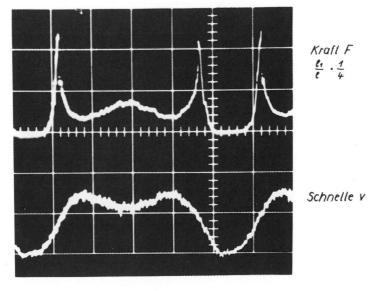

Bild 13: Elektrodynamisches Analogon der gestrichenen Saite
(nach P. Heinrich)

steht. Es würde aber schon einen wesentlichen Fortschritt in der Theorie der gestrichenen Saite bedeuten, wenn man für eine einfache Impedanz die Rechnung durchführen könnte. Sie darf allerdings nicht — wie in der Ramanschen Theorie — ein reiner Widerstand sein, sondern muß den Abfall der Nachhallzeit mit der Frequenz richtig wiedergeben, was eine Kombination von Federung und mechanischem Widerstand leistet. Lazarus versucht zur Zeit auf dieser Basis eine dynamische — d. h. die Bogenkräfte einbeziehende — Theorie der gestrichenen Saite zu entwickeln. Es ist erstaunlich, wie schwierig die Dinge werden, wenn man auch nur die schwankenden Reibungskräfte und die Stegverluste mit möglichst einfachen Annahmen erfassen will. Dabei sind die wirklichen Stegimpedanzen infolge der Resonanzen des anschließenden Geigenkörpers sicher viel komplizierter.

Daß der Geigen- oder Cellokörper auf den Streichvorgang rückwirken kann, ist einmal durch das schon von Raman [14] und neuerdings von Eggers [15] eingehend studierte Phänomen des Wolfstons erwiesen, aber auch an der jedem Spieler geläufigen Tatsache, daß es Instrumente gibt, die sich „leicht spielen" und solche, die sich „schwer spielen".

Nun mag die Suche nach und das Denken in stationären Zuständen möglicherweise der Streichmusik grundsätzlich nicht gerecht werden. Einmal ist zu erwarten, daß die Reibungsverhältnisse je nach Kolophoniumbelag von Stelle zu Stelle sich ändern. Andererseits können gerade die dadurch bedingten statistischen Schwankungen den besonderen Reiz eines Geigenklangs ausmachen.

Eine Hypothese kann dagegen als widerlegt gelten, nämlich die, daß es sich um einen Regelvorgang handelt, in den das Nervensystem des Spielers einzubeziehen ist, indem der Spieler auf Grund dessen, was er hört, seine Handhabung des Bogens beeinflußt. Dies gilt sicher für das Spielen von Melodien und anderen langfristigen Änderungen. Für Regulierungen innerhalb einer Periode sind die Laufzeiten im Nervensystem sicher zu lang. Um aber ganz sicher zu sein, haben wir die Saitenschwingungen an einem gespielten Instrument registriert, während wir dem Spieler über dicht aufgesetzte Kopfhörer sein Spiel künstlich verzögert zu Gehör brachten. Das fiel ihm nur beim Bogenwechsel auf. Auch wenn wir ihm das Hören seines Spiels durch verdeckendes Rauschen unmöglich machten, war dies den Schwingungen nicht anzusehen *).

Nun zum nächsten Problemkreis, der Funktion des Steges. Man könnte ihn bereits als Teil des Instrumentenkörpers betrachten; dies wäre sogar mit Rücksicht auf die resultierende Rückwirkung angebracht. Der Physiker aber tut gut daran, alles, was sich noch sinnvoll trennen läßt, getrennt zu untersuchen. Diese Trennung ist aber beim Steg leicht möglich, weil die Energiezufuhr jeweils an einem Punkt, in der Kerbe für die schwingende Saite, die

---

*) Ich möchte hier der Bratscherin Liselotte Schönewald und der Cellistin Brunhilde Schönewald für ihre interessierte Mithilfe meinen besonderen Dank aussprechen.

Energieabgabe an seinen Fußpunkten erfolgt. Dort interessieren nur die zur Decke senkrechten Kraftkomponenten F_1 und F_2 und die zugehörigen Geschwindigkeiten, oder — wie der Akustiker sagt — die „Schnellen" v_1 und v_2 (siehe Bild 14, oben links). Nehmen wir weiterhin an, daß die Saitenkraft F_s parallel zur Verbindung der Füße gerichtet ist, so werden die Kräfte F_1 und F_2 jedenfalls bei tiefen Frequenzen, d.h. unter Vernachlässigung vertikaler Beschleunigungen der Stegmasse, entgegengerichtet und nahezu gleich sein. In dieser Übertragung einer transversalen Saitenkraft in ein Paar zur Decke senkrechter Kräfte liegt offenbar die Funktion des Steges. Da hierbei auch Leistung übertragen werden muß, kann das nicht ohne Bewegungen geschehen. Wie groß diese im Einzelfall sind, hängt von der Rückwirkung des Geigenkörpers ab.

Nach der äußeren Erscheinung einer Geige wäre man geneigt, sie für vollständig symmetrisch zu halten. Das ist aber im Innern keineswegs der Fall (siehe Bild 14, oben rechts). Unter dem — vom Spieler aus gesehen — linken Fuß befindet sich ein mit der Decke verleimter Stab, der sogenannte Baßbalken, unter dem anderen Fuß, allgemeiner gesagt, nahe dem anderen Fußpunkt, ein zur Decke senkrechter kurzer Stab, der sogenannte Stimmstock, der Kraft und Bewegung des rechten Fußes zum Teil auf den Geigenboden überleitet.

Es sind also verschiedene „Belastungsimpedanzen" unter beiden Stegen zu erwarten und somit trotz gleicher Kräfte verschiedene Schnellen. Der Steg ist somit — in der Sprache der elektrischen Übertragungstechnik — ein „Dreitor" — wie es dort ein Kasten darstellt, der auf einer Seite zwei Eingangsklemmen, auf der anderen Seite zwei Paare von Ausgangsklemmen aufweist (siehe Bild 14, unten links). Die in der Nachrichtentechnik sehr weit entwickelte Theorie solcher Systeme kann aber für mechanische Gebilde analog übernommen werden, was für den Steg von Zimmermann [16] durchgeführt wurde. Man erhält dann zwischen den drei Kräften und den drei Schnellen drei lineare Gleichungen, die bis zu sechs verschiedene Koeffizienten enthalten können, und diese kann man experimentell bestimmen, indem man die Impedanzen an allen Toren mißt, wenn an den anderen jeweils bestimmte Zustandsgrößen zum Verschwinden gebracht werden. Im elektrischen Fall sind das Spannung und Strom, und die Zustände heißen Leerlauf und Kurzschluß. Im mechanischen Fall sind es Kraft und Schnelle, und die Zustände bedeuten freie Beweglichkeit oder starre Einspannung.

Man wäre zunächst geneigt, den Steg als eine starre Scheibe anzusehen, die lediglich durch einige hübsch geformte Unterbrechungen leichter gemacht ist. Aber selbst der frei im Schwerpunkt aufgehängte Steg zeigt, wie Steinkopf [17] an einem Cellosteg zeigen konnte, oberhalb 4000 Hz ein resonanzartiges Verhalten, das dabei auf seine Unterteilung in zwei Massen und eine verbindende Feder schließen läßt. Noch deutlicher wurde dies bei starr, d. h. auf einen Betonklotz aufgeklebten Steg. Hier zeigte sich beim hochbeinigen

Cellosteg eine Resonanz bei 1000 Hz, beim kurzfüßigen Geigensteg erhält man nach Reinicke [18] eine solche bei 3000 Hz und eine schwächere bei etwas über 6000 Hz. Reinicke zeigte, daß dieses Verhalten sich erklären läßt, wenn man das Stegoberteil von den Fußteilen durch ein verdrehbares Federnsystem trennt (siehe Bild 14, unten rechts), wie das die seitlichen Einkerbungen nahelegen. Mag ihre Form im Einzelnen — wie sicher vieles im Geigenbau — auch ästhetisch bedingt sein, es ist durchaus wahrscheinlich, daß diesen Einkerbun-

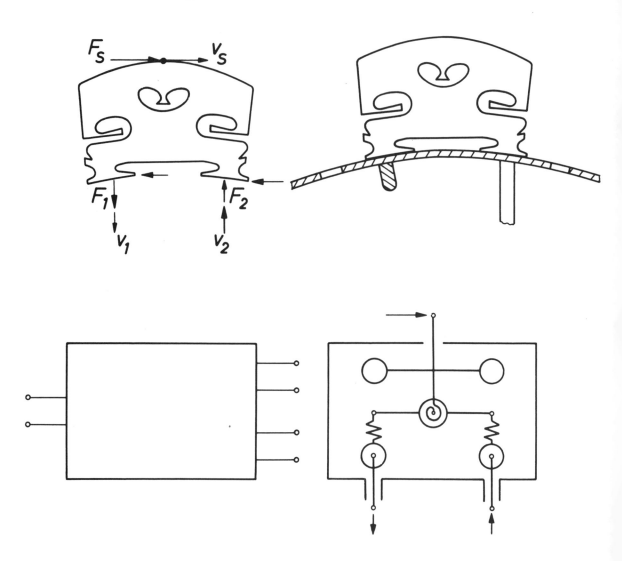

Bild 14: Geigensteg als Dreitor
oben links: Richtungen von Kraft und Schnelle; oben rechts: Steg über Baßbalken und Stimmstock; unten links: Elektrisches Ersatzschema; unten rechts: Mechanisches Ersatzschema (nach W. Reinicke)

116

gen Aufgaben zufallen, die man bei der Nachrichtenübertragung als Filterwirkungen bezeichnet.

Jedenfalls hat Reinicke beim starr aufgesetzten Steg mit Hilfe kleiner piezoelektrischer Geber in der Saitenkerbe und unter einem Stegfuß festgestellt, daß der Kraftübertragungsfaktor F_2/F_s gegen die untere Resonanzfrequenz hin, bei der das Oberteil sich gegen das Unterteil verdreht, immer größer, also günstiger wird, und dort ein Maximum erreicht (siehe Bild 15). Es wird hierbei also der Frequenzbereich um 3000 Hz angehoben.

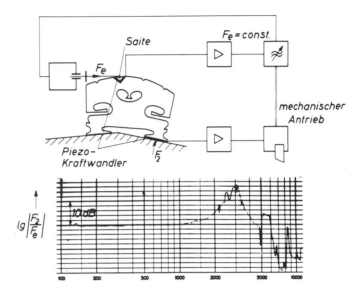

Bild 15: Messung des Kraftübertragungsfaktors am Steg
(nach W. Reinicke)

Es leuchtet ein, daß solche Filterwirkungen durch den bekannten aufsetzbaren „Dämpfer" stark beeinflußt werden, und daß es dabei, worauf schon Raman [19] hinwies, sehr darauf ankommt, ob man die Zusatzmasse am oberen Teil oder am unteren anbringt. Bild 16 zeigt die von Reinicke mit den erwähnten piezoelektrischen Gebern registrierten Zeitverläufe der von der Seite auf den Steg übertragenen Kraft (oben) und der von diesem auf die Decke an den beiden Fußpunkten übertragenen Kräfte bei aufgesetztem Dämpfer. Die Saitenkraft zeigt den zum Helmholtztyp gehörigen Sägezahnverlauf mit leichten Kräuselungen. Von allen Obertönen sind aber im Stegfuß nur noch die tieferen, diese aber relativ zum Grundton angehoben, enthalten. Man sieht übrigens auch, daß sich die Stegfußkräfte nicht nur von der Seitenkraft erheblich unterscheiden, sondern auch untereinander, daß also die Rückwirkungen (Impedanzen) der Decke neben Baßbalken und Stimmstock sehr verschieden sind.

117

Im einzelnen hängt es von den Fußpunktimpedanzen ab, wie sich die Filtereigenschaften des Steges auswirken.

Damit kommen wir zum dritten Problemkreis, den Schwingungen des Geigenkörpers. Für den Instrumentenbauer beginnt eigentlich hier erst das Reich seiner Kunst und seiner Geheimnisse.

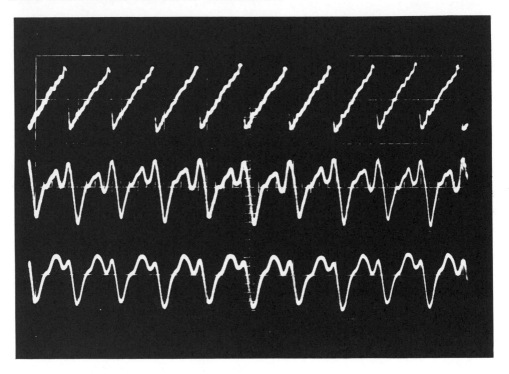

Bild 16: Oszillogramm der Kraft am Steg bei aufgesetztem Dämpfer (nach W. Reinicke) oben: Saitenkraft; Mitte: Stegfuß über dem Baßbalken; unten: Stegfuß über dem Stimmstock

In der Tat handelt es sich hier um ein sehr kompliziertes schwingungsfähiges System. Das zeigt sich bereits, wenn wir die sehr frequenzabhängigen Reaktionen des Instrumentenkörpers auf Wechselkräfte an den Stegfußpunkten betrachten, die schon von Eggers [15] am Cello, von Reinicke [18] an der Geige in Form der logarithmisch aufgetragenen Schnelle über dem Logarithmus der Frequenz bezogen auf konstante Wechselkraft registriert wurden (siehe Bild 17). Es ergibt sich ein wildes Gebirge, das viele Resonanzgipfel erkennen läßt, also starke Schwingungen, weil hier die Erregerfrequenz mit einer sogenannten Eigenfrequenz des Systems zusammenfällt.

Kein Akustiker würde einen solchen kompliziert frequenzabhängigen Sender anstreben, wenn es darum geht, ein Nutzsignal, sei es Sprache oder Musik, klangfarbentreu abzustrahlen. In der Tat hat sich auch die Verwendung von

Geigenkörpern in Verbindung mit elektroakustischen Lautsprechersystemen keineswegs bewährt, während die von den Blechblasinstrumenten übernommenen Trichter sich — in modifizierter Form — sehr wohl bewährt haben.

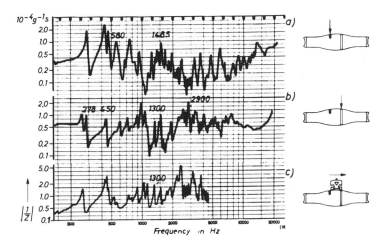

Bild 17: Frequenzgang der Schnelle der Geigendecke am Anregungsort dividiert durch die Kraft (nach W. Reinicke)

oben: Stegfußpunkt über dem Baßbalken; Mitte: Stegfußpunkt neben dem Stimmstock; unten: am Steg in Saitenkraft-Richtung

Offenbar geht es eben nicht um eine gleichmäßige Abstrahlung aller Frequenzen. Eine gewisse Ungleichmäßigkeit kann durchaus erwünscht sein. Bei Orgelregistern, wo gleichartige Pfeifen sich viel weniger unterscheiden, wird sie oft künstlich zwischen benachbarten Pfeifen herbeigeführt.

Daß uns beim Spielen und Hören diese Ungleichmäßigkeiten nicht so kraß bewußt werden, wie es den Schwankungen des Bild 17 entspricht, hängt damit zusammen, daß wir ja beim Spiel keine reinen Töne erzeugen, sondern einen obertonreichen Klang.

Die Ungleichmäßigkeit der Reaktion des Resonanzkörpers ist übrigens nicht erst eine Folge seines komplizierten Aufbaus. Auch ein der Rechnung zugängliches Gebilde, wie eine allseitig aufgestützte ebene Platte, weist bei bestimmten Frequenzen Eigenschwingungen auf, die sich untereinander dadurch unterscheiden, daß sie die Platte verschieden oft in beiden Richtungen durch in Ruhe befindliche Linien, die sogenannten Knotenlinien, unterteilen. Bekanntlich hat bereits Chladni um 1800 diese Knotenlinien dadurch sichtbar gemacht, daß er Pulver auf die Platte streute, dessen Teilchen in die Knotenlinien wanderten und dort liegen blieben.

Wir besitzen heute in der holografischen Interferometrie ein noch leistungsfähigeres Verfahren, das nicht nur die Knotenlinien erkennen läßt, sondern

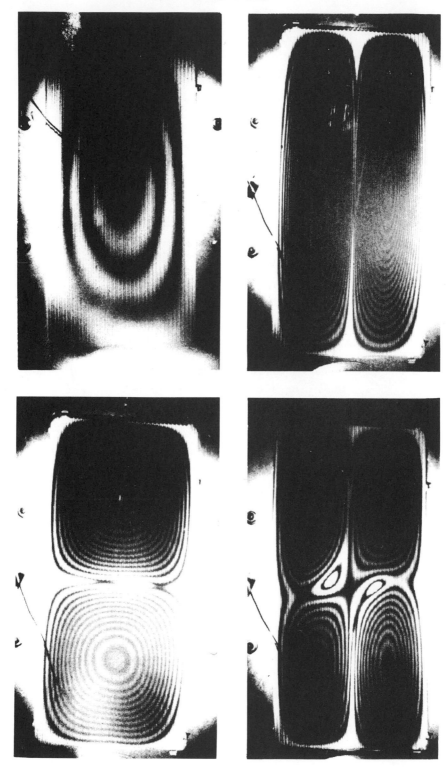

auch die Höhen der sich zwischen ihnen bildenden Berge. Bild 18 zeigt solche holografischen Interferenzbilder von Eigenschwingungen einer am Rande aufgestützten Platte [20]. Sie können diese Bilder als Aufnahmen in einem Moment größten Schwingungsausschlages betrachten, bei welchem die dunklen und hellen Linien Höhenlinien darstellen, wobei immer ein Hügel mit einer Mulde wechselt. Diese Linien entstehen durch Interferenzen kohärenter Lichtwellen, die in einem extrem frequenzstabilen Strahler, einem sogenannten Laser, erzeugt werden. Dabei bedeutet jede folgende Höhenlinie eine Ausschlagsänderung um ein Viertel der benutzten Lichtwellenlänge, das ist weniger als ein Zehntausendstel eines Millimeters. In dieser Größenordnung liegen auch die beim Geigenspiel erzeugten Verschiebungen in Boden und Decke.

Trotzdem hätte es nicht erst der holografischen Interferometrie bedurft, um die entsprechenden Schwingungsformen bei den Streichinstrumenten zu ermitteln. Backhaus [21], wie auch später Eggers [15] haben hierzu die extrem kleinen Kapazitätsänderungen herangezogen, die sich zwischen einem auf den Korpus aufgeklebten Metallblättchen und einer davor in Ruhe gehaltenen Gegenelektrode ergeben. Aber sie mußten diese punktweise erhaltenen Ergebnisse erst zu Knoten- oder gar Höhenlinien zusammensetzen, während wir diese bei den nachfolgend gezeigten, von W. Reinicke [22] aufgenommenen Hologrammen, ebenso wie bei der Platte, gleich als Ganzes überblicken (siehe Bild 19).

Daß die Schwingungsformen komplizierter sein würden als bei der rechteckigen, ringsum aufgestützten Platte, war zu erwarten, denn der Korpus ist ja in vieler Hinsicht komplizierter aufgebaut.

Schon die Randform entzieht sich jeder analytischen Erfassung. Die seitlichen Einbuchtungen sind nötig, um die Bogenbewegung auf g- und e-Saite zu ermöglichen. Auch die Abrundungen haben sicher — abgesehen von ihrer Symmetrie — nicht nur Gründe der Eleganz. Sie erleichtern das Halten am Kinn oder zwischen den Knien und erhöhen auch die Festigkeit. Zu dieser äußeren Randform kommen bei der Decke noch die f-Löcher hinzu, die ja auch einen Rand, und zwar einen freien, bedeuten.

Es ist ein — seltsam verbreiteter — Irrtum, daß komplizierte Formen das Auftreten von Eigenschwingungen behindern. Dem gegenüber hat der Mathematiker Weil nachgewiesen, daß die Zahl der Eigenschwingungen unterhalb einer bestimmten — freilich hohen Frequenz — nur von der Ausdehnung eines Kontinuums, bei räumlichen Schwingungen vom Volumen, bei zweidimensionalen, wie hier, von der Fläche abhängt.

Bild 18: Hologramm einer aufgestützten Rechteckplatte bei verschiedenen Eigenformen (nach O. König)

Die nächste Komplikation, die allein genommen mathematisch erfaßbar wäre, liegt in der sogenannten Anisotropie des Holzes, das quer zur Faser leichter zu biegen ist als in Faserebene. Sie stellt einen gewissen Ausgleich der unterschiedlichen Spannweiten in Richtung des Griffbretts und quer dazu dar.

Ferner ist die Decke ja nicht einfach eine Platte, sondern sie enthält als versteifendes Element den Baßbalken. Er sorgt bei der sogenannten Hauptresonanz des Instrumentes dafür, daß die Decke möglichst als Ganzes schwingt. Bei der hohen Frequenz von 1300 Hz, bei der Bild 19 aufgenommen wurde, markiert sich der Baßbalken deutlich als Knotenlinie.

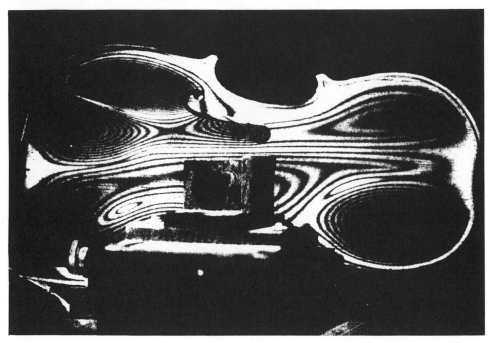

Bild 19: Hologramm einer Geigendecke bei Anregung über dem Stimmstock bei 1300 Hz, $\lambda_L/2 = 13,1$ cm, Korpuslänge 36 cm (nach W. Reinicke)

Eine weitere Komplikation, die aber auch noch rechnerisch erfaßbar wäre, stellt die unterschiedliche Dicke innerhalb von Decke und Boden dar.

Mehr Schwierigkeiten würde die Berücksichtigung der Wölbung machen. Durch sie kommt zur Biegesteife noch eine Schalenversteifung hinzu. Sie beruht darauf, daß ein über zwei Stützen gespannter Bogen nicht ausgelenkt werden kann, ohne gedehnt oder gestaucht zu werden. Deshalb kann ein schalenförmiges Dach größere Lasten aufnehmen als ein gleich dickes flaches. Bei der Geige hat diese Wölbung sicher auch statische Gründe. Wir brauchen uns nur klarzumachen, daß die gespannten Saiten über dem Steg mit einer Kraft von 20 kp auf den Instrumentenkörper drücken. (Das entspricht dem gebührenfreien Höchstgepäckgewicht eines Fluggastes!)

Hieraus wird klar, daß auch dem Stimmstock neben seiner Aufgabe als Schwingungsübertrager eine statische zukommt.

Eine schwedische Forschergruppe [23] hat übriges schrittweise mit Hilfe sehr schöner holographischer Bilder verfolgt, wie sich die Schwingungsformen im Verlaufe der Bearbeitung ändern. Bild 20a zeigt eine Schwingung der zunächst nur richtig gewölbten und ausgearbeiteten Geigendecke bei aufgestützten Zargen. Man sieht wie die „Höhenlinien" des Hologramms sich der Taillenform des Umfangs ein wenig anpassen. Die Frequenz dieser Eigenschwingung betrug 550 Hz. Sobald aber die f-Löcher eingeschnitten sind (Bild 20b) erhalten wir eine völlig andere Schwingung, die mit den f-Löchern jäh abbricht, dort aber ihre größten Ausschläge aufweist. Entsprechend der erhöhten Beweglichkeit sinkt die Frequenz der Eigenschwingung auf 420 Hz. Mit der Anleimung des Baßbalkens (Bild 20c) erhält sie eine Unsymmetrie; die Seite des Baßbalkens wird stärker „mitgenommen" als die andere, und die Eigenfrequenz steigt entsprechend der erhöhten Steife wieder etwas an, auf 465 Hz. Mit dem Aufsetzen auf den hier auf eine starre Unterlage aufgestützten Stimmstock verstärkt sich die Unsymmetrie, es kommt sogar jenseits desselben zu einer Insel gegenphasiger Schwingung. Die Eigenfrequenz liegt bei 540 Hz.

Ähnliche Bilder haben die Autoren von dem Boden angefertigt sowie von dem zusammengebauten Instrument.

Damit kommen wir zur letzten Komplikation, nämlich dazu, daß die Geige zwei senkrecht zu ihrer Oberfläche schwingende Platten aufweist, die in dreifacher Weise gekoppelt sind:

a) Einmal durch die schmalen Seitenwände, die sogenannten Zargen. Sie führen, wie Eggers [15] am Cello zeigen konnte, bei den tiefsten Frequenzen sogar dazu, daß der Instrumentenkörper als starrer Körper Drehschwingungen um seine Längsachse ausführt.

b) Sodann durch den Stimmstock, der übrigens erst später in den Instrumentenbau eingeführt worden sein soll. Da er kurz ist, bewegt er sich als Ganzes, bewirkt also auch, daß die anschließenden Teile von Decke und Boden sich immer in gleicher Richtung bewegen. Dadurch kann man die Bewegungsrichtungen in Boden und Decke einander zuordnen. Daß diese Schwingungen außerhalb der Stimmstockfußpunkte völlig verschiedene Bilder ergeben, wie das in Bild 21 zu sehen ist, leuchtet bereits bei der Verschiedenheit des Aufbaus von Boden und Decke ein. Hinzu kommt, daß die Decke und der Boden örtlich verschieden erregt werden. Bei Bild 21 erfolgte die Erregung über der Decke am Baßbalken-Fußpunkt des Steges; am Boden geht sie immer vom Fußpunkt des Stimmstockes aus. Übrigens markiert sich in dieser Aufnahme der Ort des Stimmstockes in der Decke durch eine kleine Insel.

Daß der Stimmstock überhaupt sich so leicht einschließlich der benachbarten Deckenteile bewegen läßt, hängt sicher mit der Nähe des nahen f-Lochs zu-

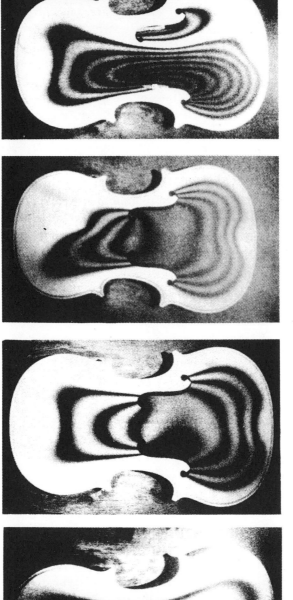

ohne f-Löcher
ohne Baßbalken
ohne Stimmstock
550 Hz

ohne Baßbalken
ohne Stimmstock
420 Hz

mit Baßbalken
ohne Stimmstock
465 Hz

komplette Decke
mit unbeweglichem
Stimmstock
540 Hz

Schwingungs-Eigenformen einer Geigendecke nach Jansson, Molin und Sundin

Bild 20: Hologramm von Geigendecken bei verschiedenen Fertigungsstufen (nach E. Jansson, N. E. Molin, H. Sundin)

sammen, dessen längliche Form diesem Deckenteil eine besondere Beweglichkeit bietet.

Es leuchtet daher auch ein, daß eine kleine Versetzung des Stimmstocks die Klangeigenschaften eines Instrumentes wesentlich beeinflussen kann.

Je höher die Frequenz ist, um so größer die Trägheit von Stimmstock und Zargen, um so mehr verharren beide in Ruhe, um so kleiner auch die Anregung des Bodens. Die bei 2900 Hz erhaltenen Aufnahmen des Bildes 22 lassen die geringere Bodenschwingung deutlich an der geringeren Zahl der Höhenlinien der dortigen Berge und Täler erkennen.

Sö kompliziert auch alle diese Schwingungsbilder sind, lassen sich doch einige allgemeine Regeln für sie aufstellen, die aus mechanischen Ähnlichkeitsgesetzen folgen: Die einfachste ist die, daß eine Vergrößerung aller Abmessungen auf das Doppelte alle Eigenfrequenzen um eine Oktave senken würden. Dem entspricht, daß die um eine Quinte tiefer gestimmte Bratsche länger ist als die Geige, aber glücklicherweise nicht im Quintenverhältnis 3 : 2, denn dann würde kein Bratscher das Instrument noch halten können. Auch Cello und Kontrabaß sind nicht so viel länger, als es ihren tiefst abgestimmten Saiten entspricht.

Diese einfachste Modellregel kann aber auch dadurch umgangen werden, daß man die Dicke von Decke und Boden nicht im gleichen Verhältnis vergrößert, sondern nach anderen Regeln variiert.

Arbeitet man beim gleichen Instrument immer mehr Material heraus, macht man Boden und Decke also immer dünner, so ergibt sich ebenfalls eine Senkung der Eigenfrequenzen. Auch hier beträgt sie bei Halbierung der Dicke — wenn die technisch möglich wäre — eine Oktave. Meinel [3] hat durch die Obertonanalyse von Geigenklängen bewiesen, daß deshalb zu dicke Instrumente zu scharf, zu dünne Instrumente zu dumpf klingen.

Sehr wichtig bei allen Eigenschwingungserscheinungen sind auch die Verluste. Sie bestimmen, wie hoch die Resonanzgipfel werden und wie steil ihre Abhänge. Sie werden hauptsächlich durch innere Verluste im Material bedingt. Die abgestrahlte Energie, auf die es letztendlich ankommt, ist viel geringer. Auch hier gibt es sicher ein Optimum. Man kann weder Metall, bei dem die inneren Verluste sehr gering sind, noch Kunststoff, bei dem sie sehr hoch wären, verwenden. Vermutlich hängt die Auswahl des „richtig klingenden" Holzes damit zusammen, wobei sicher auch sein Alter eine wesentliche Rolle spielt. Unterschiede in der Holzsteife könnte man viel eher durch Unterschiede in der Dicke ausgleichen.

c) Das Instrument aber besteht nicht nur aus festen Teilen, sondern auch aus einem Hohlraum, der über die f-Löcher sogar unmittelbar mit der umgebenden Luft verbunden ist. Auch dieser Luftraum verbindet Boden und Decke miteinander. Denn die Schwingungen der Decke erfolgen bei höheren Frequenzen so schnell, daß die durch sie komprimierte Luft im Hohlraum keine Zeit hat, über die f-Löcher zu entweichen.

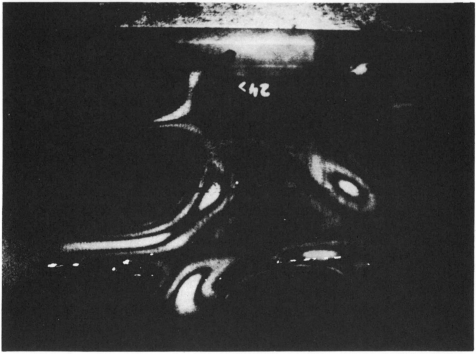

Bild 21: Hologramm von Decke und Boden bei Anregung über dem Baßbalken
$f = 1465\,\text{Hz}$, $\lambda_L/2 = 11{,}6\,\text{cm}$, Korpuslänge $36\,\text{cm}$ (nach W. Reinicke)

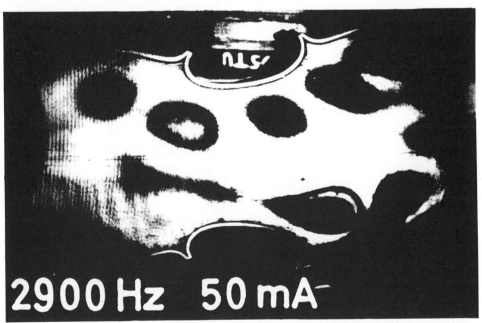

Bild 22: Hologramm von Decke und Boden bei Anregung über dem Stimmstock
$f = 2900$ Hz, $\lambda_L/2 = 5{,}9$ cm, Korpuslänge 36 cm (nach W. Reinicke)

Bei sehr tiefen Frequenzen kann dies allerdings geschehen. Wichtiger aber ist, daß es auch eine Hohlraum-Eigenschwingung gibt, bei der die kinetische Energie ihren Sitz in und an den — hierin gleichberechtigten — f-Löchern hat, wo die Luft beschleunigt hin- und herschwingt, und die potentielle Energie ihren Sitz in der im Hohlraum komprimierten oder dilatierten Luft, übrigens auch in den beiderseits nach außen oder innen gerichteten Bewegungen von Boden und Decke.

Es leuchtet ein, daß es solche beiderseits nach außen oder nach innen gerichtete Bewegungen sind, die erst die Anregung dieser Hohlraumresonanz ergeben, und daß die entgegengesetzte Kraftübertragungsrichtung vom Steg auf Baßbalken und Stimmstock solche begünstigt, wobei wir nur von dem stimmstocknahen Gebiet der Decke absehen müssen, das sich mit dem Boden in gleicher Richtung bewegt.

Diese „Atmungsbewegung" des Geigenkörpers ist nun auch besonders wichtig für unseren letzten Problemkreis, den der Abstrahlung, der einzige übrigens, der ohne Rückwirkung ist auf die bisher behandelten Teilprobleme, auf Körper, Steg und Saite. Dazu ist der Unterschied zwischen festen und gasförmigen Medien zu groß.

Abgesehen von der erwähnten Hohlraumresonanz spielt nämlich die Luftbewegung in den f-Löchern nur eine geringe Rolle. (Sonst wären die Streichinstrumente indirekte Blasinstrumente!)

Auch für die Schallabstrahlung gilt, daß sie sich infolge der komplizierten Körper und Schwingungsformen einer rechnerischen Erfassung entzieht, daß aber auch hier allgemeine Regeln angegeben werden können.

Solange die Wellenlängen groß gegen die Dimension des strahlenden Körpers sind — bei der Geige gilt das etwa bis zum g_1 auf der d-Saite — sind diese Regeln noch einigermaßen einfach. Wenn insbesondere in diesem Bereich die Instrumentenkörper im wesentlichen atmen, also allseitig gleichphasig Luft verdrängen oder heranziehen, ist der Wirkungsgrad der Strahlung besonders gut. Außerdem erfolgt die Strahlung gleichmäßig nach allen Seiten. Man spricht von einer Kugel-Richtcharakteristik.

Backhaus [21] vermutete, daß die „gut tragenden" Instrumente dadurch ausgezeichnet sind, daß diese optimale Bewegung bei noch relativ hohem Grundton, z.B. bei e_2, erfolgt. Dabei handelt es sich nicht mehr um die viel tiefer liegende Hohlraum-f-Löcher-Resonanz. Die zugehörige sogenannte Haupteigenschwingung der Decke entspricht dabei eher der tiefsten Eigenschwingung des freien Baßbalkens, welcher die Decke außerhalb des Stimmstockgebietes gleichphasig mitnimmt. Der Boden hat, wenn er dick genug ist, noch keine Veranlassung zur Unterteilung.

Aber auch wenn der Boden gar nicht schwingen würde, ist die Tatsache, daß durch ihn die Rückseite der Decke abgedeckt wird, wichtig. Würde nämlich nur eine einzige schwingende Platte vorhanden sein, würde um sie herum ein sogenannter „hydrodynamischer Kurzschluß" entstehen können, d. h. die auf der Vorderseite weggeschobene Luft könnte sich mit der von hinten herangezogenen ausgleichen, ohne daß es zu den zur Schallabstrahlung notwendigen periodischen Verdichtungen und Verdünnungen kommt. Das Ab-

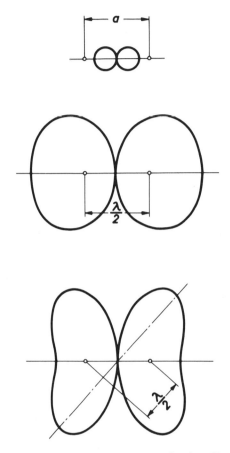

Bild 23: Richtdiagramme zweier gegenphasiger Punktstrahler
oben: $a/\lambda = {}^1/{}_{12}$ Mitte: $a/\lambda = {}^1/{}_2$ unten: $a/\lambda = {}^3/{}_4$

strahlungsverhalten entspricht dann dem eines sogenannten Dipols, worunter man genau genommen zwei sehr kleine, mit entgegengesetzter Phase aber gleich stark atmender Körper versteht (siehe Bild 23 oben). Die Verbindung dieser „Punktstrahler" heißt Dipolachse. Senkrecht zu ihr heben sich die Wirkungen beider Punktstrahler auf, in Achsrichtung bleibt eine kleine Wirkung übrig, die damit zusammenhängt, daß für einen Punkt auf der Achse die beiden Anteile entsprechend dem durch ihren Abstand gegebenen Wegunterschied nicht ganz gleichphasig ankommen. Dies gilt auch mit geringeren Phasenunterschieden für alle anderen Richtungen. Man erhält so die in Bild 23 links eingezeichnete „Achter-Charakteristik".

Eine solche ungleichmäßige Richtwirkung ist auch zu erwarten, wenn Decke oder Boden selbst in einer Richtung einmal unterteilt schwingen. Auch hierbei können sich die Strahlungswirkungen gerade in der zu Boden und Decke senkrechten Richtung aufheben.

Man erkennt an diesem Grenzfall, daß jede Unterteilung von Decke und Boden in gegenphasig schwingende Gebiete für die Abstrahlung nachteilig ist, solange die Wellenlänge groß ist. Findet die Unterteilung von Decke oder Boden noch öfter, insbesondere in beiden Richtungen statt, so kompensieren sich — bei gegebener Frequenz — die gegenphasigen Gebiete noch stärker, die Abstrahlung wird noch geringer. Die Streichinstrumente arbeiten also auch in dieser Hinsicht mit einem sehr schlechten Wirkungsgrad.

Nun verbessern sich aber die Verhältnisse, wenn die Wellenlänge im Verhältnis zum Abstand der gegenphasigen Schwingungsgebiete kleiner wird. Auch dies läßt sich am besten an den beiden gegenphasigen Punktstrahlern zeigen. Liegen sie nämlich eine halbe Wellenlänge auseinander (Bild 23 Mitte), so addieren sich ihre Teilwellen in Achsrichtung maximal, und wird die Wellenlänge noch kürzer, so findet zwar in Achsrichtung schon wieder eine Schwächung statt, es gibt aber dann eine schräge Richtung, in der die Abstrahlung optimal wird (Bild 23 unten).

Dieser Verbesserung der Abstrahlung mit wachsender Frequenz steht aber nun entgegen, daß sich der Instrumentenkörper — wie die holografischen Aufnahmen gezeigt haben — mit wachsender Frequenz immer mehr in gegenphasig schwingende Gebiete aufteilt.

Je mehr Punktstrahler wir zu berücksichtigen haben, um so krasser wird der Einfluß des Verhältnisses ihres Abstandes zur Wellenlänge auf die Strahlung. Man überblickt das leicht am Grenzfall einer unendlichen Strahlerreihe mit ständig sich umkehrenden Phasen, oder noch besser an einer unendlich ausgedehnten Platte, die durch eine in ihr von einer Stelle aus angeregten Biegewelle deformiert wird. Der Abstand gegenphasiger Gebiete ist dabei durch die halbe Biegewellenlänge gegeben. Ist er größer als die halbe Luftwellenlänge,

$$\lambda_B > \lambda_L$$

so gibt es, wie oben bei den beiden Punktstrahlern, eine Abstrahlung in der durch

$$\sin \vartheta = \frac{\lambda_L}{\lambda_B}$$

gegebenen Richtung, aber nur in dieser (Bild 24 oben). Ist dagegen

$$\lambda_B < \lambda_L \,,$$

so findet nur noch ein kompressionsfreier Ausgleich („hydrodynamischer Kurzschluß") zwischen benachbarten Bergen und Tälern statt (Bild 24 unten).

Die Abstrahlungsbedingungen für ein Streichinstrument würden sich daher mit wachsender Frequenz kaum verbessern, obschon der Abstand der Rand-gebiete immer größer im Verhältnis zur Wellenlänge wird, wenn nicht die Biegeschwingungen die glückliche Eigenschaft hätten, daß ihre Wellenlänge sich mit wachsender Frequenz weniger verkürzen, nämlich nur mit

$$\lambda_B \sim \frac{1}{\sqrt{f}},$$

als die Luftwellenlängen, für die die bekannte Regel

$$\lambda_L \sim \frac{1}{f}$$

gilt.

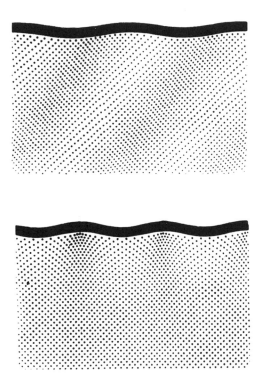

Bild 24: Abstrahlung einer Biegewelle in unendlicher Platte
oben: $\lambda_B > \lambda_L$ unten: $\lambda_B < \lambda_L$

Es gibt daher bei jeder homogenen und anisotropen Platte eine Grenzfrequenz, bei der der in Bild 24 gezeigte Übergang von Nichtstrahlung (unten) zur Strahlung (oben) erreicht wird.

Nun kann man bei der Geige infolge der Anisotropie des Holzes und der Variation der Holzdicke keine bestimmte Biegewellenlänge für eine bestimmte Frequenz angeben. Vergleicht man aber in den Hologrammen die unter ihnen jeweils angegebene halbe Luftwellenlänge $\lambda_L/2$ mit den Abständen benachbarter Hügel und Mulden, so erkennt man, daß diese gegenphasigen Gebiete jedenfalls in Querrichtung selbst bei 2900 Hz noch nicht näher aneinander liegen als jene, sich also in der Abstrahlung immer noch teilweise entgegenwirken.

Aus allen diesen Ausführungen können wir entnehmen, daß das Abstrahlungsproblem die letzten Endes interessierende Frage nach dem Schalldruck im Fernfeld nochmals kompliziert, indem hier zu der zerklüfteten Frequenzabhängigkeit der Oberflächenbewegungen noch für jede Schwingungsverteilung eine andere Richtcharakteristik hinzukommt. Da im höheren Frequenzbereich sich die Schwingungsbilder von Ton zu Ton ändern, sind auch von

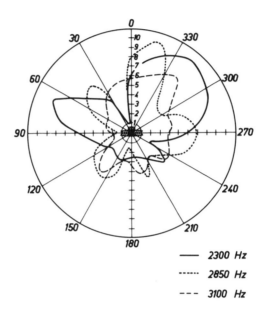

Bild 25: Richtdiagramm einer Geige
(nach H. Backhaus und W. Weymann)

Ton zu Ton andere Richtcharakteristiken zu erwarten. Dies zeigen die einer Arbeit von Backhaus und Weymann [24] entnommenen Richtdiagramme (Bild 25), die zu den benachbarten Frequenzen 2300, 2850 und 3100 Hz, also etwa zu den Tonhöhen d_4, f_4 und g_4 gehören. Es ist daher auch fraglich, ob man die Qualität eines Instrumentes bei Messung in einer einzigen Richtung beurteilen kann.

Dieser von Ton zu Ton erfolgende Wechsel der Richtcharakteristiken ist übrigens den Blasinstrumenten fremd. Um dies zu belegen, sind in Bild 26 oben die von Olsen [25] für eine Baßtuba ermittelten Richtdiagramme mit denen bei teilweise gleichen Frequenzen bei einer Geige ermittelten (unteres Bild) zusammengestellt. (Daß die Geigendiagramme hier weniger zerklüftet

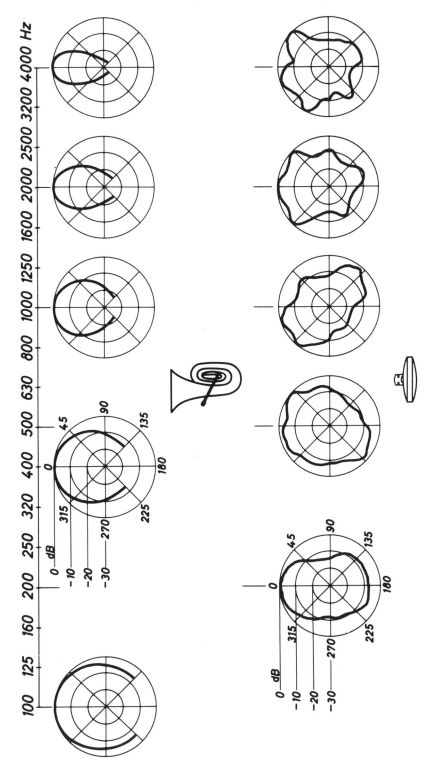

Bild 26: Richtdiagramm einer Baßtuba (oben) und einer Geige (unten) (nach H. F. Olson)

aussehen als im früheren Bild, hängt damit zusammen, daß Olsen für die Anzeige der Schalldrücke in den verschiedenen Richtungen einen logarithmischen Maßstab verwendet hat; er hat die sogenannten Schalldruckpegel in dB aufgetragen.)

Wohl ändert sich auch bei der Baßtuba die Richtcharakteristik mit der Frequenz, aber monoton und stetig. Man hört bei ihr, wie alle hohen Teiltöne

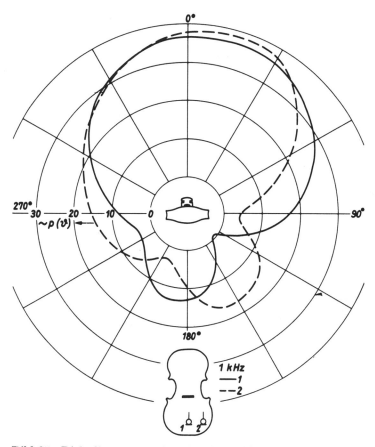

Bild 27: Richtdiagramme eines Punktstrahlers über der Geigendecke
(nach H. Ising und Y. Nagai)

bei einem Stakkatolauf von der Decke zurückkommen. Die Geige aber sendet Ton und Ton bevorzugt in andere Richtungen. Man könnte sagen, sie schillert wie ein Brillant.

Dies mag auch die Ursache dafür sein, warum Streichinstrumente im Freien nicht klingen und warum sich bei ihnen in Räumen das Fehlen mischender Reflexionen viel nachteiliger bemerkbar macht als bei Bläsern.

Wir bemühen uns zur Zeit in Berlin, aus der Schwingungsform die Richtcharakteristik zu berechnen. Man wird dazu die einzelnen Schwingungshügel
des Hologramms durch eine endliche Anzahl von Einzelsendern annähern
können. Aber man kann das Ganze nicht als eine Gruppe von Punktstrahlern
im freien Raum ansehen. Der Geigenkörper bildet ja für die ungestörte Schall-
ausbreitung ein Hindernis; dieses muß berücksichtigt werden; aber es ist
jedem theoretischen Physiker klar, daß es kein orthogonales Koordinatennetz

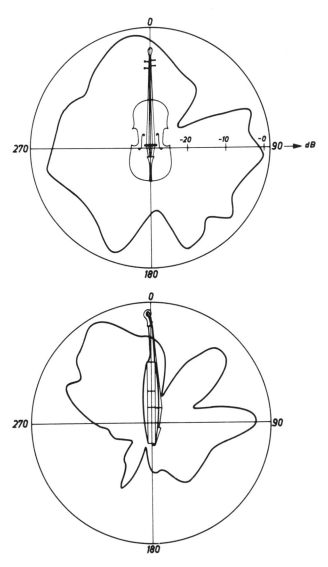

Bild 28: Richtdiagramm einer Geige mit Spieler
(nach J. Eikhold)

gibt, das in der Lage wäre, eine Geige abzubilden. Man kann aber die für die Richtwirkung benötigten Funktionen sich durch eine einfache Messung beschaffen. Statt rundherum das Schallfeld zu messen, wenn dicht über Decke oder Boden ein kleiner Lautsprecher an bestimmter Stelle strahlt, kann man ebensogut dort ein Mikrofon anbringen und die Kombination Geige—Mikrofon von verschiedenen Richtungen aus beschallen. Bild 27 zeigt zwei so von Ising und Nagai in unserem schalltoten Raum ermittelte Richtdiagramme. Es läßt klar erkennen, daß die zerklüfteten Richtwirkungen bei hohen Frequenzen auch mit den Beugungen am Instrument zusammenhängen. (Damit das Instrument nicht zusätzliche Schwingungen ausführte, mußten wir es mit Sand füllen. Es war kein wertvolles. Die Beugungserscheinungen dürften von Instrument zu Instrument sich viel weniger ändern als die Schwingungsbilder.)

Übrigens ist bei der Messung von Richtwirkungen unbedingt der Spieler mit einzubeziehen, der schon bei der Geige (siehe Bild 28), namentlich aber beim Cello die Abstrahlung in einer Richtung für die kürzeren Wellenlängen abschattet [26]. Man kann dies beobachten, wenn die Celli in Paaren hintereinander rechts vom Dirigenten sitzen und man etwa einen Rangplatz hinter ihnen hat. In der Berliner Philharmonie, in der bekanntlich die Musiker allseitig vom Publikum umringt sind, ist diese Abschattung nach der Rückseite dadurch etwas gemildert, daß die Celli über einen Viertelkreis verteilt sind.

Einer weit verbreiteten Ansicht muß der Physiker jedoch mit einer gewissen Skepsis gegenüberstehen, nämlich der, daß die Abstrahlung von Celli und Kontrabässen durch Anregung des Podiums mit dem Stachel wesentlich verstärkt wird. Einmal ist überhaupt nur bei den tiefsten Tönen eine Wirkung nachweisbar [27]. Diese besteht wiederum nur bei bestimmten Frequenzen in einer Verstärkung, bei anderen gerade in einer Schwächung, wobei es sehr darauf ankommt, an welcher Stelle der Stachel aufgesetzt wird, die sich aber der Musiker gar nicht nach akustischen Gesichtspunkten aussucht.

Die Physik der Geige wird heute in vielen Ländern quantitativ untersucht, namentlich in den USA, wo es sogar eine eigene, von Saunders gegründete „Cat Gut Acoustical Society" gibt. Carleen Hutchins [28], eine in New Jersey lebende Akustikerin hat sogar eine ganze Streichinstrumenten-Familie mit neuen Zwischenstufen zwischen Geige und Cello nach physikalischen Gesichtspunkten selbst gebaut.

Alle in dieser — naturgemäß unvollständigen — Übersicht erwähnten Beiträge des Instituts für Technische Akustik der Technischen Universität Berlin waren nur möglich infolge der großzügigen Unterstützung durch die Deutsche Forschungsgemeinschaft. Ihr gebührt dafür besonderer Dank.

Literatur

[1] H. v. Helmholtz, Die Lehre von den Tonempfindungen. Braunschweig, 1. Ausgabe 1862.

[2] H. A. Müller, Untersuchungen zur Physik der Geige (Unveröffentlichtes Manuskript der Geigenbauschule Mittenwald 1962).

[3] H. Meinel, Akustische Zeitschrift 2 (1937) 22.

[4] E. J. Völker, Diplomarbeit ITA (1961).

[5] C. V. Raman, Indian Assoc. Sci., Bull. 15 (1918) 1.

[6] L. Cremer und H. Lazarus, 6. ICA-Kongreß, Tokio N-2-3.

[7] W. Reinicke, Bericht an die DFG (1967).

[8] N. Schwarz, Studienarbeit ITA (1965).

[9] D. Ottmer, Diplomarbeit ITA (1967).

[10] W. Tippe, Studienarbeit ITA (1967).

[11] P. Eisenberg, Studienarbeit ITA (1968).

[12] P. Heinrich, Studienarbeit ITA (1967).

[13] H. Lazarus, Technischer Bericht 117 des Heinrich-Hertz-Instituts, Berlin-Charlottenburg (1970).

[14] C. V. Raman, Phil. Mag. 32 (1916) 391.

[15] F. Eggers, Acustica 9 (1959) 453.

[16] P. Zimmermann, Studienarbeit ITA (1964).

[17] G. Steinkopf, Diplomarbeit ITA (1963).

[18] W. Reinicke, Holographische Interferenzbilder zur Schwingungsanalyse am Streichinstrumenten-Steg. Vortrag auf der DAGA, Sept. 1970.

[19] C. V. Raman, Phil. Mag. 35 (1918) 493.

[20] O. König, Studienarbeit ITA (1970).

[21] H. Backhaus, Z. f. Physik 72 (1931) 218.

[22] W. Reinicke und L. Cremer, Journ. Acoust. Soc. Amer. 48 (1970) 988.

[23] E. Jansson, N. E. Molin und H. Sundin, Physica Scripta 2 (1970) 243.

[24] H. Backhaus und G. Weymann, Akustische Zeitschrift 4 (1939) 302.

[25] H. F. Olson, Musical Engineering, New York 1952, 232.

[26] J. Eikhold, Diplomarbeit ITA (1964).

[27] H. Fuchs, Studienarbeit ITA (1965).

[28] C. M. Hutchins, Scientific American, Nov. 1962.

(ITA heißt Institut für Technische Akustik der Technischen Universität Berlin. Die zitierten Studien- und Diplomarbeiten liegen dort aus.) .

6

The Violin from the Point of
View of the Physicist

LOTHAR CREMER

ABSTRACT by Lothar Cremer

This paper gives a survey of some of the basic physical problems of violins, with special reference to the contributions of the Institute for Technical Acoustics of the Technical University of Berlin. It starts with the motion of the bowed string, as first stated by Helmholtz (Figure 3, upper part), and with its registration by a description of the placement of a small part of a metallic string in a magnetic field, thus creating an electromagnetic transducer (Figures 1 and 2). In agreement with Helmholtz' description, it turns out that the maximum deviation, η_{max}, is proportional to the bow velocity v_B (Figure 4, upper part). On the other hand, the so-called bow pressure, that is, the force with which the bow is pressed against the string and which does not enter into Helmholtz description as a free motion, has no influence on the maximum deviation (Figure 4, lower part).

As already derived by Raman, the frictional force, which increases with the bow pressure, may be split up into two parts, one mean value that contributes a static deviation to be added to the Helmholtz motion (Figure 3, lower part), and a periodic part that depends on the bow velocity v_B, the length l of the string, its mass per unit length m, the distance of the bowing point from the bridge, and the energy losses outside the bow, which may be described by a decay time T. Raman calculated this alternating part under the assumptions that the velocity– time dependence is exactly that observed by Helmholtz and that the decay time is independent of frequency. (Figure 5 shows an example for $a/l = ¼$.) Although these assumptions were arbitrary, Raman could explain that the bow pressure has to be increased toward the bridge. (Figure 6 shows minimum bow pressures over a/l as measured by Lazarus.) Contrary to Raman's assumption, the decay time is decreasing with frequency (see the measurements of Reinicke, Figure 7), and the velocity– time dependence at the bow point differs from Helmholtz' idealized description, which would give sudden changes between the bow velocity and a sliding velocity (see the measurements of Lazarus in Figure 8).

Therefore, bow pressure influences the shape of the transition from bow velocity to more or less sliding. High bow pressure gives this transition steeper flanks than low bow pressure, and it follows that in the first case the time dependence contains more overtones than in the second.

It was confusing that in the first case (Figure 8, upper part) the velocity was not constant during the sticking time. It seemed possible that the bow did not always force its velocity at the frog on the string. Therefore, Ottmer measured the longitudinal impedance of the bow (see Figure 9). But since he found that this impedance is always large compared to the lateral impedance of the string, the usual assumption of a bow velocity not influenced by string motion was proved reasonable again.

But the bow touches the string at its perimeter, whereas the oscillogram presents the motion of its center of gravity. Both motions differ on account of a twisting of the string, which can be reinforced by a torsional resonance of the string. Tippe has recorded these torsional vibrations by fixing a small plate on the string, which controls the light flow picked up from a photocell (see Figure 10). Also, if transverse string waves impinge on the sticking bow, torsional waves are excited. This, together with the string stiffness against bending, results in a nontotal reflection of the transverse wave and corresponding transmission at and over the sticking bow, as measured by Eisenberg on a 30-m string (Figure 11).

Since the forces between bow and string cannot be measured directly, Heinrich tried (Figure 13) to substitute, for the excitation of the bow, an electrical feedback system in which the current, which is proportional to the electrodynamically introduced force, depends on a voltage that is made proportional to the velocity as far as possible, since the frictional force depends on the relative velocity betweeen bow and string (Figure 12). But since this analogue does not cover the discontinuous change between sticking and sliding, the force – time dependence holds only in the sliding period; it does show how this force always surpasses a maximum at the beginning and at the end of the sliding. With the same model the influence of the bridge impedance on the string motion could also be studied.

The bridge may be regarded as a mechanical transducer with one input at the string and two outputs on the body. It may be treated as a three-part or "mix pole" according to the corresponding rules of electrical circuits (Figure 14). Reinicke found that the bridge of a violin rigidly fixed works as a mechanical filter at 3,000 Hz on account of a natural vibration whereby the upper part is distorted against the lower owing to the flexibility that is given by the lateral slits. Reinicke could prove this especially by measuring the forces at the input and at one output with piezoelectric transducers (Figure 15). This natural frequency is essentially lowered if the bridge is loaded by the common sordino (mute). Figure 16 gives the forces at the input and both outputs recorded in the same way.

Modern electronics also makes possible measurement of the velocities at the body independent of frequency if a constant sinusoidal force F is acting on it. Figure 12 presents three records of the ratio v/F, the admittance. The peaks correspond to natural vibrations of the body. Such natural vibrations are well known from homogeneous plates supported at the boundaries where they may be calculated exactly. In Figure 18 those vibration patterns are represented that divide the plate by either none or just one nodal line in both directions. The patterns have been recorded by König using time-average holography. The black and white lines may be regarded as contour lines of equal deviation in a moment of maximum deviation. Figure 19 shows an example of the same method for a natural vibration of the top plate of the violin at 1,300 Hz, recorded by Reinicke, whereby the bassbar nearly determines a nodal line. Figure 20 shows examples from the excellent holograms made by Jansson, Molin, and Sundin during the production of a top plate. These show how the pattern is influenced first by the f-holes, then by the bassbar, and finally by the soundpost.

The hologram in the upper part of Figure 21 shows how the motion near the soundpost contributes an island that vibrates in antiphase to surrounding parts. This island is essentially restricted by the neighborhood of the right f-hole, which has in this respect a special importance. Finally, the hologram of Figure 22, made by Reinicke at 2,500 Hz, shows how little the back is excited in comparison to the top plate, since the back plate shows fewer contour lines than the top plate. The complicated vibration patterns result in complicated radiation characteristics. At low frequencies neighboring antiphasic areas tend to compensate each other. But if their distance is larger than half the wavelength in air, their contributions are additive for a given direction (see Figure 23). For large plates, this interference problem, on account of the peculiarity of the bending waves, results in a sharp limit between low frequencies, where the length of the bending waves in the plates λB is less than the wavelength λL in air and where no radiation in the far field exists, and higher frequencies (see Figure 24, lower part), where λB is longer than λL and the sound is radiated in a direction given by the ratio $\lambda L / \lambda B$ (see Figure 24, upper part). For string instruments such a critical frequency that covers both cases cannot be given exactly because the bending wavelengths are different in the direction of the grain and across it; but it may be said that the existence of this radiation effect is an advantage for the radiation of higher frequencies.

Taking into account all that we know about the relations between the vibrational pattern of a body and its directional characteristics, we may assume that the directional characteristics of a violin change from tone to tone, as already reported by Backhaus and Weyman, Figure 25. This is in contrast to most of the wind instruments. Figure 26 gives a comparison of a bass tuba and a violin as measured by Olson. Notice that here a logarithmic scale is used for the sound pressure in contrast to Figure 25.

The calculation of the directional characteristics of a violin is prohibited by its complicated shape, which gives reason for shadowing a point source on one of its plates. But it is possible to measure the "directional functions" regarding phase and amplitude for different sound sources (locations) on the plate surface (see Figure 27). These functions are nearly the same for all violins. By multiplying these measured functions with the special volume flows belonging to an observed pattern and adding these products, it is possible to get a special directional characteristic.

Finally, it is not the directional characteristic of the pure violin that is of interest. The shadowing by the player is also important. Figure 28 shows the directional characteristics with the player as measured by Eikhold.

II

The Bowed String

The behavior of the string is more amenable to analysis than that of any other part of an instrument played with the bow. Although the uniform flexible string may seem to be simplicity itself, no general solution has been found for its motion when the bow is used because of (1) the extreme, though necessary, nonlinearity in friction that is the seat of its entire action, and (2) the meagerness of what we know of the way in which friction varies with velocity. An idealization given by Helmholtz more than a century ago provides a good first approximation that is extremely useful but does not attempt to explain musically important subtle effects that are under control of the player. With the increasingly revealing analytical and experimental methods that are now available and the renewal of interest in the field, within the next few years important advances should be made in our understanding of the bowed string.

The first clear thought on the subject, as presently known, is a rather neglected paper by the mathematician Jean-Marie-Constant Duhamel (1797–1872) (1), known to students of calculus and of heat, in which he recognizes the need to consider two parts of a period of vibration, namely sticking and slipping. He takes pains to deride the idea that the vibrations of a bowed string are caused by the plucking action of barbs on the bow hair — a concept that still plagues us — and looks upon the behavior of the string as that of a static spring. He fails, however, to understand the fact recognized a little later by Helmholtz that the behavior that distinguishes the bowed string from other commonly observed illustrations of slip–stick action is that wave propagation along the string is not instantaneous but occurs with finite velocity. Thus the string is freed from the bow not as a result of gradual increase in stress between the two, but because the sharp kink in the string has suddenly returned after reflection from the nut.

Part II

In addition to his work on the string Duhamel was intrigued by harmonic overtones and suggested independently of Ohm (2) that a complex sound can be decomposed into a group of simultaneous vibrations. For a biography of Duhamel, see *Dictionary of Scientific Biography,* Volume 4 (1971).

Helmholtz, in his *Sensations of Tone* (3), dealt convincingly with the action of the bowed string, using a vibration microscope and mathematical analysis, a beautiful example of the power of simple apparatus and penetrating thought! The gist of it lies in the recognition of two simple facts: first, the stick – slip phenomenon; and second, the ability of the stretched string — if it is flexible and low in energy loss — to vibrate in the form of a relatively unchanging standing wave. The result is that out of numberless standing waves that are possible the string accommodates itself to a steady vibration that is consistent with alternate sticking to the bow and slipping. When the bow is held near one end of the string, the motion at that point turns out to be a slow forward motion during much of the period, followed by a quick retraction that initiates the next step. (When the string is bowed at its mid-point, the two parts of the motion are equal.)

While it would be difficult to name any subject in the field of classical acoustics that Lord Rayleigh failed to illuminate in some way, his contributions to the field of violin acoustics were general rather than specifically focused on the bowed string instruments. That his work on the vibrations of strings, membranes, plates, and shells provides an important background to the study of the violin is well known, as is his use of electrical circuit analogies in vibrating systems (4).

In addition to the writings of Helmholtz and Rayleigh referenced here, several interesting papers on acoustics by Rayleigh appear in *Acoustics: Historical and Philosophical Development* by R. Bruce Lindsay (5). Two other articles in Lindsay's book, on the early mathematical study of string vibrations, may also be of interest to the reader: Brook Taylor's "The Motion of a Stretched String" (6) and Jean le Rond d'Alembert's "Investigations on the Curve Formed by a Vibrating String" (7).

References

1. Duhamel, J. M. C. "Mémoire sur l'action de l'archet sur les cordes," *Extrait des Mémoires de l'Académie des Sciences, Savants Étrangers* **7**, Imprimerie Royale, Paris, 1841, pp. 6 – 36. In French.
2. Ohm, Georg Simon. "On the Definition of a Tone with the Associated Theory of the Siren and Similar Sound Producing Devices." Translated from Poggendorf's *Ann. Physik Chemie,* **59**, 497 ff. (1843) in *Acoustics: Historical and Philosophical Development*, R. Bruce Lindsay, ed., Dowden, Hutchinson & Ross, Inc., Stroudsburg, Pa., 1973, pp. 243 – 247.
3. Helmholtz, Hermann von. *Sensations of Tone As a Psychological Basis for the Theory of Music,* (revised 1877). Republished by Dover Publications, Inc., New York, 1954.
4. Strutt, John William (Lord Rayleigh). *Theory of Sound,* 1887. First American edition with an introduction by Robert Bruce Lindsay, Dover Publications, Inc., New York, 1945 (in 2 volumes).
5. Strutt, John William (Lord Rayleigh): "Our Perception of the Direction of a Source of Sound," "On the Theory of Resonance," "On the Application of the Principle of Reciproc-

ity to Acoustics," "On the Physics of Media That Are Composed of Free and Perfectly Elastic Molecules in a State of Motion," "On the Cooling of Air by Radiation and Conduction, and on the Propagation of Sound," all in *Acoustics: Historical and Philosophical Development*, R. Bruce Lindsay, ed., Dowden, Hutchinson & Ross, Inc., Stroudsburg, Pa., 1973, pp. 376–416.

6. Taylor, Brook. "De Motu Nervi Tensi/The Motion of a Stretched String." Translated from *Phil. Trans. Roy. Soc. London*, **28**, 26–32 (1713) in *Acoustics: Historical and Philosophical Development*, R. Bruce Lindsay, ed., Dowden, Hutchinson & Ross, Inc., Stroudsburg, Pa., 1973, pp. 96–102.

7. d'Alembert, Jean le Rond. "Recerches sur la courbe que forme une corde tendue mise en vibration/Investigations on the Curve Formed by a Vibrating String." Translated from *Hist. Acad. Sci. Berlin*, **3**, 214–219 (1747), in *Acoustics: Historical and Philosophical Development*, R. Bruce Lindsay, ed., Dowden, Hutchinson & Ross, Inc., Stroudsburg, Pa., 1973, pp. 119–123.

Additional Writings on the Bowed String

Barton, Edwin, H., and Thomas F. Ebblewhite. "Vibration Curves of Violin Bridge and Strings," *Phil. Mag.*, Ser. 6, **20**, 456 – 464 (1910). An experimental study of the simultaneous motions of a violin bridge and strings. Each string is dealt with separately and its vibration recorded with that of the bridge in seventy-two photographs, the strings being excited at various places by bowing, plucking, striking, and the like. The study was made without reference to any theory of vibration in the violin as a whole.

Bouasse, H. *Cordes et membranes*, Delagrave, Paris, 1926.

Friedlander, F. G. "On the Oscillations of a Bowed String" *Proc. Cambridge Phil. Soc.*, **49** (1953).

Keller, Joseph B. "Bowing of Violin Strings," *Comm. Pure and Appl. Math.*, **6**, 483–495 (1953). Both this paper and the preceding one present mathematically excellent methods of dealing with nonlinear systems. The motion of the string under the influence of the bow is considered for certain idealized conditions, such as absence of lossy compliance at the ends of rotation in the string. A result is the prediction of exact constancy in both the period of forward motion and that of retraction, which is contrary to results of actual experiment.

Kohut, J., and M. V. Mathews. "Study of Motion of a Bowed Violin String," *J. Acoust. Soc. Amer.*, **49**, 537 (1971).

Lazarus, Hans. "Die Behandlung der selbsterregten Kippschwingungen der gestrichenen Saite mit Hilfe der endlichen Laplacetransformation," No. D-83, Technical University, Berlin, 1973 (thesis).

Note: A listing of papers written in India following the work of E. H. Barton and C. V. Raman on the bowed string can be found at the end of the commentary on Raman, which follows.

Editor's Comments on Papers 7, 8, and 9

7 Raman: *On the Mechanical Theory of the Vibrations of Bowed Strings and of Musical Instruments of the Violin Family, with Experimental Verification of the Results: Part I*

8 Raman: *On the "Wolf-Note" in Bowed String Instruments*

9 Raman: *Experiments with Mechanically-Played Violins*

Sir Chandrasekhara Venkata Raman (1888–1970) was awarded the Nobel Prize in Physics for his discovery of the effect universally known by his name, which has contributed enormously to knowledge of the structure and dynamics of molecules and of techniques for identifying them. In 1909 he helped to found the Indian Association for the Cultivation of Science, and in 1915 the *Indian Journal of Physics.* He received many worldwide honors and awards for his research in the field of optics, but little is generally known of his early work in acoustics. During the 1930s he corresponded with F. A. Saunders about his investigations on the bowed string, and in 1968 wrote to the Catgut Acoustical Society: "My studies on bowed string instruments represent a phase of my earliest activities as a man of science. They were mostly carried out between the years 1914 and 1918. My call to the Professorship at the Calcutta University in July, 1917, and the intensification of my interests in *optics* inevitably called a halt to my further studies of the Violin family instruments."

Included here is a section from his long and distinguished monograph on the bowed string, notable for its originality and thoroughness both in theory and experiment, in which he used a kinematic method of describing seven classes of motion in the bowed string, of which only the first is significant musically. On the basis of these idealized concepts he investigated with great ingenuity many related effects. Recent researches have dealt with several aspects of string motion that the idealization ignores. Two shorter Raman papers of special interest are given in their entirety—experiments with a mechanically played violin, and his comments on the wolf-tone. Although reference is often made to the latter, Raman's denial that it is a beating effect has been disputed by J. C. Schelleng (Papers 5 and 12) and by Firth and Buchanan (Paper 14).

Selected Publications by C. V. Raman

"The Maintenance of Forced Oscillations of a New Type," *Nature,* **82, 156** (1909).
"The Maintenance of Forced Oscillations of a New Type," *Nature,* **82,** 428 (1910).
"Small Motion at Nodes of Vibrating String," *Phys. Rev.,* **32,** 309 (1911).
"Photographs of Vibration Curves," *Phil. Mag.,* **21,** 615 (1911).
"Maintenance of Vibrations by Variable Spring," *Phil. Mag.,* **24,** 513 (1912).
"Some Remarkable Cases of Resonance," *Phys. Rev.,* **35,** 449 (1912).
"Experimental Investigations on the Maintenance of Vibrations," *Bull. Ind. Assoc. Cultivation Sci.,* **6** (1912).
"Maintenance of Vibrations," *Nature,* **90,** 367 (1913).

"Maintenance of Vibrations," *Phys. Rev.*, **4**, 12 (1914).
"Combinational Vibrations Maintained by Two Simple Harmonic Forces," *Bull. Ind. Assoc. Cultivation Sci.*, **11**, 1 (1914).
"Motion in a Periodic Field of Force," *Bull. Ind. Assoc. Cultivation Sci.*, **11**, 25 (1914).
"Motions of Bowed Strings and Bow," *Bull. Ind. Assoc. Cultivation Sci.*, **11**, 43 (1914).
"On Motion in a Periodic Field of Force," *Phil. Mag.*, **29**, 15 (1915).
"On the Maintenance of Combinational Vibrations by Two Simple Harmonic Forces," *Phys. Rev.*, **5**, 1 (1915).
" 'Wolf-Note' of Violin and 'Cello." *Nature*, **97**, 362 (1916).
"Discontinuous Wave Motion" (with S. Appaswamaiyar), *Phil. Mag.*, **31**, 47 (1916).
* "Wolf-Note in Bowed Stringed Instruments," *Phil. Mag.*, **32**, 391 (1916).
"Discontinuous Wave Motion — II," (with A. Dey), *Phil. Mag.*, **33**, 203 (1917).
"Discontinuous Wave Motion — III," (with A. Dey), *Phil. Mag.*, **33**, 352 (1917).
"Maintenance of Vibrations by a Periodic Field of Force," (with A. Dey), *Phil. Mag.*, **34**, 129 (1917).
"Alterations of Tone Produced by Violin Mute," *Nature*, **100**, 84 (1917).
"Wolf-Note in Bowed Instruments," *Phil. Mag.*, **35**, 493 (1918).
"Wolf-Note in Pizzicato Playing," *Nature*, **101**, 264 (1918).
† "Vibrations of Bowed Strings and Instruments of the Violin Family," *Bull. Ind. Assoc. Cultivation Sci.*, **15**, 1 (1918).
"An Experimental Method for the Production of Vibrations," *Phys. Rev.*, **14**, 446 (1919).
"Note on the Theory of Sub-synchronous Maintenance," *Proc. Roy. Soc.*, **95** (1919).
"New Method for the Absolute Determination of Frequency," (with A. Dey), *Proc. Roy. Soc.*, **95**, 533 (1919).
"Partial Tones of Bowed Strings," *Phil. Mag.*, **38**, 573 (1919).
"Kaufmann's Theory of the Impact of the Pianoforte Hammer," (with B. Banerji), *Proc. Roy. Soc.*, **97**, 99.
"Sounds of Splashes," (with A. Dey), *Phil. Mag.*, **39**, 145 (1920).
"Mechanical Violin-Player for Acoustical Experiments," *Phil. Mag.*, **39**, 535 (1920).
"Musical Drums with Harmonic Overtones," (with S. Kumar), *Nature*, **104**, 500 (1920).
"Some Applications of Hertz's Theory of Impact," *Phys. Rev.*, **15**, 277 (1920).
* "Experiments with Mechanically-Played Violins," *Proc. Ind. Assoc. Cultivation Sci.*, **6**, 19 (1920).
"Variation of Bowing Pressure with the Pitch of the Note," *Nature*, **106**, 355 (1920).
"On Some Indian Stringed Instruments," *Proc. Ind. Assoc. Cultivation Sci.*, **7**, 29.
"The Nature of Vowel Sounds," *Nature*, **107**, 332 (1921).

Additional Indian Papers on the Bowed String

Kar, K. C. *Phys. Rev.*, **20**, 184 (1922).
Sen, B. K. *Indian J. Phys.*, **23**, 7 (1949).
Kar, K. C., N. K. Datta, and S. K. Ghosh. "Investigations on the Bowed String with an Electrically Driven Bow,"*Indian J. Phys.*, **25**, 423 (1951).
Kar, K. C. *Indian J. Theo. Phys.*, **2**, 46 (1954).
Bagchi, R. N. and Chinmayee Dutta. "The Experimental Study of the Bowed String and Verification of the Formula for the Bowing Pressure," *Indian J. Theo. Phys.*, **3**, 1 – 16 (1955).
Bagchi, R. N. *Indian J. Theo. Phys.*, **4**, 15 (1956).

---

*Included in this volume.
†Part I included in this volume.

Kar, K. C., and Chinmayee Dutta. *Indian J. Theo. Phys.,* **5**, 19 (1957).

Kar, K. C., and Chinmayee Dutta. "Extension of the Kar Formula for Minimum Bowing Pressure,"*Indian J. Theo. Phys.,* **5**(2), 31 – 37 (1957).

Dutta, Chinmayee. "On the Verification of the Kar Formula for Minimum Bowing Pressure," *Indian J. Theo. Phys.,* **5**, 105 – 110 (1957).

Dutta, Chinmayee. "On the Work Done During Bowing," *Indian J. Theo. Phys.,* **7**(1), 5 – 15 (1959).

Reprinted from *Indian Assoc. Cultivation Sci. Bull. 15*, 1–27 (1918)

On the Mechanical Theory of the Vibrations of Bowed Strings and of Musical Instruments of the Violin Family, with Experimental Verification of the Results: Part I.

By C. V. Raman, M.A.

CONTENTS.

SECTION I.—INTRODUCTION.

The vibrations of stretched strings excited by bowing and their practical application in musical instruments of the violin class present many important and fascinating problems to the mathematician and to the physicist. In the present monograph which is the first part of a more complete work on the whole subject, I propose to deal with the theory of the excitation of these beautiful and characteristic types of vibration under various conditions, and of their communication to the resonator on which the string is stretched. Experimental results in confirmation of those obtained from dynamical theory are also presented.

The problem which it is proposed to consider has formed the subject of investigation by many mathematicians and physicists. A list of the works and original papers that I have consulted is given in the Bibliographical Appendix. The present position of the subject cannot be considered satisfactory, in view of the fact that no complete and detailed dynamical theory has been put forward which could predict and elucidate the many complicated phenomena that have already been found empirically by those who have worked in the field and that could also point the way for further research. It was this defect in the present state of knowledge of the subject that induced me to undertake the investigation. Some preliminary work had already been carried out by me on the vibrations of bowed strings and the physics of bowed instruments. Reference may be made here to three papers which may be regarded as the starting points of this investigation[1]; the exposition given in the present paper is however self-contained. In the paper on " Discontinuous Wave-Motion " by myself and another that has appeared in the *Philosophical Magazine* for January 1916, it has been shown how the well-known principal mode of vibration of a bowed string discovered by Helmholtz[2] can be reproduced experimentally as a free oscillation by imposing on a stretched string a certain simple distribution of initial velocities involving a discontinuity. This experiment which was first made in September 1914, suggested my undertaking a thorough investigation of the general problem. Free use has been made of a simplified form of

---

[1] C. V. Raman, M.A., " Photographs of Vibration-Curves," *Philosophical Magazine*, May 1911.

C. V. Raman, M.A., " The Motion of Bowed Strings," Bulletin No. 11 of the Indian Association for the Cultivation of Science, 1914. See also Science Abstracts, February 1915, and *Nature*, August 13, 1914, page 622.

C. V. Raman, M.A., and S. Appaswamaiyar " On Discontinuous Wave-Motion," *Philosophical Magazine*, January 1916.

[2] " Sensations of Tone," English Translation by Ellis.

the theory of discontinuous wave-motion given by Harnack, Davis and others, which I have extended so as to cover cases not considered by these writers.[1]   The whole subject is considered in the light of dynamical theory, and an attempt has been made to divest it of empiricism as far as possible.   Emphasis is laid upon the cases which are of practical interest in music.   To make the present monograph as comprehensive as possible in respect of the matters dealt with, I shall develop the theory, step by step, in detail. A summary of the treatment and of the results obtained will be found in Section XIV.   Many illustrative diagrams, photographic curves and numerical results will be found in the paper.   Not only does the theory succeed in explaining all the known phenomena, but it has also justified itself by predicting many new relations and results which have been tested experimentally.   These are also referred to in the course of the paper and in the summary.

SECTION II.—EFFECT OF PERIODIC FORCE APPLIED AT A POINT.

It is clear that the motion of a bowed string is a case of maintained vibration, and an adequate treatment of the subject is only possible if the dissipative forces to which the string is subject are taken into account.   The dissipation may be due, (*a*) to the direct communication of energy to the surrounding medium from the string, or (*b*) to motion set up in the supports between which the string is stretched.   The forced oscillation of a string in the presence of dissipative forces of the first kind, is readily found on the assumption that each element of the string is resisted by a force proportional to its velocity.   Lord Rayleigh and others have discussed the motion that would ensue under such conditions when a periodic force is impressed at one point on the string.   In practice, however, it is known that the second source of dissipation is generally of much greater importance than the first.   The energy of the vibrating string is conducted through

---

1 Two preliminary notes on the subject have been published by me : " On Some New Methods in Kinematical Theory," Bulletin of the Calcutta Mathematical Society, Vol. IV, pages 1—4, " On the summation of certain Fourier Series involving discontinuities," Ibid., Vol. V, pages 5—8.

the bridges over which it is stretched to the sides of the box on which the bridges are fixed, and ultimately to the atmosphere as sound-waves.*

We shall now consider the motion of a string, one end of which is supposed to be rigidly fixed at the point $x=0$ and the other end of which $(x=l)$ passes over a bridge. A periodic force $E \cos mt$ is assumed to act at the point $x=x_0$. The string may be taken to be perfectly uniform and not subject to any resistance, so that the communication of energy to the surroundings takes place only through the bridge. The equation of motion of the string is

$$\mu \frac{d^2y}{dt^2} = T_0 \frac{d^2y}{dx^2}. \quad . \quad . \quad . \quad . \quad . \quad . \quad (1)$$

The solution of the equation for values of x between 0 and x_0 may be written

$$y = F_1 \sin px \sin mt + G_1 \sin px \cos mt \quad . \quad . \quad . \quad (2)$$

where

$$\mu m^2 = T_0 \, p^2. \quad . \quad . \quad . \quad . \quad . \quad (3)$$

From $x = x_0$ up to $x = l$ we may write,

$$y = D_2 \cos p \, (l-x) \sin mt + E_2 \cos p \, (l-x) \cos mt$$
$$+ F_2 \sin p \, (l-x) \sin mt + G_2 \sin p \, (l-x) \cos mt. \quad . \quad (4)$$

Since y must be continuous at the point x_0,

$$F_1 \sin px_0 = F_2 \sin p \, (l-x_0) + D_2 \cos p \, (l-x_0) \quad . \quad . \quad (5)$$
$$G_1 \sin px_0 = G_2 \sin p \, (l-x_0) + E_2 \cos p \, (l-x_0). \quad . \quad (6)$$

The discontinuous change in the value of $\frac{dy}{dx}$ at the point of x_0 is due to the force $E \cos mt$. From this we get the two equations

$$F_1 \cos px_0 + F_2 \cos p \, (l-x_0) - D_2 \sin p \, (l-x_0) = 0$$
$$G_1 \cos px_0 + G_2 \cos p \, (l-x_0) - E_2 \sin p \, (l-x_0) = \frac{E}{p \, T_0}. \quad (7,8)$$

At the point $x = l$,

$$y = D_2 \sin mt + E_2 \cos mt$$

and this motion at the bridge must be due to its yielding under the transverse periodic components of the tension. If the equation of motion of the bridge is

---

*Or dissipated as heat in the material vibrating.

$$M\frac{d^2y}{dt^2} = -T_0\frac{dy}{dx} - f^2 y - g^2\frac{dy}{dt}$$

where M is the mass of the bridge and associated parts, we obtain, by substitution, the equations

$$(f^2 - M\,m^2)\,D_2 = T_0\,p\,F_2 + g^2 m\,E_2 \quad . \quad . \quad . \quad . \quad (9)$$
$$(f^2 - M\,m^2)\,E_2 = T_0\,p\,G_2 - g^2 m\,D_2 \quad . \quad . \quad . \quad . \quad (10)$$

From the six equations numbered (5) to (10), the six unknowns F_1, F_2, G_1, G_2, D_2 and E_2 should obviously be capable of complete determination. Putting

$$\frac{T_0\,p}{f^2 - M\,m^2} = \tan\theta \text{ and } \frac{g^2 m}{f^2 - M\,m^2} = \tan\phi,$$

the equations may be solved by first eliminating D_2, E_2 and then F_1, G_1. Using for brevity the expression $\tan\psi = \tan\theta\cos^2\phi$ and $\delta = \tan\theta\sin\phi\cos\phi$, the eliminant equations obtained are

$$F_2\sin(pl + \psi) + G_2\,\delta\cos\psi\cos pl = 0$$
$$F_2\,\delta\cos\psi\cos pl - G_2\sin(pl + \psi) + \frac{E\cos\psi\sin px_0}{p\,T_0} = 0$$

Solving these two equations, we obtain

$$F_2 = \frac{-E\,\delta\cos^2\psi\cos pl\sin px_0}{p\,T_0\,[\sin^2(pl + \psi) + \delta^2\cos^2\psi\cos^2 pl]}$$

$$G_2 = \frac{-E\cos\psi\sin(pl + \psi)\sin px_0}{p\,T_0\,[\sin^2(pl + \psi) + \delta^2\cos^2\psi\cos^2 pl]}$$

If the impressed force $E\cos mt$ is regarded as an arbitrarily determined quantity, the interpretation of the preceding result is a simple matter, provided $\tan\phi$ (which involves the damping factor g^2) is regarded as very small. The second term in the denominators is then very small relatively to the first, and the maximum of F_2 is obtained when the first term in the denominator is zero, i.e. when

$$(pl + \psi) = v\,\pi \text{ or}$$
$$pl = v\,\pi - \theta,$$

ψ being then practically equal to θ. G_2 is found to be zero when F_2 has its maximum value.

When $\sin px_0$ is zero, it is found from equations (5) to (10) that F_2, G_2, D_2, E_2 and F_1 are all equal to zero. The significance

of this is that when the point of application of the force coincides
with a node of the string for the particular frequency of oscillation, the whole of the string between the bridge and the point of
application remains completely at rest.   Only the portion of the
string between the fixed end and the point of application has any
movement, this being of very small amplitude, viz., $\dfrac{E}{p\,T_0}$ sin px
cos mt.   It is thus seen that a periodic force of given magnitude
produces an effect which is insignificant when it is applied at a
node of the resulting oscillation, and which gradually increases as
the point of application is removed further and further from the
node.   This result has many applications, as we shall see later on.

Generally speaking, the angle θ may also be taken to be very
small, the quantity $(f^2 - M\,m^2)$ being either positive or negative
and large compared with $T_0\,p$.   We then find, as may have been
expected, that the vibration set up by the periodic force is a
maximum when its frequency is the same as that of the free
vibrations of the string of length l with both ends rigidly fixed.
But the case is otherwise when $(f^2 - M\,m^2)$ is small, that is, when
the free periods of vibration of the string and the bridge, taken
separately, are nearly equal to one another.   If the two periods
are nearly equal to one another, the amplitude of the vibration
of the string set up by the application of a periodic force of
given magnitude and of frequency equal to that of its free
oscillations is considerably smaller than if the natural periods
of the string and of the bridge differed appreciably.   To elicit
the same amplitude of vibration, therefore, a comparatively much
larger force would have to be applied when the frequency of the
vibration is the same as that of the free period of the bridge and
associated parts.   This is the explanation of the difficulty
noticed in bowing a string steadily when its pitch is that of
the maximum resonance of the instrument.   In Section XII, we
shall consider the special effects observable under these conditions when the pressure of the bow is insufficient to maintain a
steady vibration, and also those produced by loading the bridge.

In dealing with the motion of bowed strings, we have to
consider the effect, not of a simple harmonic force, but of a
system of forces whose frequencies form a harmonic series acting

PLATE I.

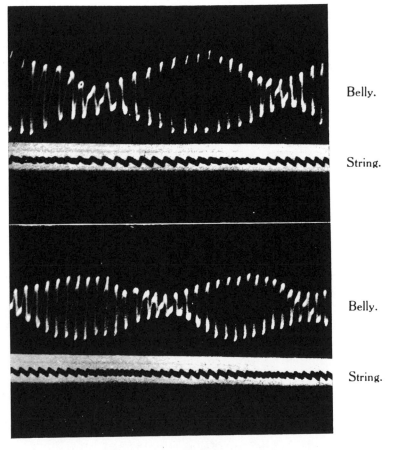

Belly.

String.

Belly.

String.

Simultaneous vibration-curves of Belly and G-Strings of
Violoncello at the " Wolf-note " pitch showing
Cyclical changes.

over a finite region of the string which may, by courtesy, be styled the " bowed point." As the bridge over which the string passes, together with its associated masses, may have several free periods of vibration, it is obvious that the formulae connecting the various harmonic components of the periodic force brought into play by the bow, with the respective components of the resulting motion, would not, in general, be of a simple character. Fortunately, however, as will be shown in the course of the paper, it is possible to build up a theory, which successfully predicts the phenomena observable under a very wide variety of conditions. The only assumptions that need be made for the present are (1), that the string is uniform and of negligible stiffness, and (2), that the yielding at the bridge is negligibly small in comparison with the motion of the string at the bowed point, or at any other point actually chosen for observation. These assumptions, which may be approximated to, in practice, as closely as desired, greatly simplify the treatment. The main result of the preceding treatment that is utilized, is that the effect of any of the harmonic components of the impressed force depends upon the point at which it is applied, vanishing when it is applied at a node, and increasing gradually as it is removed further and further from it.

The assumptions of the uniformity and flexibility of the string are made to ensure the treatment being as far as possible rigorous. Except, however, in the case of very complicated types of vibration, these assumptions are not essential, provided the frequencies of the normal modes of vibration are not so far from forming a harmonic series as to prevent the bow eliciting all the members of the series which are of importance, as components of a strictly periodic forced oscillation. Owing to this restriction and the dispersion which the waves suffer in travelling on a non-uniform string, the treatment then requires modification, as will be referred to again, later on.

Section III.—The *Modus Operandi* of the Bow.

The function of the bow as normally applied is both to elicit and to maintain the vibrations of the string. The two processes are interdependent, but it is well to remember that they should not be confused with each other, inasmuch as it may well happen that the character of the motion in its initial stages is not necessarily the same or even analogous to that maintained in the final steady state. For the present, however, we need not enter into these intricacies, but may simply assume that the motion is maintained in some perfectly periodic manner by the action of the bow, and proceed to find its character. It is obvious that on the assumptions set forth in the preceding section the period of the maintained oscillation is the same as that of the free vibrations of the string.

In a well-known paper on "Maintained Vibrations" (*Phil. Mag.*, 1883) reproduced in his Theory of Sound, Vol. I, page 81, Lord Rayleigh has discussed the general theory of vibrations elicited by generators and has shown that the supply of energy to the system through the action of the generator in any given time, may sometimes actually exceed that lost by the system in the same time through dissipative forces. When this happens, the excited vibrations continue to increase indefinitely in amplitude, until some physical limit is reached beyond which the equations of motion originally assumed cease to apply. The motion of a bowed string is evidently a case of this kind, the physical limit beyond which the vibrations cannot increase being imposed by the finiteness of the velocity of the bow. It has been suggested (with more or less definiteness) by the previous writers on the subject, that the bowed point on the string does attain or nearly attain the velocity of the bow in its movement in one direction. As will be seen presently, the question whether the forward velocity of the bowed point is *absolutely* the same as that of the bow is one of fundamental importance in the theory of the subject, and in one of my own previous publications,[1] I have shown how this identity of velocities can be brought experimentally to a test. What I propose to do here is to discuss the dynamical principles underlying the case in some detail, and then to pursue the argument to its logical conclusions.

The magnitude of the frictional force due to the bow at any instant must obviously depend upon the pressure with which it

---

[1] See Science Abstracts (Physics), Feb. 1915, page 87 (C. V. Raman).

is applied and upon the relative velocity at the point of contact. It is also clear that this relative velocity cannot ordinarily change sign during the motion, for, if it did, the entire frictional force would also change sign and the excess velocity of the bowed point would be immediately damped out. (An excellent illustration of this principle may be had by bowing a fork vigorously and then suddenly reducing the velocity of the bow. It will then be found that the amplitude of vibration of the fork also falls with practically equal suddenness.) With an efficient generator, e.g., a bow with rosined horse-hair acting on the string, the frictional force exerted would be much greater when the relative-velocity is nearly but not quite zero, than when it is large. On the other hand, when the relative velocity is actually zero, the friction ceases to be a determinate function of the relative velocity. From these premises, it is clear that, when the bow is applied with sufficient pressure and not too great a velocity, the maintaining forces brought into play would be far in excess of those required to maintain the vibration of the string (the mass and damping of the latter both being small), so long as the relative velocity at the point of contact does not actually become zero during any part of the vibration. On the other hand, we know that a steady state of vibration is only possible when the energy gains and losses balance each other, i.e., when the harmonic components of the force exerted by the bow are just sufficient to maintain the motion. The only possible inference that can be drawn under the circumstances is that the bowed point does actually attain the velocity of the bow during part of its motion and ultimately throughout the fractional part or parts of the period of vibration during which it has a forward movement. During these stages, the bow merely carries forward the point of the string with which it is in contact, and it is important to notice (accordingly to the preceding argument) that the frictional force then acting on the bowed point would actually fall below the maximum statical value; by how much it would fall below this maximum, would depend on the circumstances of the case, viz., the magnitude of the friction during the other stages of the motion as determined by the relative velocity, and the magnitude of the forces required to sustain the motion.

c

When the bowed point has the velocity of the bow in all the stages of forward movement, there is necessarily a discontinuous change of velocity[1] when it starts moving backward. The preceding argument may be pressed a little further if we assume that the forces required to maintain the motion are very small compared with the variation of frictional force due to a finite change of relative velocity. (Such an assumption would, in general, be justifiable if the pressure of bowing were sufficient and the damping co-efficient were sufficiently small). It would then follow that the frictional force exerted by the bow is practically constant throughout the whole motion, and that during all the intervals in which the relative value is not zero, it has a finite constant value which is the same for all such intervals. The relative velocity changes from this value to zero and *vice versa* in a discontinuous manner.

From a consideration of dynamical principles and of the relative order of magnitudes of the quantities involved, we thus arrive at the following two results: (*a*) during one or more intervals in each period of vibration, the bowed point has a forward movement which is executed with constant velocity exactly equal to that of the bow; (*b*) during the other interval or intervals, the bowed point moves backwards, also with constant velocity, this being the same for all such intervals (if there be more than one). The preliminary treatment of the vibrational modes given in the succeeding sections is mainly founded on these two results. It must be observed, however, that as the argument by which the second result was deduced, is rigorous only in the limiting case of a vanishingly small damping co-efficient, this particular result, viz., the constancy of velocity in the backward movement, cannot be regarded as holding good with the same completeness and generality as the first result, i.e. the constancy of velocity in the forward movement. We are thus led by the argument to anticipate the existence of cases in which the velocity of the bowed point varies in a continuous manner, particularly in the stages in which the movement is in a direction opposite to

---

[1] Equal to the velocity of the bow plus the initial speed of backward movement.

that of the bow. This is a feature which becomes of great importance in certain cases, specially in those of musical interest, and which therefore requires to be emphasised. For the present, however, it is advantageous to consider the constancy of the velocities of the bowed point as holding good rigorously, both in the forward and backward movements. This assumption serves admirably as the basis on which the kinematical theory of the various possible modes of vibration may be discussed. We shall accordingly proceed on this basis.

We may call the two velocities possible at the bowed point, v_B and v_A respectively, v_B being the velocity of the bow, and v_A the velocity of the bowed point when it travels against the bow. The intervals of time T_1, T_3, T_5 etc. in each period of vibration during which the velocity is v_A and the intervals T_2, T_4, T_6 etc. in which it is v_B are obviously connected by the equations,

$$T_1 + T_2 + T_3 + T_4 + T_5 + T_6 + + = T \text{ (the complete period of}$$
vibration).

$$(T_1 + T_3 + T_5 +) \, v_A + (T_2 + T_4 + T_6 + +) \, v_B = 0.$$

As already remarked, the argument shows that in the presence of dissipative forces, the constancy of velocity in the intervals of backward movement is not by any means so generally assured as the intervals of forward movement, and a steady state of motion in which the total of the time intervals of movement in the direction opposite to that of the bow, exceeds that of the intervals of movement with the bow, is altogether out of the question. The only cases, therefore, whose kinematics need be considered in detail are those in which v_A is numerically not less than v_B.

SECTION IV.—SIMPLIFIED KINEMATICAL THEORY.

From the general results indicating the nature of the motion at the bowed point obtained in the preceding section, we now proceed to build up a detailed kinematical theory of the motion of the bowed string. For this purpose the ordinary Fourier analysis is unsuitable, as it is neither convenient nor suggestive. I have therefore devised a simple graphical treatment which is based

upon the use of the velocity-diagram of the string and appears admirably adapted for the present investigation.

The general solution of the equation of wave propagation on
an infinite string not subject to damping is,

$$v = f_1 (x - at) + f_2 (x + at). \quad . \quad . \quad . \quad . \quad (11)$$

It is well known that this solution for the case of an infinite
string can be used to represent the configuration at any instant
of a vibrating string of finite length, by arranging the form of
the displacement waves in such manner that the motion is
periodic and satisfies the terminal condition $y = 0$ at the two ends
of the string.

Similarly, the solution obtained by differentiating (1) with
respect to time, viz.,

$$\frac{dy}{dt} = - a f_1' (x - at) + a f_2' (x + at). \quad . \quad . \quad . \quad (12)$$

can be applied to represent the velocity diagram of a finite string
at any instant during its vibration, if the periodicity of the
motion and the terminal conditions of velocity are secured. It
is obvious that solution (12) as it stands, represents the velocity
waves that travel on an infinite string without change of form
in the positive and negative directions respectively. In the case
of a finite string of length l, the reflexions that take place at the
two ends have to be taken into an account and we may write

$$\frac{dy}{dt} = \theta (x - at) + \phi (x + at). \quad . \quad . \quad . \quad . \quad (13)$$

The functions $\theta (x - at)$ and $\theta (x + at)$ represent the positive
and negative velocity waves which must be imagined as extending
to infinity in both directions. Further they must each of them
be periodic with wave length equal to twice the length of the
string, and must be so related that at the two ends of the string
$x = 0$ and $x = l$, the terminal condition $\frac{dy}{dt} = 0$ is always satisfied.
This can be secured by arranging the form of the velocity waves
in much the same way as the displacement waves would be arranged to secure the terminal condition $y = 0$, i.e. the form of the
positive velocity wave from $x = 0$ up to $x = l$ in its initial position

PLATE II.

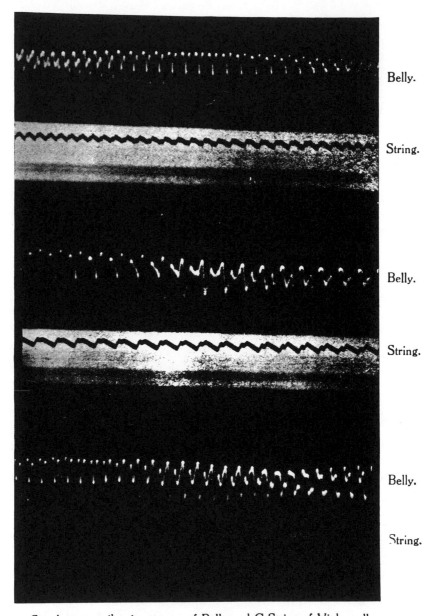

Belly.

String.

Belly.

String.

Belly.

String.

Simultaneous vibration-curves of Belly and G-String of Violoncello at half the " Wolf-note " pitch showing the resonance of the octave and Cyclical changes.

is an inverted and reflected image of the negative wave from $x=l$ up to $x=2l$, and *vice versa.*

The cases in which the positive and negative velocity waves are completely identical in their initial positions, present features of special interest. Half the initial velocity at each point on the string is then due to the positive wave, and the other half to the negative wave, and there are no initial displacements, i.e., the string is everywhere in its position of statical equlibrium. After the expiry of half a period, i.e., when the positive and negative waves have each moved through a distance equal to the length of the string, the latter is again everywhere in its position of equilibrium. This is so because one half of each wave is merely the inverted and reflected image of the other half, and the displacements resulting from the initial velocities are annulled during the same half period. During the second half period the velocity at every point on the string goes back again to its original value through exactly the same stages ; in other words, the velocity is everywhere an even periodic function of the time which when plotted gives a figure with the symmetry characteristic of such functions. It is thus seen that the positive and negative velocity waves are necessarily identical in any case in which the changes of velocity at points on the string take place in a symmetrical manner with respect to time.

We now proceed to consider cases in which the velocity changes that take place at some particular point on the string, say the point $x=x_b$, can be assumed to have a specified form. Then the form of the velocity waves $\theta\,(x-at)$ and $\phi\,(x+at)$ must be such that by their movement and superposition, the known changes of velocity at the point x_b are reproduced. For example, let us assume that the string at the point x_b moves during the vibration with a succession of constant velocities, the velocity passing in a discontinuous manner from each value to the next. Then at the point x_b, $\dfrac{d^2y}{dt^2}$ is always zero, except at certain instants in each period of vibration when it becomes $\pm$ infinity.

Differentiating (13) with respect to time, we have

$$\frac{d^2y}{dt^2} = -a\,\theta'\,(x-at) + a\,\phi'\,(x+at). \quad . \quad . \quad . \quad . \quad . \quad (14)$$

Since at the point x_b, $\dfrac{d^2y}{dt^2}$ is generally zero, we must have

$$\theta'(x_b - at) = \phi'(x_b + at). \quad . \quad . \quad . \quad . \quad . \quad (15)$$

If the velocity-waves $\theta(x - at)$ and $\phi(x + at)$ are represented graphically, equation (15) may be given a geometrical significance ; if any two points are taken, one on the positive wave and the other on the negative wave, the distances of which from the point x_b measured along the string are equal but in opposite directions, we should find the *slopes* of the waves at these two points to be equal. We have already seen that the positive and negative waves must satisfy certain other conditions, viz., that they are periodic with wave-length $2l$, and that, initially the form of the positive wave from $x = 0$ up to $x = l$ is the inverted and reflected image of the negative wave from $x = l$ up to $x = 2l$ and *vice versa*. It is a definite geometrical problem to find the configuration of the waves which would simultaneously satisfy these three conditions. By inspection, we get the following remarkably simple and significant solution : if the point x_b divides the string in an *irrational* ratio, the only possible form of the velocity waves is that in which the slope is everywhere the same, in other words, they are representable by a number of straight-lines which are all parallel to one another, a discontinuity intervening wherever one straight-line leaves off and the next begins. Velocity-waves having this form also satisfy the geometrical criteria when the point x_b divides the string in a rational ratio (i.e., in the ratio of two whole numbers), but in the latter case, this is not the only form of velocity-waves *geometrically* possible. This result is not a matter for surprise, for the point x_b would then coincide with a node of one of the harmonics of the string, and the ordinary Fourier analysis of the kinematics of the vibration shows that the motion of the string as a whole is not fully determinate, even though the motion at one nodal point on the string is fully ascertained.

It is noteworthy, however, that the result stated above was obtained solely from geometrical considerations without any reference to the methods of harmonic analysis.

The utility of the preceding discussion is obvious. For, we have seen in Section III that at the bowed point, generally speaking, the velocity alternates between two and only two constant values, once or oftener in each period of vibration. The condition $\frac{d^2y}{dt^2} = 0$ is thus generally satisfied at the bowed point, except at the instants at which the velocity changes from one value to the other and *vice versa*. At these instants, $\frac{d^2y}{dt^2}$ becomes $\pm$ infinity. The preceding arguments are thus applicable, and it follows that when the bow is applied at some point dividing the string in a *irrational* ratio, the form of the velocity-waves is that of a number of straight-lines parallel to one another, with intervening discontinuities. It can now be seen that this is the case even when the bow is applied at a point dividing the string in any *rational* ratio, i.e., at some node on the string. For, the kinematical uncertainty in the latter case is due only to the harmonic components in the motion which have a node at the point of application of the bow, and we have established from dynamical principles that such harmonics are not excited by the bow and do not therefore exist in the motion under consideration. The quantities that determine the motion at the bowed point must therefore also determine the motion at every other point on the string whose position is known. These quantities, in the case of the bowed point, are its initial velocity and the magnitudes and positions of the discontinuities in the velocity-waves. For, the slopes of the positive and negative velocity-waves passing over the bowed point in opposite directions being equal, the velocity at that point remains unaffected except when a discontinuity passes over it, the velocity then suddenly changing by a quantity equal to the magnitude of the discontinuity : the *times* at which these changes occur are determined by the initial *positions* of the discontinuities and *vice versa*. As stated above, these quantities must also completely determine the motion at all other points on the string, and this is only possible when, between the points of discontinuity, the velocity-waves consist of straight-lines that are all parallel to one another.

It is thus seen that the problem of finding the mode of

vibration of the string under the action of the bow reduces itself to one of finding the number, position and magnitudes of the discontinuities in the velocity-waves. From the mode of construction of the positive and the negative waves, it is obvious that the number of discontinuities in a wave-length of either of the two waves is the same, and is equal to the *total* number of discontinuities actually on the region of the string at any instant during the vibration. When a discontinuity travelling with the positive wave reaches the end of the string, it is reflected and returns as a discontinuity in the negative wave; moving on towards the other end, it reaches it and is again reflected and brought on to the positive wave. This process then repeats itself indefinitely. In Section III it was shown that the velocity at the bowed point is alternately v_A and v_B, changing discontinuously from one value to the other, and *vice versa*. The positions and magnitudes of the discontinuities in the velocity-waves must be such that by their passage over the bowed point, the specified changes of velocity at that point are produced. The simplest case possible is that in which the discontinuities pass in succession over the bowed point, belonging alternately to the positive and negative waves, i.e., pass alternately over the bowed point in opposite directions. It is obvious that the discontinuities must then be all of the same magnitude and sign, i.e., $v_A - v_B$. In other words, the discontinuities are all equal in magnitude to one another and to the arithmetical sum of the two speeds possible at the bowed point, i.e., to the relative velocity of the bowed point with respect to the bow during the backward movement. Other cases that may possibly arise are the following:—(a), two discontinuities of the same magnitude and sign may pass simultaneously over the bowed point in opposite directions. (b), two discontinuities differing in magnitude or sign or both may simultaneously pass over the bowed point in opposite directions. (c), two or more discontinuities may pass over the bowed point in succession in the same direction instead of alternately in opposite directions. The contingency in (c) does not however actually arise, as it is impossible to construct positive and negative waves which would give rise to it and which would at the same time satisfy the condition that the velocity at the bowed point should alternate between two values only. Further, it is found that if the bow is applied at a point dividing the string in an irrational ratio, the contingency in (*b*) is also impossible, and the discontinuities in the velocity-waves are necessarily all equal to one another and to $v_A - v_B$. This result is of great importance in the theory of the subject.

The reason why the discontinuities are all equal to $v_A - v_B$ if the bowed point divides the string in an irrational ratio, is not very difficult to see. The result has already been demonstrated for cases in which the discontinuities pass in succession over the bowed point and never simultaneously. If two equal discontinuities pass over the point in opposite directions at the same instant, the velocity of the point is left unaffected. Further, if two discontinuities thus cross at the bowed point, they cannot again pass simultaneously over the bowed point when returning after one reflexion at the ends of the string (the distance of the bowed point from the two ends being unequal). On the return journey, the discontinuities must therefore pass the bowed point either separately or else simultaneously with certain other discontinuities. In the former case their magnitudes are necessarily equal to $v_A - v_B$. In the latter case also, a precisely similar result holds good, except when all the discontinuities of a given set pass over the bowed point in twos and twos and never singly. From very simple geometrical considerations it can be shown that the discontinuities would so all pass in twos and twos only if they were situated at regular intervals equal to an aliquot part of the wave-length, and the bow were itself applied at a point of division of the string into an equal number of aliquot parts, i.e., at a point or node dividing the string in a rational ratio.[1] We thus

---

[1] The following simple model serves very effectively to picture the movement and successive reflexions of the discontinuities in the velocity-waves. Consider the motion of an endless cord which runs on two parallel axes between which it is stretched straight. A number of particles fixed to the cord at intervals may represent the discontinuities. If there are N particles fixed at equal intervals along the cord, the particles moving towards one axis would pass those moving the other way, at points dividing the distance between the axes into N equal parts. No particle would ever pass these points singly, i.e., by itself. A similar result would not be possible if the particles were fixed to the cord at unequal intervals or if any other point of observation were chosen. This model may be used for a lecture demonstration of the results given in Sections VII to X with reference to types of vibration in which there are two, three or any larger number of discontinuities.

D

arrive at the following two general results regarding the form of the velocity-waves :—1. When the bow is applied at a point dividing the string in an irrational ratio, the velocity-waves consist of straight lines that are all parallel to one other with intervening discontinuities all equal to $v_A - v_B$, and the number of such discontinuities per wave length is the same in the positive and negative waves. 2. When the bow is applied at a point dividing the string in a rational ratio, the velocity-waves also consist of parallel straight lines with intervening discontinuities, the number of which is the same per wave-length in the positive and negative waves ; the magnitude of the discontinuities is not however the same as in (1). The argument shows that in this case, the non-appearance of the harmonics having a node at the bowed point results in a number of discontinuities being present at regular intervals equal to an aliquot part of the wave-length in the positive and negative waves, viz., at intervals of $2l/s$ when the bow is applied at one of the points of division of the string into s aliquot parts.[1] The discontinuities pass in pairs (never singly) in opposite directions over the bowed point and also over the other points of rational division of the string.

From the foregoing it is seen that in any case in which the bow is applied at a point of rational division of the string, the form of the velocity waves can be derived by a very simple geometrical construction from velocity-waves of the irrational type, i.e., those in which the discontinuities are all equal to $v_A - v_B$; the construction is equivalent to the abolition or removal of those harmonics which have a node at the bowed point and leaves the resulting motion at the bowed point and at the other points of rational division unaffected. Examples of the method will be dealt with later. Its usefulness is evident from the fact that all the possible types of vibration may thus be considered as special cases of what may be termed the irrational types of vibration, the theory of which can be worked out geometrically with the greatest ease and simplicity and which we shall now proceed to discuss.

---

[1] s is taken to be the smallest possible number of aliquot parts.

PLATE III.

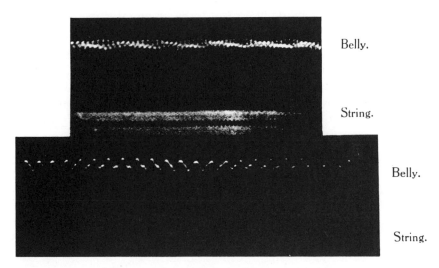

Belly.

String.

Belly.

String.

Simultaneous Vibration-Curves of Belly and G-String of Violoncello
at 264 Vibrations per second, showing Cyclical changes.

SECTION V.—CLASSIFICATION OF THE VIBRATIONAL MODES.

From the results given in the preceding section it is obvious that the vibrational modes in the cases in which the bow is applied at a point of irrational division of the string can be very simply classified according to the total number of discontinuities in the velocity waves. If there is one discontinuity, we may call it the first type of vibration of a bowed string. If there are two discontinuities, it may be called the second type of vibration, and so on. Generally speaking, each of these types of vibration includes the complete series of harmonics.

We may now proceed to deduce a few results of general application, examples which will be met with in the detailed graphical discussion of individual cases to be given later in the course of the paper.

Since the positive and negative velocity waves are representable by parallel lines separated by equal intervening discontinuities, the points at which the x – axis is cut by the parallel lines (or would be cut by them if produced) must be equi-distant from one another. If there are n discontinuities per wave length, the intercepts on the x – axis are evidently $2l/n$. If $n = 1$, the intercepts are equal to $2l$, i.e., to the wave-length, and it is obvious that in this case the positive and negative waves are necessarily of the same form (symmetrical about the x – axis) and are completely coincident twice in each period of vibration.

If the inclination of the lines to the x – axis is $\tan^{-1} c$ and there are n discontinuities per wave-length each equal to $(v_A - v_B)$, we have

$$2cl = n (v_A - v_B). \qquad \ldots \ldots \ldots (16)$$

By summation of the ordinates of the positive and negative waves, the velocities at all points on the string can be ascertained and represented graphically. The velocity graph thus obtained for the string must evidently consist of parallel straight lines inclined to the x – axis at an angle $\tan^{-1} 2c$, the maximum number of such straight lines being $(n+1)$, (there being n discontinuities on the string, some of which might be instantaneously coincident in

position). Further, these $(n+1)$ lines on the velocity diagram
pass through fixed points on the x – axis situated at equal intervals. Since the two ends of the string have always zero velocity,
these fixed points are in fact the $(n+1)$ nodes of th n^{th} harmonic,
and we thus obtain the result that the lines of the velocity diagram pass, or would pass if produced through some or all of the
$(n+1)$ nodes of the nth harmonic, if the particular type of vibration elicited by the bow is that in which there are n equal discontinuities.

From the preceding result, we may very readily deduce an
expression for the ratio v_A/v_B of the two velocities possible at
the bowed point, which besides being of perfectly general application is also valid for all points on the string, the velocity at which
alternates between two constant values only, once or oftener in
each period of vibration. Assume first that the bow is applied
at a point of irrational division of the string and the mode of
vibration elicited is that in which there are n equal discontinuities.
Consider the motion at a point on the string lying between the rth
and $(r+1)^{th}$ nodes of the nth harmonic (counting from one end)
and whose distance x from that end of the string is therefore
greater than $(r-1) l/n$ and less than rl/n. The velocity of this
point on the string at any instant during the vibration is given by
the ordinate of the velocity diagram. As we have just seen,
this velocity-diagram consists of parallel lines drawn through the
successive nodes of the nth harmonic at an inclination of $\tan^{-1} 2c$
to the x – axis, with intervening discontinuities. If the velocity
at the particular point on the string alternates between two and
only two constant values, it must be because the ordinate
drawn through it intersects alternately the two lines of the
velocity diagram passing through the two nodes of the nth harmonic on either side of it, as a result of the movement of
the discontinuities. In other words, the two velocities at the
point considered are $2c \left(x - \overline{r-1}\, l/n \right)$ and $2c \left(x - r\, l/n \right)$. The
ratio of these velocities is merely the ratio of the distances of the
point from the two nodes, and if (for brevity) the symbol x_n is
used to denote the shorter of the two distances, and ω is used to

denote the total fraction of the period of vibration in which the point moves with the larger of the two velocities, we have

$$\omega = \frac{n\,x_n}{l} \quad . \quad . \quad . \quad . \quad . \quad . \quad (17)$$

The algebraic difference of the two velocities at the point is $2cl/n$ and this is equal to $(v_A - v_B)$, vide equation (16). Since the result given in (17) is true for all points on the string at which the velocity alternates between two constant values once or oftener in each period of vibration, it applies also at the bowed point, x_n denoting its distance from the nearest node of the nth harmonic.

The result given in (17) above is noteworthy by reason of its simplicity and perfect generality as also by reason of the simplicity and perfect generality of the reasoning from geometrical considerations by which it was deduced. The result is equally applicable in cases in which the motion at the bowed point is of the simplest possible type (one ascent followed by a descent) as well as those in which the motion is one of the so-called complicated types, a succession of several ascents and descents within the period of vibration. In deducing the result, it has been assumed that the vibration is elicited by applying the bow at a point of irrational division of the string, so that the type of motion maintained is one in which the discontinuities present in the velocity waves are all equal to $(v_A - v_B)$. Even this restriction may be removed, i.e., we may also include the cases in which the bow is applied at a point of rational division of the string, the only difference being that the result given in (17) would then be applicable only at the bowed point and at some or all of the other nodes of the principal member of the missing series of harmonics, and not at any other point on the string. For, as already referred to in the preceding section, any type of vibration elicited by applying the bow at a nodal point on the string can be considered as one of the modes of vibration of the 'irrational' type with the series of harmonics having a node at the bowed point dropped out. The process leaves the motion at the bowed point and at the other nodes of the principal member of the missing series of harmonics, unaffected.

When $n = 1$, equation (17) reduces to the well-known relation discovered by Helmholtz, i.e., the ratio of the velocities of ascent and descent at the point considered is the same as the ratio of its distances from the two ends of the string. Krigar-Menzel and Raps found in their experimental work that Helmholtz's relation was satisfied when the bow was applied in the normal manner at any point very close to an end of the string or else very exactly at one of the nodal points distant $\frac{l}{2}, \frac{l}{3}, \frac{l}{4}, \frac{l}{5}, \frac{l}{6}$ or $\frac{l}{7}$ from the end. The value of ω for other points of application of the bow was also measured by Krigar-Menzel and Raps, and they state as the result of these measurements that no general algebraic relation connecting the value of ω at the bowed point with its position on the string could be found even when the motion at the bowed point was of the simplest possible type representable by a two-step zig-zag. Their deliberate conclusion on this question was that, except when the bow was applied close to the end of the string or at one of the nodes of some fairly important harmonic, the value of ω was to be regarded as a purely empirical quantity depending on the experimental conditions. It is obvious that if the value of ω is thus regarded as an arbitrary quantity determinable by experiment, no complete theoretical discussion of the kinematics of the string is possible, and in fact Krigar-Menzel and Raps did not attempt any such complete discussion. While on the experimental side their paper was a notable contribution to the subject, the treatment given by them on the theoretical side was thus obviously defective and incomplete. The general kinemetical analysis set out in the present paper shows that the value of ω in all cases (i.e. both for rational and irrational points of bowing) should satisfy the relation given in equation (17), n being given by the appropriate integral value, 1, 2, 3, 4 or 5, etc. The failure of Krigar-Menzel and Raps to discover this general algebraic relation, or rather this series of relations connecting the value of ω at the bowed point with its position on the string, must be attributed to their having adopted an almost exclusively empirical method of treatment. If, instead of relying solely on the result of the measurements which were necessarily subject to experimental error in some degree, they had investigated in detail the kinematics of some of the simpler types of vibration other than those known through the work of Helmholtz, e.g., that obtained by applying the bow at a point close to but not coincident with the centre of the string, the functional relation connecting the value of ω at the bowed point with its position on the string could have been looked for with a greater chance of success. That such a functional relation exists must indeed have been evident from the fact that the characteristic vibration-curves in such cases also are perfectly reproducible, time after time, with strings of any length, diameter or material.

The failure to establish a proper scheme of classification of the vibrational modes and to find the general form of the functional relation connecting ω at the bowed point with its position anywhere on the string was also one of the fundamental defects in the paper by A. Stephenson cited in the Bibliography. In this paper (published in 1911) only the work of Helmholtz is referred to, and a perusal of it shows that Stephenson was unacquainted with the work of Krigar-Menzel and Raps published in 1891, and that he was, indeed, unaware of many facts which anyone who has experimented with a bow and monochord could readily observe for himself. It is not a matter for surprise therefore that, though Stephenson's paper is noteworthy as an attempt to treat the motion of a bowed string as a case of maintained vibration, it takes us little beyond the work of Helmholtz. Stephenson also failed to realise that the Fourier analysis is obviously incapable of giving any useful indication of what would happen if the bow is applied at a point of irrational division of the string, i.e., at a point not exactly coinciding with any nodal point of importance, and it is precisely such indication that is required to explain the phenomena observed in experiment.

We may now pass on to consider the kinematics of the irrational types of vibration more in detail. If there are n equal discontinuities, the velocity-diagram of the string consists of not more than $(n+1)$ parallel lines passing through the $(n+1)$ nodes of the nth harmonic. As the discontinuities move one way or the other, the lines of the velocity-diagram increase in length or else shorten and sometimes vanish altogether, and given the form of the velocity-diagram at any epoch of the vibration, it is quite an easy matter to find its form at any subsequent epoch or to trace directly the succession of velocity-changes at any point on the string and thus to determine the form of the vibration curves. If the positive and negative velocity-waves are of identical form, it is obviously convenient to commence with the epoch at which they are completely coincident and the string everywhere passes through its position of statical equilibrium. At that epoch, the discontinuities are everywhere situate in pairs along the string, the odd discontinuity, if any being at one end of the string, and the lines of the velocity-diagram pass through the *alternate* nodes of the nth harmonic. In the subsequent motion, the discontinuities situate along the string separate and move off in opposite directions, the odd discontinuity, if any, situate at the end of the string moving straight off towards the other end. After half a period, the positive and negative waves again coincide and the velocity-changes at every point on the string are gone through in the reverse order, as already described in Section II of the paper. It should be remarked that the positive and negative waves are necessarily of the same form when the motion at any one point on the string is representable by a simple two step zig-zag, or by any other curve possessing a similar type of symmetry. As normally applied, the bow excites the vibrations of the string from an initial state in which the latter is everywhere in its position of equilibrium. The tendency is thus, in a large majority of cases, to set up vibrations having this characteristic type of symmetry.

SECTION VI.—THE FIRST TYPE OF VIBRATION.

Of the possible types of vibration set up by the application of the bow at a point of irrational division, the first type with only one discontinuity on the velocity-diagram is the simplest and most important. In this case, as already remarked, the positive and negative velocity-waves are necessarily of the same form, and at the instant at which they are coincident, the velocity-diagram is a straight line passing through one end of the string ($x=0$), with a discontinuity at the other end ($x=l$). As this discontinuity moves in along the string, the velocity-diagram consists of parallel lines passing through its two ends, and the velocities at

PLATE IV.

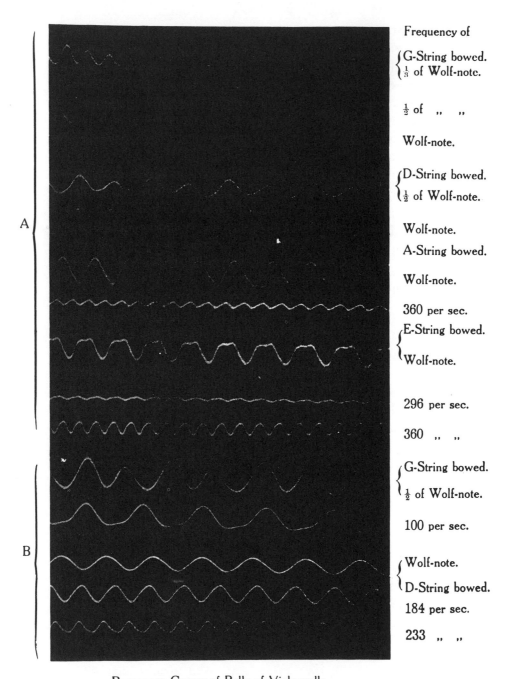

Frequency of

{ G-String bowed.
⅓ of Wolf-note.

½ of ,,   ,,

Wolf-note.

{ D-String bowed.
½ of Wolf-note.

Wolf-note.

A-String bowed.

Wolf-note.

360 per sec.

{ E-String bowed.

Wolf-note.

296 per sec.

360 ,,   ,,

{ G-String bowed.

½ of Wolf-note.

100 per sec.

{ Wolf-note.

D-String bowed.

184 per sec.

233 ,,   ,,

Resonance-Curves of Belly of Violoncello.

A.   Without any load on the bridge: Wolf-note frequency 176
vibrations per second.

B.   With a load of 40·4 grammes fixed on top of bridge.
Wolf-note frequency 137 vibrations per second.

any point before and after its passage are respectively proportional to the distances from the two ends. When the discontinuity reaches the end $x=l$, it is reflected and the velocity-diagram then passes back through the same stages to its original form.

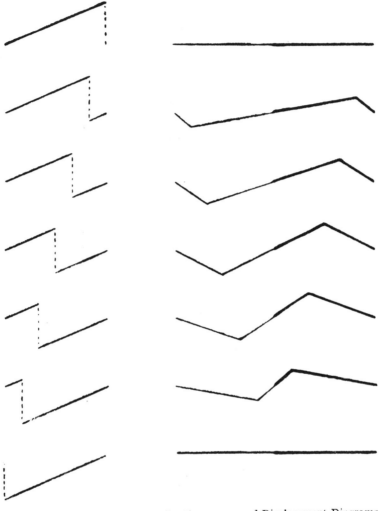

Velocity-Diagrams.   Vibration-curves and Displacement Diagrams.
Fig. 1.—First Type of Vibration.

Fig. 1 (first column) shows the successive velocity-diagrams at intervals of one-twelfth of an oscillation.

E

Fig. 1 (second column with the heavy and thin lines taken separately) also shows the displacements of the string from its initial position at similar equal intervals throughout the complete period of vibration. These configuration-diagrams, as we may call them, evidently consist of two straight lines passing through the ends of the string and meeting at the point up to which the discontinuity in the velocity-diagram has travelled at any instant. (It should be remarked here that the displacements are always measured from the position of equilibrium of the string under the steady frictional force exerted by the bow, and not from the position of equilibrium attained when the bow is removed). The second column of Fig. 1, with the heavy and thin lines taken together represents on the same scale of ordinates, the vibration curves for the complete period, of points situated on the string at successive equal intervals of one-sixth of its length, commencing from one end. When extended on either side for a number of complete periods, the vibration-curves are seen to consist of simple two-step zig-zags, the fraction ω of the complete period during which any point moves with the larger of the two velocities being given by the relation

$$\omega = x_1/l \quad . \quad . \quad . \quad . \quad . \quad . \quad . \quad (18)$$

where x_1 is the distance of the point from the nearer of the two ends of the string.

The correspondence noticed above between the configuration of the string as a whole and the vibration-curves of individual points on it is not peculiar to the present case, but may be established with generality for any possible type of vibration of a stretched string in which the positive and negative velocity-waves are of the same form. A geometrical proof is very readily given by noticing that the displacement at any point is the time-integral of the velocity and is therefore representable by the area enclosed by two ordinates drawn at equal distances on either side of the point under consideration, to intersect the velocity-wave. The following is an analytical proof. In such cases we have

$$y = \sum_{1}^{\infty} a_n \sin \frac{n\pi x}{l} \sin \frac{2\pi n t}{T} \quad . \quad . \quad . \quad . \quad . \quad (19)$$

If x is regarded as constant and t as a variable, equation (19) represents the form of the vibration-curves. On the other hand, if t is regarded as constant and x as the variable, the equation gives us the configuration of the string. By taking only half the complete period into consideration, i.e., from $t=0$ up to $t=T/2$ or from $t=T/2$ up to $t=T$, we can get identical geometrical representations for the motion at individual points on the string, and for the configuration of the string as a whole, provided that the times for the latter and the positions for the former are so chosen that $2t/T=x/l$.

Two other important consequences of equation (19) may also be noticed here. If two points are taken on the string, one on either side of the centre at equal distances from it, the form of their vibration-curves are the mirror-images of one another with respect to the centre of the string. The second consequence is that the vibration-curve at a point very close to the end of the string from $t=0$ to $t=T/2$ or from $t=T/2$ up to $t=T$ is, in the limit, of the same form as the *velocity-diagram* of the string at time $t=0$ or $t=T/2$ as the case may be. This may be regarded as a particular case of the correspondence of form noticed in the preceding para. obtained by putting the chosen values of t and x very small. For, when t is very small but not actually zero, the small displacement at any point is proportional to the initial velocity at the point.

8

Reprinted from *Phil. Mag.*, Ser. 6, **32**, 391 – 395 (Oct. 1916)

XLIII. *On the "Wolf-note" in Bowed Stringed Instruments.*
By C. V. RAMAN, *M.A.* [*]

[Plate IX.]

IT has long been known that on all musical instruments belonging to the violin family there is a particular note which it is difficult to elicit in a satisfactory manner by bowing. This is called the "wolf-note," and when it is sounded the body of the instrument is set in vibration in an

[*] Communicated by the Author.

unusual degree; and it appears to have been realized that the difficulty of maintaining the note steadily is due in some way to the sympathetic resonance of the instrument[*]. In a recent paper [†], G. W. White has published some interesting experimental work on the subject, confirming this view. The most striking effect noticed is the *cyclical* variation in the intensity of the note when the instrument is forced to speak at this point. White suggests as an explanation of these fluctuations of intensity that they are due to beats which accompany the forced vibration impressed on the resonator when the impressed pitch approaches the natural pitch of the system. The correctness of this suggestion seems open to serious criticism. For, the beats which are produced when a periodic force acts on a resonator are of brief duration, being merely due to the superposition of its forced and free oscillations, and when, as in the present case, the resonator freely communicates its energy to the atmosphere and the force itself is applied in a progressive manner and not suddenly, such beats should be wholly negligible in importance, and should, moreover, vanish entirely when the impressed pitch coincides with the natural pitch.. In the present case the essential feature is the *persistency* of the fluctuations of intensity and their markedness over a not inconsiderable range; and it is evident that an explanation of the effect has to be sought for on lines different from those indicated by White. I had occasion to examine this point when preparing my monograph on the " Mechanical Theory of the Vibrations of Bowed Strings," which will shortly be published, and the conclusions I arrived at have since been confirmed by me experimentally.

From the mechanical theory, it appears that when the pressure with which the bow is applied is less than a certain critical value, proportionate to the rate of dissipation of energy from the vibrating string, the bow is incapable of maintaining the ordinary mode of vibration in which the fundamental is dominant, and the mode of vibration should progressively alter into one in which the octave is the predominant harmonic [‡]. In the particular case in which the frequency of free oscillation of the string coincides very nearly with that of the bridge of the violin and associated masses, the mode of vibration of the string is *initially* of the well-known type in which the fundamental is dominant.

* Guillemin, " The Application of Physical Forces," 1877.
† G. W. White, Proc. Camb. Phil. Soc. June 1915.
‡ Compare with the observations of Helmholtz, ' Sensations of Tone,' English Translation by Ellis, p. 85.

But the vibrations of the string excite those of the instrument, and, as the vibrations of the latter increase in amplitude, the rate of dissipation of energy increases continually till it outstrips the critical limit, beyond which the bow fails to maintain the usual type of vibration. As a result of this, the mode of vibration of the string progressively alters to a type in which the fundamental is subordinate to the octave in importance. The vibration of the belly then begins to decrease in amplitude, but, as may be expected, this follows the change in the vibrational form of the string by a considerable interval. The decrease in the amplitude of the vibrations of the belly results in a falling off of the rate of dissipation of energy, and, when this is again below the critical limit, the string regains its original form of vibration, passing successively through similar stages, but in the reverse order. This is then followed by an increase in the vibrations of the belly, and the cycle repeats itself indefinitely. The period of each cycle is approximately twice the time in which the vibrations of the belly would decrease from the maximum to the minimum, if the bow were suddenly removed.

The foregoing indications of theory are amply confirmed by the photographs reproduced in Plate IX., which show the simultaneous vibration-curves of the belly and string of a 'cello at the wolf-note pitch. It will be seen that the form of vibration of the string alters cyclically in the manner predicted by theory, and that the corresponding changes in the vibration-curve of the belly *follow* those of the string by an interval of about quarter of a cycle. That the two sets of changes are dynamically interconnected in the manner described is further confirmed by the prominence of the octave in both curves at the epochs of minimum amplitude. The explanation of the cyclical changes given above is also in accordance with the observed fact that they disappear and are replaced by a steady vibration when the ratio of the pressure to the velocity of bowing is either sufficiently increased or sufficiently reduced. In the former case the string vibrates in its normal mode, and in the latter case the fundamental disappears altogether and the string divides up into two segments.

Effect of Muting on the "Wolf-note."

Since the pitch of the wolf-note coincides with that of a point of maximum resonance of the belly, we should expect to find that by loading the bridge or other mobile part of the body of the instrument important effects are produced.

This is readily shown by putting a mute on the bridge. The pitch of the wolf-note then falls immediately by a considerable interval. On the particular 'cello I use, a load of 17 grammes fixed at the highest point of the bridge lowers the wolf-note pitch from 176 to 160 vibrations per second. A larger load of 40·4 grammes depresses it further to 137 vibrations per second, and also causes two new but comparatively feeble resonance-points to appear at 100 and 184 respectively, without any attendant cyclical phenomena. An ordinary brass mute has a very similar effect.

The Formation of Violin-tone and its Alteration by a Mute.

The positions of the frequencies of maximum resonance of the bridge and associated parts of the belly for notes over the whole range of the scale are undoubtedly of the highest importance in determining the character of violin-tone, and the explanation of the effect of a mute on the tone of the instrument is chiefly to be sought for in the effect of the loads applied on the frequencies of the principal free modes of vibration of the bridge and associated parts of the belly. The observations of Dr. P. H. Edwards on the effect of the mute * are evidently capable of explanation on the basis of the lowering of the frequencies of maximum resonance by the loading of the bridge. But a more detailed understanding of the dynamics of the problem requires further theoretical and experimental investigation. Recently, I have secured an extensive series of photographs showing the effect on the motion of the bridge in its own plane produced by fixing a load on it at one or other of a variety of positions. The close parallelism between the effect of loading, as shown by these photographic curves and as observed by the ear, seems to show that the motion of the bridge in its own plane determines the quality of violin-tone to a far greater extent than might be supposed from the work of Giltay and De Haas †. A detailed discussion of this and other problems relating to the physics of bowed instruments is reserved for a separate communication.

This investigation was carried out in the Laboratory of the Indian Association for the Cultivation of Science, Calcutta.

20th May, 1916.

* P. H. Edwards, Physical Review, January 1911.
† Giltay and De Haas, Proc. Roy. Soc. Amsterdam, January 1910. See also E. H. Barton and T. F. Ebblewhite, Phil. Mag. September 1910, and C. V. Raman, Phil. Mag. May 1911.

Note dated the 8th of August added in proof.

Since the paper was first written, several other interesting effects have been noticed, of which the following is a summary:—

(*a*) Cyclical forms of vibration of the G-string and belly of a 'cello may also be obtained when the vibrating length is double that required for production of the wolf-note, that is, when the frequency is half that of the wolf-note. In this case, when the pressure of the bow is sufficient to maintain a steady vibration, the second harmonic in the motion of the belly is strongly re-inforced. When the pressure is less than that required for a steady vibration, cyclical changes occur, the principal fluctuations in the motion, both of the string and the belly, being in the amplitude of the second harmonic. In this, as in all other cases, the cyclical changes disappear and give place to a steady vibration, when the bow is applied at a point sufficiently removed from the end of the string. In this particular case, a large, *almost soundless* vibration may be obtained by applying the bow rather lightly and rapidly at a point distant one-fifth or more of the length from the end; the octave is then very weak in the vibration of the string, but may be restored, along with the tone of the instrument, by increasing the pressure of the bow.

(*b*) The 'cello has another marked point of resonance at 360 vibrations per second. The pitch of this is also lowered by loading the bridge.

(*c*) When the vibrating length of the G-string or A-string of the 'cello is about a fourth of the maximum or less, cyclical forms of vibration may be obtained at almost any pitch desired, by applying the bow with a moderate pressure rather close to the bridge.

(*d*) As the frequency of vibration is gradually increased from a value below to one above the wolf-note frequency, the phase of the principal component in the " small " motion at the end of the string, that is also of the transverse horizontal motion of the bridge, undergoes a change of approximately 180°. This is in accordance with theory.

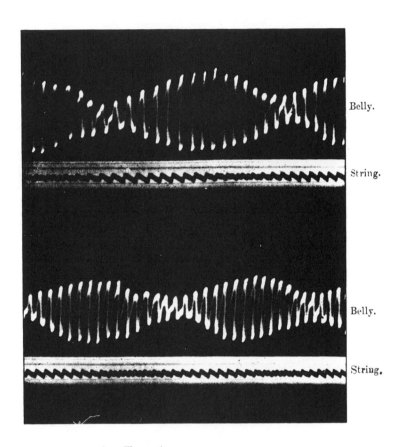

Time Axis ————·—→

Simultaneous Vibration-Curves of Belly and String of Violoncello
at the " Wolf-note " pitch.

Reprinted from *Proc. Indian Assoc. Cultivation Sci.*, **6**, 19–36 (1920–1921)

II.—Experiments with Mechanically-Played Violins.

By Prof. C. V. Raman.

(Plate II.)

CONTENTS.

Section I.—Introduction.

In the first volume (recently published *) of my monograph on the theory of the violin family of instruments, I have discussed on mechanical principles, the relation between the forces exerted by the bow and the steady vibration maintained by it, and the conditions under which the bow is capable of eliciting a sustained musical tone from the instrument. An experimental test of the results indicated by the theory on these points would obviously be of interest. Especially is this the case, as the analysis shows that the yielding of the bridge and the communication of energy from the strings through their supports into the instrument and thence into the air, play a very large part in determining the

---

* Bulletin No. 15 of the Indian Association for the Cultivation of Science, 1918, pages 1-158.

magnitude of the forces required to be exerted by the bow. An experimental study of the mechanical conditions necessary for obtaining a steady musical tone could thus be expected not merely to throw light on the *modus operandi* of the bow but also to furnish valuable information regarding the instrument itself, its characteristics as a resonator and the emission of energy from it in various circumstances. Further, a study of the kind referred to could be expected also to furnish illustrations of the physical laws underlying the technique of the violinist and to put these laws on a precise quantitative basis. The experiments described in the present paper were undertaken with the objects referred to above, and the description of the results now given in these Proceedings is preliminary to a more exhaustive treatment of the subject which it is proposed to give in the second volume of my monograph under preparation for publication as a Bulletin of the Association.

SECTION II.—DESCRIPTION OF MECHANICAL PLAYER.

As the object of the work was to elucidate the theory of production of musical tone from instruments of the violin family, it was decided that the experimental conditions should approximate as closely as practicable to those obtaining in ordinary musical practice. The general principle accordingly held in view in designing the mechanical player was to imitate the technique of the violinist as closely as possible. There was also another reason for adopting this course. It is well known that the bowing of a stringed instrument so as to elicit a good musical tone is an art requiring much practice for its perfect accomplishment. The performance of the same task by purely mechanical appliances under such conditions as would permit of accurate measurements of the pressure and speed of bowing and the discrimination by ear of the effect of varying these factors obviously involves difficulties which it was thought would be best surmounted by imitating the violinist's handling of the bow as closely as the mechanical conditions would permit. A mechanical player designed on this general idea which has fulfilled the requirements of the work is illustrated in Plate II.

As can be seen from the photograph, a violin and a horse-hair bow of the ordinary type were used in the mechanical player.

A Mechanical Violin-player for Acoustical Experiments.

Instead, however, of moving the bow to and fro, it was found a much simpler matter from the mechanical point of view to keep the bow fixed and to move the violin to and fro with uniform speed. This was arranged by holding the violin lightly fixed in a wooden cradle, the points of support being the neck and the tail piece of the violin as in the ordinary playing of the instrument. The cradle was mounted on a brass slide which moved to and fro noiselessly on a well-oiled cast-iron track. The slide received the necessary movement forward and backward from a pin carried by a moving endless chain and working in a vertical slot carried by the slide. The chain was kept in motion by the rotation of one of the two hubs between which it was stretched, this hub being fixed on the same axis as the driving wheel seen in the Plate.*

The apparatus was driven by a belt running over a conical pulley which in its turn was driven by a belt passing over the pulley of a shunt-wound electric motor controlled by a rheostat which was allowed to run without any load except the apparatus. Using the rheostat and a Weston Electrical Tachometer, a very constant speed could be maintained during the experiments. Different speeds of motion of the slide carrying the violin were obtained by putting the driving belt of the apparatus on to different parts of the conical pulley, or by adjusting the rheostat.

The mounting of the bow required special attention in order to ensure satisfactory results. As is well known, the violinist in playing his instrument handles the bow in such manner that when it is applied with a light pressure, only a few hairs at the edge touch the string. The bow is held carefully balanced in the fingers of the right hand, the necessary increases or decreases in the pressure of bowing being brought about by increase or decrease of the leverage of the fingers. The suppleness of the wrist of the player and the flaccidity of the muscles of the fore-arm secures the necessary smoothness of touch. These features are carefully imitated in the mechanical player. The violin-bow is held fixed at the end of a wooden lath, an adjustment being provided so that fewer or more hairs of the bow may be made to touch the string

---

* The whole of the apparatus was improvised in the laboratory from such materials as were to hand. The slide and cast-iron track were parts of a disused optical bench. The chain and hubs were spare parts purchased from a cycle-dealer. The ball-bearing of the axle of the lever (referred to below) was also part of a cycle. The other fittings were made up in the workshop.

of the violin. The lath itself is balanced after the manner of a steelyard, the axis of the lever being mounted on ball-bearings so as to secure the necessary solidity combined with freedom of movement. The weight of the bow is balanced by a load hung freely near the end of the shorter arm of the lever. The axis of the lever can be raised or lowered to the proper height above the violin such that when the hairs of the bow touch the string, they are perfectly parallel to the cast-iron track along which the violin slides. This is of great importance in order to obtain steady bowing, as otherwise the bow would swing up and down with the movement of the violin along the track, and its inertia would result in a variation of the pressure exerted by it. Any residual oscillations of the bow due to the elasticity of the lever or imperfection in the adjustment referred to above are checked by the damping arrangement shown in the Plate. A wire with a number of horizontal disks attached to it at intervals is hung freely from the shorter arm of the lever and dips inside a beaker of water or light oil. This effectually prevents any rapid fluctuations in the pressure of the bow and ensures a smooth movement. The pressure exerted by the bow on the string can be varied by moving a rider along the longer arm of the lever which is graduated. An adjustment is provided by which the block carrying the axis of the lever can be moved by a screw perpendicular to the track, and the distance from the violin-bridge of the point at which the bow touches the string may thus be expeditiously altered.

It will be noticed that with the arrangements described above, the pressure exerted by the bow on the string of the violin would not be absolutely constant throughout, but would vary somewhat as the violin moves along its track from the point nearest to the point furthest from the axis of the lever. This is not however a serious difficulty as the lever is fairly long and the variation of pressure is thus not excessive. Further, the observations of the character of the tone are always made for a particular position and direction of movement of the violin and no ambiguity or error due to the cause referred to above arises.

The speed of bowing may be readily determined from the readings of the Electrical Tachometer or directly by noting on a stop-watch the time taken for a number of strokes to and fro of the violin on its track.

Section III.—Variation of Bowing Pressure with the Position of the Bowed Region.

One of the well-known resources of the violinist is to bring the bow nearer to or to remove it further away from the bridge of the violin, the extreme variation in the position of the bow being from about $\frac{1}{5}$th to about $\frac{1}{25}$th of the vibrating length of the string from the bridge. In a recent paper on " The Partial Tones of Bowed Stringed Instruments" published in the " Philosophical Magazine" (November 1919), I have discussed in some detail the changes in the amplitudes and phases of the various partials brought about by these changes in the position of the bowed region. In all the cases of musical interest within these limits, the mode of vibration of the string is practically the same as in the principal Helmholtzian type * in regard to the first three partial components, but differs from it in respect of the higher components to an extent depending on the removal of the bow from the bridge. The ratios of the amplitudes and the relative phases of the first three partials remain practically the same throughout the range, the actual amplitudes for a given speed of the bow varying inversely as the distance of the bowed point from the bridge. The amplitudes of the fourth, fifth and higher partials vary in a similar way with the position of the bowed point provided this is not too far from the bridge, but deviate from this law more and more as the bow is removed further and further from the bridge. The net effect of bringing the bow nearer the bridge (its speed remaining constant) is greatly to increase the intensity of the tone of the instrument, and to make it somewhat more brilliant in character, as is of course well known. Simultaneously with these changes, the pressure with which the bow is applied has to be increased. The mechanical player described above may be used to find experimentally the relation between the bowing pressure and the position of the bow under these conditions.

The graphs in Fig. 1 (thin lines) represent the results obtained with the player on the D-string of the violin. A few words of explanation are here necessary. As a finite region of the bow is in contact with the string, it is not possible to specify the position

---

* The principal Helmholtzian type is the mode of vibration in which the time-displacement graph of every point on the string is a simple two-step zig-zag.

of the bow by a single constant. Accordingly, the positions of the inner and outer edges of the region of contact were noted in the observations. The graph therefore shows two curves connecting the positions of the two edges of the bowed region with the

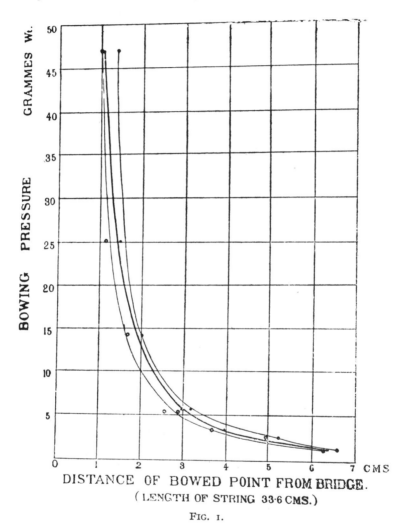

DISTANCE OF BOWED POINT FROM BRIDGE.
(LENGTH OF STRING 33·6 CMS.)

FIG. I.

magnitude of the bowing pressure, The ordinates of the graphs represent the values of the *minimum* bowing pressure found necessary to elicit a full steady tone with pronounced fundamental. [For bowing pressures smaller than this minimum, the fundamental falls off in intensity, and the prominent partial becomes its octave or twelfth. In certain cases, as for example near the wolf-note pitch, we get 'cyclical' or 'beating' tones.] It will be noticed from the graphs that the bowing pressure necessary increases with great rapidity when the bow is brought near the bridge.

The curve in Fig. 1 (heavily drawn) lying between the experimental graphs is a representation of the algebraic curve $x^2y =$ constant. (The constant was, of course, suitably chosen.) It will be noticed that the graph follows the trend of the experimental values quite closely. In other words, we may say that in the cases studied, the bowing pressure necessary varies practically in inverse proportion to the *square* of the distance of the bow from the bridge. It may be readily shown that this is the result to be expected from theory. In my monograph,* I have shown that the minimum bowing pressure P is given by the formula

$$P = \frac{P_A' - P_A}{\mu - \mu_A}$$

where P_A' is the maximum value at *any* epoch of the series

$$\sum_{n=1}^{n=\infty} k_n B_n \frac{\sin\left(\frac{2n\pi t}{T} + e_n + e_n'\right)}{\sin \frac{n\pi x_0}{L}},$$

and P_A is the value of the series at the epoch at which the bowed region of the string slips past the bow. μ is the statical coefficient of friction, and μ_A is the dynamical coefficient of friction during the epoch of slipping. B_1, B_2, etc. are the amplitudes of the partial vibrations of the string, k_1, k_2, etc. are numerical constants for the respective partials depending on the instrument, the mass, length and tension of the string, and x_0 is the distance of the bowed point from the end of the string. We have already seen that amplitudes B_n of the first few partials for a given speed of bowing vary in inverse proportion to the distance of the bow from the bridge, and their relative phases remain unaltered. In respect of these partials, the factor

$$1/\sin \frac{n\pi x_0}{L}$$

---

* Bulletin No. 15, pages 73 to 75.

also varies practically in inverse proportion to x_0, so long it is a small fraction of l.

To effect a simplification, we may proceed by ignoring the influence of all the partial vibrations except the first few, an assumption which is justifiable in the case under consideration, as by far the greater proportion of the energy of violin-tone is confined to the first few partials. Further, we may for simplicity, treat the difference $(\mu - \mu_A)$ between the statical and dynamical coefficients of friction as practically a constant quantity. This will not introduce serious error, provided the speed of the bow is not very small. For, if the slipping speed be fairly large, any changes in it due to change of the position of the bowed point would not seriously alter the dynamical coefficient of friction. On these simplifying assumptions, it will be seen from the formulæ given above that, within the limits considered, the minimum bowing pressure should vary in inverse proportion to the *square* of the distance of the bow from the bridge, exactly as found in experiment. This relation would, of course, cease to be valid when the bowed point is removed too far from the bridge or when the speed of the bow is very small.

SECTION IV.—RELATION BETWEEN BOWING SPEED AND BOWING PRESSURE.

The changing of the speed of the bow is another of the well-known resources of the violinist. The principal effect of this is to alter the intensity of tone. *Pari passu* with the change of speed of the bow, other things remaining the same, the violinist has to alter the pressure of the bow. The relation between these may be readily investigated with the mechanical player. The experimental results for the D-string and for a particular position of the bowed point are shown in Fig. 2.

The graph shows the following features : (1) for very small bowing speeds, the bowing pressure tends to a finite minimum value; (2) the increase of bowing pressure with speed is at first rather slow; (3) later, it is more rapid, the pressure necessary increasing roughly in proportion to speed, and for large amplitudes of vibration possibly even more than in proportion to the speed of the bow.

The foregoing results are, broadly speaking, in agreement with what might be expected on theoretical grounds.* This can be seen from the formula for the bowing pressure referred to in Section III. With increasing speed of the bow, the amplitudes B_n of the partial vibrations increase in proportion, so that if the difference $\mu - \mu_A$ between the statical and dynamical coefficients of friction be

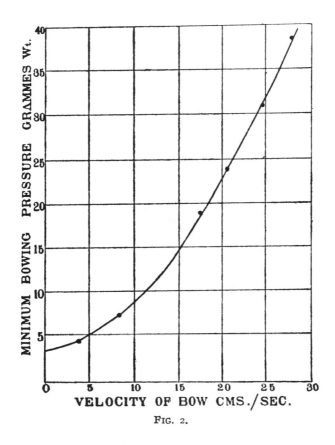

FIG. 2.

regarded as a constant, the bowing pressure necessary should vary directly as the speed of the bow. For very small speeds of the bow, however, it is not correct to take $\mu - \mu_A$ as constant, and it would be nearer the mark for such speeds to take $\mu - \mu_A$ as proportional to the velocity of slip, that is, as proportional to the speed of the bow. Thus, for very small speeds of the bow, the pressure neces-

---

* Bulletin No. 15, pages 151-153.

sary should be nearly independent of the speed, that is, should converge to a finite minimum speed. For larger speeds of the bow, it would be correct to take $\mu - \mu_{.1}$ as constant and the bowing pressure should then vary proportionately with the speed. For very large speeds, the theory of small oscillations would no longer be applicable, and the quantities k_1, k_2, k_3, etc. might increase with the speed of the bow. For such large speeds, the bowing pressure necessary might increase more than in proportion to the speed of the bow.

A more precise discussion of the experimental results would be possible on the basis of quantitative data as to the manner in which the coefficient of friction between rosined horse-hair and catgut varies with the velocity of slip at different pressures.

Section V.—Variation of Bowing Pressure with Pitch.

The pitch of violin tone depends on (1) the linear density of the bowed string, (2) its length, and (3) its tension, and may be varied by varying any one or other of these factors. In practice, the violinist varies the pitch by (1) altering the vibrating length by "stopping" down the string on the fingerboard, or (2) by passing from one string to another. The mechanical player may be used to investigate the dependence of bowing pressure upon pitch when the latter is varied in any of the ways that may be suggested. Obviously, the sequence of phenomena observed would not be exactly the same for the four strings of the violin as these are of different densities and tension, communicate their vibrations to the body of the instrument at different points of the bridge and also vibrate in considerably different planes relatively to the bridge and belly when excited by the bow in the usual way. In the experimental work now to be described, a particular string of the violin, e.g. the 4th or G-string, was used, and the pitch was varied as in the ordinary playing of the instrument by 'stopping' the string at different points. This was arranged by clamping the string down to the fingerboard, with a light but strong brass clamp shaped like an arch which could be put across the fingerboard, passed down upon it and then lightly fixed to it by two set-screws at the two ends. The inner face of the clamp was lined with leather to imitate the ball of the fingers of the violinist and to prevent damage to the strings.

A few remarks are here necessary. In actual practice, when the violinist stops down the string so as to elicit a note of higher pitch, he generally takes the bow up rather nearer the bridge so as to preserve the relationship between the vibrating length of the string and the distance of the bow from the bridge. Strictly speaking, this should also have been done in the present investigation. But as it would have been somewhat troublesome and involved the risk of errors in the adjustment of the position of the bow, it was decided to keep the bow in a fixed position somewhat close to the bridge and to find the relationship between the bowing pressure and the pitch of the string under these conditions. We have already seen that when the bow is fairly close to the bridge,

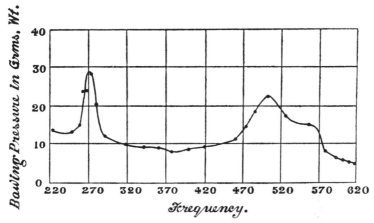

FIG. 3.—Relation between Bowing Pressure and Pitch (without Mute).

the bowing pressure necessary varies practically in inverse proportion to the square of the distance of the bow from the bridge. Accordingly, the effect of keeping the bow in a fixed position when the pitch is altered, instead of it bringing it nearer the bridge at each stage, is to decrease the bowing pressure necessary in a progressive and calculable ratio. This effect does not accordingly interfere with our observation of the characteristic changes of bowing pressure with pitch, which are connected with the changes in the forced vibration of the bridge and belly of the violin brought about by the change in the frequency of excitation.

The graph in Fig. 3 represents the relationship between bowing pressure and pitch within a part of the range of tone of the violin which includes the first three of the natural frequencies of vibration of the body of the instrument. The particular violin used was of German make, marked copy of Antonius Straduarius, the bridge being of the usual Straduarius model. The experiments were made on the 4th or G-string, stopped down to various pitches. It will be noticed that the graph for the bowing pressure shows pronounced maxima and minima. There is a strong maximum at 270, another maximum between 470 and 520, and a distinct hump between 520 and 570. These maxima pretty nearly coincide in pitch with the first three maxima of intensity of the fundamental in the tone of the violin as estimated by ear, in other words with the frequencies of maximum resonance of the instrument to the gravest component of the force exerted on it by the vibrating string. The maximum lying between 470 and 520 is specially interesting as this region exhibits the well-known phenomenon of the 'wolf-note.' In the ascending part of this portion of the graph, and especially at and near the peak of the curve, it is found that when the pressure of the bow is less than the minimum required to elicit a steady tone with a well-sustained fundamental component, we get 'cyclical' or beating tones of the kind described and illustrated by me in previous papers.* The rapidity of the beats depends on the pitch of the tone which it is attempted to elicit. and also on the pressure and speed of the bow. A similar tendency to production of a 'wolf-note' though not so striking, is also manifested in the part of the graph between 520 and 570. The maximum in the region of 250 to 285 does not show a similar tendency, at any rate to any appreciable extent. It would appear that the gravest resonance of the violin chiefly involves a vigorous oscillation of the air within the belly of the instrument, but not so vigorous an oscillation of the bridge and belly as in the second and third natural modes of vibration, which show the wolf-note phenomenon. Further evidence on this point is furnished by experiments on the effect of putting a load or mute on the bridge of the violin as will be referred to in the following section.

The formula for the bowing pressure quoted on page 25 enables the variation of bowing pressure shown in Fig. 3 to be explained. In the series

---

* Bulletin No. 15, also Phil. Mag. Oct. 1916.

$$\sum_{n=1}^{n=\infty} K_n B_n \frac{\sin\left(\frac{2n\pi t}{T} + e_n + e_n'\right)}{\sin \frac{n\pi x_o}{L}},$$

of which the graph practically determines the bowing pressure required, the B_n's stand for the amplitudes of the different partial vibrations of the string, and the K_n's are quantities which are practically proportional to the corresponding partial components of the forced vibration of the bridge transverse to the string at its extremity. In the case of a string bowed near the end, $B_n / \sin \frac{n\pi x_o}{L}$ varies nearly as $1/n^3$ and thus decreases rapidly as n increases. Further, in the part of the range of violin-tone covered by the graph in Fig. 3, the fundamental component of the vibration of the bridge should obviously be well marked, and K_1 would therefore be of the same order of quantities as K_2, K_3, etc., or even larger. Hence the value of the series given above would be principally determined by its leading term proportional to K_1, and the variation of bowing pressure with pitch would practically follow the fluctuations of K_1, in other words would follow the variations in the amplitude of the fundamental component in the forced vibrations of the bridge and pass through a series of maximum values at the successive frequencies of resonance of the instrument. This is practically what is shown by the experimental results for the bowing pressure appearing in Fig. 3, and the graph gives us an idea of the *sharpness* of the resonance of the instrument at each of the frequencies referred to. It must be remembered, however, that the bowing pressure required is also influenced by the terms proportional to K_2, K_3, etc., that is by the resonance of the instrument to the second, third and higher partial components of the vibration, and some evidence of this also appears in the graph in Fig. 3. For instance, though the first resonance of the instrument was actually found to be at a pitch of 284, the peak of the curve for the bowing pressure is at about 270 as can be seen from Fig. 3. This appears to be a consequence of the fact that the course of the curve is to some extent modified by the resonance of the *octave*. It is obvious that corresponding to the resonance of the instrument to the fundamental tone in the range of pitch from 470 to 570, the octave should be strongly reinforced when the pitch lies within the range 235 to 285 ; hence the peak of the curve for the bowing pressure instead of being at 284 is actually shifted towards a lower frequency (270) as is seen from Fig. 3.

Obviously, the experiments described in this section may be extended in various directions. The curves for the three other strings of the violin, and especially over the whole of the possible range of pitch of the tone of the instrument, and the differences between the curves obtained with the different strings would deserve investigation. The differences between different violins could obviously be studied by this method, and the constants K_1 K_2, etc., for any particular violin and string and for various pitches may be found experimentally by study of the free and forced oscillations of the bridge, and used for a theoretical calculation and comparison with experiment of the bowing pressure required for exciting the tone of the instrument.

SECTION VI.—EFFECT OF MUTING ON BOWING PRESSURE.

Perhaps the best illustration of the close relation existing between the forces required to be exerted by the bow and the

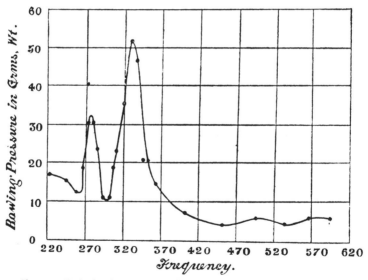

FIG. 4.—Relation between Bowing Pressure and Pitch (with Mute).

communication of vibrations from the string to the bridge and belly of the instrument and thence to the air, is furnished by the effect of putting a mute on the bridge on the bowing pressure required for eliciting a musical tone. Very striking results on this point may be obtained with the aid of the mechanical player. Fig. 4 illustrates the relation between bowing pitch and bowing pressure observed experimentally, the G-string of the violin being used as in Fig. 3, and the results shown in Fig. 3 and Fig. 4 being obtained under the same conditions except that in the latter case, a brass mute weighing 12·4 grammes was clamped to the bridge, while in the former case, the bridge was unmuted. The great difference between the two cases is obvious, and the change in the form of the graph for the bowing pressure shows a very close analogy with the change in the character of the forced vibration of the instrument and the intensity of the tone of the violin over the whole range of pitch produced by application of the mute.

In view of the discussion of the graph for the bowing pressure contained in the preceding section, and the theoretical treatment of the effect of muting already given by me in previous papers * it is perhaps not necessary to enter here into a detailed examination of the subject, and it may suffice briefly to draw attention to some of the features appearing in the graph in Fig. 4. It will be noticed that there is a peak in the curve at a frequency of about 280. This is the pitch of the first resonance of the instrument, the fundamental component of the vibration being re-inforced, and in this case, the position of the peak in the curve for the bowing pressure is not appreciably influenced by the resonance of the octave as in Fig. 3. It is clear that the pitch of the first resonance of the instrument is hardly influenced at all by the application of a load of 12·4 grammes to the bridge, and the bowing pressure required at the first peak of the curve is nearly the same as in Fig. 3. The first natural mode of vibration of the violin does not therefore appear to involve any very large vibration of the bridge. Following the peak at 280, we have in fig. 4 a very high peak near 330 which is the pitch of the 'wolf-note' as lowered by the mute of 12·4 grammes. The great lowering of pitch (from 490 to 330) shows that the second mode of vibration of the body of the violin involves a very large vibration of the

---

* Phil. Mag. October 1916, Nature October 1917, Phil. Mag. June 1918, and Bulletin No. 15, pages 143 to 151.

bridge, and the enormous increase of bowing pressure at the peak of the curve is also noteworthy. This can no doubt be explained on dynamical principles as due to the very greatly increased amplitude of the forced vibration of the bridge due to the loading. At higher pitches, the bowing pressure necessary falls off very rapidly, though one or two minor maxima (due to the resonance of the instrument in its higher modes as altered by the loading) are also obtained. The tone of the instrument in the higher ranges of pitch when muted is extremely feeble.

Further investigations which are worthy of being carried out would be the study of the effect of gradually increasing the mass of the mute on the graph for the bowing pressure, and also of putting the mute at different places on the bridge. In view of what has been stated above, it is clear that we may expect the changes in the form of the graph to follow closely the changes in the pitch of resonance of the instrument in its various modes produced by the loading.

SECTION VII.—OTHER APPLICATIONS OF THE MECHANICAL PLAYER.

The investigations described in the preceding sections may be extended in various directions. Some indications have already been given on these points, and it will suffice here to suggest some of the other possible applications of the mechanical player. As the instrument affords a means of bowing the violin at precisely measurable speeds and pressures, it furnishes a means by which the intensity of violin-tone and its variations with pitch may be quantitatively determined and compared with the indications of mathematical theory. Various questions, such as for instance the effect of heavier or lighter stringing, the effect of varying the pressure, speed, and the width of the region of contact of the bow, and the position of the bowed region on the tone-quality of the violin may be quantitatively studied with a degree of accuracy that cannot be approached in manual playing with its undetermined conditions. Further, the study of the tone-intensity, of the bowing pressure curves, and of the vibration-curves of bridge and belly of the instrument under quantitative conditions made possible by mechanical playing may be expected speedily to clear up various structural problems relating to the construction of the violin, e.g. the effect of the peculiar form of the Stradinarius bridge, the influence of its position, the function of the sound-post and base-bar, the shape of the air-holes, the thickness, curvature and shape of the elastic plates composing the violin, and the influence of various kinds of varnish. The dynamical specification of the constants determining the tone-quality of any violin over the whole range of pitch may be regarded as one of the aims towards which these investigations are directed.

Section VIII.—Synopsis.

The paper describes the construction of a mechanical violin-player intended for study of the acoustics of the instrument, and some of the investigations in which it has been applied. The principal feature in the player which is worthy of notice is that the conditions obtaining in ordinary musical practice are imitated with all the fidelity possible in mechanical playing, and the results obtained with it may therefore be confidently regarded as applicable under the ordinary conditions of manual playing. The following is a summary of the results obtained in the four investigations described in the present paper: (1) *Effect of the position of the bowed region on the bowing pressure*: it is shown that provided the speed of the bow is not too small, the bowing pressure necessary within the ordinary musical range of bowing varies inversely as the *square* of the distance of the bow from the bridge. (2) *Relation between bowing speed and bowing pressure*: it is shown that for very small bowing speeds, the bowing pressure necessary tends to a finite minimum value, and the increase of bowing pressure with speed is at first rather slow, but later becomes more rapid. (3) *Variation of bowing pressure with pitch*: the graph for the bowing pressure for different frequencies shows a series of maxima which approximately coincide in position with the frequencies of resonance of the instrument. (4) *Effect of muting on bowing pressure*: it is found that the mute produces profound alterations in the form of the graph. The bowing pressure necessary is increased in the lower parts of the scale and decreased in the higher parts of the scale. The peaks in the graph shift towards the lower frequencies in consequence of the alteration in the natural frequencies of resonance of the violin produced by the loading, and the change in the form of the graph is closely analogous to the change of the intensity of the fundamental tone of the instrument produced by the muting.

Some further possible applications of the mechanical player are also indicated in the paper.

Editor's Comments on Papers 10 and 11

10   Bladier: *Contribution à l'étude des cordes du violoncelle*

11   Bladier: *Sur les phénomènes transitoires des cordes vibrantes*

Author of numerous works in electrophysiology, the acoustics of music and musical instruments, architectural acoustics, and the study of complex sounds, M. Benjamin Bladier (1908–   ) holds the Diplôme d'Études Supérieures de Sciences Physiques, and is Ingénieur au Centre National de la Recherche Scientifique and Chef de Service du Laboratoire d'Acoustique des Salles du Centre de Recherches Physiques à Marseille. In 1955 he was Lauréat du Prix des Sciences de l'Association des Arts, Lettres et Sciences de Provence, and holds the order of Chevalier dans l'Ordre des Palmes Académiques.

His papers presented here (Papers 10, 11, and 16) represent three of his published works on the bowed string and on the vibrational characteristics of the bridge of the cello.

Selected Publications by Benjamin Bladier

"Nouvelle étude sur la vibration des cordes dans les instruments de musique," *J. Phys. Radium*, **15**, 65–67 (1954).

"Effect of Bow Speed and Pressure on the Vibration Speed of Catgut Strings," *Compt. Rend.*, **240**, 1868–1871 (May 1955).

"Sur l'état transitoire résultant de l'ébranlement d'une corde par archet," *Compt. Rend.*, **242**, 2704–2707 (1956).

"Sur la caisse sonore, l'âme et le chevalet du violoncelle," *Compt. Rend.*, **245**, 791–793 (1957).

"Contribution à l'étude du violoncelle," *Colloques Internationaux du Centre National de la Recherche Scientifique*, LXXXIV (May 1958), Acoustique Musicale, Marseille.

*"Sur le chevalet du violoncelle," *Compt. Rend.*, **250**, 2161–2163 (1960).

†"Contribution à l'étude des cordes du violoncelle," *Acustica*, **11**(6), 373–384 (1961).

‡"Sur les phénomènes transitoires des cordes vibrantes," *Acustica*, **14**(6), 331–336 (1964).

"Sur les réponses transitoires en acoustique des salles," *Acustica*, **18**(6), 309–322 (1967).

"Visualisation et critères de phase de sons complexes," *Compt. Rend.*, **276**, 401–404 (1973).

---

*Included in this volume as Paper 16.
†Included in this volume as Paper 10.
‡Included in this volume as Paper 11.

10

Reprinted from *Acustica*, **11**(6), 373–384 (1961)

CONTRIBUTION A L'ETUDE DES CORDES DU VIOLONCELLE

par B. Bladier

Centre de Recherches S.I.M. Marseille

Sommaire

Dans cette étude l'auteur précise d'abord les accessoires et le dispositif expérimental utilisés; il élimine d'emblée instrument et instrumentiste: le premier, car les caisses sonores des instruments sont différentes de l'une à l'autre; le deuxième, eu égard à la conduite de l'archet qui doit, dans les expériences, être constante en ce qui concerne les paramètres: vitesse, pression et distance archet-chevalet. Ces restrictions conduisent à réaliser un dispositif simulant le jeu de l'archet, et permettant de faire varier indépendamment chacun des trois paramètres ci-dessus.

Après avoir repris, dans quelques cas, des expériences d'Helmholtz, de Bouasse, et de Charron, sur la vibration des cordes, il étudie d'autres grandeurs, et fixe son choix sur la vitesse de vibration qui permet d'accroître la quantité d'information et de préciser certaines particularités. Il expose en détail l'influence de différents paramètres en régime stationnaire.

Zusammenfassung

In dieser Arbeit wird zunächst die verwendete Apparatur und das benutzte experimentelle Verfahren beschrieben. Es wird von vornherein der Einfluß des Instruments und des Spielers ausgeschaltet; der des Instruments, weil die Klangkörper der Instrumente von einem zum anderen variieren; der des Spielers im Hinblick auf die Bogenführung, die bei den Versuchen hinsichtlich der Parameter: Geschwindigkeit, Druck und Abstand Bogen-Steg, konstant sein muß. Diese Forderungen lassen den Bau einer Anordnung, die die Bogenführung nachahmt und es ermöglicht, jeden der drei Parameter unabhängig voneinander zu variieren, als sinnvoll erscheinen.

Zunächst werden einige Versuche von Helmholtz, Bouasse und Charron über die Schwingungen der Saiten wiederholt, dann werden andere Größen untersucht, vor allem die Schnelle der Schwingungen, die es ermöglicht, neue Informationen zu gewinnen und einige Einzelheiten zu bestimmen. Der Einfluß verschiedener Parameter im stationären Bereich wird ausführlich dargelegt.

Summary

In this study the author describes at first the apparatus and the experimental system utilised. He eliminates initially the instrument and the player, the first because the sound boxes of instruments are different from one to another, the second in regard to the guidance of the bow which should, in the experiments, be constant in so far as it concerns the parameters, velocity, pressure and the distance between bow and bridge. These restrictions lead to the realisation of a system which simulates the playing of the bow and permits the independent variation of each of the three parameters given above.

After having repeated, in several cases, the experiments of Helmholtz, of Bouasse and of Charron, on the velocity of vibration to provide more information and to specify certain features. He shows in detail the influence of the various parameters on the stationary state.

1. Introduction

Si l'on demande à un instrumentiste d'essayer une dizaine de cordes (jugées satisfaisantes du point de vue physique et semblables) sur son instrument, son choix, après un travail délicat, long et fastidieux, se fixe en général sur deux ou trois cordes qu'il trouve les mieux adaptées à son instrument. L'objet de ce travail a été tout d'abord de rechercher une méthode pour connaître dans quelle mesure on pourrait différencier, par quelques critères physiques, ces cordes apparemment semblables. Après avoir essayé d'exploiter le son de la corde sans caisse, nous avons utilisé sa vitesse de vibration. Chemin faisant nous avons étudié les paramètres qui avaient une influence marquée sur la vitesse de vibration des cordes, en régime stationnaire.

Pour faire cette étude nous avions comme antécédents les travaux d'Helmholtz [1], de Bouasse [2] et de Charron [3].

Notre souci majeur, dans ce travail, a été d'accroître l'information recueillie, de la mieux conserver, de choisir des appareils d'analyse l'altérant le moins possible, d'effectuer leur vérification, de connaître l'influence de petites erreurs commises pendant les expériences pour, ensuite, nous attacher à ne pas dépasser les bornes au-delà desquelles les influences de ces erreurs sont marquées.

2. Accessoires et dispositif expérimental

2.1. Cordes

Nous avons étudié des cordes variées, boyau, nylon, métal, mais nous avons plus particulièrement porté notre attention sur les cordes filées sur boyau. L'âme de celles-ci, constituée de lanières de boyau de mouton, présente la forme d'un toron, recouvert d'un sous filage textile, par dessus lequel on dispose le filage métallique ou trait, dont le pas est inverse de celui du boyau. Certaines cordes sont filées à double trait, parfois avec des métaux de densités différentes, dans le but de modifier le timbre propre de la corde.

Pour une longueur d'environ 700 mm et selon les matériaux utilisés, la tension nécessaire pour amener une corde ut_1, filée sur boyau, à son accord (66 Hz) varie entre 10,5 et 12,5 kg. Pour la corde sol_1 (99 Hz) elle est généralement comprise entre 10 et 11 kg.

Avec un jeu de cordes métalliques, pour réaliser les accords: $ut_1 = 66$ Hz; $sol_1 = 99$ Hz; $ré_2 = 148,5$ Hz; $la_2 = 220$ Hz; nous avons noté respectivement les tensions suivantes: 13,95; 15; 14,55 et 14,5 kg.

La corde est tendue à l'aide d'un petit treuil de contre-basse. L'accord, vérifié par un capteur couplé à un analyseur électronique, est ajusté à mieux qu'un Hertz près. La longueur de la partie vibrante utile de la corde (entre chevalet et sillet) utilisée dans cette étude est de 706 mm.

2.2. Archet

L'état de surface de l'archet a une influence sensible sur les résultats des mesures. C'est donc l'archet classique en crin qu'il faudrait utiliser, mais celui-ci est incommode pour une longue durée des expériences. Après plusieurs essais nous avons adopté un archet circulaire: roue de vielle en ébonite. Cet archet doit permettre d'obtenir une vibration bien stable de la corde; si le fuseau qu'elle forme en vibrant «respire» trop, les résultats des mesures sont erratiques. Le phénomène n'est pas reproductible. Il faut donc une roue de vielle bien rectifiée et qui de plus porte également sur la corde sur toute sa largeur (10 mm). Sa largeur ou jante est en permanence légèrement colophanée. Un berceau mobile permet de modifier: a) sa distance par rapport au chevalet, b) la force exercée ou pression de l'archet sur la corde. La mesure de cette pression s'effectue à l'aide d'un pont à jauges. La vitesse de l'archet est mesurée et maintenue constante avec un stroboscope. Le moteur d'entrainement de l'archet possède un variateur de vitesse; ses pertubations propres, vibrations mécaniques et dans certains cas bruits acoustiques, sont minimisées par l'utilisation de matériaux adéquats.

2.3. Colophanage de l'archet

L'influence du colophanage, pour des vitesses d'archet très lentes, a fait l'objet d'une publication [4].

Pour une vitesse de 150 mm/s et une pression de l'archet sur la corde ($ut_1 = 66$ Hz) de 1600 g, la répartition spectrale du phénomène vitesse de vibration (voir plus loin), obtenue pour un point pris au milieu de la corde, montre qu'en l'absence du colophanage il n'y a pas disparition d'harmonique mais seulement, par rapport à un colophanage normal, une diminution générale des amplitudes des différentes composantes. Cette diminution, pour les harmoniques impairs de rang 1 à 11, évolue entre 12 et 35%. Il importe de noter qu'il s'agit ici d'un ordre de grandeur, car suivant le type de corde, pour une pression faible, en l'absence de colophanage de l'archet, le fuseau de la vibration de la corde ne se forme pas ou est erratique; les mesures sont alors impossibles. Enfin, un colophanage trop important apporte lui aussi des troubles dans la vibration de la corde. Nous avons utilisé un léger colophanage permanent.

2.4. Chevalet

Pour conserver plus pur le mouvement de la corde nous avons adopté un chevalet en plomb massif de 10 kg environ. Pour quelques expériences seulement, des chevalets du type classique en bois, pesant environ 13 g, ont été utilisés. L'angle dont se brise la corde après le chevalet a été fixé à environ 25°. Le sillet, en plomb massif, est semblable au chevalet.

2.5. Massif de montage

L'étude de la vibration d'une corde montée sur un banc d'essai métallique d'une seule pièce, sur lequel se trouvaient disposés: cordier, chevalet, sillet et chevillier ne nous a pas donné des résultats satisfaisants. Eu égard aux différentes interactions cordier, chevalet, sillet, nous avons dû scinder en trois parties non solidaires le massif de montage. Celui-ci est réalisé à l'aide de trois blocs de béton reposant chacun sur une semelle disposée à même le sol, les trois semelles étant séparées entre elles par du granité de liège et de la soie de verre. De cette façon, cordier, chevalet, sillet, reposant chacun sur un bloc de béton, sont nettement séparés.

Fig. 1. Vue du massif de montage montrant: la séparation des blocs de béton; le berceau et sa roue de vielle; la corde disposée sur le chevalet en plomb reposant sur le bloc central; le bloc de gauche supportant le dispositif jouant le rôle de cordier, celui de droite soutenant le sillet (également en plomb) et le tendeur (petit treuil permettant d'accorder la corde).

La Fig. 1 donne une vue du massif de montage montrant: la séparation des blocs de béton; le berceau et sa roue de vielle; la corde disposée sur le chevalet en plomb reposant sur le bloc central; le bloc de gauche supportant le dispositif jouant le rôle de cordier, celui de droite soutenant le sillet (également en plomb) et le tendeur: petit treuil permettant d'accorder la corde.

3. Son aérien

Nous avons cherché tout d'abord à étudier le son aérien produit par la corde, excitée par l'archet, lorsque celle-ci repose seulement sur les deux chevalets en plomb. L'expérience montre que ce son est perceptible par l'oreille placée à quelques centimètres de la corde en vibration. L'oreille remplacée par un microphone suivi d'un amplificateur, conduit à des résultats aberrants; le son étant mêlé aux bruits de fond de l'appareillage, aux pertubations sonores du moteur et au crissement de l'archet sur la corde.

Cet échec nous a conduit à l'étude de la vibration puis de la vitesse de vibration de la corde; le mode de vibration de celle-ci étant lié au son émis, c'est-à-dire à son fondamental et à ses harmoniques, donc au timbre.

4. Vibration en régime stationnaire

4.1. Travaux d'Helmholtz

HELMHOLTZ [1] a observé la forme de vibration d'un point d'une corde de violon, dont le déplacement était perpendiculaire à celui d'un microscope animé d'un mouvement sinusoïdal. La composition des deux mouvements lui a permis de donner la forme de la vibration: deux segments rectilignes réunis par un point anguleux, et de préciser que chaque point de la corde va et vient avec une vitesse constante entre les positions limites de sa vibration. Sauf pour un point pris au milieu de la corde, la vitesse de la phase ascendante est différente de la phase descendante. La corde se déplace à l'intérieur d'un fuseau délimité par deux arcs de parabole.

4.2. Vérifications expérimentales

a) La corde étant excitée en régime stationnaire par la roue de vielle disposée à 25 mm du chevalet, nous avons tout d'abord cherché à obtenir la forme de la corde sur toute sa longueur vibrante ($l = 706$ mm).

La Fig. 2 donne le résultat de deux instantanés du déplacement, à l'aller et au retour de la corde au moment où les points anguleux A et C, peu nets sur notre cliché, sont sensiblement en face. Apparemment la corde se déplace à l'intérieur d'un fuseau délimité par deux arcs de parabole et les parties 0A, 0C et AB, BC peuvent être assimilées à des segments de droites. En première approximation ces résultats confirment les travaux de HELMHOLTZ.

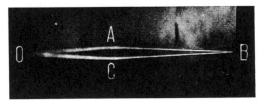

Fig. 2. Forme de la corde sur toute sa longueur vibrante; elle se déplace à l'intérieur d'un fuseau délimité par deux arcs de parabole; les parties 0A, 0C et AB, BC peuvent être assimilées, en première approximation, à des segments de droites.

b) Ensuite nous avons enregistré, en différents points de sa longueur ($l = 706$ mm), l'élongation en fonction du temps du mouvement de la corde. La Fig. 3, donne les graphiques du déplacement d'un point de la corde situé, par rapport au chevalet, Fig. 3 a, à environ 150 mm, lorsque l'archet est tiré; puis, Fig. 3 b, de nouveau à 150 mm, pour archet poussé; enfin Fig. 3 c, au milieu de la corde, pour archet poussé. L'inversion du mouvement de l'archet

ACUSTICA
Vol. 11 (1961)

fait subir aux phénomènes un retournement de 180°. Ces résultats confirment les travaux antérieurs cités précédemment.

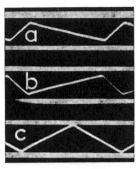

Fig. 3. Graphiques du déplacement d'un point de la corde situé par rapport au chevalet à environ 150 mm: en «a» archet titré, en «b» poussé. En «c» archet poussé pour un point pris milieu de la corde. L'inversion du sens de l'archet fait subir aux phénomènes un retournement de 180°.

Toutefois cette méthode, pour le but à atteindre, ne nous a pas permis de différencier distinctement les cordes, ni non plus d'obtenir nettement sur les diagrammes les petites dentelures signalées par HELMHOLTZ. Nous avons pu y parvenir, dans une certaine mesure, en considérant, la vitesse de vibration de la corde, dont l'enregistrement approximativement carré ou rectangulaire nous paraissait contenir le maximum d'information [5].

4.3. Montages dérivateurs

Pour obtenir la vitesse de déplacement de la corde d'une part, pour ne pas perturber le mode de vibration d'autre part et enfin pour contrôler nos résultats d'expériences, nous avons utilisé deux méthodes différentes: l'une utilisant les variations de capacité, l'autre les courants induits dans la corde se déplaçant dans un champ magnétique. Les diagrammes de la vitesse, obtenus avec ces deux méthodes, sont semblables, Fig. 4, aussi bien pendant le régime de vibration stationnaire, que pendant l'état transitoire de la corde.

4.3.1. Dérivateur à capacité

Pour que la corde, dont le filage ou trait métallique constitue l'armature mobile d'un condensateur, provoque des variations de capacité approximativement proportionnelles à ses propres déplacements, nous avons donné aux deux armatures fixes la forme de deux petits triangles isocèles. La corde se meut à l'intérieur de ces deux surfaces triangulaires de 5 cm² chacune, disposées l'une au dessus de l'autre. On localise ainsi les déplacements d'un petit élément de longueur de corde qui peut être choisi en un point quelconque de la partie vibrante.

L'épaisseur du diélectrique air séparant ces deux armatures[1], réunies électriquement, est de 5 mm. La capacité de l'ensemble varie, suivant la position de la corde, autour de 1 pF. Cette faible capacité C, du fait de son montage en série, avec la résistance R_g du circuit d'entrée, Fig. 5, fournit une tension

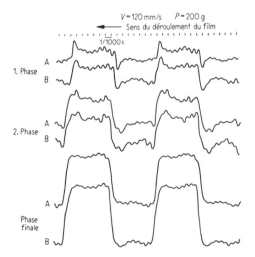

Fig. 4. Graphiques de la vitesse de vibration, obtenue à $l/2$, pendant l'état transitoire, puis stationnaire (phase finale), avec deux méthodes: A) électromagnétique; B) électrostatique, qui mettent en évidence une forme d'onde sensiblement la même. Humidité = 65%, $V = 120$ mm/s, $P = 200$ g.

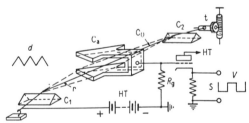

Fig. 5. Dérivateur à capacité schématisé, sur lequel on a volontairement exagéré les dimensions du capteur C_a;

C_o	=	corde;
C_1 et C_2	=	chevalets en plomb;
d	=	forme d'onde du déplacement de la corde;
r	=	emplacement de la roue de vielle;
R_g et S	=	respectivement résistance de fuite, et sortie de la triode de couplage;
t	=	treuil utilisé pour l'accord de la corde;
V	=	forme d'onde de la vitesse de vibration obtenue en S.

[1] Ces deux armatures sont nécessaires pour compenser le «balancement» de la corde qui, nous le verrons plus loin, ne vibre pas dans un seul plan.

voisine de la dérivée, par rapport au temps, de la tension engendrée par les déplacements d'un point de la corde, soit la vitesse de vibration.

4.3.2. Dérivateur par courants induits (électromagnétique)

La corde condutrice (son filage est le plus souvent en argent ou en cuivre) se déplaçant dans un champ magnétique de façon à couper des lignes de force, est le siège de forces électromotrices induites. Celles-ci, recueillies sur la corde en deux points situés en dehors de la partie vibrante, sont données en volt par la formule $e = - \dfrac{d\varphi}{dt} \cdot 10^{-8}$; la dérivée par rapport au temps du flux total d'induction, pour un point de la corde à l'instant considéré caractérise sa vitesse de vibration.

Pour que l'induction se localise sur un petit élément de longueur de corde, choisi en un point quelconque de la partie vibrante, nous avons réalisé les pièces polaires de l'aimant permanent en forme de prisme triangulaire; les deux arêtes, nord-sud, sont situées dans un même plan perpendiculaire au plan de déplacement de la corde.

Le montage est donné Fig. 6. Il comprend deux aimants permanents montés en parallèle, reliés entre eux par deux armatures de fer doux, au milieu desquelles sont disposées les pièces polaires nord-sud, et dont la distance (entrefer où se meut la corde) est réglable.

Nous avons vérifié, à l'aide d'un stroboscope, que l'introduction de l'aimant, pendant la vibration de la corde, n'amenait pas de pertubation sensible sur son déplacement.

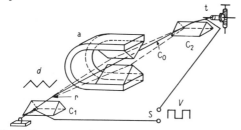

Fig. 6. Dérivateur électromagnétique schématisé, sur lequel on a exagéré les dimensions de l'aimant "a" qui, pour la clarté du dessin, ne comporte pas deux aimants en parallèle que nous avons utilisé en réalité.

Co = corde;
C_1 et C_2 = chevalets en plomb;
d = forme d'onde du déplacement de la corde;
r = emplacement de l'archet circulaire;
S = sortie reliée à l'amplificateur-analyseur;
t = petit treuil utilisé pour l'accord;
V = forme d'onde du signal vitesse de vibration recueilli en S.

5. Résultats obtenus

5.1. Vitesse de vibration à·l/2

La Fig. 7, diagramme A – B – C, donne l'élongation d'un point pris au milieu de la partie vibrante de la corde. Sur la même figure le diagramme α, α', β, β', γ, γ', correspond à la vitesse de vibration, au même point milieu de la corde. Ce relevé permet, en augmentant la précision de l'analyse et en mettant en valeur les harmoniques supérieurs, de faire les remarques suivantes:

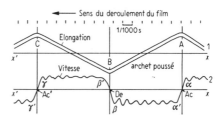

Fig. 7. Elongation et vitesse de vibration d'un point pris au milieu de la partie vibrante d'une corde.

a) La répartition des vitesses n'est pas uniforme; les dentelures, ici très nettement mises en évidence, montrent que AB et BC ne sont pas rectilignes; de plus la partie α'β diffère de β'γ, donc AB n'est pas identique à BC, comme l'avait suggéré le diagramme C de la Fig. 3.

b) La concavité tournée vers le bas, en A correspondant à αα', n'est pas semblable à la concavité tournée vers le haut en B, correspondant à ββ'.

c) La corde, quel que soit le sens du mouvement de l'archet, se comporte d'une façon différente selon qu'elle est accrochée par l'archet (Ac, α', β, De) ou qu'elle se décroche (De, β', γ, Ac'). Si l'on inverse le sens de mouvement de l'archet, les phénomènes, vibration et vitesse en un point donné, subissent un retournement de 180° autour de leur axe xx'; on note cependant de légères modifications sur le diagramme de la vitesse, mais toutes les remarques précédentes relatives à archet poussé, restent valables pour archet tiré.

5.2. Analyse harmonique

La forme d'onde abrupte (approximativement carrée ou rectangulaire) du signal «vitesse de vibration» à analyser, nous a procuré quelques inquiétudes quant au choix, et aux performances que devait présenter l'appareil électronique, analyseur de fréquence.

Nous avons tout d'abord effectué nous-même, par le calcul, plusieurs analyses harmoniques (partiels de rang 1 à 24). Pour contrôler nos propres résultats d'une part, nous avons fait exécuter ces calculs

ACUSTICA
Vol. 11 (1961)

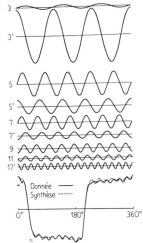

Fig. 8. Analyse harmonique d'un oscillogramme du type Fig. 7. Expression mathématique:
$y = -216,6 \sin x + 145,7 \cos x + 6 \sin 3 x + 83,7 \cos 3 x + 36,2 \sin 5 x + 21 \cos 5 x + 20,4 \sin 7 x - 10,56 \cos 7 x - 12,55 \cos 9 x - 7,45 \sin 11 x + 9,6 \cos 17 x.$

Le fondamental n'a pas été représenté. Les nombres indiquent le rang harmonique (les termes en sinus sans accent, ceux en cosinus accentués). La synthèse des premiers harmoniques impairs ($n \leq 17$) diffère peu de l'oscillogramme original.

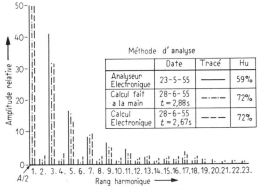

Fig. 9. Comparaison des analyses. Deux expériences ont été faites dans les mêmes conditions à un mois d'intervalle. Archet: vitesse = 120 mm/s, pression = 200 g. Corde P 4-AC = 66 Hz.

Le spectre de la première est obtenu très rapidement au moyen d'un appareil à filtres; ceux de la deuxième relatifs à $t = 2,88$ s et à $t = 2,67$ s, ont été calculés (l'un à la main, l'autre à la machine). La différence entre les deux derniers, donne une idée de la «respiration» du phénomène; la comparaison entre ceux-ci et le premier permet d'accorder sa confiance au filtre analyseur au sujet duquel nous avions quelque inquiétude.

Les amplitudes des partiels de rang 1 sont réduites de moitié. Cette réduction est notée par $A/2$, comme elle le sera sur d'autres figures.

par machines électroniques et d'autre part nous avons comparé donnée et synthèse de nos résultats calculés; la Fig. 8 illustre ceux-ci et montre que la recomposition de la forme d'onde est satisfaisante. Une autre vérification a consisté à intégrer la forme d'onde «vitesse de vibration»; la courbe obtenue coïncide bien avec la forme d'onde du déplacement. Enfin la Fig. 9 donne pour une même corde, dans les mêmes conditions d'expérience, les répartitions spectrales obtenues par le calcul et par l'analyseur de fréquence électronique. La concordance des trois spectres n'est pas parfaite mais seulement satisfaisante. On y remarque un maximum ou pointe aux environs de l'harmonique 17, étudié plus loin. On s'assure de la reproductibilité des phénomènes à quelques jours d'intervalle après avoir démonté puis remonté la corde et la roue de vielle. Les fluctuations maxima observées (affectant l'amplitude des partiels et non leur rang) sont notables, ±14% sur deux partiels, mais insuffisantes pour jeter un doute sur l'existence de la pointe signalée (Fig. 10).

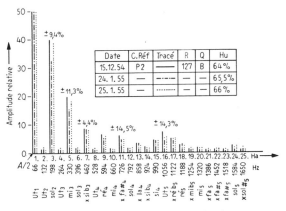

Fig. 10. Analyse harmonique de la vitesse de vibration. Archet: $v = 120$ mm/s, $P = 1600$ g, Ac = 66 Hz. Les fluctuations observées dans la reproductibilité des phénomènes, après avoir démonté puis remonté la corde et la roue de vielle sont notables, mais pas au point de jeter un doute sur l'existence d'un maximum ou pointe (harmoniques 15, 16 et 17). En ordonnée, amplitude relative; en abscisse, rang harmonique Ha; la fréquence en Hz et les notes correspondantes. Devant le nom de quelques notes la croix indique que la fréquence est seulement voisine de la note considérée.

5.3. Vitesse de vibration à l/n

Les analyses effectuées en des points de la corde pris à $l/2$, $l/3$, $l/7$, respectivement distants du chevalet de 353, 235, et 101 mm, donnent pour un même accord, $ut_1 = 66$ Hz, des spectres différents. Les harmoniques pairs, jusqu'à l'harmonique 10, sont pratiquement éliminés pour un point pris à $l/2$; pour

$l/3$ c'est l'harmonique trois et ses multiples qui sont altérés: partiels, de rang 3, 6, 9, etc. ... Cependant, si pour un point pris à $l/2$ on compte dans la période du phénomène un nombre pair de dentelures, le spectre correspondant met en évidence sans ambiguïté un maximum ou pointe, comprenant deux ou trois partiels successifs, dont un au moins est pair. Le milieu de la corde n'est donc pas exactement un point nodal pour tous les partiels harmoniques pairs. Suivant le type de corde utilisée le même cas peut se présenter à $l/3$ pour un partiel impair; à $l/4$ pour un nouveau partiel pair etc.; toutefois, pratiquement limité à l'harmonique 18 avec nos moyen actuels d'investigation.

Pour un point pris à l/n de la longueur de la corde un, au moins, des partiels multiples de n, n'est pas exactement un point nodal.

5.4. Dentelures

Les diagrammes de la vitesse Fig. 11, obtenus à $l/2$ sont relatifs à des cordes filées avec un trait d'aluminium, d'argent et de tungstène; elles pèsent respectivement 5, 11, et 22 g; elles ont des diamètres voisins, mais des raideurs différentes. On remarque que l'amplitude des dentelures paraît d'autant plus grande que la densité du métal utilisé est plus petite. Inversement, leur nombre paraît, d'autant plus grand que la densité du métal est plus grande.

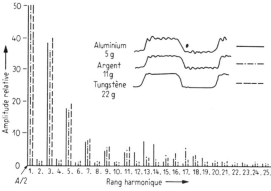

Fig. 11. Analyse harmonique de la vitesse de vibration. Archet: $v = 120$ mm/s, $P = 1600$ g, Ac = 66 Hz. Diagrammes et spectres relatifs de trois cordes filées d'un trait différent. La différenciation est nette. Les masses linéiques sout différentes.

Sur cette même figure, les spectres, relatifs aux trois diagrammes sont différents entre eux, plus particulièrement par leur maximum ou pointe correspondant aux dentelures.

On remarque: 1) dans les trois cas l'importante diminution des partiels pairs jusqu'à l'harmonique 10; 2) l'importance de l'amplitude des partiels de rang 12, 13 et 14 pour la corde aluminium 5 g,

ainsi que celle des partiels de rang 16, 17 et 18 de la corde argent 11 g. A notre échelle d'observation, la faible amplitude des dentelures de la corde tungstène 22 g échappe à l'analyse. La différenciation des trois cordes est cependant nette.

5.5. Modification des dentelures

Compte tenu de l'influence de la pression de l'archet, étudiée plus loin, et de la distance archet-chevalet qui modifie le nombre et la forme des dentelures, nous avons recherché si ces dentelures étaient modifiées sensiblement par:

a) Le sens du mouvement de l'archet,
b) l'état de surface de l'archet,
c) la raideur des cordes,
d) l'humidité relative,
e) la masse linéique [2] de la corde suivant le trait utilisé.

a) Sens du mouvement de l'archet

Les diagrammes obtenus avec un mouvement de la roue de vielle, correspondant à archet tiré puis poussé, ne montrent pas, semble-t-il, de modifications sensibles de l'amplitude et du nombre de dentelures. Cependant, le spectre de l'analyse du phénomène global met en évidence 5 à 10% de fluctuations sur l'amplitude de quelques partiels. L'influence du sens du mouvement de l'archet est donc négligeable en première approximation pour le but qui nous occupe.

b) Etat de surface de l'archet

Le remplacement de l'archet en ébonite par un archet en crin conduit à la main appelle les mêmes remarques. Toutefois, dans ce cas l'analyse n'a pu être effectuée eu égard à la difficulté de maintenir constantes: pression, vitesse et distance archet-che-valet. A en juger par l'aspect des diagrammes les modifications ne peuvent être, ici aussi, que faibles.

c) Raideur des cordes

Nous appelons «raideur» d'une corde accordée le résultat des mesures suivantes effectuées sous la corde à l'emplacement de l'archet: Une jauge de contrainte est placée au contact de la corde sur laquelle on n'exerce aucune pression; on règle le zéro du pont, relié à la jauge, ensuite on note la valeur du courant résultant du déséquilibre du pont, pour des pressions de plus en plus importantes exercées sur la corde à l'emplacement de l'archet. La Fig. 12 donne le résultat de ces mesures pour une corde raide, trois cordes normales et une souple. On

---

[2] Par "masse linéique" nous entendons la masse rapportée à l'unité de longueur.

mesure donc l'abaissement de la corde provoqué par une même force; le nombre résultant R est d'autant plus grand que la corde est plus souple.

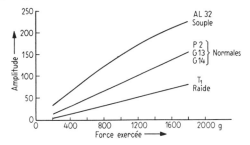

Fig. 12. Résultats des mesures de la «raideur» sur des cordes ut_1 diverses, accordées sur 66 Hz.

Nous n'avons pu préciser l'importance et l'influence de ce paramètre sur les dentelures et les résultats de l'analyse; les cordes classiques semblables ayant des raideurs du même ordre.

d) Humidité relative

Pour un type de corde donné, les expériences effectuées entre 55% et 77% d'humidité relative, ne montrent pas, sur le spectre, de fluctuations d'amplitude nettement supérieures à celles observées après avoir démonté puis remonté la corde et la roue de vielle. Sur d'autres types de cordes on remarque qu'une humidité importante affecte un peu les dentelures, partiels de la pointe qui d'une manière générale, deviennent un peu plus amples. Si la corde est bien séchée près d'un radiateur, le trait n'adhère plus sur l'âme, la corde «frise» elle est devenue défectueuse.

Nous nous sommes efforcés d'effectuer mesures et analyses dans les limites de $66 \pm 8\%$ d'humidité relative, zone où l'influence parait négligeable en première approximation.

e) Masse linéique

Pour connaître si l'influence de la masse linéique mise en jeu était plus importante que le type de métal utilisé comme trait, nous avons fait réaliser des cordes ayant un trait, en aluminium, en cuivre, en tungstène et présentant approximativement le même poids (respectivement 11; 11,4 et 11,02 g) et des raideurs voisines, mais entre celles-ci les diamètres sont très différents (respectivement 2,72; 1,84 et 1,53 mm). Les diagrammes obtenus et les spectres de ceux-ci, Fig. 13, ne sont pas identiques mais très voisins entre eux, aussi bien pour les analyses effectuées à $l/2$ qu'à $l/7$ de la longueur de la corde vibrante.

Pour vérifier ces résultats, nous avons utilisé des cordes filées d'un trait de tungstène et d'un trait de

cuivre, présentant des raideurs différentes, (respectivement $R = 53$ et 74), mais d'un poids égal (22 g, double des précédentes) et des diamètres différents (respectivement 1,9 et 2,33 mm). Les résultats concordent bien: les spectres des cordes sont semblables entre eux, aussi bien à $l/2$ qu'à $l/7$. Sur les deux diagrammes — comme sur la Fig. 11, corde 22 g — les dentelures peu marquées en amplitude paraissent plus resserrées.

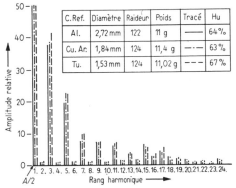

Fig. 13. Influence de la masse linéique. Distance archet chevalet $= 25$ mm. Archet: $v = 120$ mm/s, $P = 1200$ g. Les spectres de trois cordes filées d'un trait différent ayant des masses linéiques semblables, sont très voisins; la différenciation, faite sur la Fig. 11, n'est plus possible.

f) Conclusion

Dans les limites de nos expériences et en première approximation, on peut conclure que les caractéristiques des dentelures (amplitude, nombre et forme) dépendent surtout de la masse linéique de la corde mise en jeu. Les autres paramètres, tels que le type de métal ou alliage utilisé comme trait, le diamètre de la corde, la raideur, l'état de surface et le sens du mouvement de l'archet n'ont qu'une influence secondaire dans nos conditions d'expériences: corde montée sur des chevalets en plomb, reposant sur des blocs de béton.

5.6. Influence de l'accord

On compare les spectres obtenus avec deux cordes ayant des traits différents (l'un en aluminium, l'autre en cuivre) d'une longueur égale, $l = 706$ mm et d'un poids égal (5 g) dans les mêmes conditions d'expérience, sauf toutefois l'accord. La corde aluminium ut_1, est accordée sur 66 Hz; la corde cuivre sol_1, est accordée sur 99 Hz. Dans ces conditions les deux spectres sont différents, non seulement par leur maximum ou pointe, mais encore par l'amplitude relative de nombreux harmoniques. Sans modifier la longueur de la corde aluminium, si l'on fait passer son accord de 66 Hz à 99 Hz, le nouveau spectre obtenu est alors semblable à celui de la corde cuivre accordée sur 99 Hz. Le maximum ou pointe, situé aux environs de l'harmonique 14 pour un accord de 66 Hz, passe à l'harmonique 15 pour un accord de 99 Hz; il coïncide avec celui de la corde cuivre; d'une manière générale les amplitudes relatives, à chacun des harmoniques des deux cordes, sont voisines.

Une vérification de l'influence de l'accord sur une même corde plus lourde (11 g) conduit à des résultats semblables. Le maximum ou pointe situé sur l'harmonique 15 pour un accord de 50 Hz passe à l'harmonique 17 pour un accord à 66 Hz.

Des variations importantes de la tension de la corde modifient le spectre de la vitesse de vibration; on note des modifications de l'amplitude relative de nombreuses composantes ainsi qu'un déplacement de la pointe dans le rang harmonique.

5.7. Influence de la longueur

1) Si l'on modifie la longueur de la corde à tension constante on agit simultanément sur: l'accord, la masse vibrante mise en jeu, la répartition des nœuds et des ventres sous l'archet pour un même emplacement. Du moment qu'on ne peut faire agir une de ces variables à la fois ces mesures perdent un peu de leur intérêt. Sur l'instrument, lorsque l'on passe d'une corde à vide à la même corde doigtée, ces mêmes paramètres, augmentés de la réponse de la caisse, varient bien simultanément.

2) Si l'on maintient l'accord constant, pour une longueur initiale de la corde de 706 mm, les répartitions spectrales montrent:

a) une variation de ± 2 mm de longueur à une influence négligeable,

b) une variation de ± 28 mm de longueur à une influence sur les amplitudes et sur le rang harmonique plus particulièrement marquée sur le maximum ou pointe. Cette dernière située sur les harmoniques de rang 15, 16, 17 pour $l = 678$ mm, se déplace d'un rang soit 16, 17, 18 pour $l = 734$ mm. Ces résultats sont obtenus dans les deux cas pour une distance chevalet-archet de 25 mm. D'autre part ce sont seulement les harmoniques qui présentent un nœud sous la largeur de l'archet (de 25 à 35 mm) qui ne sont pas exaltés. Soit pour $l = 678$ mm de l'harmonique 20 à 27 et pour $l = 734$ mm de l'harmonique 21 à 29. Or, dans ces deux cas, les harmoniques minimisés par l'archet se trouvent en dehors de la pointe de 15 à 18. La nature des nœuds pris sous l'archet peut donc être négligé. C'est donc seulement la longueur qui est responsable des modifications signalées.

5.8. Conclusions

En ce qui concerne le maximum ou pointe, on peut conclure en première approximation et dans les limites des nos expériences, que cette pointe se déplace vers les harmoniques plus élevés si:

a) La masse linéique, pour une même longueur et un même accord, est plus importante.

b) A longueur égale, la tension (accord) est plus élevée.

c) A accord constant, la longueur et simultanément la masse vibrante [3] augmentent.

5.9. Question non résolue

Recherchons une longueur de corde, pour une attaque par l'archet à 25 mm du chevalet, dont les harmoniques de rang 14 à 18 présenteraient un nœud sous l'archet. Soit une corde $ut_1 = 66$ Hz de $l = 706$ mm poids 12 g, réduite de $1/3 = 470,6$ mm, poids 8 g, son accord passe à $sol_1 = 99$ Hz; les harmoniques présentant un nœud sous l'archet sont alors au nombre de 5, (de l'harmonique 14 à 18) ceux-ci sont bien situés dans la plage du maximum ou pointe signalée précédemment. L'analyse montre qu'effectivement les harmoniques de rang 14 à 18 sont minimisés [4]. Il en est bien de même pour une autre corde de même longueur, $sol_1 = 99$ Hz, poids 5 g, qui réduite de $1/3$ devient $ré_2 = 148,5$ Hz, poids 3,34 g. Toutefois, de ces résultats on ne peut conclure que seul le paramètre distance archet-chevalet [5] est responsable de l'altération des harmoniques de rang 14 à 18, car la réduction de la masse et de la longueur ont, nous l'avons vu, un effet dans le même sens. La question reste posée.

5.10. Déductions

Des expériences réalisées avec une distance archet-chevalet constante on peut, d'une manière générale, déduire que le timbre s'appauvrit et l'énergie s'amenuise lorsque l'on réduit notablement la longueur de la corde.

5.11. Influence de la vitesse de l'archet

L'accroissement de la vitesse de l'archet (entre 60 et 150 mm/s) à pression constante (Fig. 14), rend plus abrupte la montée et la descente du diagramme, et augmente l'amplitude moyenne du phénomène

[3] Par masse vibrante nous entendons la masse totale del a corde en mouvement.

[4] Le maximum semble se situer alors sur les harmoniques 11 et 13.

[5] En ce qui concerne l'influence de la variation de la distance archet-chevalet sur une corde de longueur constante, nous jugeons nos résultats encore trop sommaires pour en faire état.

global, sans modifier sensiblement le nombre, la forme et l'amplitude des dentelures [6].

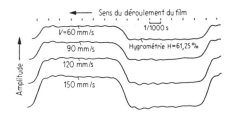

Fig. 14. Augmentation de l'amplitude moyenne du phénomène global due à l'accroissement de la vitesse de l'archet. On remarque que les caractéristiques des dentelures ne se trouvent pas sensiblement modifiées. Corde P_1, $P = 200$ g.

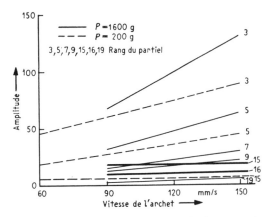

Fig. 15. Influence de l'accroissement de la vitesse de l'archet sur l'amplitude des partiels (pour des pressions P de 1600 g et 200 g). Cette influence est d'autant plus faible que le rang du partiel est plus élevé, et elle s'annule pour ceux de la pointe. C'est le fondamental $n = 1$, non figuré ici, qui possède la plus grande pente. Les courbes $n = 15$ et $n = 16$ (partiel de la pointe) sont en trait gras.

La corde filée sur boyau (ut$_1$ accordé sur 66 Hz, archet à 25 mm du chevalet en plomb, pression 200 g, longueur vibrante de corde $l = 706$ mm) est étudiée à $l/2$ (emplacement du dérivateur) de façon à obtenir seulement les partiels harmoniques de rang impair. Dans ces conditions d'expériences l'amplitude globale est sensiblement une fonction linéaire de la vitesse; il en est de même de l'amplitude des partiels (Fig. 15). L'influence de la vitesse est d'ailleurs d'autant plus faible que le rang du partiel est plus élevé, et elle s'annule pour ceux de la pointe. L'accroissement de la vitesse se traduit surtout par une diminution de l'amplitude relative des partiels de la pointe (les 14ème, 15ème et 16ème pour la corde étudiée, Fig. 16), donc appauvrit le timbre d'une façon particulière.

5.12. Influence de la pression de l'archet

Dans les conditions d'expériences citées, mais à vitesse d'archet constante, l'augmentation de la force exercée par l'archet sur la corde (suivant l'usage appelé pression), fait apparaître les dentelures (Fig. 17), leur amplitude croît, leur forme se modifie légèrement, mais leur nombre semble rester constant [6]. L'amplitude moyenne du phénomène global varie peu; cependant la descente et la montée du diagramme (accrochage et décrochage de la corde par l'archet) deviennent légèrement plus abruptes.

L'augmentation de la pression de l'archet agit d'une manière générale dans le sens de l'enrichissement du timbre (Fig. 18). Elle augmente principalement l'amplitude des partiels mis en évidence par les dentelures; cette accentuation étant d'ailleurs d'autant plus marquée, en valeurs relatives, que la vitesse de l'archet est plus faible [7].

5.13. Balancement en régime stationnaire

Une corde ne vibre pas dans un seul plan car, perpendiculairement à sa vibration principale (trans-

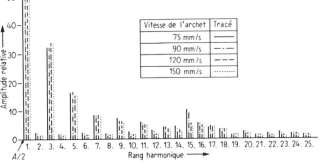

Fig. 16. Appauvrissement particulier du timbre provoqué par l'accroissement de la vitesse de l'archet. Cette influence est surtout importante sur l'amplitude relative des partiels mis en évidence par les dentelures, les 14 e, 15 e, et 16ème. Corde P_2, Ac = 66 Hz, H = 58%. Archet: $P = 1600$ g.

Vitesse de l'archet	Tracé
75 mm/s	——
90 mm/s	— · —
120 mm/s	— — —
150 mm/s	········

---

[6] Quelques cordes ont présenté avec 1800 g de pression une dentelure de moins qu'avec des pressions plus faibles, par suite l'analyse montre un glissement de la pointe vers le partiel voisin de rang inférieur.

versale), on peut en enregistrer une autre, bien qu'imperceptible à l'œil nu. La dérivée de cette dernière vibration, que nous appelons «vitesse de ba-

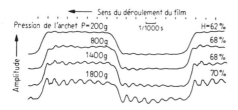

Fig. 17. Modification des caractéristiques des dentelures due à l'augmentation de la pression de l'archet. On remarque que l'amplitude moyenne du phénomène global varie peu. Corde P_1; archet: $v = 120$ mm/s.

lancement», est étudiée lorsque la vibration de la corde correspond à la nuance «fortissimo» sur l'instrument, soit 10 à 12 mm d'amplitude transversale (largeur du fuseau à $l/2$), obtenue avec une vitesse d'archet de 140 mm/s et une force pressante sur la corde d'environ 800 g; distance archet-chevalet 25 mm.

Les diagrammes de la Fig. 19 donnent pour trois cordes, un aperçu de la «vitesse de balancement», enregistrée en disposant un capteur à capacité sous la corde en son milieu.

Ces diagrammes a, b, c, sont respectivement relatifs à des cordes filées avec un trait d'aluminium, d'argent et de tungstène; elles pèsent respectivement

5, 11 et 22 g. On remarque que le nombre, la forme, l'amplitude des dentelures et celle du signal global, sont nettement différents entre eux; ceux-ci sont donc liés aux caractéristiques physiques de la corde (âme, sous-filage, filage etc.). On note que la balancement paraît d'autant plus grand que la densité du métal utilisé comme trait, est plus petite. Sur la

Fig. 19. Forme d'onde de la vitesse de «balancement» pour trois cordes filées d'un trait différent.

Fig. 19 la courbe «a» relative à la corde aluminium 5 g a été relevée avec un gain moitié de celui utilisé pour les deux autres cordes. La recherche des composantes par l'analyse harmonique de ces diagrammes, Fig. 20, montre comparativement entre eux des spectres différents; on y remarque une pointe qui, respectivement pour les cordes aluminium, argent et tungstène, se situe sur les harmoniques de rang 13, 16 et 17. Cette pointe qui correspond sensiblement au nombre de dentelures par période varie dans le même sens que précédemment (pointe vitesse de vibration): plus la masse linéique de la corde étu-

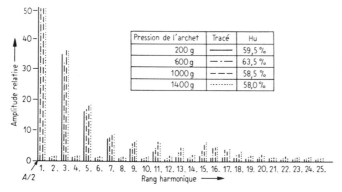

Pression de l'archet	Tracé	Hu
200 g	——	59,5 %
600 g	—·—	63,5 %
1000 g	— — —	58,5 %
1400 g	·········	58,0 %

Fig. 18. Enrichissement du timbre provoqué par l'augmentation de la pression de l'archet. La présence des partiels 14, 16 et 18, d'amplitude non négligeable, démontre que le milieu de la corde n'est pas exactement un point nodal pour tous les partiels pairs. Sur ces spectres on a réduit de moitié les amplitudes des partiels de rang 1. Corde B_3; archet: $v = 120$ mm/s.

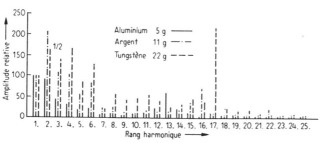

Fig. 20. Spectres relatifs aux trois cordes filées d'un trait différent. Sur ces spectres on a réduit de moitié l'amplitude du partiel de rang 2 de la corde tungstène.

diée est importante, plus le rang harmonique de la pointe est élevé. On note que le milieu de la corde n'est pas un point nodal, dans ce type de vibration, pour les partiels pairs. L'étude de la vitesse de balancement permet de différencier les cordes entre elles d'une façon plus nette que ce qu'avait permis l'analyse de la vitesse de vibration transversale; comparer à ce sujet les Fig. 20 et 11, relatives aux mêmes cordes.

5.14. Résumé sur le régime stationnaire

On étudie tout d'abord les accessoires et les montages expérimentaux indispensables utilisés. Les dispositifs d'appareillages que nous avons réalisés ont eu pour but d'accroître la quantité d'information contenue dans les grandeurs étudiées. Ceci nous a permis de connaître, si de petites erreurs commises dans les expériences avaient une importance sur les résultats de l'analyse; nous nous sommes donc attachés à ne pas dépasser les bornes au-delà desquelles les influences des erreurs sont marquées. Dans ces conditions, entre autres résultats, nous avons pu mettre en évidence des dentelures sur les diagrammes de la vitesse de vibration, puis préciser que celles-ci sont liées non seulement aux caractéristiques physiques des cordes, mais encore à divers paramètres: tension de la corde; pression de l'archet; distance archet-chevalet; masse linéique; longueur de la corde. Nous avons pu préciser également que le milieu de la corde $l/2$, n'est pas un point nodal pour tous les partiels pairs. Il peut en être de même pour un point pris à $l/3$ en ce qui concerne un partiel impair, ainsi qu'à $l/4$ pour, de nouveau, un partiel pair. On détermine ensuite que l'accroissement de la vitesse de l'archet correspond à un appauvrissement du timbre; et que l'augmentation de la pression de l'archet correspond à un enrichissement particulier du timbre. Enfin on met en évidence une vibration perpendiculaire à la vibration transversale, qui permet de différencier nettement les cordes et qui, bien qu'imperceptible à l'œil nu, montre que la corde ne vibre pas dans un seul plan.

Remerciements

Nous adressons, nos vifs remerciements à M. G. DELARUOTTE Technicien du C.R.S.I.M., pour son efficace collaboration à ce travail; notre témoignage de reconnaissance aux Etablissements Babolat, Maillot, Witt, de Lyon, fabricants des différents types de cordes utilisées; notre gratitude à M. CH. MAILLOT pour ses suggestions et conseils sur celles-ci.

(Reçu le 19 Juin 1961.)

Bibliographie

[1] HELMHOLTZ, H., Théorie physiologique de la musique (Trad. fr.). V. Masson, Paris 1868.
[2] BOUASSE, H., Cordes et membranes. Delagrave, Paris 1926.
[3] CHARRON, E., Théorie de l'archet. Thèse, Gauthier-Villars, Paris 1916.
[4] BLADIER, B., J. Phys. Radium **15** [1954], 65.
[5] BLADIER, B., C. R. Acad. Sci., Paris **238** [1954], 570.
[6] BLADIER, B., J. Phys. Radium **16** [1955], 108.
[7] BLADIER, B., C. R. Acad. Sci., Paris **240** [1955], 1868.

Nous avons consulté aussi:

FLEURY, P. et MATHIEU, J., Vibrations mécaniques, Acoustiques. Eyrolles, Paris 1955.
FOCH, A., Acoustique. A. Colin, Paris 1942.
GRANIER, J., Les systèmes oscillants. Dunod, Paris 1936.
LUCAS, R., Acoustique, Elasticité. C. Hermant, Paris 1958.

11

Reprinted from *Acustica*, **14**(6), 331 – 336 (1964)

SUR LES PHÉNOMÈNES TRANSITOIRES DES CORDES VIBRANTES

par B. Bladier

Centre de Recherches Physiques, Marseille

Sommaire

Dans ce travail l'auteur étudie les phénomènes transitoires qui amènent une corde filée sur boyau, initialement au repos, jusqu'à son régime stationnaire, puis inversement de ce régime stationnaire au repos. Il expose dans le détail l'influence des paramètres de l'archet: vitesse et pression, puis de la corde: longueur, tension et masse linéique. Il donne la répartition spectrale due à l'ébranlement, précise évolution du spectre en fonction du temps, présentant un timbre plus riche pendant l'état transitoire que pendant le régime permanent.

Summary

In this work the author studies the transitory phenomena which arise from the bowing of a gut string, intially at rest, until it attains the stationary regime, then in reverse from the stationary regime to rest. It shows in detail the influence of the bow parameters, velocity and pressure, and of the cord, viz. its length, tension and mass per unit of length.

It also gives the spectrum distribution due to the motion, in particular the evolution of the spectrum as a function of time, giving a tone much richer in the transitory state than during the steady regime.

Zusammenfassung

In dieser Arbeit wird untersucht, wie eine Darmsaite aus der Ruhelage in den stationären Schwingungszustand beziehungsweise aus dem stationären Schwingungszustand in die Ruhelage übergeht. Hierbei werden im einzelnen die Einflüsse der Bogenführung (Schnelligkeit und Andruck) und der Saite (Länge, Spannung und Massenbelegung pro Längeneinheit) untersucht. Die spektrale Verteilung beim Einsetzen der Schwingung und die genaue Entwicklung des Spektrums in Abhängigkeit von der Zeit wird angegeben. Es zeigt sich, daß der Klang beim Ein- oder Ausschwingen voller als im stationären Bereich ist.

1. Introduction

Les vibrations des cordes, avec les nombreux paramètres liés à ces vibrations, ont fait l'objet d'importantes recherches: v. Helmholtz [1] observe et décrit la forme de vibration d'un point d'une corde en mouvement. Bouasse [2] expose et commente les résultats des expériences fondamentales sur les cordes vibrantes.

Charron [3] étudie l'entretien de la vibration de la corde et les frottements de l'archet. Young [4] examine la justesse de la série harmonique des fréquences propres d'une corde de piano. Leipp [5] précise la vibration longitudinale et l'influence de certains paramètres sur le spectre d'une corde. Lottermoser et Meyer [6] étudient le spectre du bruit de frottement de l'archet sur la corde. Nous avons,

nous-même, donné les résultats des expériences, effectuées sur la vitesse de la vibration transversale des cordes en régime stationnaire [7]; nous étudions ci-après les phénomènes transitoires.

2. Technique expérimentale

Nous utilisons la technique, mise au point pour l'étude de la vitesse de vibration transversale des cordes en régime stationnaire, qui a fait l'objet d'une publication antérieure [7]. Nous prions le lecteur de se rapporter à cette publication, pour tout ce qui concerne: la description des accessoires, le dispositif expérimental, les montages dérivateurs réalisés pour obtenir la vitesse de vibration, les diverses vérifications effectuées sur les résultats et les analyses.

Dans nos expériences, instruments et instrumentistes sont éliminés. La corde, filée sur boyau, le plus souvent ut_1 de violoncelle accordée sur 66 Hz, montée sur des chevalets en plomb reposant sur des blocs de béton, est attaquée à 25 mm du chevalet par un archet circulaire en ébonite (roue de vielle) de 10 mm de largeur. La vitesse de l'archet et la pression que celui-ci exerce sur la corde sont mesurées, réglées et maintenues constantes au cours des expériences. Enfin, le relevé de la vitesse de vibration s'effectue le plus souvent au milieu de la corde, sans contact matériel, pour ne pas perturber le mode de vibration.

3. Durée du phénomène transitoire

Nous avons tout d'abord cherché à connaître si ces phénomènes transitoires étaient bien reproductibles. Plusieurs expériences, sur une même corde excitée dans les mêmes conditions, montrent que la durée du phénomène transitoire fluctue autour d'une valeur moyenne; compte tenu de ces fluctuations, l'évolution de la forme reste semblable d'un tracé à l'autre.

4. Influence de la pression et de la vitesse de l'archet

Pour l'étude des paramètres vitesse et pression de l'archet, nous avons effectué dans certains cas 52 mesures et dans d'autres 104, pour éliminer l'effet des fluctuations de durée. Le résultat de ces mesures, obtenues avec une humidité relative sensiblement constante de 55%, est donné par le Tableau I ci-après:

Tableau I.
Résultat des mesures de la durée du transitoire en fonction de la pression et de la vitesse de l'archet.

Nombre de mesures effectuées	Archet pression	vitesse	Durée du transitoire Marge[1]	Moyenne
	g	mm/s	s	s
52	1000	150	1,1 à 2,5	1,8
52	1000	120	1,1 à 1,9	1,51
52	1000	90	0,7 à 1,7	1,13
52	1000	60	0,5 à 1,3	0,9
52	600	150	1,7 à 3,3	2,28
52	600	120	1,3 à 2,5	1,76
52	600	90	0,9 à 2,1	1,39
52	600	60	0,7 à 1,3	0,95
104	200	150	2,5 à 4,9	3,4
104	200	120	1,9 à 3,9	2,82
104	200	90	1,7 à 3,3	2,32
104	200	60	1,1 à 2,5	1,75

[1] Différence entre les mesures extrêmes de la distribution.

En valeur absolue la dispersion des mesures autour de la moyenne est d'autant plus grande que la moyenne est plus élevée; mais sa valeur relative n'est pas très différente dans toutes les conditions expérimentales. Ces résultats, sur la durée du transitoire, donnent une répartition qui n'est pas gaussienne mais dissymétrique et bimodale dans deux cas (vitesse 150 mm/s et pression 600 g et 200 g), ce que nous avons vérifié en effectuant un plus grand nombre de mesures pour chaque cas [8].

La Fig. 1 montre, dans les limites de nos expériences, que la durée moyenne de l'état transitoire est une fonction sensiblement linéaire de la vitesse de l'archet; l'influence de cette vitesse paraît d'ailleurs d'autant plus faible que la pression de l'archet est plus grande.

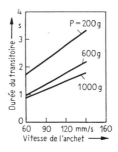

Fig. 1. Influence de la vitesse et de la pression de l'archet sur la durée de l'état transitoire.

La Fig. 2 reproduit quelques uns des enregistrements obtenus sur une corde ut_1 lorsque, à pression constante, $P = 200$ g (Fig. 2 a, b, c, d), la vitesse d'archet passe de 60 mm/s à 165 mm/s, et lorsque à vitesse constante, $V = 120$ mm/s, (Fig. 2 m, n, o, p), la pression augmente de 200 g à 1600 g.

On a ensuite vérifié l'influence de la pression sur une corde sol_1 accordée sur 98 Hz (Fig. 2 h, i), et de nouveau sur la corde ut_1 réduite de 1/3 de sa longueur et donnant par suite le sol_1 (Fig. 2 r, s); les résultats confirment les précédents: la durée de l'état transitoire augmente d'une part lorsque la pression de l'archet diminue et, d'autre part lorsque la vitesse de l'archet croît.

On peut avoir une idée des fluctuations sur la durée et sur l'évolution de l'état transitoire en comparant entre eux les Figs. 2 c et m, puis s et t; enfin la Fig. 2 j montre un début aberrant du transitoire, comparativement à la Fig. 2 h obtenue dans les mêmes conditions. On note encore que l'augmentation de la pression modifie l'évolution du transitoire et le rend plus fluctuant.

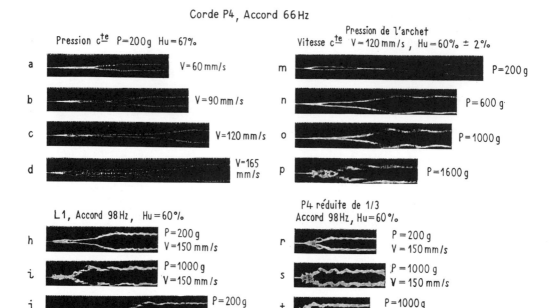

Fig. 2. Evolution de l'état transitoire avec la vitesse V et la pression P de l'archet.

5. Longueur, tension et masse linéique de la corde

En ce qui concerne l'influence de la longueur, de la tension et de la masse de la corde sur la durée du transitoire, on note que cette durée diminue lorsque l'on raccourcit la corde et encore lorsque l'on augmente la tension; l'influence de la masse n'a pu être nettement précisée; cet échec serait peut-être explicable par l'écart important qui existe entre les «raideurs» de cordes ($R = 101$, 46 et 32, ayant respectivement des masses de 5, 13 et 22 g), utilisées pour ces expériences [9].

6. Détail de l'évolution transitoire initiale (ébranlement)

La Fig. 3 donne, de l'état transitoire, le détail de ses formes mouvantes au cours du temps. L'évolution des phénomènes périodiques qui composent cet état, est sensiblement reproductible lorsque, sur les enregistrements comparés, les durées des transitoires sont du même ordre. Les diagrammes, de la Fig. 3 a à i, représentent quelques uns des aspects de l état transitoire pendant son évolution. Sur le haut des tracés sont inscrits des repères chronologiques espacés de 0,001 s.

Sur la Fig. 3 a, la corde au repos reçoit une première impulsion dont la durée d'environ 1 ms (correspondant à l'harmonique 15) est légèrement su-

périeure à celle d'une dentelure (0,8 ms correspondant à l'harmonique 17). Après cette première phase, où la période n'est pas très bien marquée et le phénomène quelque peu erratique, il y a une ébauche périodique en Fig. 3 b d'où se détachent semblable à l'impulsion initiale, une première impulsion dirigée vers le haut; puis en Fig. 3 c une deuxième impulsion, située au milieu de la période, et dirigée ver le bas. A ces impulsions qui croissent en amplitude, viennent s'en ajouter d'autres: tout d'abord vers le haut en Fig. 3 d, puis vers le bas en Fig. 3 e, modifiant ainsi graduellement la durée de l'impulsion initiale et la forme du phénomène périodique (Fig. 3 f, g, h), pour finalement arriver à former le créneau du régime stationnaire de la phase finale en Fig. 3 i.

Dans l'ébranlement, on note que c'est un harmonique élevé (pour la corde étudiée l'harmonique 15, voisin des dentelures correspondant à l'harmonique 17) qui prend naissance le premier et non le fondamental.

7. Vérification sur une corde plus courte

Pour connaître dans quelle mesure la reproductibilité de cette évolution était possible, nous avons réduit cette même corde de 1/3 de sa longueur vibrante en conservant les mêmes pression, vitesse et distance archet-chevalet. La seule réduction de la longueur, nous l'avons vu précédemment, entraine

Temps   Sens du déroulement du film ⟶

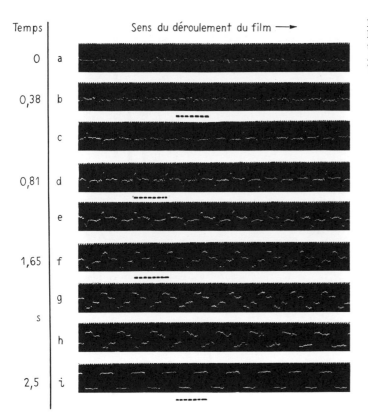

Fig. 3.
Détail de l'évolution de l'état transitoire au cours du temps. Corde P_4, $P = 200$ g, $V = 120$ mm/s, $Hu = 72\%$, accord 66 Hz.

la variation simultanée de: la masse vibrante, la raideur, l'accord (quinte) et aussi l'emplacement des nœuds et des ventres qui correspondent à l'endroit où la corde est attaquée par l'archet de 10 mm de largeur. La variation simultanée de ces paramètres a surtout affecté la durée du transitoire qui a notablement diminué et, bien que le nombre de dentelures [2] soit lui aussi plus faible (11 au lieu de 17), on remarque que le détail du processus de l'évolution du phénomène transitoire reste semblable au cas précédent.

8. Evolution transitoire finale (amortissement)

L'étude de l'amortissement de la vitesse de vibration (obtenu à partir du régime vibratoire stationnaire, lorsque la vitesse initiale de l'archet diminue jusqu'à s'annuler en 3 s), montre une diminution progressive de l'amplitude globale du phénomène obtenu en régime stationnaire. Le créneau et ses

[2] Pour cette corde réduite de 1/3 de sa longueur, l'analyse en régime stationnaire montre un appauvrissement du timbre plus particulièrement net sur la pointe de rang 17, par rapport à l'analyse de la corde de longueur totale.

dentelures sont modifiés lorsque l'amplitude résiduelle n'est plus qu'environ 1/4 de l'amplitude initiale. Toutefois, comme pour l'ébranlement, l'augmentation de la pression de l'archet modifie cette évolution et la rend plus fluctuante.

Dans l'amortissement on note que fondamental et harmoniques subsistent et persistent avant de laisser place seulement aux harmoniques correspondant aux dentelures, puis à des impulsions erratiques.

9. Répartition spectrale à l'ébranlement

L'analyse harmonique a été effectuée pour chacune des périodes situées au-dessus du trait ponctué de la Fig. 3. Ces diagrammes ont été reproduits en haut de la Fig. 4 qui donne les spectres au temps $t = 0,38$ s; $0,82$; $1,65$ s; $2,5$ s après l'ébranlement de la corde. Au temps $t = 0,38$ s, les composantes principales sont les harmoniques: 3 (sol_2); 5 (mi_3); 7 ($\approx$ si $\flat_3$) et le fondamental ut_1; sur la pointe ce sont les harmoniques: 15 (si 4); 17 ($\approx$ ré $\flat$ 5); 18 (ré 5) et 19 ($\approx$ mi $\flat$ 5). Au temps $t = 0,82$ s, il y a un très net appauvrissement du timbre (seul l'harmonique 5 reste important); le timbre au temps $t = 1,65$ s est encore un peu plus pauvre, et ainsi jusqu'à la phase finale $t = 2,5$ s.

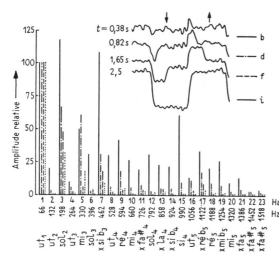

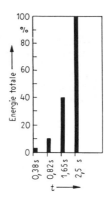

Fig. 5. Energie de la vibration au cours de l'état transitoire.

Fig. 4. Variation du spectre lors de l'attaque de la corde P_4. En abscisse on a porté: le rang harmonique Ha, la fréquence en Hz, et les notes correspondantes; devant le nom de quelques notes la croix indique que la fréquence est seulement voisine de la note considérée. $V = 120$ mm/s, $P = 200$ g, $Hu = 72\%$.

10. Conclusion

D'une façon générale, le régime stationnaire a un timbre plus pauvre que la période transitoire. Au cours de celle-ci le milieu de la corde n'est pas exactement un nœud de vibration pour tous les harmoniques pairs, résultat déjà signalé pour le régime stationnaire [7].

11. Etude des phases des composantes

Pour chacun des phénomènes analysés aussi bien dans l'état transitoire que dans le régime stationnaire, on a déterminé les phases des composantes; il n'a été possible de dégager de loi ni en ce qui concerne les groupements des phases des différentes composantes à un instant donné, ni en ce qui concerne leur variation dans le temps. Peut-être de telles lois apparaîtraient-elles si le nombre de périodes analysées était notablement plus grand.

12. Variation de l'énergie

L'énergie de la vibration croît pendant la période transitoire (Fig. 5) d'environ 3% de la valeur finale ($t = 2,5$ s) au temps $t = 0,38$ s; elle passe à 10% pour t = 0,82 s, puis à 40% pour $t = 1,65$ s.

Le rapport de l'énergie de chacun des harmoniques à l'énergie totale (Fig. 6) montre que sur les fondamentaux, l'énergie croît en fonction du temps; elle décroît sur les harmoniques et plus particulièrement sur ceux de rang 3, 7, 9, 11.

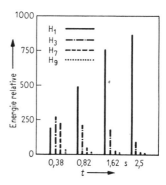

Fig. 6. Rapport de l'énergie de chacun des harmoniques à l'énergie totale.

13. Résumé

On vérifie par deux méthodes (électrostatique puis électromagnétique) la vitesse de vibration transversale pendant l'état transitoire de la corde. La durée du phénomène fluctue autour d'une valeur moyenne; pour éliminer cette fluctuation on effectue un grand nombre de mesures. Dans les limites de nos expériences la durée moyenne de l'état transitoire est une fonction sensiblement linéaire de la vitesse de l'archet; l'influence de cette vitesse paraît d'ailleurs d'autant plus faible que la pression de l'archet est plus grande.

La durée de l'état transitoire augmente d'une part, lorsque la pression de l'archet diminue et, d'autre part, lorsque la vitesse de l'archet croît. Pour une même longueur de corde cette durée diminue si la tension (accord) augmente. De même, cette durée diminue si l'on raccourcit la corde. L'influence de la masse n'a pu être précisée.

On donne dans le détail l'évolution transitoire initiale, avec une vérification sur une corde plus courte, ainsi que l'évolution finale (amortissement). On note pour l'ébranlement que c'est un harmonique élevé qui prend naissance le premier et non le fondamental; dans l'amortissement, fondamental et harmoniques subsistent et persistent simultanément avant de laisser place aux dentelures.

De la répartition spectrale, on dégage que le régime stationnaire a un timbre plus pauvre que celui de la période transitoire. Au cours de celle-ci le milieu de la corde n'est pas exactement un nœud pour tous les partiels pairs.

En ce qui concerne les groupements des phases des composantes à un instant donné, ainsi que leur variation dans le temps, il n'a pas été possible de dégager de loi.

On donne la variation de l'énergie pendant l'état transitoire et le rapport de l'énergie des premiers harmoniques impairs à l'énergie totale.

Remerciements

Nous adressons nos vifs remerciements à MM. J. Basso et G. Delaruotte, Techniciens du C.R.P., pour leur efficace collaboration à ce travail; notre témoignage de reconnaissance aux Etablissements Babolat, Maillot, Witt, de Lyon, fabricants des différents types de cordes utilisées; notre gratitude à M. Ch. Maillot pour ses suggestions et conseils sur celles-ci.

(Reçu le 6 Janvier 1964.)

Bibliographie

[1] Helmholtz, H., Théorie physiologique de la musique (Trad. Franç.). V. Masson, Paris 1868.
[2] Bouasse, H., Cordes et membranes. Delagrave, Paris 1926.
[3] Charron, E., Théorie de l'archet. Thèse, Gauthier-Villars, Paris 1916.
[4] Young, R. W., Inharmonicity of piano strings. Acustica 4 [1954], 259.
[5] Leipp, E., Les paramètres sensibles des instruments à cordes. Thèse, Paris 1960.
[6] Lottermoser, W. et Meyer, J., Über das Anstrichgeräusch bei Geigen. Instrumentenbau-Z. 12 [1961], 382.
[7] Bladier, B., Contribution à l'étude des cordes du violoncelle. Acustica 11 [1961], 373.
[8] Bladier, B., Sur l'état transitoire résultant de l'ébranlement d'une corde par l'archet. C. R. Acad. Sci. Paris 242 [1956], 2704.
[9] Bladier, B., Evolution des phénomènes transitoires dans la mise en vibration des cordes. J. Physique Radium 17 [1956], 57 S.

Nous avons consulté aussi:
Fleury, P. et Mathieu, J., Vibrations mécaniques, Acoustique. Eyrolles, Paris 1955.
Foch, A., Acoustique. A. Colin, Paris 1942.
Granier, J., Les systèmes oscillants. Dunod, Paris 1936.
Lucas, R., Acoustique, Elasticité, C. Hermann, Paris 1958.
Mercier, J., Acoustique, Vol. III. Presse Univ. de France, Paris 1962.

Editor's Comments on Paper 12

12 Schelleng: *The Bowed String and the Player*

The viewpoints of performer and acoustician are so different that unless they are combined in one and the same person a no-man's land tends to develop between them. While the artist cannot be solving problems in physics as he plays, he must always be aware, subconsciously if not consciously, of the possibilities and dangers through which he is passing in his unrelenting search for optimum effects. A knowledge of the meaning of this terrain can certainly do no harm, and for some players it can definitely help. It is the purpose of the following article to unite some aspects of these two points of view.

Information about the life and work of John C. Schelleng can be found in "Editor's Comments on Paper 5."

Reprinted from *J. Acoust. Soc. Amer.*, **53**(1), 26– 41 (1973)

The bowed string and the player

JOHN C. SCHELLENG

301 Bendermere Avenue, Asbury Park, New Jersey 07712

(Received 14 December 1971)

Relations between bowing parameters, i.e., force applied, bow position, and velocity, are derived in terms of load impedance presented by bridge to string, characteristic impedance, and frictional coefficients. The range between least applied force needed to couple bow to load during sticking and maximum permitting uncoupling following sticking provides the generous tolerance, variable but typically in the ratio of 1 to 10, that makes the bowed string so flexible in performance. Domains of string behavior recognized by players are related graphically to bowing parameters. An electromagnetic method for observing particle velocity in the string reveals small but significant ripples caused by forces at the bow that the idealized explanation ignores. In one very flexible string, force exerted on the bridge varied approximately inversely with frequency out to the 15th harmonic, whereas for a string equivalent to a gut G for violin force became zero near the 7th. Elastic effects are considered, and it is suggested that in strings, either solid or wound, inharmonicity of 0.1 cent per square of mode number will not perceptibly degrade bowed-string performance.
SUBJECT CLASSIFICATION: 6.7.

INTRODUCTION

The work of Helmholtz[1] and that of Raman[2] contain information on the bowed string that is important to the player. Helmholtz, besides elucidating general aspects of string behavior, observed, for example, the effect when the bow applies abnormally low "bow pressure," the musician's term for the force applied normally to the string. Raman in another classic of musical acoustics, extended Helmholtz's work and clarified many aspects of the bowed string, including a calculation of frictional driving force, the manner in which velocity and position of the bow as well as the frequencies of resonance of the instrument affect minimum bow pressure, and the spectral differences that can be controlled by bowing. Since Helmholtz's work appears in his book, it is widely disseminated. That of Raman, while included in every bibliography, is mainly read "by title," in part for the reason that it is out of print and unavailable even in some large libraries.

During the past 40 years, with the rapid development of acoustics, there have been many important additions to our understanding of the violin family, in considerable detail with respect to the vibrations of the body but with less attention to the strings. Particularly informative in relation to strings is the well-known experimental paper by Saunders,[3] in which he extended Raman's work on minimum bow force and its relation to body resonances, and called attention to the existence of a maximum force that should be investigated. It is one aim in the present paper to draw existing strands together by use of simple circuit concepts and a new

method of observation in the hope that, if not actually useful to the player, the synthesis may at least interest him as the physical setting behind the string behavior with which he has long contended. This involves knowing the difference between the mechanisms of minimum and maximum bow force, applying available data to their calculation and relating the results to various domains of bowing, good and bad, that he has always recognized; the factors controlling volume of sound and the rather subtle differences in timbre controlled by bowing; the efficiency of the stick-slip process and the effects of negative resistance at the contact between bow and string; conditions occurring at the beginning of vibration and the equivalent circuit of the string and related matters. Naturally in a field so diverse there can be no pretense of comprehensiveness. Concern is mainly with sustained tones. The important subject of dynamics in such techniques as staccato and spiccato is left untouched. In the last part of the paper, the effects of longitudinal and flexural stiffness and of torsional compliance in the string are examined.

I. MOTION IN THE BOWED STRING

The primary features of motion in the bowed string as outlined by Helmholtz are (1) beneath the bow the period of vibration is broken into two parts whose durations have the same ratio as the lengths of the string either side of the bow, and in which velocities are constant but different in magnitude and sense; (2) the string is broken into two straight sections by a discontinuity that shuttles end to end and back once in one period of vibration; and (3) the discontinuity travels around a closed lens-shaped path having parabolic curvature, direction of circulation reversing with direction of the bow. These conditions are shown graphically in Figs. 1(a) and 1(b). Conditions become more complicated when the bow is located precisely at a node of a normally active mode. When the string is plucked, two such waves are set up, equal in amplitude and opposite in circulation.

Raman's work begins with an elegant kinematic method of analysis, which he applies to a wide range of types, only the simplest of which is important in music. He chooses to view the waves in terms of their transverse particle velocity rather than displacement; these oppositely directed waves have the form on the string shown in Fig. 1(c). Their sum, in Fig. 1(d), gives the two velocities possible at any point x' along the string. When viewed in time at any point, the two slopes (accelerations) are equal and opposite, together representing a velocity that is constant except when the

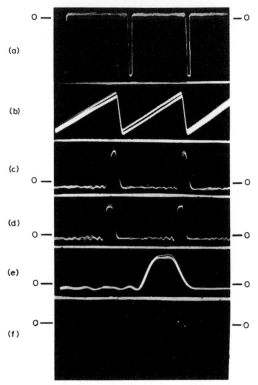

FIG. 2. (a) Particle velocity at bow, very flexible string on monochord; (b) displacement at bow; (c) velocity near minimum bow force, C string mounted on cello; (d) same but with increased force; (e) same as (d), expanded; (f) cello A string on monochord, moderately high bow force.

discontinuity passes, as in Figs. 1(e) and 1(f). The result is identical with that of Helmholtz, both agree with experiment qualitatively, commonly with considerable accuracy. The traveling velocity waves will be referred to here as "Raman waves."

A. Oscillograms of String Motion

The motion of the string beneath the bow is illustrated by oscillograms of Fig. 2. These were obtained by a method in which a restricted magnetic field is brought to bear on the string at the point anywhere along the string where motion is to be studied. An emf proportional to velocity is thus induced in the conductive string, amplified and displayed oscillographically.[4] Figure 2(a), a photograph of such a display, shows the particle velocity of a string precisely beneath the bow. To obtain Fig. 2(b), which plots displacement instead of velocity, an integrating circuit was interposed in the amplifier chain.[5] Conditions were chosen to illustrate experimentally what happens under the idealized conditions assumed by Helmholtz, though as

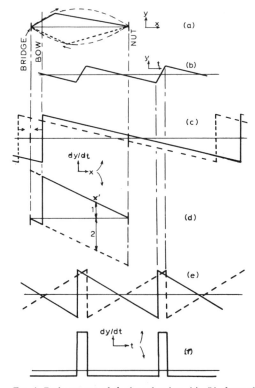

FIG. 1. Basic concept of the bowed string; (a), (b) shape of string and displacement at bow according to Helmholtz; (c) the two Raman waves at release; (d) their sum along the string and the two velocities at any point x'; (e) separately as seen at the bow; (f) together as seen at bow.

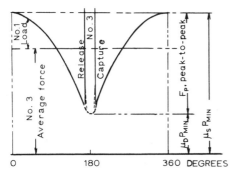

FIG. 3. Frictional forces exerted by bow on string, through one cycle.

will be remarked later the correspondence is not perfect; a thin steel string under high tension was mounted on a monochord employing heavy steel blocks at the ends of the string. The distance from the "bridge" to the bow divided by total length of string we shall call β. In this case $\beta = 1/20$, or 0.05. As a result, the duration of the velocity pulse in Fig. 2(a) at its half-amplitude is also about 0.05 of the period of vibration, an equality merely signifying the fact that waves are propagated back and forth at a fixed velocity. The displacement curve of Fig. 2(b) is made up of two straight portions, during the longer of which the string travels with speed of the bow, snapping back at a speed 19 times greater during the shorter period. Fiddle strings act about the same, but the curves lack the sharp corners shown here and display differences that are musically significant. Figures 2(c) and 2(d) illustrate the behavior of a cello C string as normally mounted on the instrument and bowed by hand. The only difference between Figs. 2(c) and 2(d) is in bow force, which for the first is near the minimum and in the second well above it (minimum force is discussed in Sec. II-A). Increased force obviously sharpens the high-frequency "edge" of the note, both by squaring the rounded shoulders of the main pulse and by accentuating the ripple that occurs during the long period of nominally constant velocity.[6]

II. ON THE MECHANICS OF PLAYING

In playing a bowed-string instrument there are certain limits in the speed of bow, its location on the string, and the normal force applied that must not be overstepped in a specific musical situation. For the performer, this is usually a matter of choosing more or less subconsciously from among familiar patterns of action rather than analyzing into the particular parameters mentioned. In extreme cases, however, such as in making a long tone as loud as possible, conditions are marginal and the effort is probably always intensely conscious. Even in circumstances where the limits are

wide enough to prevent disaster for the mediocre player, the artist always seeks the best compromise. The extreme limits of bow force are wider than most people would probably expect. In the rapid changes of condition involved in virtuoso performance each note cannot be optimized; if it were not for the wide tolerance in fine instruments, modern violin playing would be impossible.

From the viewpoint of the player, the subject is best approached through a study of bow force and its limits. There are two important questions: When is it so small that the fundamental vibration fails, and when is it so large that the wave on the string is unable to trigger a clean repetition of similar pulses? The result of the first is that a "solid tone" recognizable as the fundamental vibration has degenerated into a mere "surface sound" comprising mainly higher modes; the result of the second is a raucous scratching. The mechanical requirements for avoiding these failures are that (1) throughout the period in which the bow clings to the string and particularly toward the middle, bow force must exceed a certain minimum value, $P_{\min}$, and (2) *at the end* of the period force must be less than the value $P_{\max}$ that interferes with release of the string. The frictional force that the blow exerts on the string may be thought of as comprising three components: first, one that provides the power that vibrates the bridge and sustains other losses, such as those in the string itself; second, one associated with release and capture; and third, a steady unvarying component in the direction of the bow's motion. These are indicated in Fig. 3. The one that vibrates the bridge has its greatest backward value in the phase when slipping occurs, and this value, in combination with the third or average component gives the forward resultant that we associate with sliding friction. The sum of the three is at all times in the direction of the bow.

A. Minimum Bow Force

Raman found that when bowing speed v_b is small, $P_{\min}$ tends to a finite minimum, increases rather slowly at first, but later roughly in proportion to speed (or possibly faster). He explained this in terms of the change in $(\mu_s - \mu_d)$ with speed, these terms being respectively the coefficients of static and dynamic friction. When speed was held constant and the distance from the bow to bridge changed, minimum force was found to vary inversely as the square of distance.

Experiments by Kar, Datta, and Ghosh[7] confirmed the variation with speed but indicated variation with distance to be the inverse first rather than second power. These findings were confirmed by their theory, which unfortunately is based on assumptions for friction at variance with those commonly found for one solid sliding over another. Besides being proportional to the force applied normally and to a coefficient of friction, frictional force was assumed to be proportional to relative velocity, thus acting like a viscous resistance. While elsewhere Kar and his associates have considered the effect of the bridge load on minimum pressure, at this point all loss except that in friction at the contact has been implicitly assumed absent. With this assumption, however, no force is needed to drive a perfectly flexible string once a Helmholtz vibration is set up, and the bowing pressure may be made zero.

In considering minimum bow force, we are concerned with the positive and negative extremes of the first force component in Fig. 3. A simplified circuit which gives a basis for computation of this force is shown in Fig. 4. Ideally the impedance into which the bow delivers power is a pure resistance R_n at fundamental and harmonics, an assumption sufficiently good for present purposes. X_n is the reactance of the short section between bow and bridge, and for fundamental and a few harmonics may be regarded as a simple compliance. The bridge impedance is represented by a high reactance X_b in parallel with a high resistance r_n. Bridge resistance r_n will be taken to represent all sinks of vibrational power referred to the end of the string at the bridge. Each of these adds a component of conductance at that point, r being the reciprocal of their sum. We may think of r as composed of two components in parallel, one the resistance of the body of the instrument, the other representing the string itself—its internal friction, those losses at the contact with the bow that may be considered parasitic to the vibration (due perhaps to the finite width of the bow) and the finger at the other end of the string. Of the first, we know something of its magnitude from measurements by Eggers[8] for a cello and Reinicke and Cremer[9] for a violin. The second probably has the same order of magnitude as the first, and equals $2Z/\delta$, δ being the resultant logarithmic decrement per fundamental cycle of the string without the body. R_n is inversely proportional to the square of the order of harmonic so that components higher than the fifth have little effect. The effect of reactance X in the long section of string is to reduce to negligibility total reactance into which the bow operates when, as we are now assuming, the string is perfectly flexible; otherwise its effect on minimum bow force may exceed that of resistance.

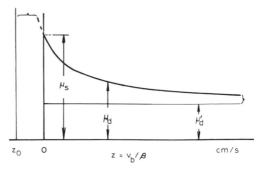

FIG. 5. Hyperbolic frictional characteristic.

The difference between the extremes of the first force component enters into an estimate of minimum bow force because, as Raman points out, unless it is less than or equal to the range between static friction and dynamic, slipping will set in within the long period of sticking. It is shown in Appendix A that this peak-to-peak force is $F_p \doteq 0.5Z^2 v_b / r_1 \beta^2$. ($Z$ is the characteristic impedance of the string, square root of tension times mass per unit length.) As at the minimum this equals $P_{\min}(\mu_s - \mu_d)$, we have

$$P_{\min} \doteq (0.5Z^2/r_1) \cdot v_b / (\mu_s - \mu_d)\beta^2. \tag{1a}$$

Since, in general, μ_d is a function of v_b, it is helpful to consider a simple friction characteristic having qualitatively the variation that actually occurs. For simplicity, we choose a rectangular hyperbola as shown in Fig. 5, raised above the velocity axis by amount μ_d' and moved to the left by velocity z_0, so that it intersects the vertical axis at μ_s. We have then for the coefficient along the curve to the right of the axis

$$\mu = \mu_d' + K/(z+z_o)$$

from which, since

$$\mu_s = \mu_d' + K/z_o,$$
$$\mu = (\mu_d'z + \mu_s z_o)/(z+z_o)$$

and

$$(\mu_s - \mu_d) = z(\mu_s - \mu_d')/(z+z_o).$$

Hence from Eq. 1a, since $z = v_b/\beta$ during the period of slipping,

$$P_{\min} \doteq \frac{0.5Z^2}{(\mu_s - \mu_d')r_1} \cdot \frac{v_b + \beta z_0}{\beta^2}, \tag{1b}$$

which reduces to Eq. 1a when $z_0 = 0$. Equation 1b indicates that when v_b becomes sufficiently large P_m tends to rise in proportion, as both Raman and Kar *et al.* have found experimentally. It also agrees with experiment that when v_b becomes small, minimum pressure tends toward a finite minimum rather than zero. (Though the calculation indicates a linear variation of P_m with v_b, a curve of μ other than hyperbolic

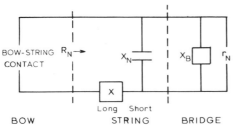

FIG. 4. Load circuit as seen at bow–string contact, simplified.

228

could be chosen to give curvature in P_{min}.) Similarly it is consistent with experiment that when v_b is large, P_{min} tends to vary inversely with the square of β, and when small inversely with the first power. Admittedly, however, this is not the whole story.

The order of magnitude predicted is consistent with experiment without the help of arbitrary constants. If for the cello A string we assume r_1 to be 25×10^3 cgs units,[8,9] Z to be 450 cgs units, $(\mu_s - \mu_d) = 0.4$; $\beta = 0.1$, and v_b to be 20 cm/sec, then P_{min} is indicated by Eq. 1a to be 20×10^3 dynes or 20 g weight, about one-third the wieght of a cello bow. The corresponding figure for a violin E string would be somewhat less than 10 g (i.e., 10 g wt.). If Eq. 1b for the friction curve of Fig. 5 had been used, somewhat higher values would have been indicated.

Raman calculated the waveform of the force involved in minimum bow force, assuming loss to occur in the string itself. The waveform implied here is of the same general shape, except that for mathematical convenience Raman assumed bowing at the node of a normally active partial, with the result that the shape instead of being smooth has steps corresponding in number to the order of harmonic and is obtained without approximation. Figure 3 shows one cycle of the waveform obtained by the present method using fundamental and all harmonics up to and including the sixth. It is characterized by a broadened top and a narrowed bottom. There is loss in detail throughout owing to neglect of higher harmonics. The two spurs, labeled "release" and "capture," assume the continuous frictional characteristic with maximum friction corresponding to "static" friction occurring at a very small but finite velocity; the magnitude of the spurs corresponds to minimum force.

B. Maximum Bow Force

When force greater than the minimum is applied, the frictional grasp of the bow at the transition points rises above the curve in Fig. 3 and the load impedance is no longer a danger to vibration. It is now necessary to ensure that this grasp with rising applied force does not exceed the second condition that must be satisfied. When ultimately it does interfere the condition of maximum bow force is reached. This condition depends on all the quantities in Eq. 1a except bridge resistance r. The reason for the exception is that when failure occurs, the frictional force changes suddenly from that of sliding friction at the minimum to that of "static" friction, and the smaller peak-to-peak force produced by the load is irrelevant because of both size and timing.

When vibration is prevented by too much bow force, the limiting frictional force equals or exceeds $\mu_s P_{max}$—static coefficient times maximum pressure. In the slip region, the force is $\mu_d P_{max}$, and the difference in order to detach the string with regularity must be supplied by the discontinuity arriving from the nut. These waves

produce whatever dislodging force is dictated by bow pressure, but there is a limit to this capability, namely $Z v_b / \beta$, which ignoring string rotation is characteristic impedance times magnitude of the velocity discontinuity.[10] It follows that

$$P_{max} = Z v_b / (\mu_i - \mu_d) \beta. \qquad (2)$$

With the same numerical assumptions used for P_{min}, we obtain for a cello A string the value $P_{max} = 225 \times 10^3$ dynes or 225 g wt, several times the weight of a cello bow. A corresponding figure for a violin E string is 80 g wt. Dividing Eq. 2 by Eq. 1, we obtain

$$P_{max} / P_{min} = 2 \beta r / Z. \qquad (3)$$

C. Volume of Sound

In the simple concept of the bowed string, the vibrational force exerted by the string on the bridge is proportional to the amplitude of the incident Raman wave and thus has the same sawtoothed waveform. The force discontinuity thus equals $2 Z v_b / \beta$ and, for a given string, sound pressure is proportional to v_b / β. The same volume is therefore maintained when change in bow velocity is proportional to the change in distance of the bow from the bridge. It is significant that no mention need be made of bow force, which as indicated above can change by a large factor from its minimum to its maximum with but little effect on volume. Bow force is important primarily as the catalytic agent that makes possible a correct reaction between bow speed and bow position. Usually one increases volume by increasing both velocity and force or by bowing closer to the bridge and increasing force. There is no rule to fit all cases. Procedure depends on length of stroke. With the bow close to the bridge, the force needed may become excessive. Loudest possible sound can only be obtained in a stroke of short duration. To keep bow force within reason the bow then tends to be moved away from rather than toward, the bridge.

D. Spectrum of Sound

It is the shape of the force wave at the bridge—primarily the shape of the Raman wave—rather than the motion of the string at the bow, that gives directly the string's contribution to the timbre of a bowed–string instrument. In the idealized sawtooth waveform, this is particularly simple: the force exerted by the nth harmonic is that of the fundamental divided by n, entirely without regard to bow position provided that it is not nodal. This statement has to be modified for actual vibrations, which follow the rule for the lower several harmonics but deviate in higher orders according to bow position and bow force. The extremely flexible string of Figs. 3(a) and 3(b) in one test with moderate bow force exhibited a good inverse frequency character up to about the 15th harmonic as shown in Fig. 6,

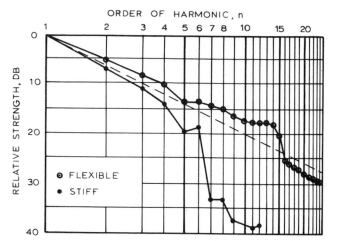

FIG. 6. Spectrum of a very flexible and a very stiff string; inharmonicity of the latter comparable with that of a violin gut G string.

beyond which response fell off more rapidly. As the force is lowered to the minimum, the higher frequencies are reduced in response. In strings normally used, this reduction is greater than in the flexible string.[11]

In obtaining Fig. 6, the string, mounted on a monochord, was bowed continuously by machine, the same device described and used by Saunders.[3] In order to measure the relative forces exerted on the bridge, all that is necessary is to move the magnet to an antinode of the harmonic in question, since velocity at an antinode is proportional to force at the end of the string. The midpoint of the string serves for all odd orders. A simple commercial sound analyzer was used.

The disappearance of the nth harmonic when the bow approaches a point an nth submultiple of string length away from the bridge, i.e., a node of the nth partial, must cause some change in timbre, but its musical importance as a means of intentional musical expression is doubtful. At best, this subtle effect, superposed on the more complex spectrum of the instrument itself, would require the bow to be placed

with exactness at a nodal point, a distance from the bridge that changes with fingered length of the string. With the violin, tolerance to mislocation would be less than a few millimeters. This point one nth of string length away from the bridge is of course a forbidden position for the bow when a "harmonic" is fingered by a light touch at some other node since it is then a point of zero motion for all desired components of vibration. Best positions are of course much closer to the bridge.

E. Graphical Representation for Sustained Tones

Owing to the many resonances in the instrument, the player on even a fine instrument must adapt bowing technique to variations both in minimum bow force and acoustical response. Raman found that minimum bow force can be increased by a factor of the order of two as the fundamental frequency is swept through the air resonance or the principle wood resonance. Variations in acoustical responsiveness are still greater. Long acquaintance with a particular instrument develops

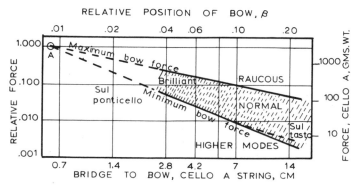

FIG. 7. Typical normal and abnormal playing conditions in the violin family related to bow force and bow position at constant bow velocity for sustained tones. A second set of coordinates refers to a cello A string bowed at 20 cm/sec. Dotted branch suggests effect of curvature in Fig. 5.

automatic compensation, facilitating performance on that instrument but adding difficulty in playing one that is unfamiliar. The effect of these variations is fortunately diluted by a lack of correlation with response by harmonics, but the total effect as shown in the Saunders "loudness" test is still a pronounced one. The effect of a pronounced resonance is to narrow the gap between minimum and maximum bow force.

The limits of bow force can be related to concepts of tone production by constructing a chart as in Fig. 7 showing the range of variables used in Eqs. 1 and 2. It is intended to express general conditions. Strictly speaking, it applies to a single fingered position on the instrument and a single bow speed. Qualitatively, it applies to any position with suitable shift of the diagram on the coordinates. Relative bow forces are plotted logarithmically against bow position β, so that separate straight lines represent maximum and minimum force applied. Intersection A of the lines lies so close to the bridge (about 7 mm for the cello A string) as to be within the width of the bow, but it is of significance as a starting point in the construction of the diagram. Its force coordinate is $P = 2rv_b/(\mu_s - \mu_d)$, independent of the nature of the string. The line of maximum force on the other hand is independent of the impenance of the body of the instrument. Its abscissa is $Z/2r$, or one-fourth the logarithmic decrement (under conditions assumed) of the freely vibrating string as damped by total losses. The two lines radiate from A with slopes corresponding to β^{-1} and β^{-2}. Generality is made impossible by the fact that r is too variable from note to note and from harmonic to harmonic to permit a single quantitative representation of the minimum pressure for an instrument, a statement not applying to maximum bow force. Nevertheless, in a qualitative sense, the picture is confirmed by experience.

Normal playing employs the cross-hatched area. With the bow well away from the bridge there is a large range of forces that can be used for long-sustained tones without the tone breaking. To play louder or softer, the player changes speed of bow or its position or both. To keep within safe limits force also usually changes, but change of force alone does not suffice to control loudness.

Far from the bridge but still within the crosshatched area one plays *sul tasto*, the soft quality with reduced high-frequency content when the bow lies "over the fingerboard." Closer to the bridge, the tone tends toward increased brilliance and power. Exceed maximum bow force and the result is unmusical. Fall short of the minimum and the fundamental mode of vibration is lost. The closer the bow is to the bridge, the less generous is the ratio of maximum to minimum and the steadier is the hand needed to preserve acceptable tone. The experienced player prizes this domain for its nobility of tone; the beginner finds it prudent to play nearer to the fingerboard. Closer still to the bridge as

bow force mounts prohibitively, the gap between maximum and minimum narrows and the solid fundamental disappears. Bowing still closer yields the eerie effect without predominant fundamental known as "sul ponticello," on or rather near to the "little bridge."

The applied forces indicated refer to the cello A string and are in the correct order of magnitude. Still greater force is needed on the lower strings. For the violin bow, force will be less than for the cello by a factor of about 2.5. Incidentally one might reverse the procedure, determining minimum force experimentally and therefrom calculating an equivalent resistance r.

What we have called minimum bow force for a given velocity v_b and bow position β might also have been called minimum β for given force P and v_b, or maximum velocity for given P and β. Similarly, our maximum force for given β and v_b corresponds to maximum β for given P and v_b, or minimum velocity for given P and β.

In Fig. 7, velocity is held constant in the interest of simplicity but a plot in which it is variable might be useful. A three-dimensional plot (P, v_b, and β) can be visualized from Fig. 7 merely by raising (or lowering) anchor point A in the same proportion as velocity is increased (decreased). The extent of the cross-hatched area indicating normal conditions will then have to be reconsidered.

F. Forms of "Attack"

There are two conditions in which a string can conceivably be set into vibration by the bow, first while it is slipping, and second by release from sticking. In the first, the moving bow makes a "soft landing" on the string, and the gradualness of force increase permits growth of vibration by virtue of "negative resistance" (to be discussed later). The vibration expands continuously into the abruptness of capture and release. Frequency is not necessarily that of the fundamental at the beginning. The importance of this form of beginning seems open to question.

The second condition begins so to speak with a plucking or quasipizzicato. Whatever the initial velocity, whether zero or finite, force is more than enough to ensure that the string clings to the bow for a moment, that is, until friction can pull the string aside no longer. At this moment, oppositely moving discontinuities set out from the bow, but the one toward the bridge, since it is in a position normal to slipping, is the one presumably to grow into equilibrium. This second type of beginning is unquestionably the easier one to perform. It can be harsh or gentle depending on musical intention and skill.

Clean enunciation at the beginning of a note is so important that cello players use certain percussion effects produced by the fingers of the left hand to ensure it. These fingers may for example start vibration by acting like the tangents of a clavichord. A finger corresponding to one frequency but otherwise unengaged may support the start of a tone lower in the scale by an actual plucking action. The latter method is particularly useful for attack on an open (unfingered) string, which without benefit of the absorption that the soft finger provides at very high harmonics is more liable to a false beginning than a fingered note.[12]

The beginning of a note or a rapid crescendo requires more force on the bow than is thereafter needed to maintain it because in addition to the power needed for steady vibration it must during that moment provide the energy that is stored in the vibrating string; otherwise the bow "loses traction" as do the driving wheels of a motor car when acceleration demanded is excessive.

Thus in a simple tuned circuit in which mass, resistance, and stiffness are M, R, and S, force over velocity is

$$F/V = Mp + R + S/p,$$

in which in exponential growth, amplitude is proportional to e^{pt}, where $p = \alpha + j\omega$. Substituting this value of p in the previous equation and assuming F and V to be in the same phase during growth, we find that

$$F/V = R + 2M\alpha. \tag{4}$$

This is the resistance offered to the growing oscillation in the string: to the usual resistance of dissipation it adds one of storage that depends only on the mass and the exponential rate of growth α. If decay in free vibration contains the factor $e^{-\alpha_o t}$, it follows that resistance to the growing oscillation contains the factor $(1 + \alpha/\alpha_o)$. As an illustration, consider the cello A string whose equivalent load at the bridge is the same as that assumed in previous examples. If it is required to increase volume at the rate of 70 dB (about 8 nepers)/sec, a moderate rate in music, the factor $(1 + \alpha/\alpha_o)$ is 2 and minimum force will double. It may be necessary either to increase force or to move the bow further from the bridge in order to preserve a healthy tone.

G. Power Efficiency in the Stick-Slip Process

Of the power delivered to the bow-string contact, the fraction converted into power of vibration is of some interest. Extending the procedure of Appendix A to that end, one arrives at the following approximations, which assume width of bow to be negligible:

EFFICIENCY AT MIN. FORCE $= E \doteq (1 - \gamma)/(1 + 0.5\gamma)$,

EFFICIENCY AT MAX. FORCE

$$= E' \doteq 1/[1 + (1/E - 1)P_{\max}/P_{\min}], \tag{5}$$

where $\gamma = \mu_d/\mu_s$. If as previously assumed $\mu_d = 0.2$ and $\mu_s = 0.6$, the value of E is 55%, and if $P_{\max}/P_{\min} = 10$, E' becomes 11%. These approximate values suggest that if the overall mechanical-acoustical efficiency of the instrument is as low as is commonly supposed (1% or less), the chief cause is to be sought elsewhere than in the stick-slip process. There is reason to believe that at the principle resonance, the radiation efficiency can

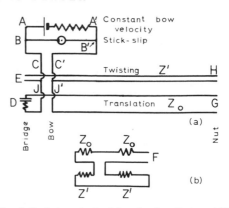

FIG. 8. Equivalent circuit of bowed string. During sticking, B–B' is an open circuit. DG and EH, though shown in electrical analogy as two conductors each, propagate in only one mode.

be better than 20%, a unique frequency in which overall efficiency at minimum bow force might be about 10%, and at maximum pressure 2%. For the range as a whole, however, it will be much less, at top and bottom frequencies in particular. It is well-known that at the lowest fundamental frequency radiation is almost nonexistent.

H. Equivalent Circuit of Bowed String

Figure 8(a) represents the bowed string in terms of hypothetical single-mode balanced transmission lines, i.e., lines the two sides of which never act in parallel. The constant velocity of the bow (branch A–A') divides into three parts: (1) slipping velocity in branch B–B'; (2) string translational velocity in series lines D and G; and (3) string rotational velocity in lines E and H. Rotation will be discussed in a later section. If all string motion is blocked, all bow motion appears in branch B as slipping. If, on the other hand, slipping is prevented, bow motion appears as string translation and rotation, and as translation alone if torsional admittance is zero.

During the period when the string clings to the bow, branch B is a very high impedance; branch A, likewise being a source of constant velocity, has very high impedance. When therefore a Helmholtz discontinuity of amplitude $D = v_b/\beta$ approaches the bow from the right (translation), the equivalent circuit is that of Fig. 8(b); F by Thevenin's theorem then equals $2Zv_b/\beta$. The characteristic impedance of rotation Z' will be defined later. Hence the transverse force which that section of string exerts when the discontinuity is incident at the bow is $F(Z + 2Z')/(2Z + 2Z')$, or F if Z' is infinite.

As seen by the discontinuity approaching from G the torsional mode is in series with the translational mode. As seen by the changing force of friction in B the two are in parallel. If Z' were zero the discontinuity would

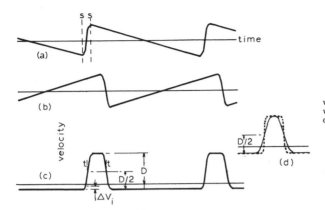

Fig. 9. A more realistic shape of Raman wave as seen at bow: (a), (b) the two waves separately; (c) their sum; (d) effect of increasing bow force.

pass the bow from G into D without reflection and the translational mode would be isolated from C–C'.

I. Dependence of Timbre on Bow Force and Placing

The vibrational force at the bow–string contact is the product of a particular spectrum of string velocities at that point multiplied by the impedance spectrum offered by the string. The impedance is independent of bow velocity and force, but by experiment we know that the string velocity is not. It is interesting to inquire what will happen in a hypothetical instrument, for which the impedance at the bow, unlike that for the violin, is a pure resistance independent of frequency as at C in Fig. 8(a), and the frictional force during slipping is independent of velocity, with instantaneous transition at capture and release as conceived in the simple concept of friction. Assuming string rotation negligible in effect, the velocity during sticking will be the constant speed of the bow, and the vibrational force on it must therefore be constant because the load is a pure resistance. Similarly during slipping, since force is constant, velocity must be constant, again because of the simplicity of the resistance. The result is therefore a simple Helmholtz rectangular wave in an arrangement for which the load is finite. Unlike that of the violin, the timbre of such an instrument will be independent of the way the bow is handled. The chief reason for the variation of timbre in actual instruments (and this of course is a musically useful feature) lies in the complexity of impedance.

In the hypothetical instrument just discussed, temporal sharpness in velocity change at the bow will be paired with spacial sharpness in the apex that shuttles back and forth on the string. In a real instrument rounding of the apex will pair with an S shape in the velocity curves, a lengthening in the interval of slipping and a shortening of time during which slipping velocity is more' or less constant. Figure 9 illustrates how the incident Raman waves vary at the bow as a function of time; (a) is the wave from the nut and (b) is its retarded reflection from the bridge, while (c) is their sum. For a given position of the bow, the period of constant slip rate equals that in the Helmholtz construction if the S-wave has negligible duration s–s, but if s–s grows to the time of transit from bow to bridge and back, pulselength doubles and the period of constant slipping disappears.

These extreme conditions are related, respectively, to the conditions of maximum and minimum bow force. In any case, since the apex is never actually discontinuous, a finite time is needed for the frictional force of release to build to the breaking point. Let the change of incident velocity that leads to release at minimum force be ΔV_i in Fig. 9(c). Therefore

$$P_{\min}(\mu_s - \mu_d) = 2\Delta V_i \cdot Z$$

and from Eq. 1a,

$$\Delta V_i/D \doteq 0.25 Z/r\beta, \qquad (6)$$

where D is the greatest difference in velocity between the two periods. The same numerics used following Eq. 1b give $\Delta V_i/D = 0.045$ for a cello A string. The bow at minimum bow force has so precarious a hold on the string that even though the front of the velocity wave is gently continuous, escape is almost immediate.

When the velocity pulse is watched oscillographically while the bow is suddenly raised, the region previously identified as that of slip spreads into the flat region where sticking formerly occurred. This overflow occurs because of inharmonicity and inequality in decay rates among the natural modes of the string. Previously, it was held in check by the damlike action of static friction. This restraint by friction is increased when the bow is pressed more firmly on the string. Escape is now delayed, but when it does come it is more abrupt. Similarly the end of slipping is hastened by the greater effect of negative resistance close to zero velocity. Since in typical fiddle strings the rapid rise of velocity and its later fall in Fig. 9(c) represent substantially the same unchanged Raman wave before and after reflection, the two points of half-velocity (relative to the bow) have the same origin and their time separation is that of the Helmholtz mode regardless of bow force. The force brought into play at escape and also that at capture enter into the equilibrium shape of the Raman wave and therefore they influence both the beginning and the end of slipping. The squeezing together of the bottom of this peculiar "scissors" at points of low slip, therefore, spreads the top and makes the wave more nearly rectangular, as shown in Fig. 9(d). Certain aspects of string motion discussed here, including the reduction in duration of the slip period that accompanies increase in bow force, have been mentioned previously by Lazarus.[13]

J. The Ripple

The ripple in the long period of sticking in Fig. 3 indicates that the string, while still retaining "traction," is rolling under the hair of the bow. It is equal to the transverse force exerted by the bow multiplied by the torsional admittance at its contact with the string. More vividly it may be thought of in terms of reflection from the bow of the fronts of the two Raman waves with their rounded corners. If the bow were immediately transparent to the Raman waves these reflections would not arise. However, a finite time is needed and during that time new wave components proportional to the force are set up, not only in one direction away from the bow, but in both, this being another way of saying that with regard to the force the two sections of string are in series.

For example, the Raman wave approaching from the nut (the end opposite the bridge) cannot because of its rounded corners immediately release the string. During the finite time in which force is building to the breaking point the wave is reflected rather than transmitted. The reflections from the bow travel to the nut and are reflected back toward the bow, arriving there just before the next release and causing a momentary rolling. Thus, the reflection produced at a given first release, as in Fig. 2(d), returns to the bow as the first ripple to the left of the third release, and so on. This accounts for the seeming growth during the long period that is particularly evident in Figs. 2(c) and 2(d). Actually, within one long period, age in this series increases to the *left* while time increases to the *right*.

The force existing during capture is likewise the source of reflections, but since the discontinuity now moves toward the nut, the reflections move toward the bridge and so are trapped in the short section of string. They are commonly smaller than those set up at release. This series can be observed to decay during the long period in an opposite sense: age increases to the right as with time. A tendency for a minimum at the middle of the period is seen in Fig. 2(f) in which both series occur.

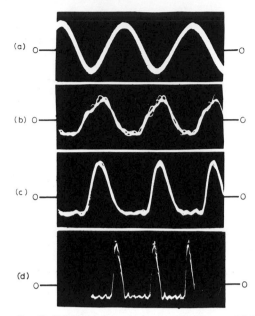

FIG. 10. Velocity curves at bow, very stiff string, $\beta = 1/7.5$: (a) well below minimum force; (b) force near minimum; (c) increased force; (d) heavy force.

As it travels, the string reflection is much larger than the ripple produced when it returns to the bow. This ripple is merely the fraction that is transmitted past the bow. In the violin, steel E string calculation shows the fraction to be about 0.2. This is the reason why the ripple goes through a minimum as the magnet passes the position of the bow. Initially, the reflection can with sufficient bow force be in the order of half the main velocity discontinuity. Figure 2 indicates ripples much lower than this.

The ripple in the region of *slip* [e.g., in Fig. 2(f)] is to be identified with the residue of the reflections that shuttle between the ends of the string and the bow, observed at the moment when the bow is at last transparent. They are now transmitted in entirely past it without the reflection loss previously encountered.

It seems likely that these reflections play a part in the peculiarities observed at the beginning of a note.[14] Their damping by the finger contributes to the difference between the timbre of fingered notes and that of the open string.

K. Starting Conditions and Negative Resistance

It is conceivable that when a bow is slowly drawn across a fiddle string the string will move a little to one side and stay without vibration. The fact that it never does this means that the condition is one of instability. The vibration, however, can be one of very low ampli-

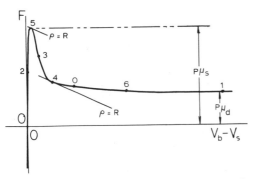

FIG. 11. Frictional force as function of relative velocity; point 1 represents dynamic, point 5 static friction.

tude. In the oscillogram in Fig. 10(a) vibration is essentially sinusoidal. Here we have vibration without discontinuities: sticking does not necessarily alternate with slipping.

Qualitatively, the friction for rosined surfaces will be similar to Fig. 11, that is, continuous.[15] The vibration just discussed might have occurred at point 0, the requirement being that the slope of the force–velocity diagram be negative and greater than the resistance of the load. Consider a bow moving across a taut string and for simplicity ignore energy dissipation in the string itself and string rotation. At the bow, velocity of the string V_s has two components, that of incoming waves from both directions and that produced by the force F of the bow:

$$V_s = V_i + F/\bar{Z}. \tag{7}$$

Since F produces waves in both directions, $\bar{Z}$ is twice characteristic impedance Z.

Let F be a function of $z = V_b - V_s$, subscripts meaning "bow" and "string." For small changes in velocity

$$dV_s = dV_i + 1/\bar{Z} \cdot \delta F/\delta z \cdot dz,$$

where $\delta F/\delta z$ is the slope ρ in Fig. 10. Since $dz = -dV_s$,

$$dV_s = \frac{dV_i}{(1 + \rho/\bar{Z})}. \tag{8}$$

When $\rho = 0$, as at 1 in Fig. 11, $dV_s = dV_i$ and the bow is "transparent" to V_i. With ρ positive, $dV_s < dV_i$, and if ρ is great enough, motion is negligible, as in sticking at point 2. If ρ is negative, $dV_s > dV_i$, and amplification occurs, as between 4 and 6; the contact has become a "regenerative receiver." When V_i is part of a standing wave system, it will have suffered a decrement during the past cycle, apart from the effect of F. If this equals the increment produced by amplification, steady oscillations can occur in the system as a whole.

When $\rho/\bar{Z} = -1$, as at 4 and 5, dV_s/dV_i becomes infinite and instability local to the contact develops,

leading to capture or to release. The system as a whole however, remains stable provided losses continue to be made good by the overall action of negative resistance.

L. Unsteadiness in Vibration

Measurements of the frequency and amplitude of vibration within short intervals have shown that they are not constants as is ideally supposed, but that there is sometimes a small variation which might be large enough to impart warmth of tone color somewhat like that of the vibrato.[14] A. H. Benade[16] remarks that while physically one can point to probable reasons for "jitter" in bowed strings, analogous causes do not exist in the wind instruments, and he points out the greater difficulty in obtaining beats between bowed strings than between woodwinds. A first-order change in bow velocity would produce a first-order change in amplitude, perhaps a second-order change in phase. A first-order change in bow force or in friction might produce a significant change in phase by delaying release or hastening capture.

When the generated wave is strictly periodic in time τ, V_s and V_i in Eq. 7 are repeated with exactness at times t, $t+\tau$, $\cdots$, $t+n\tau$, $\cdots$, as is their difference $V_s - V_i = f/\bar{Z}$. This does not hold true at a wolf tone, in which there exists no waveshape that satisfies conditions for a wave periodic in time τ. This is one case in which our trust that Nature will find conditions for a periodic wave is beside the mark, and it is to be noticed that the situation occurs without assuming vagaries in bow motion or friction. One naturally wonders whether there are less drastic situations in which some subtle failure of periodicity occurs, such as when an upper mode in the vibration of the string falls on or near a body resonance.

III. ELASTIC EFFECTS

In its simplest terms, the conception of the ideal string contains somewhat contrary requirements. The string must withstand tension but be without significant longitudinal or flexural stiffness: longitudinal stiffness must be low enough to ensure that tension is not increased by the extension in length accompanying finite vibration, and flexural stiffness must be low enough to ensure a harmonic relation between natural modes. But finite Young's modulus means finite rigidity modulus, and their customary numerical relationship guarantees a finite amount of twisting in the bowed string that, at least for purposes of explanation, we should prefer to be without.

A. Longitudinal Stiffness

On the lowest string of an instrument, as vigorous bowing shows, the amplitude of vibration at its midpoint can be large, on a cello almost 1 cm. When the string is first mounted, there is a first-order stretching

Δ depending on the tension applied, length, and cross-sectional area, and the Young's modulus of the core. As one now pulls the string aside a distance d at the midpoint, it is stretched further by a second-order length δ equal to $2d^2/L$, L being the length of string. As a result, the second-order increase in tension divided by the first-order tension is $2d^2/L\Delta$, the relative sharpening in frequency being $d^2/L\Delta$. Such a rise can easily be observed in a commercial steel-core string if it is tuned down $\frac{1}{2}$ oct and bowed vigorously. Approximate theory indicates that with bowing to amplitude d, assuming the Helmholtz type of vibration, the relative frequency rise is $\frac{2}{3}d^2/L\Delta$. In a properly designed string, the effect is small. Thus a certain cello C string (the string of lowest frequency) has a core of three steel wires about 0.008 in. (0.020 cm) in diameter, so that δ=0.029 cm and Δ=0.44 cm. When the open string is bowed at normal tuning to an amplitude $d=1$ cm, it will sharpen 36 cents.[17] Fingered notes can, of course, be corrected. When tension is relaxed below normal, the effect can become very pronounced for two reasons: (1) the decrease in Δ and (2) the increase in amplitude d for a given speed of bow. With gut, the effect is less than with steel because Δ is greater.

The loss of brilliance of an instrument when strings are tuned down a few semitones is a matter of common experience. While in part this is a result of decrease in vibrational force that the relaxed string can exert on the bridge, it is sometimes thought that the static force of the strings on the body actually changes its responsiveness to the force of vibration, perhaps by changing elastic modulus or by imparting firmness to glued joints —in any case because of some nonlinearity in elasticity. Since the hypothesis seems not to have been checked by objective measurements, it is worth noticing that another influence needs to be considered, namely, the effect of the longitudinal stiffness of the strings. Resonances might be studied in two ways, first by the use of strings having tension and high compliance (e.g., something like rubber), and second, strings with or without tension and having low compliance (e.g., steel). Even if the elasticity of the violin body were perfect in its linearity, the addition of longitudinal stiffness in the strings would alter modes of the ensemble. The effect will be increased by decreasing the obtuse angle of the strings at the bridge. Steel strings will call for an angle different from that for gut strings, even though tensions are the same.

Some measurements by Leonhardt[18] are consistent with the idea that it is stiffness rather than tension in the strings that is accountable for the difference: the acoustical responsiveness of the body of a viola to force electromagnetically applied at the bridge was found to depend on whether the strings were steel or gut.

B. Flexural Stiffness

The beautiful simplicity of the flexible string is endangered by flexural stiffness. The periodic motion produced in bowing precludes inharmonicity such as that found among the freely vibrating modes of the piano string, in which higher modes are sharper than the corresponding integral multiples of the lowest. Instead, one looks for difficulties in intonation, deterioration in tone quality, reduction in amplitude of harmonics, the need for abnormal bow force, and so on.

In a flexible string, resonance occurs not only at the fundamental frequency but also at all of its integral multiples up to a high order. The string is naturally adapted for response to a periodic wave however complex. In a string flexurally stiff and under tension, this remains true for the lower orders but fails for the higher ones. Consider for example the G string of a violin. Prior to about 1700, when wound strings became available, all string were of gut, but the G string was unsatisfactory. The reason why it is never used today is not hard to see. In bowing, the frequency of vibration is close to that of the lowest natural mode of the string, but its seventh harmonic falls midway between the sixth and seventh natural modes, that is, completely out of resonance. Regardless of bow force, vibration at this harmonic must be negligible.

The lowest mode of vibration has a natural advantage in controlling frequency in the bowed string in that it is twice as large as the second, three times as large as the third, and so on. The lowest few modes have the additional advantage over the upper in that they are essentially synchronous (except for their integral relationship), since, in going from upper to lower, the perturbation in frequency falls in proportion to the square of the order. In bowing, frequency remains near that of the lowest natural mode.

Figures 10(b), 10(c), and 10(d) illustrate an extreme case in which the steel string used in obtaining Fig. 10(a) was subjected to increased bow force so as to bring in a period of sticking. With each increase, the duration of this period increased and the angularity of the slip pulse became greater. At no time, however, within the range of conditions shown, was duration of the flat region as great as one would expect merely on the basis of bow position ($\beta=1/7.5$). Increased force increased the strength of higher harmonics, but the lower ones remained of greatest importance in controlling frequency.

An expression for inharmonicity given by Young provides a convenient starting point.[19] The ratio of frequency of the nth mode in a string with flexural stiffness to that without it is

$$\nu_n/n\nu_o = 1 + Bn^2 = 1 + (\pi^2 Q\kappa^2 S/2L^2 T)n^2, \quad (9a)$$

and inharmonicity in cents is

$$\delta_n = bn^2, \quad (9b)$$

where b is inharmonicity in the nth mode in cents per square of mode number n, and equals $1731B$. Here Q is Young's modulus, κ the radius of gyration (equals

TABLE I. Inharmonicity due to various string materials.

Material	Young's modulus, $Q/10^{11}$ (dyn/cm²)	Density (g/c³)	Inharmonicity parameter/10^{10} Q/ρ^2
(Silver)	7.5	10.5	0.68
Brass	9.2	8.6	1.24
German silver	10.8	8.4	1.53
Gut	0.39	1.37	2.1
Steel	19.0	7.8	3.1
Aluminum	7.0	2.7	9.6

$\frac{1}{4}$ diam for a circular cross section and for rotation about a diameter), S the area of section, L the length of string, and T the tension.

In the choice of string for a given string position (i.e., in the violin, E, A, D, or G), acoustical balance requires that tension remain unchanged. In the following convenient form for a fiddle, the first parenthesis comprises values required by the instrument, the second the pertinent constants of the string material:

$$b = 42.5 (T/\nu_o{}^4 L^6) \cdot (Q/\rho^2). \tag{10a}$$

Table I lists values of Q/ρ^2 for various materials. Silver is thus the best of these materials if tensile strength does not come into question. Aluminum is the worst. Gut and steel fall midway in the table. It is noteworthy that the difference between gut and steel is not large.

As an example, in the violin steel E string, T is about 16 lb (7.6×10^6 dyn), $L \doteq 32.5$ cm, and $\nu_o = 660$ Hz. Using the value of Q/ρ^2 in Table I, $b = 0.042$. In the piano,[19] the strings having the same fundamental frequency have b typically equal to 1.0, so that in the fifth mode inharmonicity is about 25 cents compared with 1 cent in the violin E string.

The following is an alternative form[19]:

$$b = 133.5 (d^2/\nu_o{}^2 L^4) \cdot (Q/\rho). \tag{10b}$$

Assuming that all four strings of a fiddle are of the same material, the inharmonicity coefficient in the mth string ($m = 1$ in the highest) is

$$b_m = b_1 \cdot (T_m/T_1) \cdot (\nu_{01}/\nu_{0m})^4.$$

Assuming typical values of tension T, coefficients for solid strings for the violin are as given in Table II.

When the nth harmonic of the repetition rate falls about midway between the nth and the $(n-1)$th natural mode, the corresponding component of frictional force at the bow encounters an antiresonance and is almost completely suppressed. Nonlinear friction can generate such a force, but its effect will be small. A slight shift in repetition above the assumed rate is likely but will not change the result qualitatively.

The situation may alternatively be stated as follows. In one cycle of the wave during bowing, there will be a perturbation in phase in the nth mode that can be

shown to be approximately $2\pi B n^3$. The shift most destructive to resonance in this mode is π rad. Hence the value of n that gives that condition will be $n_{min} = (1/2B)^{\frac{1}{3}}$.

If we apply this result to the stiff string used in obtaining Fig. 10, we find that $b = 2.2$ or $B = 1.27 \times 10^{-3}$, so that $n_{min} = 7.3$. The spectrum shown by the solid circles in Fig. 6 applies to this string. The low values of the seventh and eighth harmonics are consistent with the prediction.

The following experimental method gives a basis for determining b for a string of unknown physical constants or composite structure having a frequency ν_o and length L. A short length L_{cant} is mounted as a cantilever and the rate of vibration in its lowest mode measured. According to a standard method, this rate for a solid string is

$$\nu_{cant} = 0.140 (Q/\rho)^{\frac{1}{2}} \cdot d/L_{cant}{}^2.$$

Substituting in Eq. 10b the experimental value of $(Q/\rho) \cdot d^2$ obtained above, we obtain for the unknown string

$$b = 6820 (\nu_{cant}/\nu_o)^2 \cdot (L_{cant}/L)^4. \tag{11}$$

Since this requires no knowledge of structure, it permits the estimation of b for wound strings. It assumes that flexural stiffness is the same with or without tension.

Coefficient b was estimated according to Eq. 11 for a few wound strings. The results, along with calculations for some solid strings, are given in Table III. The quantity n_{min} giving the order of harmonic suffering a phase reversal may be regarded as a figure of merit of a given string in respect to stiffness. It lies above the upper limit of usefulness of the string.

When n is small, ϕ_n is a decrement in the same sense as the usual logarithmic decrement per fundamental cycle, in that it represents the change in unit vector of the harmonic in that interval, the change being at 90° to the change used by energy losses. In both cases, it represents a change to be countered by the negative resistance mechanism. The usual logarithmic decrement as previously noted equals $2Z/r$ and has a value that while variable is in the order of 0.05. Phase distortion will, therefore, have effects comparable with those of resistance losses when $2\pi B n^3 = 0.05$ and $n = n'$ $\doteq 0.2 B^{-\frac{1}{3}}$. This means that for orders higher than n',

TABLE II. Inharmonicity coefficient b in solid strings for violin.

m	String	Tension (lb)	Inharmonicity b in cents/square of mode number, n		
			Gut	Steel	Aluminum
1	E	18	0.027	0.04	0.124
2	A	11	0.084	0.124	0.38
3	D	11	0.43	0.63	1.95
4	G	11	2.3	3.2	9.9

stiffness is a more important limitation than are resistive losses and is more a determinant of bow force than are resistive losses. The actual importance will of course depend on the ability of the negative resistance mechanism to contend with ϕ; however when $\phi = \pi$, resonance is completely destroyed. Harmonics near n_{min} will also be adversely affected. Values for n_{min} are given in Table III.

● A method of rating flexibility of bowed strings according to the order of harmonic that suffers a phase shift of π rad in one fundamental cycle and therefore gives negligible response is suggested. The useful limit is still lower, perhaps two-thirds as great. Flexural effect begins to be of consequence at about one-fourth of that order.

● It is suggested that any string having b equal to 0.1 or less, i.e., n_{min} not less than 21, will have flexibility satisfactory for bowing. The limit of acceptability, however, cannot be stated with exactness on the basis of present information.

● A solid steel string is somewhat more inharmonic than one of solid gut, but not much.

● Of the materials considered, solid silver is the least inharmonic, solid aluminum most, while steel and gut lie midway between them.

● Inharmonicity in solid strings becomes rapidly worse in going to lower strings of a fiddle. In the violin, the lowest solid strings that seem acceptable are the A for gut and the E for steel. The E for aluminum seems to fall on the border line of acceptability.

● One cello C string wound on steel cable equaled the violin solid-steel string in flexibility.

TABLE III. Inharmonicity in various solid and wound strings.

String	Relative frequency perturbation per n^2 $B \times 10^3$	Perturbation cents per n^2 b	n_{min}
0.006-in. rocket wire	0.0013	0.0023	73.0
Violin steel E	0.023	0.04	28.0
Violin gut D	0.25	0.43	13.0
Violin gut G	1.3	2.3	7.5
Violin D, alum. on gut	0.12	0.2	16.0
Violin G, metal on gut	0.075	0.13	19.0
Cello G, on gut	0.046	0.08	22.0
Cello C, on gut	0.13	0.2	16.0
Cello C, 2 layers on steel cable	0.23	0.04	28.0
Suggested aim	0.058	0.1	21.0

TABLE IV. Torsional/transverse velocities and impedances for violin and cello strings.

	Density ρ (grams/cm³)	$G \times 10^{-10}$ (dyn/cm²)	$V' \times 10^{-5}$ $(G/\rho)^{\frac{1}{2}}$ (cm/sec)	$V \times 10^{-5}$ (cm/sec)	V'/V	Z'/Z
Violin						
Steel E	7.9	75.0	3.2	0.43	7.5	3.8
Gut E	1.36	1.0	0.86	0.43	2.0	1.0
Cello						
Gut A	1.36	1.0	0.86	0.31	2.8	1.4
Metal on steel A			1.5	0.31	4.9	2.5
Metal on steel C			0.5	0.09	5.9	3.0

● In an aluminum-wound gut D for violin, b equalled 0.2 as compared with 1.95 for solid aluminum. A winding of different metal would probably be preferable.

C. Torsional Compliance

Since the frictional force exerted by the bow is applied along the circumference of the string, the resulting motion includes a degree of twisting. E. Leipp[20] seems to have been the first to show by simple stroboscopic observation that the string rotates somewhat. However, in his view a torsional relaxation vibration occurs independently of the transverse vibration, its frequency being the lower of the two. He identifies it with the vibration found above maximum bow force, apparently in the region of Fig. 7 labeled "raucous." This erratic vibration does have a lower "frequency" than the normal transverse one, and torsion doubtless enters as a complication, but the reason for the lowering is the lengthening of the period of sticking through failure of release rather than the existence of a torsional resonance having lower frequency. As shown in Table IV, the torsional velocities of the strings studied, rather than being less than those of the transverse, are typically five times greater.

Figure 12 indicates a simple method for measuring pertinent parameters in any string either solid or wound as normally mounted on an instrument. The bow is placed on the string with sufficient normal force to ensure sticking and moved to a position at the right, thus producing transverse displacement and angle of rotation ϕ. A light-weight index AB attached to the string moves to position $A'B$, B being the stationary point where rotation annuls translation. Length $AB = R$ and radius a are the measurements to be made, or if preferred, a, z, and ϕ.

In the following, V is velocity of transverse propagation and Z the corresponding characteristic impedance, while V' is velocity of torsional waves and Z' their characteristic impedance. Z' is defined in terms of linear motion at the circumference of the string rather than in angular measure, thus permitting the particle velocities in the two modes to be considered in parallel.

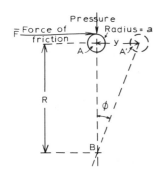

FIG. 12. Static measurement of torsional properties: rotation versus translation.

In Appendix B, it is shown that for any string thus measured

$$V/V' = (k/a) \cdot (a/R)^{\frac{1}{2}},$$
$$Z/Z' = (a/k)^2 \cdot (V/V') = (a/k) \cdot (a/R)^{\frac{1}{2}}. \quad (12)$$

For solid strings,

$$\frac{Z'}{Z} = \frac{\frac{1}{2}V'}{V} = \frac{(G/\rho)^{\frac{1}{2}}}{4Lf}. \quad (13)$$

Table IV gives velocity data for certain strings thus estimated. Shear modulus G for gut was measured in a torsion pendulum.

● Referring to Fig. 8(b), the fraction of force F that is exerted across torsional impedance $2Z'$ and therefore available to release the string is $(1+Z/Z')^{-1}$. If Z'/Z is large compared with unity, almost all of F can be thus applied. For the steel E string for violin, $Z'/Z = 3.8$ and the force ratio 0.8, whereas for gut $Z'/Z = 1.0$ and the force ratio is 0.5. Hence maximum bow force is significantly reduced by torsional compliance.

● Torsional compliance probably alters waveform, but the nature and degree of the effect are difficult to foresee. So far no significant acoustical effect seems to have been traced to it.

● The ripples in motion of the string at the bow during the long period of sticking could obviously not exist except for torsional compliance.

● Effects of twisting are greatest at maximum bow force and probably entirely negligible at minimum force.

● Within one set of metal strings for cello wound on steel cable, the measured ratio Z'/Z and the presumable effects of rotation vary less from string to string than would be expected from a set of solid strings.

● The steel E string for violin and the obsolete gut E represent extremes between which the other strings measured fell. This suggests, to the extent that both

are considered acoustically satisfactory, that the effect of rotation is in general unimportant.

● In Eq. 13 for solid strings, Z'/Z depends only on the parameters of the material and the velocity of transverse propagation. For a given material, the highest strings suffer the greatest rotational effects, since Z'/Z is least. For a given string position, the effect is independent of diameter.

ACKNOWLEDGMENT

It is a pleasure to acknowledge valuable discussion with W. S. Gorrill and C. M. Hutchins, and help in procurement of apparatus from Frank Fielding, Stewart Hegeman, C. M. Hutchins, Mrs. F. A. Saunders, and C. E. Schelleng.

APPENDIX A. DERIVATION OF EQUATION 1

The string is a series resonant circuit at fundamental and harmonics as seen at its contact with the bow, and its reactance when bowed is substantially zero. Referring to Fig. 5, resistance at the nth (low) harmonic is $R_n = X_n^2/r_n$, where $X_n = T/\beta Ln\omega$.

Confining attention to the case when $\beta \ll 1$, the velocity pulse is short and the first several harmonics have peak values equal to or near $2v_b$. Therefore, peak force for the nth harmonic is

$$F_n = 2v_b R_n,$$

and if $r_n \doteq r_1$, which appears true enough for present purposes, peak-to-peak force on the string is

$$F_P \doteq 4v_b R_1 \sum_1^3 \frac{1}{(2k-1)^2},$$

since all components have maximum values of the same sign at the middle of the period of slip and only odd harmonics enter into peak-to-peak force. Since $(Tm)^{\frac{1}{2}} = Z$ and by Mersenne's law

$$\omega = \frac{\pi(T/m)^{\frac{1}{2}}}{L},$$

it follows that $T/L\omega = Z/\pi$ and $R_1 = Z^2/(\pi^2\beta^2 r_1)$. From this it follows that

$$F_p \doteq 0.5Z^2 v_b/r_1\beta^2.$$

The fundamental is responsible for about 80% of the peak-to-peak force. The higher harmonics are more important in their effect on timbre than on minimum bow force.

APPENDIX B. TWISTING AT MAXIMUM BOW PRESSURE

Referring to Fig. 12, let a represent the radius of string; k the radius of gyration; ϕ the radians of twist, static; $\bar{F}$ the tangential static force; $S_y = \bar{F}/y =$ stiffness to transverse deflection; $S_\phi = \bar{F}a/\phi =$ torsional stiffness, angular measure; R the distance AB; V' the torsional propagational velocity; and V the translational propagational velocity $= (T/m)^{\frac{1}{2}}$.

From foregoing, $S_y/S_\phi = \phi/ya$. For small ϕ, $y = R\phi$ and $S_y = T/\ell$; therefore torsional stiffness of unit length is $S_\phi \ell = TaR$. For simplicity of explanation, the long section of string is ignored, both for transverse and torsional deflections, but no error is thereby introduced into the final result. Since V' equals the square root of torsional stiffness of unit length in angular measure divided by the moment of inertia per unit length, mk^2, it follows that

$$(V'/V)^2 = Ra/k^2. \qquad \text{(B1)}$$

When force is applied circumferentially to a disk so as to produce rotation about its center, one may replace the actual mass distributed over its surface by a mass m' distributed around the circumference such that $m'a^2 = mk^2$; i.e., $m' = m(k/a)^2$. Interpreting m and m' as masses per unit length, characteristic impedance in rotation Z', as seen in linear motion at the circumference, i.e., by the bow, is

$$Z' = V' \cdot m' = V' \cdot m(k/a)^2.$$

Also, since $m/m' = (a/k)^2$,

$$Z/Z' = (a/k) \cdot (a/R)^{\frac{1}{2}}. \qquad \text{(B2)}$$

[1] H. L. F. Helmholtz, *On the Sensations of Tone* (Dover, New York, 1954), Part 1, Chap. 5, pp. 80–88.

[2] C. V. Raman, *On the Mechanical Theory of Vibrations of Bowed Strings, etc.* (The Baptist Mission Press, Indian Association for the Cultivation of Science, Calcutta, 1918), Part 1, pp. 1–158; Proc. Assoc. Cultivation of Science 6, 19–37 (1920).

[3] F. A. Saunders, J. Acoust. Soc. Amer. 9, 81–98 (1937).

[4] Catgut Acoust. Soc. Newsletter, No. 10, 18–21 (1968).

[5] Note loss of detail in passing from Fig. 3(a) to Fig. 3(b). One of the two lines in Fig. 3(b) should be ignored. The integrating

circuit, inserted between the preamplifier (output impedance 0.05 MΩ) and the oscilloscope (input impedance 3.3 MΩ) consisted simply of a series resistance (0.25 MΩ) and a shunt capacitance of a value to give a 3 dB loss at 12 Hz. Repetition rate in the string was 440 Hz.

[6] In a flexible lossless string the motion at a point for which the time of propagation to the nut and thence back to the bow is half of a period is the same in form and magnitude as at the bow. For practical purposes, the distance of this image point from the nut equals that from bow to bridge. Oscillograms for the C string were made in this way on the cello rather than the monochord. When the magnet strays away from the bow or the image point even a small distance, the ripple increases because it no longer represents pure rolling of the string beneath the bow.

[7] K. C. Kar, N. K. Datta, and S. K. Ghosh, Ind. J. Phys. 25, 423–432 (1951); K. C. Kar and C. Dutta, Ind. J. Theoretical Phys. 5, No. 2, 31–37 (1957).

[8] F. Eggers, Acustica 9, 453–465 (1959).

[9] W. Reinicke and L. Cremer, J. Acoust. Soc. Amer. 47, 988–992 (1970).

[10] This is approximately right, but to know that it is really the limit of musical acceptability would require experiment. With pressure so great as to cause complete reflection, the force would be doubled.

[11] In a plucked string, waveform of force is rectangular rather than sawtoothed, therefore tending to give equality to the lower harmonics rather than an inverse-frequency decline. This is the reason for the "metallic" sound when the string is plucked close to the bridge.

[12] M. Eisenberg, *Cello Playing of Today* (The Strad, London, 1957), pp. 5–6.

[13] H. Lazarus, J. Acoust. Soc. Amer. 48, 74(A) (1970); see also L. Cremer and H. Lazarus, "Influence of the Normal Force between the Bow and String on a Steady Vibration of the Bowed String," Sixth Int. Congr. Acoust., 6th, Tokyo (1968), paper N2-3.

[14] H. Fletcher, E. D. Blackham, and O. N. Geertsen, J. Acoust. Soc. Amer. 37, 851–863 (1965); J. C. Risset and M. V. Mathews, Phys. Today 22, No. 2, 23–30 (1969); J. W. Beauchamp, Tech. Rep. No. 15, Univ. of Ill. (1967) (unpublished).

[15] E. Rabinowicz, Sci. Amer. 194, No. 5, 109–118 (May 1956); F. Palmer, Sci. Amer. 184, No. 2, 54–58 (Feb. 1951).

[16] A. H. Benade (personal communication, 1967).

[17] For example, by bowing the open C string 7 cm from the cello bridge with a velocity of 50 cm/sec.

[18] K. Leonhardt, *Geigenbau and Klangfrage* (Verlag Das Musikinstrument, 6 Frankfort am Main, 1969), pp. 68–69. In a review of this book, J. Acoust. Soc. Amer. 48, 638–639 (1970), I too ignored the possibility set forth here.

[19] R. W. Young, J. Acoust. Soc. Amer. 24, 267–273 (1952).

[20] E. Leipp, Ann. Télécom. 17, 99–106 (1962).

Editor's Comments on Papers 13A and 13B

The changing force that the player applies to the string is one of the most important parameters in bowing an instrument. It is of particular interest to the player because of its limits and the effects within those limits on tone quality and loudness. To the physicist it suggests an interesting study in stability of oscillation and the changes in waveform necessary if stability is to be achieved when extreme nonlinearity in frictional force exists in an otherwise linear situation. The following papers offer a solution to this problem.

Information about the life and work of Lothar Cremer can be found on page 100.

The Influence of "Bow Pressure" on the Movement of a Bowed String: Part I

LOTHAR CREMER

The very simple and clear description of the motion of a bowed string which Hermann von Helmholtz gave in his "Lehre von den Tonemfindungen" based on measurements and a theoretical treatment as a free motion neglecting losses and stiffness may even today be regarded as a good first approximation. This free motion is depicted as a triangle with connection of the supports of the string as the base and a constant top angle whose apex propagates to and fro inside two enveloping parabolae, changing its sign with direction of propagation (Figure 1a). We may call this the "Helmholtz motion," though in fact Helmholtz himself observed other more complicated motions. At the bowing point at distance a from the

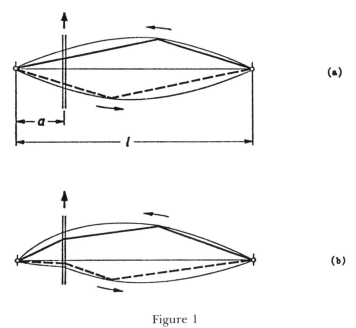

(a)

(b)

Figure 1

bridge it results in a displacement that increases linearly with time during the "sticking time" t_+ according to the constant bow velocity v_B and decreases, here again linearly, during the "sliding time" t_-, so that the following equation holds:

$$\frac{t_-}{t_- + t_+} = \frac{a}{l} \tag{1}$$

where l = length of string.

Moreover, the velocity jumps discontinuously from bow velocity $v_B = v_+$ to a sliding velocity v_-, i.e., by

$$\Delta v = v_+ + |v_-| \tag{2}$$

and since

$$|v_-|t_- = v_+t_+ \tag{3}$$

we get

$$\Delta v = v_B \frac{1}{a} \tag{4}$$

On the other hand, Δv is proportional to the lateral force at the bridge, which excites the body, and is in turn proportional to the sound pressure in the room. So Helmholtz's description allows the conclusion that this pressure increases with the bow velocity and is inversely proportional to the distance a from the bridge. Both are in accordance with the experience of the player.

But the Helmholtz treatment does not include the influence of the frictional force between bow and string that increases with the so-called "bow pressure" (this means the force with which the bow is pressed against the string), because Helmholtz treated the string motion as a free one.

This kind of treatment is often a good approximation in self-exciting systems; for instance the forces which keep the pendulum of a clock vibrating are very small compared with the forces of gravity and inertia at the pendulum.

At the bowed string the frictional force is certainly not negligible. This becomes evident if we compare the reverberation of the plucked string when free with the rather short damped sound that we get when we pluck it when it is loaded by the weight of the bow.

But the frictional force which thus appears would not change the Helmholtz motion if during the full period it were independent of time. Then the energy introduced by the bow would be the same as that which friction transforms into heat during the sliding time, and we would have only to add to the Helmholtz displacement a constant triangular deformation with the apex at the bowing point, as already stated by Raman (see Figure 1b). We may therefore omit this additional

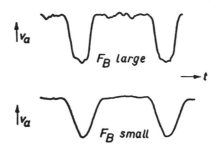

Figure 2

constant deformation in Figures 5 and 6 that follow. But, for the compensation of losses other than those of friction at the bow, we need additional friction forces which must be dependent on time because they must be larger in the sticking than in the sliding interval.

Raman calculated these forces under the assumption that the time dependence of the velocity corresponds to that of the free motion of the string and that the losses are the same for all frequency components. Neither assumption is satisfied physically. Furthermore, in this way one gets one special frictional force, which corresponds to but one value for the bow pressure.

Indeed, change in bow pressure is of importance to the player as his third means for changing the sound of his instrument, especially in that it changes the timbre. High bow pressure produces more overtones and thus produces loudness sensation although the fundamental remains unchanged.

This means that changes in bow pressure actually modify the shape of string motion at the bow. That this is so may be demonstrated by recording the transverse velocity of a metallic string at the bowing point by placing this point in a concentrated magnetic field. Figure 2 shows two oscilloscopic registrations by H. Lazarus (1) where the upper corresponds to a large "bow pressure" and the lower to a small one. Similar oscillograms have been published by J. Schelleng (2).

Especially for low applied force we get a rather smooth change from v_B to v_- instead of a jump. It is physically clear that the jump of the Helmholtz motion cannot appear in reality. Similarly, the corresponding sharp corner in the displacement of the Helmholtz displacement curve cannot exist. In fact the resistance of the string to bending would alone prevent this.

But there is another reason that makes it unnecessary to introduce the rather complicated influence of bending in the treatment. If one watches the changing motion of a plucked string (same coordinates as Figure 2), it becomes evident that its time dependence, starting with many overtones, becomes increasingly sinusoidal; that is, the higher harmonics are damped sooner than the lower ones, in contradiction to Raman's assumption for his first calculation of the forces of friction.

Figure 3 shows quantitative results of the energy-decay time in its dependence

244

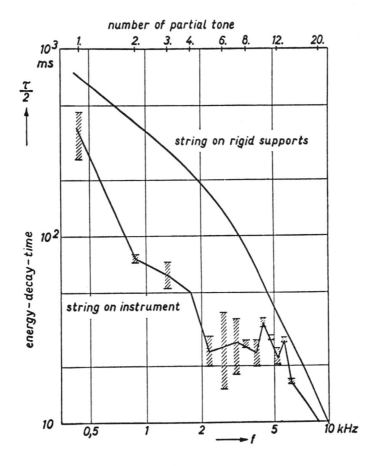

Figure 3

on frequency as found by W. Reinicke (3). Even on rigid supports a string shows this decreasing tendency. But with the string mounted on an instrument, in this case a violin, a similar but even more rapid decrease in decay time is found in going to the higher frequencies.

Any lack of high-frequency components must smooth the corner in the displacement and the jump in the velocity. If, for instance, we multiply the displacement of the Helmholtz motion by a damping term

$$e^{-n\,\delta_1 t}$$

where n is the number of the harmonic and δ_1 the decay rate of the fundamental, we

245

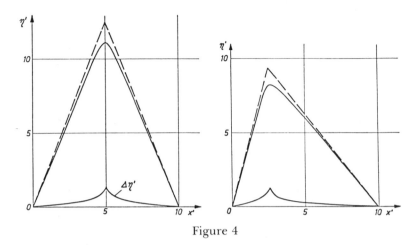

Figure 4

get after only a few periods of the fundamental a displacement with rounded corners instead of the original angular shape. In Figure 4 this is demonstrated with extremely exaggerated ordinates compared with the abscissae. This is done in order to show the difference, original displacement minus that with rounded corners, which propagates as a "correction wave" together with the corner to and fro.

If we now register the transverse velocity of the string at the bow point under the assumption that the additional force of friction is negligible. that is, under the limiting case of a very small bow pressure, we get the derivative of the displacement with respect to time. If we assume that the curvature may be represented adequately by a parabola, the velocity decreases linearly from v_B to v_-.

This is represented in Figures 5a to 5c.* At the left the displacement of the vertically extended string is to be seen, the Helmholtz line dashed and the rounded corner by a solid line, while at the right velocity is plotted versus time up to the instant depicted by the displacement curve.

In Figure 5a the curved part of the string has not yet reached the bow. Therefore, the velocity of the string equals the constant velocity of the bow. In Figure 5b the first half of the curved part has passed the bow. This is the instant when the jump of the velocity in the original Helmholtz representation appears, whereas the solid line shows the first part of the linear decrease for the rounded corner. In Figure 5c the rounded part has passed the bow. This means that we must get the same velocity v_- that we have to expect in the original Helmholtz motion. This value remains constant until the reflected rounded corner reaches the bow from the bridge side (Figure 5d). After that instant the velocity begins to increase linearly in symmetrical correspondence to Figures 5b and 5c. Figure 5e thus corresponds to 5b, and 5f to 5c. The total dependence on time from 5a to 5f forms a trapezium which may be regarded as a first approximation for the velocity course in the lower oscillogram in Figure 2.

---

*Figures 5 and 6 are taken from a 16-mm movie film which was made for the Institute of Technical Acoustics of the Technical University of Berlin.

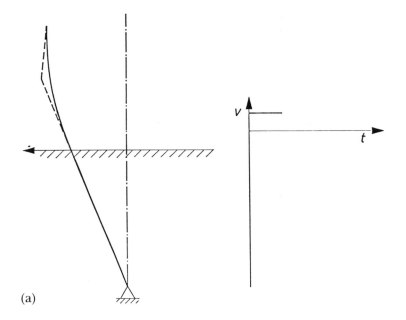

(a)

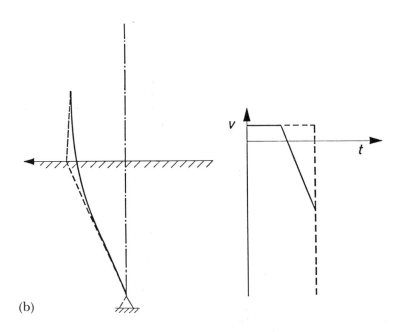

(b)

Figure 5

247

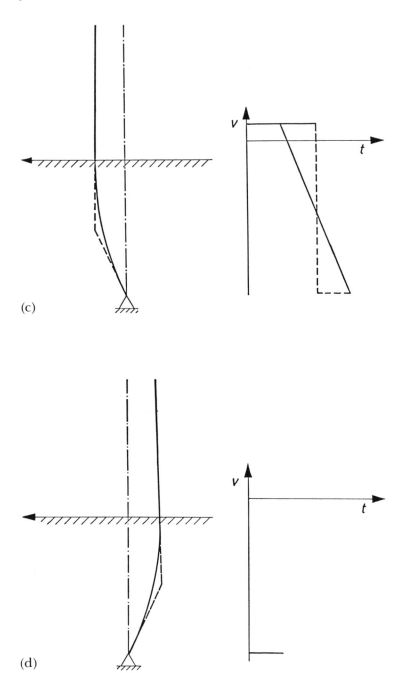

(c)

(d)

Figure 5 *(continued)*

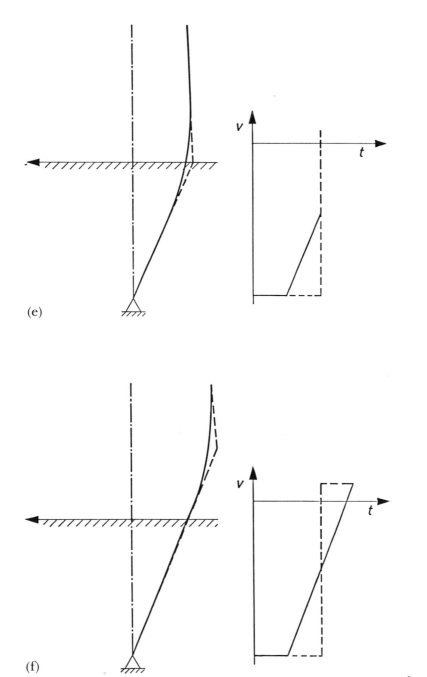

(e)

(f)

Figure 5 *(continued)*

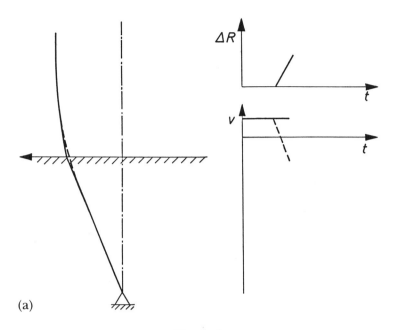

Figure 6

The rounded Helmholtz motion now also allows an explanation of the influence of bow pressure.[†] This may be seen from Figure 6. Here again the displaced string is shown in the left part as a solid line. At the right we have again the velocity dependence on time up to the instant presented at the left and in the diagram above it the frictional force ΔR which has to be added to the minimal frictional force that produces the static deformation of Figure 1b, but which is ignored in the figure.

In Figure 6a, which is different from Figure 5a, the beginning of the rounded corner of the Helmholtz motion has already passed the bow point. But the sticking friction between bow and string tends to further sticking and so to hold the velocity v equal to the bow velocity and hence to straighten the string. This needs an additional frictional force, which increases with the difference δv between the velocities which are to be expected with and without this force:

$$\Delta R = 2Z \, \delta v \tag{5}$$

Z is the characteristic impedance for transverse waves on the string and is the ratio of the lateral force in such a wave to the lateral velocity at that point. The factor 2 in (5) appears because the force always excites such "secondary correction waves" in both directions. With the simple assumption of a parabolic curvature, ΔR increases linearly with time.

[†]This method was presented at the ICA Congress in 1968 in Tokyo (1) but qualitatively only and with an arbitrary instant for the Helmholtz velocity jump.

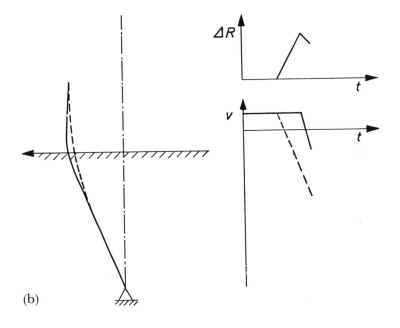

(b)

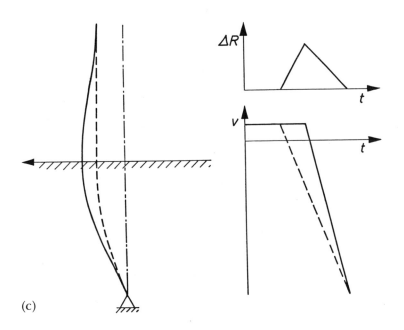

(c)

Figure 6 *(continued)*

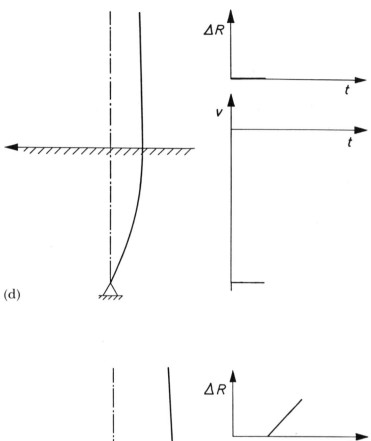

(d)

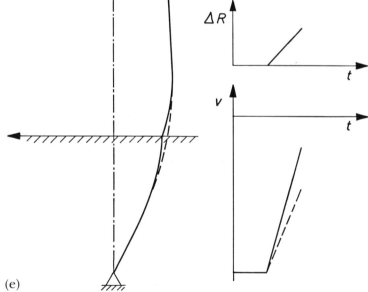

(e)

Figure 6 *(continued)*

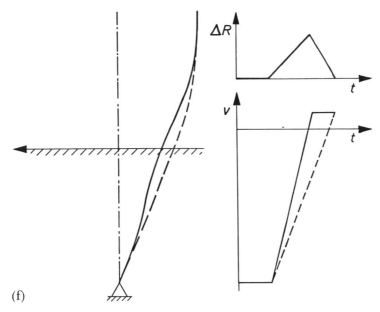

(f)

Figure 6 *(continued)*

But this increase is restricted by a maximum value $\Delta R_{\max}$ which depends on, i.e., is proportional to, the bow pressure. When this maximum value is reached, sliding begins. It is thus evident that this delay in the beginning of sliding is increased by increasing bow pressure.

The details of the sliding process are governed by the so-called friction characteristic, i.e., the dependence of the sliding frictional force on the relative velocity between bow and string. This characteristic is a concave curve (see Figure 7) with decreasing slope. Since v_{rel} is the difference between bow velocity v_B and string velocity v, and v may be thought of as a given incident quantity v_0 plus a secondary correction δv, we have

$$v_{\mathrm{rel}} = v_B - (v_0 + \delta v) = F \, \Delta R \tag{6}$$

and since that relative velocity is a function of ΔR we have two conditions from which we may eliminate ΔR and get a relation between the given v_0 and the desired δv. In any case this may be done by graphical means. But for our purposes to get a principal insight into the physical behavior only, it is reasonable to simplify the characteristic.

One way would be to substitute for the sloping part of the frictional curve just described a horizontal straight line that lies below the maximum value of sticking friction. It is known from simpler systems that even then we succeed in getting a self-exciting vibration.

For the present purpose, however, it is more convenient and mathematically simpler to replace the characteristic by a straight line (dashed in Figure 7) which

253

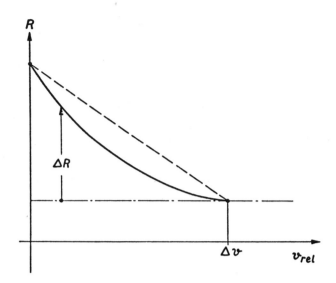

Figure 7

corresponds to the main slope and is given by Δv, i.e., the bow velocity and the point of excitation (Eq. 4), and by the maximum value of ΔR (Figure 7) and thus by the bow pressure. From

$$v_B \ - \left[\ v_0(t) \ + \ \delta v(t) \right] = \Delta v \left[1 \ - \ \frac{\Delta R(t)}{\Delta R_{\max}} \right] \tag{7}$$

and from (5) we get the relation

$$\delta v(t) \ = \ \left[\frac{\Delta v \ - \ v_B + \ v_0(t)}{- \ 1 \ + \ (2Z \ \Delta v)/\Delta R_{\max}} \right] \tag{8}$$

When $v_0(t)$ reaches its limiting value $(v_B \ - \ \Delta v)$, $\delta v(t)$ becomes zero, and if $v_0(t)$ is given by a straight line, $\delta v(t)$ is also a straight line which starts at the instant of the beginning of sliding with v_B and ends at v_- at the same instant when v_0 reaches this value. This is to be seen in Figures 6b and 6c, which correspond to Figures 5b and 5c. During the transition of the velocity v_{rel} from 0 to Δv the additional friction force decreases from the maximum value to zero.

Finally, Figures 6d, 6e, and 6f, which correspond to 5d, 5e, and 5f, show the influence of the bow at the transition from sliding to sticking. This transition starts with additional sliding frictional forces which occur as soon as the curvature passes the bow point (see 6e), and we get the beginning of sticking as soon as the linear but steeper increasing velocity approaches the bow velocity. We see that this instant is advanced in time by an interval proportional to the steepness of the slope in Figure 7, which in turn is proportional to bow pressure.

In the sticking period we get the constant velocity v_B and a decreasing force ΔR as long as the original displacement passing the bow point was curved.

Summarizing Figures 6a to 6f, the bow increases the slopes of the flanks of the original trapezium and shortens the sliding time. Both are in accord with the experimental records given in Figure 2.

Certainly this treatment is far from a complete solution because the Helmholtz motion is for a free string. Several problems remain.

For instance, if we were to follow the further fate of the correction waves we would have to take into account the secondary correction waves that propagate in a direction opposite to the primary correction wave. They may influence the next start of sliding and sticking when they just coincide with the waves which travel together with the primary correction wave. But if so the proper point of excitation would influence the motion much more than is the case with real instruments. So perhaps we may treat these secondary waves as a kind of statistical noise level with respect to their influence on sticking and sliding.

More important seems to be the problem of how the movement of the wave passing the bow and its reflections at the ends of the string, especially at the bridge, enter into the formation of a steady periodic cycle. Till now we have only shown that in passing the bow the curvature acquires a smaller radius. But we may assume that the opposite always happens ·at the ends since in reflection there are higher frequency components which tend to lose more energy than the low ones, as we have already explained in connection with Figures 3 and 4.

The exact treatment of these reflections, which are influenced not only by the bridge but by the body, seems to be the most difficult problem in the theory of bowed strings.

References

1. Cremer, L., and H. Lazarus, 6th Intern. Congr. Acoust., Tokyo, Aug. 1968, N-2-3.
2. Schelleng, J. C. *Catgut Acoust. Soc. Newsletter*, No. 12, Nov. 1969, p. 12.
3. Reinicke, W., *Bericht an die DFG*, 1967

The Influence of "Bow Pressure" on the Movement of a Bowed String: Part II[1]

LOTHAR CREMER

It is reasonable to start with the simplest bridge impedance which may be regarded as a first approximation for the frequency dependence of the energy-decay time $\tau/2$ observed by Reinicke (see Figure 3). It consists in a spring and a resistance whose forces are to be added. (See Figure 8, left side.)

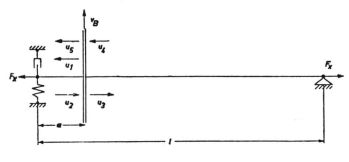

Figure 8

This bridge impedance

$$\underline{Z}_{Br} = \frac{s}{j\omega} + w \tag{9}^2$$

(s = stiffness, $j = \sqrt{-1}$, ω = radian frequency, and w = resistance) results in an "absorption coefficient" (i.e., nonreflected energy divided by incident energy) of

$$\alpha = \frac{4wZ}{(w + Z)^2 + (\frac{s}{\omega})^2} \tag{10}$$

[Z = characteristic impedance for transverse waves on the string, already introduced in Eq. (6).] By defining the nondimensional parameter which is especially suitable later

$$\epsilon = \frac{2Z}{w + Z} \quad < 1 \tag{11}$$

and the relaxation time

$$\tau_{Br} = \frac{Z + w}{s} \tag{12}$$

which characterizes the decay of a deflection at the end of a string infinite in the other direction, we may simplify (10) in

$$\alpha = \frac{(2\epsilon - \epsilon^2)}{1 + (1/\omega\tau_{Br})^2} \tag{13}$$

Since the free transverse wave of a plucked string loses the part α of its energy always after having passed the way $2l$, i.e., the full period $T_0 = 2l/c$, we get the energy decay time $\tau/2$ from the comparison

$$e^{-2t/\tau} = (1 - \alpha)^{ct/2l} \approx e^{-\alpha ct/2l}$$

$$\tau/2 \approx 2l/ac = \frac{2l}{(2\epsilon - \epsilon^2)c} \left[1 + (1/\omega\tau_{Br})^2 \right] \tag{14}$$

This means, in a diagram with logarithmized scales in ordinate and abscissa, a straight slope at low frequencies and an asymptotic constant value at high frequencies. Both may be regarded as a first approximation of the measured frequency dependence given in Figure 3.

We may therefore also assume this bridge impedance as the only reason for losses and may disregard equivalent losses at the other end or during the propagation of the transverse wave at this first attempt of a theory of the influence of bow pressure.

A good matching to the slope and the asymptotic value of $\tau/2$ in Figure 3 yields the values

$$\epsilon = 0.08_5 \quad \text{and} \quad \tau_{Br} \approx 10^{-4}\ \text{s}$$

If we later choose higher but still physically reasonable values for ϵ, this is done to get clearer figures. It is essential that the τ_{Br} value be not only small compared with the period of 23.10^{-4} s in Reinicke's experiment but even smaller as the tenth of that time which a transverse wave needs for the way $2a$ from the bowing point to the bridge and back.

Apart from this statement we may generalize our following calculations sometimes by dividing the time by τ_{Br}, so introducing a dimensionless time parameter

$$\vartheta = t/\tau_{Br} \tag{15}$$

then ϵ remains the only construction parameter, and we may keep in mind that higher ϵ mean higher losses at the bridge.

We now have to bring in the problem of the time dependence of the reflected wave at that bridge for a given time dependence of the incident wave, and we may describe both by the velocities v_1 (incident) and v_2 (reflected).

At the ideal Helmholtz motion v_1 and v_2 show an ideal sawtooth form (see Figure 9, upper diagram). Since the bridge is assumed to be rigid there, i.e., to have

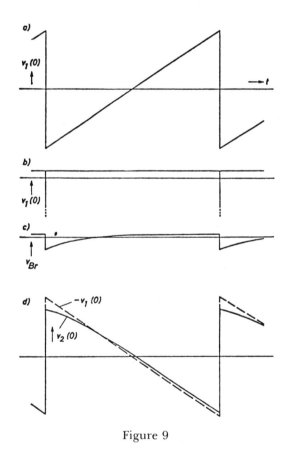

Figure 9

an infinitely large impedance, the resultant velocity there is

$$v_{Br} = v_1(0) + v_2(0) = 0 \tag{16a}$$

But in our case, where the bridge moves, v_1 and v_2 differ at the point $x = 0$ by v_{Br}:

$$v_2(0) = -v_1(0) + v_{Br} \qquad (16b)$$

In the same way the force acting on the bridge may be composed by two forces

$$F_{Br} = F_1(0) + F_2(0) \qquad (17)$$

one belonging to the incident wave according to the equation

$$F_1 = Zv_1 \qquad (18a)$$

and the other belonging to the reflected wave

$$F_2 = -Zv_2 \qquad (18b)$$

The negative sign corresponds to the propagation of this wave in the negative x-direction.

So we may express by (16) to (18) the force acting on the bridge by the velocity $v_1(0)$ of the incident wave and the velocity of the bridge:

$$F_{Br} = Z\big[v_1(0) - v_2(0)\big] = Z\big[2v_1(0) - v_{Br}\big] \qquad (19)$$

If we enter with (19) in that equation which defines the impedance in the frequency domain:

$$\underline{F}_{Br} = \underline{Z}_{Br}\underline{v}_{Br} = (w + \frac{s}{j\omega})\underline{v}_{Br} \qquad (20)$$

we may derive v_{Br} immediately from a given value of $v_1(0)$:

$$2Z\underline{v}_1(0) = (Z + w + \frac{s}{j\omega})\underline{v}_{Br} \qquad (21a)$$

or

$$\epsilon\underline{v}_1(0) = (1 + \frac{1}{j\omega\tau_{Br}})\underline{v}_{Br} \qquad (21b)[3]$$

But the problem of the excitation of the string by bowing is primarily a problem in the time domain because the periods of sticking and sliding are initiated by the surpassing of thresholds in $v(a)$ and ΔR which appears during the time dependence of these quantities.[4]

Therefore, we had better transfer (21b) in the time domain, getting:

$$\epsilon v_1(0) = v_{Br} + \frac{1}{\tau}\int v_{Br}\,dt \qquad (21c)$$

259

or avoiding the integral

$$\epsilon \dot{\underline{v}}_1(0) = \dot{v}_{Br} + \frac{v_{Br}}{\tau} \tag{21d}$$

If we first assume that the incident wave corresponds to an ideal Helmholtz motion, we have to introduce

$$\dot{v}_1(0) = \frac{\Delta v}{T} - \Delta v \; \delta(0) \tag{22}$$

or in words: from a horizontal line shifted by $\Delta v/T$ over the abscissa are to be subtracted ideal impacts with the impact strength Δv corresponding to the velocity step in $v_1(0)$ (see Figure 9, second diagram). These periodic impacts cause periodic free relaxation motions of the bridge, and the constant term in (22) causes another constant term:

$$v_{Br} = -\epsilon \Delta v e^{-t/\tau} + \frac{\epsilon \tau}{T} \; \Delta v \tag{23}$$

(Figure 9, third diagram).
   So for the reflected wave, we are interested in

$$v_2(0) = -v_1(0) - \Delta v (\epsilon e^{-t/\tau} - \frac{\epsilon \tau}{T}) \tag{24}$$

(Figure 9, fourth diagram).
   The constant term is physically essential because it guarantees that the integral of v_{Br} over a full period vanishes. Otherwise we would get a permanent shifting of the bridge in one direction. But numerically this term is so small that we could make it evident in the third diagram of Figure 9 only by assuming an unusual short period T compared with τ. In general we may neglect this term in comparison to Δv, but also in comparison to v_+ and v_- at the bowing point given in the Helmholtz motion.
   Furthermore, since the difference $v(a) - v_-(a)$ determines the friction force during the sliding period, we may simplify our equations by substituting $v(a)$ by

$$u = v(a) - v_-(a) \tag{25}$$

We are now looking for a stationary cycle for this quantity $u(t)$ where $t = 0$ always shall mean the time point at which the first deviation from the constant value of u (i.e., Δv or 0) occurs.
   We may begin with a reasonably chosen function $u_1(t)$ which appears when the motion of the string changes from sticking to sliding. This function $u_1(t)$ characterizes a modified step wave which travels from the bow to the bridge. (This direction of propagation is always indicated by the arrows over the u in Figure 8.)

u_1 now causes a reflected step wave u_2 at the bridge which moves back toward the bow point. Here u_2 gets transformed, first by the sliding and then by the sticking friction in a step wave u_3 which travels toward the other end of the string, which we assume to be ideally reflecting. By this reflection we get the step wave u_4 which approaches the bow from the right side, and here u_4 is changed in u_5 on account of the friction forces.

The condition for a stationary motion is now that

$$u_5 = u_1 \tag{26}$$

We know from our discussion of Figure 6 that it would be inadequate to assume for u_1 the ideal step of the Helmholtz motion. But the bridge motion given in (23) which "answers" an impact of $u_1(t)$ of the strength 1,

$$\phi(t) = \epsilon e^{-t/\tau} \tag{27}$$

is suitable to compose the corresponding "answer" of any time function $u_1(t)$ by differential contributions:

$$v_{Br} = \int_0^t \dot{u}_1(t')\phi(t - t')\, dt'$$

$$= \epsilon \int_0^t \dot{u}_1(t') e^{-(t-t')/\tau}\, dt' \tag{28}$$

Remembering that the constant slopes we treated in Figures 5 and 6 gave reasonable results, we may now start with

$$u_1 = \Delta v \left(1 - \frac{t}{t_1}\right) \tag{29a}$$

respectively with

$$\dot{u}_1 = -\frac{\Delta v}{t_1} \tag{29b}$$

Figure 10a shows this assumed function u_1 with its constant slope. Hereby Δv is given by the bow velocity and the distance between bow and bridge, but t_1 is to be regarded as dependent on the bow pressure and has been chosen so that the condition (26) for a stationary cycle was best approximated.

With the given u_1 we get from (28) for v_{Br}, for the time section $0 < t < t_1$,

$$v_{Br} = -\frac{\epsilon \tau \Delta v}{t_1}\left(1 - e^{-t/\tau}\right) \tag{30a}$$

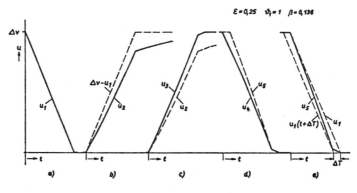

Figure 10

This means we get at the end of this section the minimal value

$$v_{Br}(t_1) = -\frac{\tau \, \Delta v}{t_1}(1 - e^{-t_1/\tau}) \tag{30b}$$

In the adjacent time $t > t_1$ this deviation decreases exponentially:

$$v_{Br} = -\frac{\epsilon\tau \, \Delta v}{t_1}(1 - e^{-t_1/\tau})e^{-[(t-t_1/\tau]} \tag{30c}$$

Correspondingly we get for

$$u_2 = \Delta v - u_1 + v_{Br} \tag{31}$$

during the time section $0 < t < t_1$,

$$u_2 = \Delta v\left[t/t_1 - \epsilon\tau/t_1(1 - e^{-t/\tau})\right] \tag{31a}$$

and during the time $t > t_1$,

$$u_2 = \Delta v\left[1 - \epsilon\tau/t_1(1 - e^{-t/\tau})e^{-[(t-t_1)/\tau]}\right] \tag{31b}$$

Figure 10b shows u_2 besides the ideal reflected u_1 (dashed) for a rather high value of $\epsilon = 0.25$. As expected, the slope of u_2 is smaller at each time point as $\Delta v/t_1$; this means the bridge tends to decrease the curvature of the string at the Helmholtz corner. But this smaller slope is far from being constant. It diminishes asymptotically; and so the "mean slope" is zero.

262

But this asymptotic part of u_2 is eliminated by the next step, the transforming of u_2 in u_3 by the sliding and sticking friction force. We simplify again the dependence of the friction force on the relative velocity between string and bow by a straight line and we may so get u_3 from u_2 by a linear equation as in (8):

$$u_3 = u_2 \frac{\Delta v}{\Delta v - \Delta R_{max}/2Z} = \frac{u_2}{1 - \beta} \tag{32}$$

We keep in mind that

$$\beta = \frac{\Delta R_{max}}{2Z \, \Delta v} = \frac{\Delta R_{max} \cdot a}{2Z v_{Br} l} \tag{33}$$

is directly proportional to the bow pressure and, therefore, may be called "bow pressure parameter."

Figure 10c shows the transition from u_2 to u_3 according to eq. (32). As soon as u_3 surpasses Δv the relative velocity between string and bow vanishes and we get sticking. It is to be seen from Figure 10c, that this happens at a time

$$t_3 = t_1 + \Delta t_3 \tag{34}$$

which is given by the condition

$$(\epsilon \tau) \frac{\Delta v}{1 - \beta} (1 - e^{-t/\tau}) e^{-\Delta t_3/\tau} = \Delta v \left(\frac{1}{1 - \beta} - 1 \right) \tag{34a}$$

i.e., by

$$\Delta t_3/\tau = \Delta \vartheta_3 = ln \frac{\epsilon \tau}{\beta t_1} (1 - e^{-t/\tau}) = ln \frac{\epsilon}{\beta \vartheta_1} (1 - e^{-\vartheta_1}) \tag{34b}$$

This transition from 0 to Δv restricted to a finite time t_3 is not so far from a mostly constant slope $\Delta v/t_3$. But it would be wrong to now adjust this slope to that of $-u_1$. We shall get another increasing of the slope if the bow is passed the second time by the velocity step. This means that the valley given in the time dependence of $v(a)$ cannot be symmetric, not even with respect to the slopes only as it appeared in Figures 5 and 6, at least not if we have different reflections at both ends of the string.

This dissymmetry may be especially pronounced if we neglect all losses during the propagation and if we regard the other end of the string as ideally reflecting.

But under these simplifying conditions we get for u_4 that

$$u_4 = \Delta v - u_3 \tag{35}$$

and finally we get u_5 from u_4 in the same way as u_3 followed from u_2 on account of the increasing sliding friction force:

$$u_5 = \frac{u_4}{1 - \beta} \tag{36}$$

This means again a restriction in time for the transition from $u = \Delta v$ to $u = 0$ by an amount Δt_5:

$$t_5 = t_4 - \Delta t_5 \tag{37}$$

which now on account of the turning given by (35) appears at the other end of the transition function. From Figure 10d we may derive for this time restriction

$$\frac{\Delta v}{(1 - \beta)} \, 2 \left[\frac{\Delta t_5}{t_1} - \frac{\epsilon \tau}{t_1} (1 - e^{-\Delta t_5/\tau}) \right] = \Delta v (\frac{1}{1 - \beta} - 1)$$

or

$$\frac{\Delta t_5}{t_1} - \frac{\epsilon \tau}{t_1} (1 - e^{-\Delta t_5/\tau}) = \beta(1 - \beta)$$

or with the time parameters

$$\frac{\Delta \vartheta_5}{\vartheta_1} - \frac{\epsilon}{\vartheta_1} (1 - e^{-\Delta \vartheta_5}) = \beta(1 - \beta) \tag{38}$$

It is not possible to transform this transcendental equation in a form which presents $\Delta \vartheta_5$ explicitly.

The appearance of Δt_5 means that the fundamental condition (26) can be fulfilled only if we allow u_5 to start later with sliding as it would correspond to the period

$$T_o = 2 \, l/c \tag{39}$$

which holds for rigid ends of the string. This extension of (26) seems to be a doubtful one if we regard the fact that the fundamental tone of a bowed string is given at least with very good approximation by (39). But on the other hand such a deviation of the period T from that given in (39) has to be taken into account already for the free motion of the plucked string. We may express this increment ΔT_0 in the time period by an equivalent increment in string length.

$$\frac{\Delta T_0}{T_0} = \frac{\Delta l}{l} \tag{40}$$

If the tension of the string is given by the force F_x, then a string element of the length Δl before a rigid end works just as a stiffness

$$s = \frac{F_x}{\Delta l} \tag{41}$$

and therefore may be replaced by a bridge with this stiffness. So we find for the plucked string with such a bridge that

$$\Delta T_0 = \frac{T_0}{l} \cdot \frac{F_x}{s} = \frac{2F_x}{c \cdot s} = \frac{2Z}{s} = \epsilon \tau \tag{42}$$

Besides this time shift, u_5 departs only so little from a transition with constant slope that it seems reasonable to restrict (26) for a first approximation to equal mean slopes, and this means to equal the slope times:

$$t_5 = t_1 \tag{43}$$

But since

$$t_5 = t_4 - \Delta t_5 = t_3 - \Delta t_5 = t_1 + \Delta t_3 - \Delta t_5$$

we get the sample condition

$$\Delta t_5 = \Delta t_3 \tag{44}$$

In Figure 11 both time increments are plotted over the bow pressure parameter β

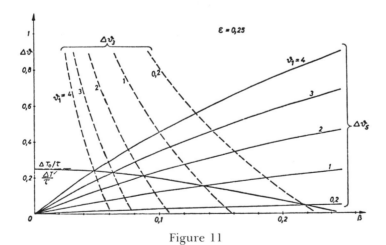

Figure 11

265

for different values of the "slope parameter" ϑ_1. The intersections of these curves for equal ϑ_1 deliver the increment of the time period ΔT:

$$\Delta T = \Delta t_5 = \Delta t_3 \qquad (44a)$$

If we compare these values with that of ΔT_0, marked at the ordinate, we see that ΔT is always smaller and has the tendency to diminish with the bow pressure. This means that we have to expect a small increase in the frequency of the bowed string with bow pressure. But we may also state that this increase can be only very small because the upper limit ΔT_0 is already very small. From Reinicke's measurements of the decay time we obtained the values

$$\epsilon = 0.085, \quad \tau_{Br} = 10^{-4}\,s, \quad T = 23.10^{-4}\,s$$

So we may estimate for T/T_0 a value in the order of 0.22 percent. Compared with the nearly 6 percent which corresponds to a half tone, we can say that this frequency shift we find for different bow pressures is negligible. But just this result is in agreement with musical praxis.

In this respect the string instruments are in a much better position than most of the wind instruments. The fact that the changes in the fundamental frequency are negligible does not mean that the same holds for all changes which happen at different bow pressures. The changes in the slopes of the v-transitions are rather large; they mean much larger deviations between the functions $u_1(t)$ for different bow-pressure parameters as they appear between u_1 and u_5.

These slopes may also be deduced from the intersections in Figure 11. They are represented by the slope parameter ϑ_1, which is plotted in Figure 12 versus the corresponding bow-pressure parameter for two different values for ϵ. As expected from Figure 6, ϑ_1 decreases; this means the slope increases with increasing bow pressure.

The curve end at the limiting case $\vartheta_1 = 0$, which means a vertical jump as in the Helmholtz motion. This limiting case defines a maximum of the bow-pressure parameter.

Hereby the initial value of the bridge velocity in (23) which marks the initial difference between u_1 and u_2 has to be compensated by the first action of the friction forces and this leads to

$$\frac{1 - \epsilon}{1 - \beta_{\max}} = 1 \qquad (45)$$

or to

$$\beta_{\max} = \epsilon \qquad (45a)$$

It is interesting that

$$(\Delta R_{\max})_{\max} = 2Z\epsilon\,\Delta v = \frac{2Z\epsilon l}{a}\,v_B \qquad (46)$$

not only increases inversely with the distance of the bowing point from the bridge in agreement with experience but also increases with the loss parameter ϵ. $(\Delta R_{max})_{max}$ is much smaller than the value

$$(\Delta R'_{max}) = 2Z \, \Delta v \qquad (47)$$

which would be necessary to prohibit the jump Δv at passing the bow, and this difference is in agreement with measurements.

It is disappointing that our treatment does not give immediately also the minimum bow pressure. We may only say that the given theory assumes that the transition covers the full difference Δv. This means that t_1 cannot exceed the time the step wave needs to travel from the bowing point to the bridge and back:

$$t_1 < \frac{2a}{c} \qquad (48)$$

$$\vartheta_1 < \frac{2a}{c\tau} = \frac{a}{l} \cdot \frac{T}{\tau} \qquad (48a)$$

We may assume that at least near this limit the conditions change so essentially that no modified Helmholtz motion can exist.

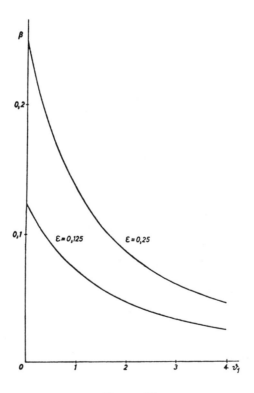

Figure 12

267

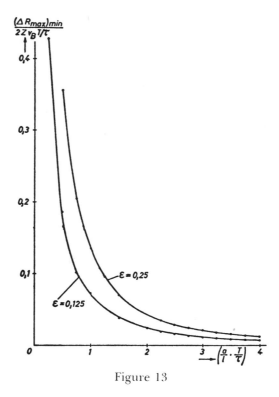

Figure 13

If we so regard $\vartheta_{1\,max}$ as an approximation of this limit, we may regard the abscissa in Figure 12 as proportional to the distance a, but we have to take into account that the ordinate β also depends on a:

$$\beta = \frac{\Delta R_{max}}{2Z\,\Delta v} = \frac{\Delta R_{max}}{2Zv_{Br}} \cdot \frac{a}{l} \tag{49}$$

If we now divide this ordinate β by a, we get the dependence of $(\Delta R_{max})_{min}$; this means also of the minimal bow pressure, on the distance bow bridge a (see Figure 13). In accordance with experience this minimal bow pressure is larger for higher losses ($\Sigma = 0.25$) than for smaller losses ($\Sigma = 0.125$) and it increases with diminishing distance more than inversely proportional.[5] So the difference between maximum and minimum bow pressure decreases as the player is approaching the bridge, which according to Schelleng is to be expected.[6]

It is possible to improve the similarity of u_1 and u_5 by now regarding u_5 as the given function and deriving after performing the same process a new $u_5{}'$ as the result, which has to be compared with u_5 and so on. To do this analytically would be rather hard. But with a computer it may be handled.[7] With such means it also would

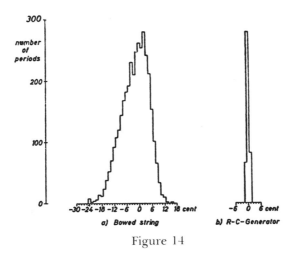

Figure 14

be possible to introduce the real dependence of the sliding friction on the relative velocity between bow and string.

But it is doubtful if it makes sense to be very exact in some respect as long as we must make simplifying assumptions with respect to the bridge impedance and as long as we neglect the possible influence of the secondary correction waves on the cycle of the regarded step wave.

Finally it is even doubtful if an exact periodic wave motion is what we have to look for. It would be sufficient to prove that the time periodicity appears between given limits.

Indeed, if we register the sound of a bowed string on a tape and if we study the distribution of the time periods on it, we get a population versus frequency which shows a maximum at a "mean frequency" but as well a remarkable variance (see Figure 14a).[8] If we do the same with an electronic tone generator, the variance is very small (see Figure 14b). This variance at bowed-string sound is not surprising. Already the rosin will not everywhere cause the same maximum sticking friction. Furthermore, the secondary correction waves may sometimes delay, sometimes speed up, the starts of sliding or sticking. And it may be that this variance even belongs to the characteristic charm of "violin sound."

This statement does not make the present investigation worthless. In spite of all its simplifying assumptions, it shows — to some extent even quantitatively — that the bow pressure modifies the Helmholtz motion, especially that increasing bow pressure shortens the finite transitions and so increases the content of overtones, that it influences the time period but only very little, that there are limits for the bow pressure as well with respect to too high as to too small values, and that these limits depend on the losses and on the bowing place, all this in principal accordance with experience.

Notes and References

1. The first part was published in *Catgut Acoust. Soc. Newsletter*, No. 18, page 13. The numbers of figures and equations continue those of the first part.
2. All complex quantities are underlined.
3. In the following equations we may write τ instead of τ_{Br}.
4. However it may be mentioned that H. Lazarus has treated this problem successfully in the frequency domain by applying the final Laplacian transformation to a chain system with 20 lumped elements instead of the continuous spring (dissertation Technical University of Berlin, 1972). But in this way the physical insight into the problem is little.
5. This is in agreement with measurements of H. Lazarus (1).
6. J. C. Schelleng, *Catgut Acoust. Soc. Newsletter*, No. 13, May 1970, p. 24.
7. Thomas Hermes of the Institut for Technical Acoustics in Berlin has treated this procedure. His results deliver transit functions which differ more from a constant slope but which obey the same tendencies as those given here as a first approximation.
8. The same fact was already observed by B. L. Cardozo and P. M. A. S. van Noorden (3rd Annual Progress Report of the Instituut voor Perceptie Onderzoek, Eindhoven, 1968).

Editor's Comments on Paper 14

14 Firth and Buchanan: *The Wolf in the Cello*

Dr. Ian M. Firth (1936 –   ), whose career has taken him through studies of microwave masers, low temperature physics, and superconductivity to musical acoustics, obtained his doctorate from the University of St. Andrews in 1962, after which he went on to a postdoctoral fellowship at the University of British Columbia. Since 1964 he has been a Lecturer at the University of St. Andrews. His present research interests center on stringed instruments — in particular, on impedance measurements and on the objective assessment of instruments.

With a B.Sc. (Honors) in experimental physics, Mr. J. Michael Buchanan (1947–   ) is now a research student in the M.R.C. muscle biophysics research unit, Department of Biophysics, King's College, University of London, working on optical and numerical methods of electronmicrograph image analysis and reconstruction.

Firth and Buchanan's study presented here disputes Raman's contention (Paper 8), which has been widely circulated, that the persistent fluctuation in intensity at the wolf note is not the result of beating between two adjacent frequencies. That the accepted explanation by Raman is not the correct one was suggested by Schelleng (Paper 5) on the basis of elementary circuit theory.

14

Reprinted from *J. Acoust. Soc. Amer.*, **53**(2), 457 – 463 (1973)

The wolf in the cello

IAN M. FIRTH AND J. MICHAEL BUCHANAN*

School of Physical Sciences, University of St. Andrews, St. Andrews, Fife, Scotland

(Received 10 October 1971)

The occurrence of the wolfnote in the cello is shown by direct tonal analysis to be caused by the beating of two equally forced oscillations. The fundamental sinusoidal component of vibration of the string is shown to split at the wolfnote into a pair of oscillations separated by a frequency interval equal to that of the stuttering frequency of the wolfnote. In one of the two cellos investigated, higher partials up to the third are also split into pairs, their separation being the multiple of the harmonic number and the separation of the fundamental pair. Further evidence is obtained to support the interpretation of the wolf as beating between two equal forced vibrations from a frequency analysis of the bowed-string vibration for different stopped string lengths. The frequencies of vibration of the string bear a close resemblance to those obtained in coupled electrical resonant circuits.

SUBJECT CLASSIFICATION: 6.7.

INTRODUCTION

In the musical world, the term "wolf" is attributed to unpleasant tonal effects, but it is most well known when referred to the cello. The wolfnote occurs in most cellos, to be found about F or F♯ on the G string, or at the same note on the C string, that is, on the two lowest and heaviest strings of the cello. The wolfnote occurs at the same frequency as that of the main body resonance. It has been described as a jarring note, an impure and wheezy sound, a cyclic stuttering response to the bow, or a cyclic fluctuation of intensity of imperfect quality. It is a note which is avoided by the cellist, and precautions are taken to suppress it by use of vibrato, or by employing a wolfnote eliminator attached to one string, usually the G string, between bridge and tailpiece.

There have been two main previous experimental investigations of the wolfnote in the cello. In the first, G. W. White[1] investigated the motion of the front plate by means of a light lever and concluded that, at the wolfnote, the instrument showed "the phenomenon of 'beats' which explain the fluctuations in intensity of the sound." Although no explanation was given for the creation of two instantaneous tones that would produce beating at this particular frequency, White recognized the wolfnote as occurring at the main body resonance of the cello.

In a second set of experiments, C. V. Raman[2] recorded both front-plate and string motion at the wolfnote by optical levers. This experiment led to criticism[2] of White's conjecture of beating. Raman argued that as beating is a transitory effect in forced vibrations of

linear systems, the persistent fluctuation in intensity at the wolfnote is not the result of beating between two adjacent frequencies. From the observed motion of the string, Raman suggested that the sympathetic resonance of the cello body takes energy from the string more rapidly than it can be replaced without a marked increase in bowing pressure. At this limit, the bow cannot maintain the string in its normal mode of vibration, in which the fundamental is dominant. The form of the vibration of the string changes, therefore, and the second mode of the string takes over as the bow is still moving. As the body is no longer forced to vibrate, its vibrations in turn die away, whereupon the string reverts to its fundamental and the cycle is repeated. Energy at the fundamental frequency passes from string to body, with a rapid alternation of fundamental and octave in the sound produced. This explanation is now frequently accepted[3] as the physical interpretation of the wolfnote.

That the accepted explanation by Raman is not the correct one was suggested by Schelleng[4] on the basis of elementary electrical-circuit theory applied to bowed-string instruments. Schelleng considered the equivalent electrical circuit for the cello at the wolfnote as a resonant transmission line (the string) coupled through a transformer (the bridge) to a series resonant circuit (the body). At the wolfnote, the two resonant circuits have identical frequencies, or nearly identical frequencies, for the wolfnote can be played over a short range of stopped string lengths. The coupled circuits are excited by a generator (the bow) placed in the first circuit close to the transformer.

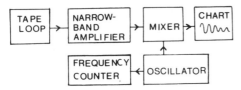

FIG. 1. Block diagram of frequency-analyzer circuit.

At the wolfnote, the circuits present to the bow an impedance for which there is not a single frequency, but three at which the reactance is zero. Steady oscillation can conceivably occur at any of these, but it is likely that the resistance of the circuit is sufficiently small only at the outer two for oscillations to be excited. This explanation is identical to that for coupled electrical resonant circuits.[5] The two steady oscillations are equally forced, so that it is only by exciting both together that the bow can elicit a note from the cello. Schelleng stresses that neither steady oscillation can exist alone at the wolfnote, but by cooperation they can produce together an instantaneous waveform acceptable to the string. In terms of the acoustical parameters of the cello, Schelleng gives a criterion for the occurrence of the wolfnote.

That this explanation is not at variance with the motion of the string as recorded by Raman was also shown to be the case by Schelleng. The recorded sawtooth motion of the string at the wolfnote, brought about by a slip-stick motion forced by the bow, can be interpreted as being caused by the superimposition of two steadily excited sawtooth waves. The two waves have the same triggering rate f_s, the resonance frequency of the string, but have a phase difference of 180°, and also are amplitude modulated at a frequency of $\Delta f/2$ in phase opposition to each other so that when one is a maximum the other is zero. A Fourier analysis[4] of the full cycle of the motion of the string indicates that at the wolfnote the fundamental sinusoidal vibration of the string at frequency f_s splits into a pair of oscillations (called the fundamental pair) separated by a frequency interval Δf, the beat or stuttering frequency of the wolfnote. Further, higher odd resonance sinusoidal components are split in frequency by the

same amount (harmonic pairs also appear) so that the frequencies present in the resultant of the two sawtooth waves are

$$\left(f_s - \frac{\Delta f}{2}, f_s + \frac{\Delta f}{2}\right), \quad 2f_s, \quad \left(3f_s - \frac{\Delta f}{2}, 3f_s + \frac{\Delta f}{2}\right), \quad 4f_s \cdots.$$

The present frequency analysis of the wolfnote confirms White's and Schelleng's suggestion that this note in the cello is the result of beating caused by the superimposition of sine waves in the string. The analysis indicates that the wolf in certain cellos is more complicated than in others. Nevertheless, wolfnotes investigated in two cellos may be explained in terms of beating. No evidence has been obtained in favor of Raman's argument of rapid cyclic alternation of fundamental and octave.

I. FREQUENCY ANALYSIS

The wolfnote in two cellos was analyzed into its frequency components. Tape recordings of each wolfnote were made in an anechoic room. Each recording was scanned by ear and by oscilloscope, and sections in which the cyclic variation in the wolfnote was most regular were made into tape loops for continuous replaying. The analysis of each wolfnote was made from such loops, the loop containing about 3 sec of recording. The tape recorder could be played back at one-half and one-quarter of the recording speed if required.

In attempting the frequency analysis of the wolfnote, it was quickly realized that even in selected tape loops the wolfnote was not sufficiently constant in phase to allow normal use of a lock-in amplifier with selective resistance–capacitance (RC) filtering.

The method used to detect closely spaced frequency components was as follows. The signal from the tape loop was amplified by a narrow-band amplifier, $Q=25$, adjusted to a particular harmonic frequency of the wolfnote (Fig. 1). The amplified signal was passed through a lock-in amplifier, the reference signal of which was variable, and then narrow-banded further with an RC filter of a time constant 1 sec. The filtered output was recorded on a chart at sequential settings of the reference frequency. From the chart, the beat

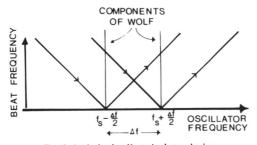

FIG. 2. Analysis of wolfnote by heterodyning.

TABLE I. Frequencies, in hertz, of components of wolfnote on G string, and linear density of strings used.

	Harmonic		1	2	3	4	5
Cello 1							
String	Pirastro Yellow Label	f Hz	168.9	336.2	500	666.8	840
	Silver on gut		173.8	342.4	510.6	679.4	
	5.2 g/m	Δf Hz	4.9	6.2	10.6	12.6	0
Cello 2							
String	Thomastic Rope—	f Hz	182.2	375	558.4	742	
	cored Chromsteel		192.2				
	7.0 g/m	Δf Hz	10	0	0	0	

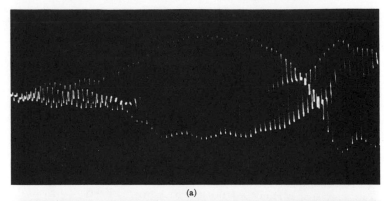

(a)

FIG. 3. String motion of typical wolf on Cello 1, G string. (a) Start of up-bow. (b) Constant motion of bow. (c) Bow leaves string at end of motion. Note interpenetrating sawtooth waves and amplitude modulation of envelope. Intervals on figure: 100 msec.

(b)

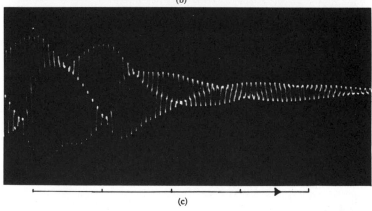

(c)

rate between signal and reference averaged over the time taken to play back the loop once could be measured. From the chart recording, both decreasing and increasing beat rates are obtained for any component frequency contained in the tape loop. For example, if the wolfnote fundamental is split into two components as suggested by Schelleng, then during the scan the chart recorder will show two signals with decreasing frequencies as the pair are approached by the oscillator,

a signal with increasing and a signal with decreasing frequency between the pair, and two signals with increasing frequency when the oscillator is moving away on the other side of the pair (Fig. 2). In a trial experiment, the beating signal between two pure ac sinusoidal oscillator signals was interpreted in this way with success, the discrimination in the frequency interval between the two being 0.1 Hz in a beat frequency of 5 Hz. Intercepts showing zero beat frequency indicate

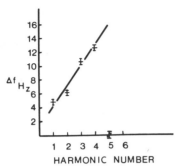

Fig. 4. Separation (Δf) of harmonic pairs at wolfnote in Cello 1 on G string. See Table I.

split components of the wolfnote. All partials of the wolfnote can be investigated in a similar fashion.

II. RESULTS

Recordings of the wolfnote on the G string were made on two cellos, and each was subjected to analysis with the chart recorder. Results from fundamental and partials up to the fifth are shown in Table I.

A. Cello 1

Cello 1, strung with a light G string, shows splitting of each partial up to the third into two component oscillations. This is confirmed by the complicated sound pattern observed on the oscilloscope and also by the complicated motion of the string. The latter was recorded in a separate experiment in which the voltage induced across the wire-covered G string when moving in a magnetic field placed close to the point of bowing was integrated to show string displacement. String motion was recorded in a camera with a film transport attachment (Fig. 3). Such motion shows no simple beat pattern as can be discerned in Raman's recording,[2,4] but a complicated motion in which phase varia-

tion in the fundamental and in higher partials is present, together with a form of amplitude modulation within the main cyclic pattern. From the complexity of the string motion, it is not surprising that both even and odd harmonics split into pairs of simultaneously forced oscillations.

The conclusion is therefore that, in this cello, the wolf occurs because the normal harmonic series of oscillations of the string alters to become a series of pairs of oscillations, the two components of each pair being equally forced by the bow. Pairs are situated about the expected frequency of the relevant harmonic of the string. The frequency separation of the pairs increases monotonically with harmonic number (Fig. 4). The wolfnote is the resultant waveform produced by the pairs beating. There is a constant phase relation between the pairs in the series; Fig. 3 shows a cyclic repetition of the resultant string motion.

B. Cello 2

The wolfnote is caused by the splitting of the fundamental only into a fundamental pair of frequencies. Although no recordings of the motion of the G string on this cello were made, the fact that the wolfnote is apparently caused by the production of a single fundamental pair of components is confirmed by the more simple motion of the heavy C string of Cello 1 (Fig. 5). The motion of this string is similar to that obtained by Raman, showing little harmonic content in the wolf.

At the beat minimum, which does not exist in the G string of Cello 1 (Fig. 3), the sawtooth apparently goes to the octave, but Schelleng has demonstrated that this is exactly what is expected from beating. An epoch of maximum string motion undergoes a phase shift of π relative to the preceding one leading to an apparent doubling of the frequency in the sawtooth motion where the phase shift is $\pi/2$, at an epoch of minimum string motion.

It is noted from the present recordings, nevertheless, that the envelope of the motion of the string is not

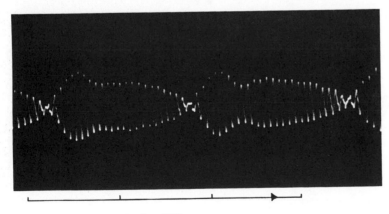

Fig. 5. String motion of complicated wolf on Cello 1, C string. Note amplitude modulation on envelope which has zero minima. Intervals on figure: 100 msec.

sinusoidal even on heavy strings (Fig. 5), and some recordings show distinct amplitude modulation (Fig. 6). In these, components other than the fundamental pair should be present in the wolfnote.

C. Acoustical Parameters

Various acoustical parameters were measured for both cellos in order to evaluate the criterion for the wolfnote introduced by Schelleng. This criterion is based on a plot of the logarithmic decrement of vibration of the cello body against the dimensionless factor $K/(SM)^{\frac{1}{2}}$, where K is the characteristic impedance of the string ($K=(T\mu)^{\frac{1}{2}}$, T is string tension, μ is string linear density, S is the stiffness of the body, and M is the mass of the body. The graph obtained can be divided into a "safe" and a "dangerous" region for propensity to wolf. The method suggested by Schelleng for measuring both S and M was followed leading to the data of Table II for the cellos.

These results put both cellos inside the "danger" region for the wolf. The logarithmic decrement for Cello 2 was not measured. If it had been similar to Cello 1, the conclusion from the wolfnote criterion would have been that Cello 2 was more susceptible to wolf. Both in terms of regularity and attainability of the wolf, this was indeed the case.

The method used for measuring S and M involves loading the bridge to lower the body resonance. Lead shot mixed with plasticine was used for this purpose. Since the wolf could be produced over an increasing range of finger positions when the load was increased, the results for S and M are not accurate, but this fact led to another way of confirming Schelleng's suggestion of the production of a fundamental pair of frequencies at the wolf.

D. Frequency Analysis by Ear

With the bridge of Cello 1 loaded with 14 g, the interval on the fingerboard over which the wolf could be elicited by the bow was about 4 cm. At the extremi-

TABLE II. Acoustical parameters of cello.

	Cello	1		2		Value from Schelleng[4]		Unit
Series stiffness S	1.2	10^8	0.5	10^8	1.13	10^8	dyn/cm
Series mass M	107		36		92		g
$(SM)^{\frac{1}{2}}$	1.12	10^5	0.4	10^5	1.02	10^5	cgs Ω
Log decrement	0.04				0.08		
Quality factor Q	34				25		
String linear density μ	5.2		7.0				g/m
String tension T	8.8	10^6	12.8	10^6			dyn
$K=(T\mu)^{\frac{1}{2}}$	6.8	10^2	9.5	10^2			cgs/Ω
Wolfnote criterion $K/(SM)^{\frac{1}{2}}$	0.006		0.024		0.009		
Body resonance f_b	168		184		176		Hz

ties of this range, the wolf had frequencies of 152 and 141 Hz. The frequency of each component of the fundamental pair of the wolf elicited was easily measured by listening for zero beating between the wolf and a pure tone produced by feeding an ac signal to a loudspeaker. The condition of zero beating could be determined to better than 2 beats/sec, and the frequency of the ac signal was measured to 1 Hz by a frequency counter. The ear acted as an efficient, real-time frequency analyzer with bandwidth of $\pm\frac{1}{2}$ Hz. In this simple fashion, the "fundamental" pair of vibrations of the wolfnote could be resolved and their frequencies measured over the interval on the finger board in which the wolf could be elicited. Outside this range, only one frequency could be determined for the fundamental of the cello note. Figure 7 shows the fundamental frequency elicited by the bow on the G string of Cello 1, and shows the splitting into a fundamental pair of frequencies over the range of finger positions at which the cello wolfed.

Figure 7 can be interpreted on the basis of the reactance presented to the bow according to Schelleng's theory. With changing string length, the reactance alters diagrammatically according to Fig. 8 (see also Schelleng,[4] Fig. 6). In this figure, at positions A high

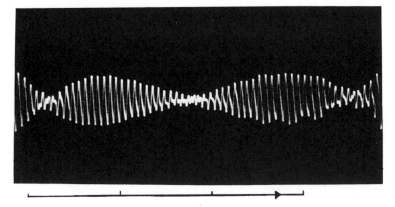

FIG. 6. String displacement against time of typical "simple" wolf on Cello 1, C string. Intervals on figure: 100 msec.

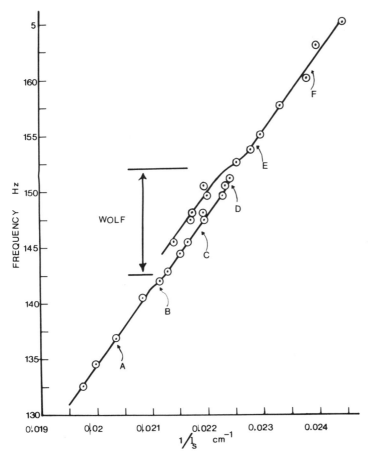

FIG. 7. Frequency components in bowed string against $1/l_s$ of stopped string. Frequency components detected by ear by adjusting pure tone from loudspeaker to zero beats. Points A–F refer to Fig. 8 (measurements on Cello 1, G string, bridge loaded with 14 g).

on the fingerboard, the reactance is zero at a frequency much lower than that of the body. Coupling between the two is negligible. As the length of the string is shortened, the influence of the body increases. At B there is still only one resonance frequency. This single frequency is lower than that expected from the inverse relation, $f_s \propto 1/l_s$, represented by the straight line AF in Fig. 8. Between C and D the bow elicits a fundamental pair of oscillations as the reactance cuts the axis at two points where $dX/d(1/l_s)$ is positive, so that both oscillations are stable. At E, a single component is produced again, its frequency being somewhat higher than expected from the string alone. Beyond E the string produces one frequency only, being proportional to $1/l_s$. Figure 7 bears out this description of the change in reactance, the two limbs of the curve at the wolf being displaced from the expected straight line relation for string alone, with no indication that they join together at the extremities of the range of the wolf.

The simplicity and accuracy of frequency analysis by ear is a method that could be exploited in further investigations of sounds containing closely spaced components. Acting in real time, with a bandwidth of $\pm\frac{1}{2}$ Hz, makes it an efficient method of investigating sounds from string instruments in which notes are not produced completely regularly. This method could have been used to investigate the harmonic content of the wolf in addition to that of the fundamental pair.

III. CONCLUSION

The wolfnote in the cello is the result of beating between pairs of equally forced vibrations. In the words of Schelleng,[4] "Neither of the pair can exist without the other because its frequency is so different from a recurrence rate possible on the string. On the other hand by cooperation they can produce an instantaneous frequency acceptable to the string equal to half the trigger rate. Like Siamese twins, they can

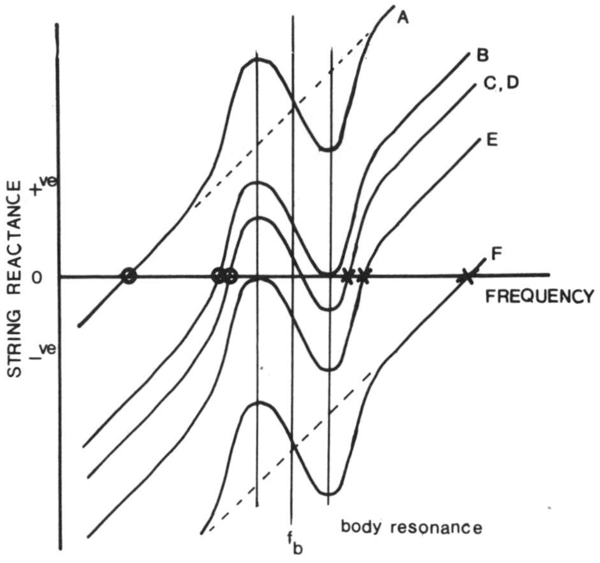

FIG. 8. Diagrammatic behavior of reactance (X) seen by bow against frequency based upon result of Fig. 7 and Schelleng.[4] Oscillation of string possible when $X = 0$ (points 0 and X). Positions A–F refer to Fig. 7. Length of stopped string decreasing from A to F.

exist as a pair, but not otherwise." The conclusion of Raman that the wolf is caused by a cyclic alternation between fundamental and octave is incorrect.

The effect can be explained in terms of the acoustical impedance presented to the bow which is that of two coupled resonators. The wolfnote produced in a cello with heavy strings is apparently more simple than that produced by a cello with light strings. In the former, the wolf occurs because of a pair of frequencies produced about the fundamental of the string, whereas in the latter, partials at least up to the third are split into pairs, each pair beating together to give a resultant waveform recognized as the wolf. This difference in species of wolf could be further investigated on one cello restrung successively with sets of strings of various weights.

ACKNOWLEDGMENTS

Our sincere thanks are due to T. W. Parnaby for lending Cello 2 for testing, and to Mrs. H. B. Buchanan for the loan of a cello the harsh wolf of which defied our attempted investigation.

* Present address: Univ. of London, King's College, Dept. Biochemistry.
[1] G. W. White, Proc. Cambridge Phil. Soc. **18**, 85–87 (1915).
[2] C. V. Raman, Phil. Mag. **32**, 391–395 (1916); Nature **97**, 363 (1916).
[3] E. Blom, Ed., *Groves Dictionary of Music and Musicians* (Macmillan, London, 1954), see entry under Wolf; C. A. Taylor. *The Physics of Musical Sounds* (The English Universities Press Ltd., 1965), p. 160; A. Wood, *The Physics of Music* (Methuen, London, 1944), p. 100.
[4] J. C. Schelleng, J. Acoust. Soc. Amer. **35**, 326–338 (1963).
[5] F. E. Terman, *Radio Engineering* (McGraw Hill, New York, 1937), p. 85.

III
The Bridge

The bridge of the violin serves to transmit vibrations from the strings to the wood of the top and in so doing acts as a filter for various wavelengths of sound. This means that the quality of the wood, its density, stiffness, mass, and characteristics of acoustical transmission, can have a marked effect on the sound of the violin. The expert selection and cutting of a bridge can enhance the desired tone qualities in a given instrument, especially in the hands of a skilled violin maker. The best bridges are cut from seasoned maple, which has a fairly high frequency of sound. Various bridges of similar dimensions can be compared by rubbing them in the fingers and listening to the sounds. Once a bridge blank has been selected, the violin maker cuts away some of the wood, particularly in the areas where bending takes place. In the cello bridge, the legs are a critical area for removal of wood, for as Reinicke (Paper 17) shows in his holograms, there is considerable bending here. The practice of removing wood from the inside of the cello bridge feet has occasioned considerable interest, but most skilled violin makers know that they can alter the sound of the instrument by careful shaping of the bridge legs and the critical areas near the heart-shaped hole. A bridge that brings out the best qualities in a given violin or cello may not be at all suited to another.

Explanation of such empirical knowledge of the violin maker is given to a certain extent in the 1937 paper of Minnaert and Vlam, included here (Paper 15), although no attempt is made to relate motions of the bridge to those of the body of the instruments. The studies of Bladier and Reinicke provide considerably more information on the motions of bridges.

Editor's Comments on Papers 15, 16, and 17

15 Minnaert and Vlam: *The Vibrations of the Violin Bridge*

16 Bladier: *Sur le chevalet du violoncelle*
English translation: *On the Bridge of the Violoncello*

17 Reinicke: *Übertragungseigenschaften des Streichinstrumentenstegs*
English translation: *Transfer Properties of String-Instrument Bridges* (abstract only)

The study of "The Vibrations of the Violin Bridge" by Minnaert and Vlam, reported in 1937, has long been the definitive work on this subject. Working at the University of Utrecht in Holland, they used an ingenious optical method to observe the various motions of the violin bridge. Now, with the more sophisticated techniques of hologram interferometry, there will undoubtedly be refinements of their findings.

Professor M. G. J. Minnaert (1893–1970), a native of Belgium, obtained a doctorate in biology in Ghent, and later a second doctorate, this time in physics with Professor Julius in Utrecht, Netherlands, where he was Professor of Physics and Director of the Observatory from 1937 to 1963. Founder of qualitative solar spectroscopy, Professor Minnaert received honorary professorships in Heidelberg, Moscow, and Nice.

After leaving the grammar school of his native Hoorn in north Holland, Dr. C. C. Vlam (1916–   ) studied physics and mathematics at the University of Utrecht and attended lectures on musicology. This combination of interests led to his research on the violin bridge. He was awarded his doctorate in 1953 for investigations on the shape of the emission bands of luminescent solids. He has published many articles on musical subjects as a contributor to the *Algemene Muziek-Encyclopedie* (Zuid-Nederlandse Uitgeverij, Antwerp–Amsterdam, 1957); *Die Musik in Geschichte und Gegenwart* (Barenreiter Verlag, 1949); *Documenta et Archivalia ad Historiam Musicae Neerlandicae,* edited by C. C. Vlam and M. A. Vente (Vereniging voor Nederlandse Muziekgeschiedenis, 1965).

Benjamin Bladier's *Sur le chevalet de violoncelle* is reproduced here in its entirety with an English translation. A discussion of his work and a list of selected papers may be found in "Editor's Comments on Papers 10 and 11."

Dr. Walter Reinicke (1939 –   ) received his training in electrotechnics and acoustics at the Technical University of Berlin, obtaining his doctorate in 1972. He is at present a staff member in the consultant office of acoustics of Müller BBN-GmbH in Munich.

Reprinted from *Physica*, **4**(5), 361–372 (1937)

THE VIBRATIONS OF THE VIOLIN BRIDGE

by M. MINNAERT and C. C. VLAM

Physisch Laboratorium der Rijks-Universiteit te Utrecht

Zusammenfassung

Die Bewegungen eines Violinstegs beim Spielen der verschiedenen Töne werden mittels eines Spiegelchens untersucht, das auf dem Steg befestigt ist, und durch welches ein Lichtstrahl reflektiert wird. Ausserdem wird auch direkte mikroskopische Beobachtung angewandt. Die Schwingungen des Stegs sind sehr kompliziert und für jeden Ton verschieden, aber sie lassen sich genau reproduzieren und sind unabhängig von der Saite auf welcher der bestimmte Ton gespielt wird. Wir unterscheiden: *a*) eine Schwingung in der Ebene des Stegs; *b*) eine Biegungsschwingung senkrecht zu dieser Ebene; *c*) eine Torsionsschwingung, gleichfalls senkrecht zu dieser Ebene. Bei bestimmten Tönen erscheinen charakteristische Maxima der einen oder der anderen dieser Schwingungen, von denen man experimentell zeigen kann, dass sie zu bestimmten schwingenden Systemen gehören. Wegen der Einkerbungen des Stegs werden nur die Schwingungen (*a*) an die Decke des Instruments mitgeteilt, welche eine Drehung um dem Punkte *A* ausführt (Fig. 1).

1. *The method of observation.* In the sound production of a violin an important role is played by the bridge, which transmits the vibrations to the belly; this on its turn causes the surrounding air to vibrate. The bridge is supported by two feet; under one of these, but slightly displaced towards the tailpiece, the small wooden „sound post" is clamped between the belly and the back of the instrument (fig. 1). The older investigators [1]) supposed that the bridge vibrates in its own plane, oscillating about the foot at the *E*-side. By the experimental work of d e H a a s and G i l t a y [2]), B a r t o n and E b b l e w h i t e [3]), R a m a n [4]), it became clear that the bridge vibrates also in a direction perpendicular to its plane.

We proposed to investigate these different movements by an optical method, and preferred to determine the vibration by simple visual inspection, because this allows to survey rapidly the complicated phenomema, and to detect any modification in the vibrating system. Our violin was an ordinary commercial instrument, fitted with a bridge marked „Bausch" of a weight of 2.1 gram. It was so clamped as to reproduce more or less the normal conditions in which the instrument is played: a cushion of cotton wool under the neck, a clamp with two pieces of cork at the place of the edge ordinarily occupied by the chinrest .In some cases, where any movement of the violin had to be excluded, it was also supported in different places by

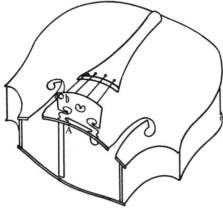

Fig. 1. Transverse section of the violin, showing bridge, sound-post, and galvanometer mirrors *a* and *b*; the point *A* corresponds to the center of rotation of the bridge (cf. 3).

Fig. 2. Photographs of vibration curves of the bridge (arrangement *a*). From each pair of photographs, this to the left is observed when a mute is used, this to the right is observed without a mute (§ 5).

wooden rods: we always found, however, that the vibrations of the bridge were unaffected by such changes in the support. The vibrations were observed by means of a galvanometer mirror, fixed to a point of the bridge with soft wax. A pencil of light, reflected by this mirror, was directed into a telescope, where the image of some very narrow pin-holes was formed. When a string was sounded, each luminous point described a curve, which did not alter its shape as long as the sound remained constant; some of these curves have been photographed, and are reproduced in fig. 2. One or more auxiliary mirrors were arranged, in such manner that it was possible to bow the strings and at the same time to look through the telescope. Because the deviations of the ray of light are mostly of the order of 1′ and seldom exceed 10′, the optical quality of the mirrors and telescope had to be very good, otherwise the motion of the points of light could not be distinguished with sufficient precision. The amplitude of the curves was determined with an ocular micrometer, disposed in the focal plane of the telescope. The galvanometer mirror had a weight of only 0.025 g; by fixing a second mirror of the same dimensions at different points of the bridge, it was found that such a small mass has no perceptible influence on the vibration, whereas a mirror of 0.150 g modifies already the curves of some determinate tones. In tuning and playing the violin, the pitch was compared by ear with a normal tuning fork; a still more sensitive criterium was found in the form of the luminous curves, observed in the telescope.

2. *The influence of the frequency of the sound on the vibrating bridge.* The vibrations of the bridge appear to depend on the frequency of the tone which is played, but to be independent of the string on which this is done. They may be reproduced with great accuracy. Generally, an interval of half a tone makes already a great difference in the form of the curves; in many cases a displacement of the finger of which the influence cannot be discerned by the ear, modifies already the optical image in a marked degree. On the contrary, only conspicuous changes in the timbre of the sound can be optically observed, e.g. the effect of a mute. It makes no difference whether the tone under investigation is played on one string or on another, even when one of the two is used open or when one of the two central strings is compared with one of the two outer strings; or when the strings are tuned somewhat too heigh or too low. By strong bowing, the dimensions of the figures increase, but small modifications only appear when the timbre is considerably modified by a too great amplitude. For harmonic tones, the vibration of the bridge is the same as for ordinary tones of the same pitch. In bowing upwards, the same curves are observed as in bowing downwards, only they are inverted (upside down and right to left).

In order to get a preliminary survey of the vibrations of the bridge as a function of the frequency, they were examined through the whole chromatic scale which can be obtained on the violin, and this with two different arrangements (cf. fig. 1):

a) the mirror fixed on the vertical narrow side of the bridge, with its plane perpendicular to that of the bridge, the ray of light being in this plane;

b) the mirror fixed in the plane of the bridge, near one of the upper edges; the ray of light perpendicular to this plane.

By discussing the curves observed, it appears that the motion can be decomposed in the following vibrations (fig. 3):

A) A vertical motion of the luminous point, observed in the arrangement (*a*); this corresponds to a vibration in the plane of the bridge. We will admit that the bridge executes this movement nearly as a rigid body (see however the end of § 3).

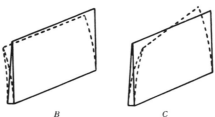

B   C

Fig. 3. Schematic representation of „flexural" and „torsional" deformation of the bridge.

B) A vertical motion of the luminous point, observed in the arrangement (*b*); in the simplest case, this corresponds to a flexural vibration perpendicular to the plane of the bridge (fig. 3B).

C) An horizontal motion of the luminous point, observed in the arrangements (*a*) or (*b*); this corresponds to a torsional vibration, perpendicular to the plane of the bridge (fig. 3C).

The vibrations *B* and *C* of the bridge cannot be considered as independent, as a matter of fact the torsion is nearly always accompanied by a bending of the edges; and so also a pure flexion as represented in fig. 3B is exceptional.

A rough comparison of the relative importance of these different vibrations was obtained by measuring their amplitudes through the whole chromatic scale, the loudness of the sound being kept approximately the same (fig. 4). The mean amplitude of the first vibration was about 2 times greater than that of the two other vibrations, which were about equal in strength. This agrees with the result of most of the earlier authors, who observed that the bridge vibrates both in its plane and perpendicularly to it; however the existence of a torsional vibration does not seem to have been remarked up to now. The vibrations of the bridge show some typical maxima, near determinate frequencies, different for each of the three types of vibration. In fig. 4 the amplitude of the vibrations is plotted against the frequency; though the pressure of the bow and the loudness can hardly be compared for distant parts of the scale, the existence of the maxima is doubtless thrustworthy. It is curious, that the vibration in the plane of the bridge shows maxima in harmonic proportion to each other; we are not sure if this is altogether fortuitous.

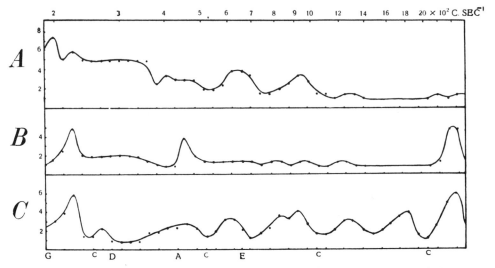

Fig. 4. Amplitude in minutes of arc of the different vibrations of the bridge as a function of frequency.
A = longitudinal vibration; B = flexural vibration; C = torsional vibration.

In analogy with the wellknown experiment of M e l d e, one could expect the vibration in the plane of the bridge to have the frequency of the string, and the transverse vibration a double frequency; the curves observed should then have the general form of the L i s s a-j o u s curve 1 : 2. In agreement with B a r t o n and E b b l e-w h i t e, we found that this is only the case for some particular tones (e.g. for a), while the curves are ordinarily more complicated and even sometimes show the form 1 : 1 (a' with a mute). It is clear that the movement of the bridge cannot be treated as a statical problem; the resonances with the proper vibrations of the different vibrating systems present in the violin play by far the chief role (see § 6).

Generally speaking, the curves produced by the movement of the bridge become more complicated in proportion to the number of partial tones given by the strings; the intricate curves corresponding to the tones *g—c'* on the *G*-string confirm the wellknown fact, that for these tones the fundamentals are weak compared with the partials.

3. *Detailed investigation of the deformations of the vibrating bridge.* In the preceding paragraph, we only considered the movement of the two upper edges, where our mirror was fixed. We will now try to find how the bridge as a whole is deformed during the vibration. In the first place we investigate the torsional vibration. Two possibilities present themselves to the mind, according to the form taken by the upper edge of the bridge, which may correspond to (I) or to (II) in fig. 5. In order to decide which of these two forms occurs, we select the arrangement (*a*), but fix our mirror first at the narrow side near *E*-string, secondly at that near the *G*-string (fig. 8, *E* and *G*). We may assume that in its own plane the bridge moves nearly as an indeformable body (cf. however the end of this paragraph); thus for the first position of the mirror, the vertical motion which is composed with the horizontal torsional vibration is directed in one sense, and for the second position of the mirror in the other sense. The curves at the *E*-and at the *G*-side must therefore by their symmetry

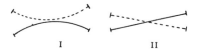

Fig. 5. Two possible forms of the upper edge of the bridge during the torsional vibration.

Fig. 6. Comparison between the curve, seen in the mirrors *E* and *G*.

properties give a decision between the torsional vibration of type (I) and or type (II). Figure 6 shows two typical images, as seen by the observer who is looking in each of the two mirrors; from this, one easily concludes that case (II) occurs.

The method here described can only be applied to a small number of figures which are asymmetric and yet not too complicated. More direct and general is the following observation: we make use of the arrangement (*b*), but we fix two galvanometer mirrors *E* and *G* on the bridge, one at each edge; by means of an auxiliary mirror *H*, the pencil of light from the lamp L_1 is reflected successively against both small mirrors (fig. 7). If these mirrors vibrate in the same plane, the angular displacements given by them to the ray of light will reinforce each other; in the reverse case they will partly cancel each other. Wishing to compare the figures for two mirrors with these of the original arrangement (*b*), we mounted a second lamp L_2 so, that it was only reflected by one of the mirrors; by means of a weak lens, cut into the halves Z_1 and Z_2, the two pencils of light could be made nearly parallel before they entered the telescope, and brought

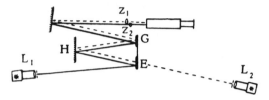

Fig. 7. Arrangement for comparing the phases
of the vibration near *E* and *G*.

to a focus in the same plane. The result of this experiment was perfectly convincing: in our arrangement, the angular movements of the edges always reinforce each other, the horizontal amplitude is sometimes increased by a factor 2, sometimes by a somewhat smaller factor.

The normal on the bridge vibrates not only in a horizontal but also in a vertical plane. We investigate how these movements are compounded, and this all along the upper edge of the bridge, by displacing the mirror from point to point. It appears, that mostly the shaded parts of fig. 8 execute a flexion, turning about the lines MP, MR. In accordance with our previous results, one must imagine the point *E* coming towards us at the same time when *G* is moving backwards. Only for the tones *f'—gis'* on the *D*-string are the movements more complicate.

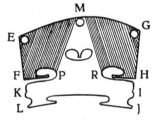

Fig. 8. Deformation of the bridge by the transverse vibrations.

However the deformations of the bridge cannot be wholly described by these two flexions. The existence of a real torsion is proved by fixing a mirror at *M*, which reveals considerable angular motions of the normal in a horizontal plane, and shows the same characteristic maxima which we found with the mirror near E and G. Generally speaking, the transverse movement of the bridge is best described as a combination of a torsional vibration with a flexure of the edges (fig. 9). Curiously, in this arrangement some very high tones on the E-string (near ais''') show a nearly pure vertical movement of the luminous spot, which proves the existence of a maximum of simple flexural vibration.

When we not only investigate the upper edges, but when we

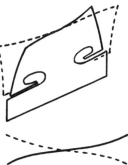

gradually displace the mirror towards the lower parts of the bridge along the sides EF, GH (fig. 8), we find that the curves are hardly modified, only the torsional vibration is somewhat smaller; we conclude that these parts of the bridge are moving more or less as straight lines. When we come under the side cuts of the bridge, near KL and IJ, the torsional and the flexural vibration disappear at once, which proves that there is no

Fig. 9. Combination of flexural and torsional vibration of the bridge.

question of a balancing of the bridge as a whole about the axis LJ (as assumed by G i l t a y and D e H a a s). The feet of the bridge move only in the plane of the bridge, and this is therefore the only movement transmitted to the belly of the violin.

In order to get more ample information about the movement of the bridge and to verify the conclusions already reached, the bridge was examined also under the microscope with a magnifying power of about 55. On the wood we had rubbed out a trace of aluminium powder; in the beam of a lamp, directed on it, a number of particles were seen shining brightly, and when the strings were bowed, one could observe directly how they were drawn out into luminous lines. In looking more or less perpendicularly to the bridge (so far as this is not hindered by the strings), one nearly always saw straight lines, corresponding to the movement of the bridge in its plane. These lines were observed at different parts of the bridge, their inclinations carefully noted and compared, so that the imaginary centre of rotation could be determined, about which the bridge rotates. This point A (fig. 1) is situated near the belly, about midway between the symmetry plane of the violin and the point supported by the sound post; however, for the different tones it has not exactly the same position. From this we infer that when the violin is played, the sound post does not remain motionless, but that it vibrates slightly up and down in its own direction, though with a small amplitude; so it can be understood that the vibration is communicated to the back of the instrument.

R a m a n remarks [5]) that the bridge during the vibration in its plane does not move as a rigid body, but that there occur elastic deformations in it. This was confirmed, by fixing two mirrors on the narrow vertical side of the bridge, one in F and one in K (fig. 8). The comparison of their movements showed that the mirror above the cut rotated over a smaller angle than that beneath the cut, but the difference seldom exceeded 10 or 20%.

4. *Investigation of bridges of special construction.* For comparison with the results obtained, we ordered a second bridge from the same factory which had manufactured the first. This second one showed entirely the same modes of vibration.

Then we procured a bridge in which no cuts had been made. The main peculiarities of its vibration are summarized in fig. 10, which is to be compared with fig. 9: the chief difference is that the flexural and torsional vibrations of the massive bridge are transmitted as low down as the feet, so that one may speak of a real rotation about the axis LJ. Obviously, the cuts in the bridge are not made for esthetical reasons only, but in order to secure that no vibrations, except these in the plane of the bridge, will be transmitted to the belly.

Some observations were also made with a brass bridge, similar to the ordinary wooden bridge in form and dimensions, but without cuts. The curves are now much simpler, often they approximate straight lines, especially the torsional vibration has diminished: the bridge is moving mainly as a rigid body in its plane. Still it is striking that such a small brass plate with a thickness of several millimeters bends

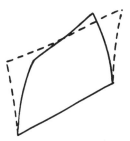

Fig. 10. Deformation of a bridge without cuts by the transverse vibrations.

sufficiently under the pull of the strings in order to show traces of the torsional maxima (at a slightly higher frequency than for the wooden bridge).

5. *The influence of a mute on the vibrations of the bridge.* By placing on a normal wooden bridge an ebony mute of 3,5 g, vibration curves were obtained of remarkable simplicity, and which therefore could be easily studied in detail (fig. 2). However, this only applies to the interval g—d''; on the E-string, the curves were either too small to be observed, or they showed a complicated form; a corresponding phenomenon in the sound emitted by the violin has been also found by E d w a r d s [6]) and explained by R a m a n [7]). Moreover, in the interval g—d'', the curves have an amplitude which is at least equal to that without a mute, and sometimes even greater, while the sound gives the impression of being damped.

In order to investigate if these characteristic modifications are due to the mass of the mute or rather to its rigidity, a mute was sawn in three parts, and these were clamped separately on the bridge. The greater simplicity of the curves was now also very striking, so that we conclude that the influence of a mute is chiefly due to its mass; still a slight effect of the rigidity was observed.

6. *The different vibrating systems of which the bridge is a part.* The characteristic maxima shown by the vibrations at determinate tones indicate the existence of separate vibrating systems with proper frequencies. In order to find out which systems are concerned in producing some of the most important maxima, we slightly modified the instrument, trying to change the vibrating mass or the elastic force; of the effect of such modifications we judged by looking at the change in the frequency of the maximum involved.

fis' and b'. These tones correspond to maxima of the motion in the plane of the bridge when a mute is used. They were selected, because they were so very conspicuous and well isolated. They were considerably lowered by a weight of 10 g, fixed on the bridge above the cuts; the effect was less when the weight was placed beneath the cuts, which is easily understood by the diminished distance to the axis of rottation (see § 3).

These maxima were also lowered when the mass was clamped to the belly of the violin in the prolongation of the plane of the bridge, at the inner side of the f-holes. So it is clear that the vibrating mass consists of the bridge and the belly between the f-holes. By relaxation of the strings, the frequency of the maximum was lowered, and this for every separate string which was tuned down. So the elastic force for the vibration of the bridge in its plane is furnished by the tension of the four strings, which are so strongly coupled for this movement that they have to be treated as one single chord. To this is added the elasticity of the belly, which has about an equal importance as that of the strings; for by lowering the pitch of the strings by 2 tones, the maxima are lowered by 1 tone only.

a'. This maximum of the torsional vibration does not change by charging the belly, which confirms that the vibration is not transmitted to this. By charging the bridge with a mass of 10 g, the maximum is slightly displaced, on the contrary it is much lowered when the tailpiece is charged. That this. part of the instrument participates in the motion, was also optically observed by fixing a galvanometer mirror to it; obviously the broad extremity of the tailpiece is oscillating with a considerable amplitude, while the narrow extremity makes only small linear excursions. So the tailpiece will determine the frequency of the torsional vibration in a much higher degree than the bridge, not only because its mass is greater, but also because of its greater amplitude. For this vibrating system, the elastic force is furnished by the strings, considered again as a whole; this is proved by observing that the maximum is lowered for each separate string which is tuned down.

b and b'. These maxima of the flexural vibration observed when a mute is used, are lowered by charging the bridge, and also by tuning down the strings; this proves, that the vibrating mass is to be found in the bridge, the elastic force in the bundle of the strings.

It is wellknown, that the air volume in the violin has a proper frequency, which in most of the violins corresponds to about 270 Hertz. In order to ascertain, whether the vibrations of this air have a perceptible influence on the movement of the bridge, we filled the whole violin with gas; wishing to keep the instrument intact in all other respects, we left the *f*-holes open, and conducted a continuous current of gas through them. In these circumstances, the proper frequency of the volume inside the instrument was lowered by one tone; however, no effect either on the vibration curves of the bridge or on the position of the maxima could be observed.

7. *The vibration of the bridge and the violin sound*. Because only vibrations in the plane of the bridge are transmitted to the belly, these vibrations must have a particular importance for the sound of the instrument. As a matter of fact, our curve fig. 4*A* confirms in a

striking way the resonance curves obtained by B a c k h a u s [8]). We have found, how the position of the maxima may be modified, and by this we have some possibility of changing the pitch of the characteristic resonances of the violin sound.

It is now a generally accepted opinion, that the brilliant sound of the best old violins is closely connected with the presence of high resonances ("formants"), especially near 3000 c. sec^{-1} Perhaps this could be obtained, by using wood of a high elasticity for the belly. Moreover, when all the strings are screwed up over one tone, we have shown that the pitch of a resonance near say 2800 will be heightened to about 3000 c. sec^{-1}, which must affect the character of the sound in a marked degree; so this effect, of which sometimes use is made in musical performances, becomes entirely comprehensible.

8. *Conclusion.* The general result of our investigation is, that the bridge executes different complicate movements, as well in its own plane as perpendicularly to it. These correspond to the proper vibrations of several vibrating systems, consisting of the whole of the strings, the tailpiece, the belly or the bridge itself, in some respects strongly coupled to each other. Thanks to the cuts in the bridge, only the vibration in its plane is transmitted to the feet, and by them to the belly of the instrument.

We thank Prof. Dr. L. S. O r n s t e i n for his kind interest in this investigation.

Received March 19, 1937.

REFERENCES

1) See the older literature in: G i l t a y, De strijkinstrumenten uit een natuurkundig oogpunt beschouwd (Leiden 1916).
2) Versl. Akad. Amst. **18**, 462, 1909; Proc. Amsterd. **12**, 513, 1910.
3) Phil. Mag. **20**, 456, 1910.
4) Phil. Mag. **35**, 493, 1918. Hdb. d. Physik **8**, 370, 1927.
5) Phil. Mag. **35**, 495, 1918.
6) Phys. Rev. **32**, 23, 1911.
7) Phil. Mag. **35**, 493, 1918 and Hdb. d. Phys. **8**, 382, 1927.
8) Z. techn. Phys. **8**, 509, 1927.

16

Reprinted from *Compt. Rend.*, **250**, 2161–2163 (Mar. 1960)

ACOUSTIQUE MUSICALE. — *Sur le chevalet du violoncelle.* *
Note (*) de M. Benjamin Bladier, présentée par M. Gustave Ribaud.

On a étudié sur plusieurs chevalets, disposés tour à tour sur un bloc de béton puis sur leur instrument, la réponse à une excitation mécanique, à l'aide de capteurs piézoélectriques fixés sur le chevalet. Ensuite, sur l'instrument, le niveau de pression sonore provoquée par deux excitateurs électrodynamiques, travaillant en série et en opposition de phase, disposés tout d'abord en deux points de la table (l'un à 3 cm en arrière du pied droit du chevalet, l'autre à 3 cm en arrière du pied gauche), enfin, de part et d'autre sur le haut du chevalet parallèlement à la table.

Le processus opératoire diffère de celui décrit antérieurement (¹), (²) par le mode d'excitation, et la durée de l'exploration continue de la plage de fréquence (20 à 5 000 Hz) du générateur, portée à 6 mn. Excitateurs et capteurs, étudiés deux à deux pour chaque type, ont des sensibilités inégales, mais conservent sensiblement la même phase en fonction de la fréquence; il en est de même des divers amplificateurs; la reproductibilité des phénomènes est satisfaisante si l'on a soin de remettre correctement en place les excitateurs, et de rendre les capteurs bien solidaires du chevalet. Il n'a pas été tenu compte de l'influence propre de chaque capteur (dont le poids est de 24 g) ni de leur surface de contact avec le chevalet (0,8 cm²).

Dans les limites de nos expériences on peut résumer ainsi les résultats obtenus :

1º Dans la transmission des vibrations des cordes à la caisse, le chevalet ne joue pas uniquement le rôle d'un organe de couplage passif.

2º Une excitation alternative exercée sur le haut du chevalet dans le plan de celui-ci donne naissance à deux forces alternatives exercées sur la table par chacun des pieds du chevalet. Nous considérons pour chacune de ces forces les trois composantes situées respectivement : deux dans le plan du chevalet (l'une X perpendiculaire à la table, l'autre Y latérale) et une troisième Z perpendiculaire au plan du chevalet.

3º D'une manière générale le niveau de vibration, recueilli sur les deux pieds, est maximal dans le plan X, suivi de près par celui du plan Y, le minimum se trouvant dans le plan Z. Les vibrations recueillies sous la découpure centrale appelée cœur, ainsi que celles obtenues de part et d'autre de ce dernier, sont maximales dans le plan Y.

4º La réponse dans le plan X de chevalets étudiés sur un bloc de béton met en évidence, en fonction de la fréquence d'excitation (20-5 000 Hz), un comportement complexe : Passif dans la transmission d'ailleurs inégale des basses fréquences (20 à 200 Hz), il devient actif entre 200 et 1 100 Hz (résonances propres d'un niveau moyen de 20 dB), enfin, il est inerte de 1 200 à 5 000 Hz (affaiblissement net de 20 dB au-delà de ces résonances).

Editor's Note: An English translation follows this article.

(2)

5⁰ La réponse du pied droit n'est pas la même que celle du pied gauche. Les chevalets classiques (13 g) sont nettement différenciés. Sur des chevalets sans découpures les résonances, moins larges et moins amples, sont d'autant plus déplacées vers des fréquences plus élevées que le chevalet est plus lourd.

6⁰ La modification de l'impédance de couplage, par interposition de caoutchouc entre les pieds et le bloc, déporte la réponse vers des fréquences plus graves.

7⁰ Sur l'instrument les trois parties distinctes de la réponse sont moins évidentes; en plusieurs points les interactions, dues vraisemblablement aux résonances de la caisse, modifient son comportement.

8⁰ Certaines résonances de la caisse sont modifiées non seulement par la présence du chevalet, mais encore suivant le type adopté. Il se comporte d'une façon différente suivant l'élasticité de la table; par suite il est capable de régir et de modifier le timbre de la production sonore dans certaines plages de fréquences.

9⁰ Si l'excitation se propageait dans le chevalet sans changer de phase (comme dans un conducteur), la différence entre l'état du sommet et celui des pieds, 9 cm plus loin, ne pourrait être notable pour les longueurs d'onde en question. Or, l'expérience ne vérifie pas toujours cette prévision : les interactions, dues vraisemblablement aux vibrations dans les plans Y et Z, semblent responsables des modifications dans le plan X, jusqu'à faire varier le déphasage entre les deux pieds, d'environ 0⁰ à $\pm$ 180⁰, et cela pour plusieurs fréquences. Donc sous les pieds du chevalet les deux sources sonores de l'instrument ([2]) ne sont pas toujours excitées alternativement ([3]).

10⁰ Le chevalet se comporte comme un amplificateur (leviers), de gain le plus souvent égal à 2 (6 dB entre ut_1 et mi_1 de 66 à 660 Hz), ensuite inférieur ou au plus égal à 1 (entre mi_1 et ut_6 de 660 à 2 112 Hz).

11⁰ Les réponses dans les plans Y et Z, montrent que le chevalet oscille approximativement autour d'une droite du plan X ([4]); le calcul situe celle-ci aux deux-tiers de sa hauteur, soit aux environs du cœur; les vérifications par l'exploration du chevalet confirment, pour une plage de fréquences comprise entre 66 et 180 à 220 Hz, un minimum de vibration au point situé par le calcul, aussi bien lorsque l'excitation s'effectue par excitateurs que lorsque ceux-ci sont remplacés par la corde et l'archet ([5]), (notes jouées : ut_1, mi_1, sol_1, sur la corde ut et sol_1, si_1, $ré_2$, sur la corde sol).

12⁰ Ce minimum persiste si le chevalet, sans corde, est collé sur l'instrument, il disparaît si le chevalet est collé sur le bloc de béton. L'effet paraît dû surtout à la dissymétrie de l'instrument (âme-barre).

13⁰ Compte tenu de nos réserves concernant l'axe d'oscillation, celui-ci ne passe ni par le pied droit (qui n'est pas fixé à travers la table par l'âme), ni par le pied gauche (également non fixé par la barre); d'une manière générale (de ut_1 à environ la_2) le niveau de vibration maximal est sur le

(3)

pied gauche, plus faible sur le pied droit et minimal aux environs du cœur. Ce minimum est d'autant mieux marqué que la force d'excitation est plus faible. De la$_2$ à ut$_3$ le niveau prépondérant est sur le pied droit (côté âme); l'effet d'axe n'existe plus.

14° Le chevalet réduit (au-delà de mi$_3$) les possibilités sonores du violoncelle, dans la plage de fréquences correspondant au maximum de sensibilité de l'oreille. Il ne rend donc plus sonore qu'une partie des vibrations.

(*) Séance du 14 mars 1960.

(¹) B. BLADIER, *Comptes rendus*, 245, 1957, p. 791.

(²) B. BLADIER, *Contribution à l'étude du violoncelle*, C. N. R. S., Paris, 1958.

(³) Les réponses d'un même instrument excité en deux points de la table en opposition de phase, puis en phase, ne sont pas les mêmes, mais les niveaux sonores sont assez souvent voisins.

(⁴) Question évoquée pour le violon par Savart (⁵), discutée par Bouasse (⁶), signalée par Rimsky-Korsakov et N. A. Diakonov (⁷).

(⁵) F. SAVART, *Ann. Chim. et Phys.*, 2ᵉ série, 12, 1819.

(⁶) H. BOUASSE, *Cordes et membranes*, Delagrave, Paris, 1926.

(⁷) RIMSKY-KORSAKOV et N. A. DIAKONOV, *Instruments de musique*, Moscou, 1952.

(⁸) Dans ce dernier cas la forme de la réponse est complexe; alors qu'en utilisant des excitateurs celle-ci est sensiblement sinusoïdale.

(*Centre de Recherches scientifiques, industrielles et maritimes, Marseille.*)

16

On the Bridge of the Violoncello

BENJAMIN BLADIER

*This article was translated expressly for this
Benchmark volume by R. Bruce Lindsay, Brown University,
from* Compt. Rend., **250**, *2161–2163*
(Mar. 1960)

Study has been made of the response to mechanical excitation of several bridges placed in turn on a block of concrete and then afterward on the instruments to which they belong. This was done by means of piezoelectric transducers fastened to the bridge. When the bridge was placed on the instrument, the level of sound pressure produced by two electrodynamic oscillators was measured — with the oscillators working in series and in opposite phase, placed at first at two points of the belly of the instrument (the one 3 cm behind the right foot of the bridge and the other 3 cm behind the left foot). Finally the two oscillators were placed on both sides of the top of the bridge parallel to the belly.

The operating procedure differs from that described earlier (1, 2) through the method of excitation and the duration of the continuous exploration of the frequency range of the generator (20 to 5,000 Hz), carried up to 6 minutes. Oscillators and transducers, studied in turn for each type, have unequal sensitivity, but maintain approximately the same phase as a function of the frequency. It is the same with the different amplifiers. The reproducibility of the phenomenon is satisfactory if one takes care to put back the oscillators in the proper place and to fasten the transducers firmly to the bridge. No account has been taken of the specific influence of each transducer by itself (the weight of each is about 24 g) nor of their area of contact with the bridge (0.8 cm²).

Under the limitations imposed by our experiments, we can summarize the results obtained:

1. In the transmission of the vibrations of the strings to the box the bridge does not play uniquely the role of an organ of passive coupling.

2. An alternative excitation applied to the top of the bridge in its plane gives rise to two alternative forces excited on the belly by each of the feet of the bridge.

For each of these forces we consider the three components situated as follows: two in the plane of the bridge (one, X, perpendicular to the belly, the other, Y, along the belly) and a third, Z, perpendicular to the plane of the bridge.

3. Speaking generally, the level of vibration measured at the two feet is a maximum in plane X, followed closely by that of plane Y; the minimum is found in plane Z. The vibration measured on the central cutout, called the heart, as well as those obtained from both sides of the latter, are a maximum in plane Y.

4. The response of the bridge in plane X as examined on the block of cement demonstrates a complex behavior as a function of the excitation frequency (20 to 5,000 Hz). Passive in the transmission of the bass frequencies (20 to 200 Hz), it becomes active between 200 and 1,100 Hz (the proper resonance of a mean level of 20dB). Finally, it is ineffective from 1,200 to 5,000 Hz (a distinct diminution of 20dB above these resonances).

5. The response of the right foot is not the same as that of the left foot. The classical cello bridges (13 g) are definitely different. For the bridges without cutouts the resonances, less large and less broad, are in addition more displaced toward the higher frequencies in proportion as the bridge is heavier.

6. The change in impedance of the coupling produced by the insertion of rubber between the feet and the block moves the response toward lower frequencies.

7. On the instrument itself the three distinct parts of the response are less evident. In many cases interactions, due no doubt to the resonances of the box, modify its behavior.

8. Certain resonances of the box are modified not only by the presence of the bridge, but also by the type used. The behavior of the box depends on the elasticity of the belly; consequently, it is capable of governing and modifying the timbre of sound production in certain frequency ranges.

9. If the excitation is propagated in the bridge without change of phase (as in a conductor), the phase difference between the top and the feet of the bridge (9 cm apart) should not be very perceptible for the wavelengths in question. But experience does not always verify this prediction. Interactions, due probably to vibrations in planes Y and Z, appear responsible for changes in plane X, producing a phase difference between the two feet in the neighborhood of 0 to $\pm 180°$, and this indeed for several frequencies. Thus, under the feet of the bridge, the two sound sources of the instrument (2) are not always excited one after the other (3).

10. The bridge behaves as an amplifier (acoustical lever) with a gain most often equal to 2 (6 dB between ut_1 [c_1] and mi_4, that is 66 and 660 Hz), and thereafter with a gain less than, or at most equal to 1 (between mi_4 and ut_6, that is between 660 and 2,112 Hz).

11. The responses in the Y and Z planes show that the bridge oscillates approximately about a straight line in the X plane (4). "Calculation" locates this line at about two thirds the height of the bridge in the neighborhood of the heart. Exploratory tests of the bridge at various points, for a range of frequencies between 66 and 180 to 220 Hz, confirm a minimum of vibration at the point located by calculation. This

is the case when the excitation is produced by oscillators as well as when the oscillators are replaced by the string and the bow (8) (notes played: ut_1, mi_1, and sol_1 on the string ut; and sol_1, si_1, re_2 on the string sol).

12. The minimum persists if the bridge, without the string, is glued to the instrument. It disappears if the bridge is glued to the cement block. The effect appears to be due to the lack of symmetry of the instrument (soundpost–bassbar).

13. With due account taken of our reservations, the axis of oscillations does not pass either through the right foot of the bridge (which is not rigidly supported on the belly by the soundpost) or through the left foot (supported with equal lack of rigidity by the bassbar). Speaking generally (concerning ut in the neighborhood of la_2), the vibration level is a maximum at the left foot, weaker at the right foot, and a minimum in the neighborhood of the heart. This minimum is more marked as the excitation force becomes weaker. From la_2 to ut_3 the preponderant level is on the right foot (near the soundpost); the effect of the axis no longer exists.

14. Above mi_4 the bridge reduces the sound-producing potential of the violoncello in the frequency range corresponding to the maximum sensitivity of the ear. Thus only a fraction of the vibrations are turned into sound.

Notes and References

1. Benjamin Bladier, *Compt. Rend.*, **245**, 791 (1957).
2. Benjamin Bladier, *Contribution à l'étude du violoncelle*, C.N.R.S., Paris, 1958.
3. The responses of a given instrument excited at two points of the belly, first in the phase opposition and then in phase agreement, are not the same, but the sound levels are often close to each other.
4. A question raised concerning the violin by Savart[5], discussed by Bouasse[6], and mentioned by Rimsky-Korsakov and N. A. Diakonov[7].
5. F. Savart, *Ann. Chim. Phys.*, 2^{e}sér., 12 (1819).
6. H. Bouasse, *Cordes et membranes*, Delagrave, Paris, 1926.
7. Rimsky-Korsakov and N. A. Diakonov, *Instruments de musique*, Moscow, 1952.
8. In the last case the form of the response is complex; only with the use of the oscillators is it reasonably sinusoidal.

Übertragungseigenschaften des Streichinstrumentenstegs *

WALTER REINICKE

1. Funktionsprinzip und Übersicht

In der Literatur ist verschiedentlich, besonders deutlich aber von L. Cremer /1/, darauf hingewiesen worden, daß das Streichinstrument aus 3 wichtigen Teilen besteht, welche prinzipiell unterschiedliche physikalische Funktionen erfüllen: Der für die Abstrahlung der Schalleistung entscheidende Teil ist der Corpus. Die Erzeugung dieser Schalleistung übernimmt das System Saite–Bogen. Der Steg überträgt die Leistung zwischen beiden.

Da insbesondere der *Streich*instrumentensteg an beiden "Füßen" quasi punktweise mit dem Corpus einerseits und mit der Saite andererseits verbunden ist, kann man ihn nachrichtentechnisch als "Mehrtor" behandeln, um seine Übertragungseigenschaften zu ermitteln. Dieser "black box" geben wir aus folgenden Gründen 2 Eingangstore und 2 Ausgangstore: Am Eingang kommen wir, selbst wenn wir nur *eine* Saite in der Mitte annehmen, nicht mit einem Tor aus, wenn wir beliebige Schwingungsrichtungen der Saite zulassen wollen. Wir zerlegen die Ein-

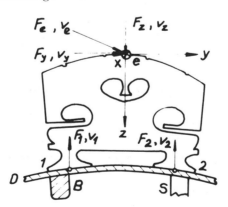

Bild 1: Koordinaten und Wirkungsrichtungen der Torgrößen

Editor's Note: An abstract in English follows this article.

gangskraft F_e in ihre Komponenten F_y und F_z, vgl. Bild 1. Trotz der Stegsymmetrie können wir im Betriebsfall auch die beiden Ausgangstore nicht zu einem zusammenfassen, weil der Corpus durch seinen unsymmetrischen Aufbau für die Stegausgänge zwei unterschiedliche Belastungsimpedanzen darstellt. Der Steg wird so zum "*Viertor*".

In Bild 2 erkennt man, daß die Abklingzeit der Stegeingangskraft F_e einer a^1-Violinsaite (4. Teilton) unterschiedlich ist, je nachdem, ob sie nicht-polarisiert schwingt (a) oder genau senkrecht zur Instrumentendecke erregt wurde (b). Mit Hilfe eines sehr dünnen Perlonfadens, an dem die Saite aus der Ruhelage gezogen wurde, bis dieser riß, konnte eine genaue Schwingungsrichtung eingehalten werden. Der steilere Teil der Abklingkurve (a) entspricht der kürzeren Abklingzeit der y-Komponente, der flachere Abfall parallel zu (b) entspricht der längeren Abklingzeit der z-Komponente. Umfassende Messungen zeigten, daß der Saite auch im Mittel in y-Richtung mehr Energie entzogen wird als in z-Richtung.

Bild 3 gibt als schraffierte Säulen den Streubereich gemessener Energie-Abklingzeiten $\tau/2$ der Teiltöne einer a^1-Saite wieder, die auf der Violine in y-Richtung gezupft wurde. Die stark schwankende ausgezogene Kurve ist ein Rechenergebnis nach Abschnitt 5. Die glatte Kurve verbindet die Meßwerte, die bei dieser Saite auf *starren* Stegen gefunden wurden.

Erste *Dreitor*-Messungen sind bereits von G. Steinkopf /2/ am Cello-Steg vorgenommen worden, jedoch ohne die höheren Eigenfrequenzen zu erfassen. P. Zimmermann /3/ hat in einer theoretischen Untersuchung diese Eigen-Frequenzen und -Formen in einem Modell erfaßt. Allerdings beschreibt dieses Modell den Cello-Steg nicht richtig, wenn man eine z-Komponente der Eingangskraft zuläßt. Es wird deshalb hier ein abgewandeltes und sogar einfacheres Funktionsmodell für den Cello-Steg behandelt. Die davon abweichenden Übertragungseigenschaften des Violinstegs werden zuvor mit einem eigenen Modell beschrieben.

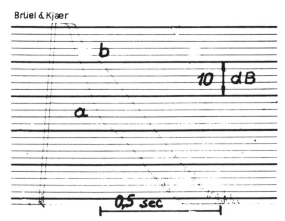

Bild 2: Abklingkurven des 4. Teiltones einer a^1-Saite
    (a) nicht polarisiert
    (b) senkrecht zur Decke

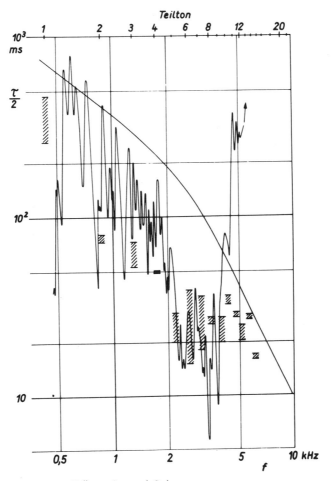

Bild 3: Abklingzeiten der Teiltöne einer a^1-Saite
— Messung bei starren Stegen
z Messung (Streubereich) und
ᴍᴍ Rechnung für den betriebsmäßig abgeschlossenen Steg

Zur Beschreibung der Übertragungseigenschaften wird die Impedanzmatrix Z gewählt, die nach Gl. (1) die Kräfte F an den 4 Toren mit den Schnellen v verbindet:

$$
\begin{bmatrix} F_1 \\ F_2 \\ F_y \\ F_z \end{bmatrix} = \overbrace{\begin{bmatrix} z_{11} & z_{12} & z_{1y} & z_{1z} \\ z_{12} & z_{11} & -z_{1y} & z_{1z} \\ z_{1y} & -z_{1y} & z_{yy} & 0 \\ z_{1z} & z_{1z} & 0 & z_{zz} \end{bmatrix}}^{Z} \begin{bmatrix} v_1 \\ v_2 \\ v_y \\ v_z \end{bmatrix} \tag{1}
$$

Weil das passive Viertor übertragungssymmetrisch ist ($Z = Z'$) und weil der Steg als geometrie-symmetrisch zur z-Achse angenommen wird, enthält die Z-Matrix statt 16 nur 6 unbekannte z-Elemente.

Man könte deswegen den Steg auch durch 2 Zweitor-Gleichungen beschreiben; man erhält sie, wenn man die Differenz bzw. die Summe der beiden ersten Zeilen von Gl. (1) bildet und sie mit je einer der Zeilen 3 und 4 zusammenfaßt. Aber der Betriebsfall am Instrument kann damit nicht beschrieben werden, vgl. /3/. Am Instrument kann z. B. die Kraft F_y eine Schnellekomponente v_z am Stegeingang hervorrufen.

2. Hologramm-interferometrische Untersuchungen

Um ein mechanisches Funktionsmodell des Steges zu finden, kann man versuchen, in die "black box" hineinzuschauen, indem man z. B. die Eigenschwingungen sichtbar macht, die der Steg bei teilweise festgebremsten ($v = 0$) Toren hat. Als berührungsfreies Meßverfahren ist hierzu die Hologramm-Interferometrie bestens geeignet. Das Verfahren von K. A. Stetson und R. L. Powell /4/ ist an anderer Stelle ausführlich beschrieben. Mit Bild 4 soll hier nur gezeigt werden, daß es mit Hilfe eines Spiegels Sp gelingt, den schwingenden Steg in einem einzigen Hologramm sowohl schräg (Richtung B) als auch normal (Richtung B^s) zu betrachten, wenn das Laserlicht fast streifend unter dem Winkel α auffällt. Der Zweck ist, mit Hilfe des

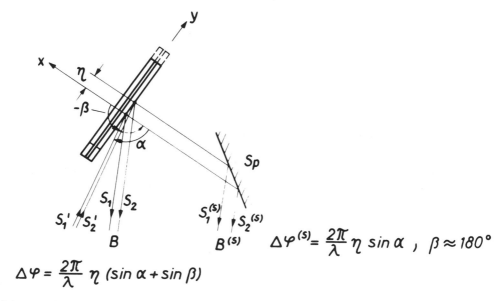

$$\Delta\varphi^{(s)} = \frac{2\pi}{\lambda}\,\eta\,\sin\alpha\ ,\quad \beta \approx 180°$$

$$\Delta\varphi = \frac{2\pi}{\lambda}\,\eta\,(\sin\alpha + \sin\beta)$$

Bild 4: Strahlenverlauf für die Signalwelle S bei Aufsicht auf den seitlich beleuchteten Steg, der in y-Richtung um $x_y = \eta$ ausgelenkt wird.

Bild 5: Zeitmittelhologramm nach Bild 4, y-Komponente bei $f_{yy} = 3000$ Hz

schräg beobachteten "Kontrollbildes" nachzuprüfen, ob eventuell Bewegungskomponenten in x-Richtung, z. B. Biegeschwingungskomponenten, das Schwingungsbild beeinflussen. Ist dies nicht der Fall, so erhält man gleichzeitig 2 Bilder, wie in Bild 5, von denen das linke (schräge Richtung B) etwa doppelt so viele Interferenzstreifen aufweist wie das rechte (normale Richtung B^s). Der Steg wird dabei am Eingang mit der Kraft F_e angeregt und steht mit beiden Füßen auf einem starren Fundament. Wenn man mit der Auswertung von Hologramm-Interferogrammen vertraut ist, erkennt man, daß der Steg in diesem Fall eine Eigenschwingung vollführt, die in Bild 7(1) als eine Drehschwingung des Steg-Oberteils um den Drehpunkt D gegen das ruhende Unterteil erkannt wird. Die Auslenkungen bei einer solchen Drehschwingung haben eine y-Komponente, welche die Interferenzstreifen in Bild 5 erzeugt, und eine z-Komponente, welche die Streifen in Bild 6 erzeugt, wenn man die Beleuchtungs- und Beobachtungsrichtungen zum Steg aus der x-y-Ebene in die x-y-Ebene verlegt, vgl. die Festlegung des Koordinaten-Systems in Bild 1. Während die Eigenform, die von den Bildern 5, 6 und 7(1) beschrieben wird, bei einer Frequenz um ca. 3000 Hz auftritt, zeigt sich eine zweite Eigenform bei ca. 6000 Hz.

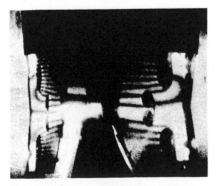

Bild 6: Zeitmittelhologramm nach Bild 4, z-Komponente bei $f_{yy} = 3000$ Hz

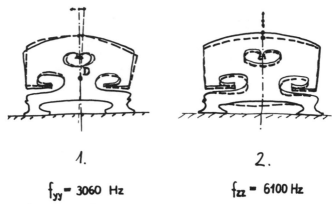

$f_{yy} = 3060$ Hz $f_{zz} = 6100$ Hz

Bild 7: Eigenformen des starr aufgesetzten Violinsteges nach den Bildern 5 und 8

Die z-Komponente erzeugt dabei das Interferogramm Bild 8, welches besagt, daß sich die Querbalken zwischen Stegmittelteil und Füßen wie Biegefedern verhalten, auf denen die restliche Stegmasse vertikal schwingt, vgl. auch Bild 7(2).

Eine Reihe von Hologramm-Interferogrammen bei anderen Anregungs- und Abschlußbedingungen für die Tore haben dieses prinzipielle Verhalten bestätigt und berechtigen dazu, den Violinsteg durch ein Funktionsmodell nach Bild 9(a) zu ersetzen. Dort ist am Eingangspunkt die Eingangskraft F_e mit ihren Komponenten für das y- und das z-Eingangstor eingezeichnet. Diese Kraft wird über den Hebelarm d auf den Drehpunkt der Drehsteife D_2 geleitet (die am Steg etwas unterhalb des "Herzens" konzentriert zu denken ist). Sie spannt D_2 und beschleunigt die Stegoberteilmasse m_2 mit ihrem Trägheitsmoment $m_2 \, i_{20}^2$. Der Massen-Schwerpunkt liegt um d_{20} über dem Drehpunkt von D_2. Dort ist D_2 an der punktförmigen Stegmittelteil-Masse befestigt. Zusammen mit m_2 kann m_1 in der Mitte der Stange von der Länge $2a$ über den Fußsteifen s_1 eine vertikale Schwingung gemäß Bild 8 und 7(2) ausführen, wenn die Füße festgebremst sind. Schließlich ist eine punktförmige Masse m_3 der Füße berücksichtigt. m_1 und m_3 können nicht

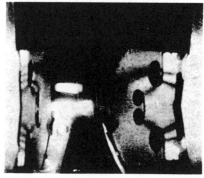

Bild 8: Zeitmittelhologramm nach Bild 4, z-Komponente bei $f_{zz} = 6000$ Hz

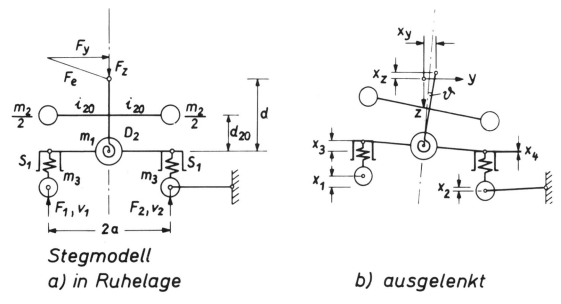

Stegmodell

a) in Ruhelage   **b) ausgelenkt**

Bild 9: Funktionsmodell des Violinsteges

horizontal (y-Richtung) ausgelenkt werden, weil die Fußmasse m_3 mittels eines masselosen Gestänges horizontal festgelegt ist. Damit ist berücksichtigt, daß die Stegfüße fest auf der Instrumentendecke sitzen und dort nur vertikal ausgelenkt werden, vgl. /3/. Alle Auslenkungen müssen sehr klein sein. Bild 9(b) zeigt das aus seiner Ruhelage ausgelenkte Modell. $x_k = v_k/j\omega$, $k = 1. 2 \ldots y$, z sind die Auslenkungen mit sinusförmigem Zeitverlauf in komplexer Schreibweise.

3. Berechnung und Messung der Übertragungseigenschaften des Violinstegs

Setzt man für dieses Modell die Lagrange'schen Bewegungsgleichungen an, so erhält man die Gleichungen für die Kräfte an allen vier Toren in Form der Gl. (1):

$$
\begin{bmatrix} F_1 \\ F_2 \\ F_y \\ F_z \end{bmatrix} = \begin{bmatrix} j\omega m_3 + \dfrac{1-\sigma}{j\omega}s_1 & \dfrac{\sigma s_1}{j\omega} & -\dfrac{\tau s_1}{j\omega} & \dfrac{s_1}{j\omega} \\[2ex] \dfrac{\sigma s_1}{j\omega} & j\omega m_3 + \dfrac{1-\sigma}{j\omega}s_1 & \dfrac{\tau s_1}{j\omega} & \dfrac{s_1}{j\omega} \\[2ex] -\dfrac{\tau s_1}{j\omega} & \dfrac{\tau s_1}{j\omega} & j\omega\mu m_2 + 2\dfrac{\sigma D_2}{j\omega d^2} & 0 \\[2ex] \dfrac{s_1}{j\omega} & \dfrac{s_1}{j\omega} & 0 & j\omega(m_1+m_2)+\dfrac{2s_1}{j\omega} \end{bmatrix} \begin{bmatrix} v_1 \\ v_2 \\ v_y \\ v_z \end{bmatrix} \quad (2)
$$

Es bedeuten dabei:

ω = Kreisfrequenz

j = $\sqrt{-1}$

$$\sigma = \frac{s_1}{2s_1 + \dfrac{D_2}{a^2}} = 0{,}38$$

$$\tau = \frac{D_2/ad}{2s_1 + D_2/a^2} = 0{,}19$$

$$\mu = \frac{d^2_{20} + i^2_{20}}{d^2} = 0{,}74$$

Die Impedanzen sind rein imaginär, also Reaktanzen, weil im Modell nach Bild 9 keine inneren Verluste im Holz berücksichtigt wurden. Diese sind klein im Vergleich zur Verlustenergie im Corpus. Die Zahlenwerte von σ, τ und μ ergeben sich aus Messungen und der Geometrie des Steges. Man erkennt, daß in der Hauptdiagonalen von Z in Gl. (2) die Impedanzen einfacher Schwinger stehen mit folgenden Resonanzfrequenzen:

$$\omega^2_{11} = \frac{(1 - \sigma)s_1}{m_3}$$

$$\omega^2_{22} = \omega^2_{11}$$

$$\omega^2_{yy} = \frac{2\sigma D_2/d^2}{\mu m_2} \qquad \text{gemessener Wert} \quad f_{yy} = \frac{\omega_{yy}}{2\pi} = 3060 \ \text{Hz}$$

$$\omega^2_{zz} = \frac{2s_1}{m_1 + m_2} \qquad \text{gemessener Wert} \quad f_{zz} = \frac{\omega_{zz}}{2\pi} = 6100 \ \text{Hz}$$

ω_{yy} und ω_{zz} sind die Eigenfrequenzen zu den Schwingungsformen nach Bild 7. Aus Gl. (2) läßt sich z.B. die Impedanz $F_e/v_e = F_t/v_t$ berechnen, die der starr aufgesetzte Steg einer tangential eingeleiteten Kraft mit den Komponenten F_y und F_z entgegensetzt. Hierbei ist die tangentiale Richtung an den Steg im Auflagepunkt der Saite a^1 gemeint, wo—abweichend von der Rechnung—die Messung nach Bild 10 durchgeführt wurde. Die Kraft F_t wurde mit einem winzigen piezoelektrischen Empfänger und die Auslenkung $x_t = v_t/j\omega$ kapazitiv an der Stegkante abgenommen. Der elektrodynamische Erreger wurde von einem Tongenerator mit Regeleingang so gespeist, daß v_t konstant blieb. Dann ist die aufgezeichnete Kraft F_t proportional

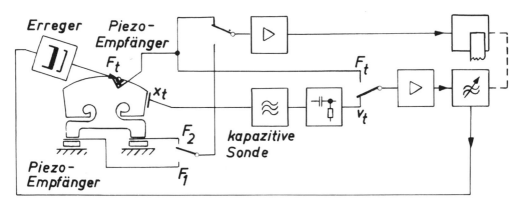

Bild 10: Aufbau für die Messungen am starr aufgesetzten Steg

zur Impedanz $|F_t/v_t|$. Ihren Verlauf über der Frequenz gibt Bild 11 wieder (ausgezogene Linie), dazu eingezeichnet ist der rechnerische Verlauf (gestrichelte Linie). Man erkennt die beiden Eigenfrequenzen bei ca. 3000 und 6000 Hz, zu denen die Hologramm-Interferogramme Bild 5 und 8 gehören bzw. die Eigenformen nach Bild 7.

Aus Gl. (2) kann man aber z. B. auch die komplexen Kraftübersetzungsverhältnisse F_1/F_t bzw. F_2/F_t berechnen, die beide bis auf das Vorzeichen gleich sein müssen. Ihre Messung wurde ebenfalls nach Bild 10 durchgeführt: Unter den Stegfüßen wurden mit piezoelektrischen Empfängern die Fußkräfte F_1 bzw. F_2 abgenommen und aufgezeichnet, während nun F_t am Regelverstärker des Tongenerators konstant gehalten wurde. Meß- und Rechenergebnisse sind in Bild 12 als ausgezogene bzw. gestrichelte Kurven wiedergegeben. Auch hier treten beide Eigenfrequenzen wieder auf. Bei tiefen Frequenzen sind die Kraftübersetzungsverhältnisse konstant, der Steg verhält sich wie ein 2-armiger Hebel. Sehr gut

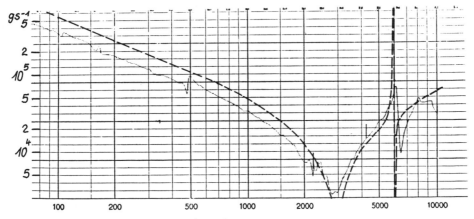

Bild 11: Verlauf der Eigenimpedanz $|F_t/v_t|$
————gemessen — — — —berechnet

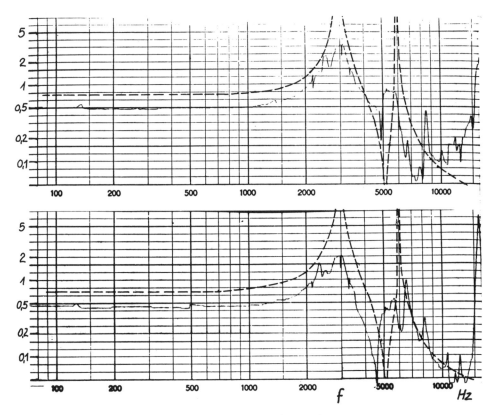

Bild 12: Verlauf der Kraftübersetzungsverhältnisse $|F_{1,2}/F_t|$
——————gemessen — — — —berechnet

konnte dabei nachgewiesen werden, wie sich f_{yy} zu tiefen Frequenzen verschiebt, wenn man ein Sordino auf den Steg setzt, und wie sich f_{yy} erhöht, wenn man die Stegschlitze mit kleinen Holzkeilen verstopft. Weitere Ergebnisse finden sich in /5/.

4. Untersuchungen am Cello-Steg

Ganz entsprechende Untersuchungen wurden am Cello-Steg vorgenommen. Die Hologramm-Interferogramme Bild 13, 14 und 15 für die Auslenkungen in der y-Richtung zeigen 3 wichtige Eigenschwingungsformen am starr aufgesetzten Steg. Sie werden mit den Bildern 16(1)–16(3) interpretiert. In der ersten Form bei $f_1 = 985$ Hz schwanken Steg-Oberteil und -Mittelteil als Ganzes über den biegeweichen (im Vergleich zum Violinsteg langen) Füßen. Bei der zweiten Eigenform bei $f_l = 1450$ Hz ist die größte Auslenkung in der Stegmitte, sie nimmt zu den Füßen und zum Eingang hin auf Null ab. Dabei ist die Durchbiegung der Füße ähnlich wie bei

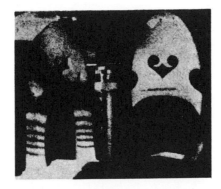

Bild 13: Zeitmittelhologramm, y-Komponente bei f_1 = 985 Hz

Bild 14: Zeitmittelhologramm, y-Komponente bei f_1 = 1450 Hz

Bild 15: Zeitmittelhologramm, y-Komponente bei f_2 = 2100 Hz

der 1. Form, nur geringer, während das Oberteil um den Eingangspunkt eine Drehung ausführt. Die dritte Form bei f_2 = 2100 Hz nach Bild 15 ist überlagert von einer störenden Biegeschwingung mit Auslenkungen in der x-Richtung. Diese wird durch das linke Kontroll-Interferenzbild erkannt, das nicht mehr die doppelte

309

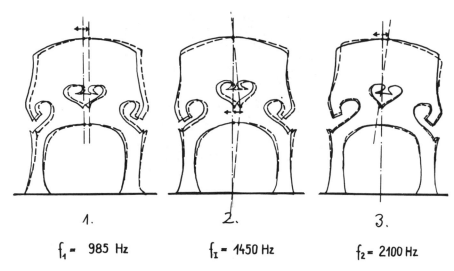

1.

2.

3.

$f_1 = 985\ Hz$ $f_I = 1450\ Hz$ $f_2 = 2100\ Hz$

Bild 16: Eigenformen des starr aufgesetzten Cellosteges nach den Bildern 13 bis 15

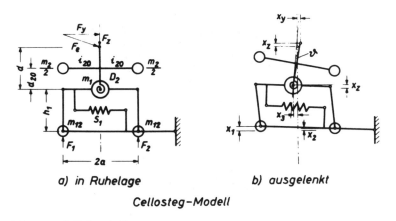

a) in Ruhelage   b) ausgelenkt

Cellosteg-Modell

Bild 17: Funktionsmodell des Cellosteges

Streifenzahl aufweist. Auch müßten die Streifen parallel verlaufen, wenn man erwarten darf, daß das Stegoberteil sich wie eine starre Scheibe verhält. Im Prinzip entspricht diese Form aber der 1. Violinsteg-Eigenform, also einer Drehung des Stegoberteils um eine Achse in Höhe des "Herzens", vgl. Bild 16(3).

Bild 17 zeigt das danach entworfene Funktionsmodell, das sich nur durch den unteren Teil vom Violinsteg-Modell unterscheidet. Der parallelogrammförmig verschiebbare, an den Eckpunkten gelenkige und von der Feder s_1 in die Ruhelage zurückgestellte Rahman erlaubt die translatorische Bewegung, die besonders in der ersten Eigenform nach Bild 13 bzw. 16(1) ausgeprägt ist, vgl. auch /3/. Auch hierfür sind die Lagrange'schen Bewegungsgleichungen aufgestellt und damit die Im-

310

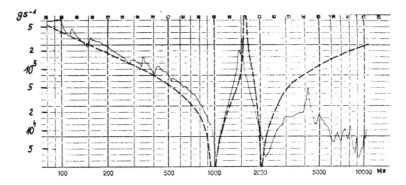

Bild 18: Verlauf der Eingangsimpedanz $|z_{yy}| = |F_y/v_y|$
——— gemessen – – – – berechnet

pedanzelemente aus Gl. (1) berechnet worden. Im Unterschied zum Violinsteg ist aber $x_z = \frac{1}{2}(x_1 + x_2)$, und es ist nicht nötig, die Variable v_z beizubehalten. Der Cellosteg wird deshalb durch 3-Torgleichungen beschrieben, für F_z existiert eine unabhängige Gleichung.

$$
\begin{bmatrix} F_1 \\ F_2 \\ F_y \end{bmatrix} = \begin{bmatrix} z_{11} & z_{12} & z_{1y} \\ z_{12} & z_{11} & -z_{1y} \\ z_{1y} & -z_{1y} & z_{yy} \end{bmatrix} \begin{bmatrix} v_1 \\ v_2 \\ v_y \end{bmatrix} \tag{3}
$$

Die Elemente z_{11} bis z_{yy} sind etwas kompliziertere Reaktanzfunktionen als beim Violinsteg und sollen deshalb hier nicht näher beschrieben werden. Sie haben *alle* eine Polstelle (Antiresonanz) bei $\omega_l = 2\pi f_l$ und 2 weitere Nullstellen (Resonanzen). Als Beispiel ist in Bild 18 der gemessene und berechnete Verlauf von $|z_{yy}| = |F_y/v_y|$ am starr aufgesetzten Steg ($v_1 = v_2 = 0$) gezeigt. Zu den Resonanzfrequenz-

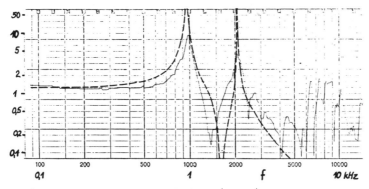

Bild 19: Verlauf des Kraftübersetzungsverhältnisses $|F_1/F_y|$
———gemessen – – – berechnet

311

en f_1 und f_2 gehören die Schwingungsformen von Bild 13 und 15 bzw. 16(1) und 16(3), zur Antiresonanz das Bild 14 bzw. 16(2). Entsprechend weist auch der Verlauf des Kraftübersetzungsverhältnisses $|F_1/F_y|$, vgl. Bild 19, diese Resonanzfrequenzen auf. Wie der Violinsteg verhält sich auch der Cellosteg bei tiefen Frequenzen wie ein Hebel mit konstanter Kraftübersetzung.

5. Der betriebsmäßig abgeschlossene Steg

Die für die Physik des Streichinstruments entscheidende Frage, wie der Steg im Betriebsfall zwischen Saite und Corpus als Übertragungsglied wirkt, ist am Beispiel der Geige untersucht worden. Dabei kann man die Belastung des Stegeingangs durch die Saiten vernachlässigen, während der Corpus den Stegfuß-Bewegungen einen großen Widerstand entgegensetzt. Diese durch die Schnellen v_1 und v_2 beschriebenen Fuß-Bewegungen sind nicht unabhängig voneinander. Das Corpus-Verhalten muß vielmehr durch eine 2-Tor-Impedanzmatrix W beschrieben werden:

$$
\begin{bmatrix} -F_1 \\ -F_2 \end{bmatrix} = \begin{bmatrix} w_{11} & w_{12} \\ w_{12} & w_{22} \end{bmatrix} \begin{bmatrix} v_1 \\ v_2 \end{bmatrix} \tag{4}
$$

Wird dieses 2-Tor an die Ausgänge des Steg-Viertores "angeschaltet", so kann man die Steg-Eingangs-Impedanz in y-Richtung, $z_y = F_y/v_y$, oder in z-Richtung, $z_z = F_z/v_z$, mit Hilfe eines Computers berechnen. Dazu braucht man nur die Matrix Z in dem homogenen Gleichungssystem Gl. (5), das aus der Addition der Gleichungen (1) und (4) hervorgegangen ist, gleich Null zu setzen:

$$
0 = \overbrace{\begin{bmatrix} z_{11} + w_{11} & z_{12} + w_{12} & z_{1y} & z_{1z} \\ z_{12} + w_{12} & z_{11} + w_{22} & -z_{1y} & z_{1z} \\ z_{1y} & -z_{1y} & z_{yy} - \dfrac{F_y}{v_y} & 0 \\ z_{1z} & z_{1z} & 0 & z_{zz} - \dfrac{F_z}{v_z} \end{bmatrix}}^{Z} \begin{bmatrix} v_1 \\ v_2 \\ v_y \\ v_z \end{bmatrix} \tag{5}
$$

Dies setzt die Messung der Impedanzelemente $w_{11} \ldots w_{22}$ in Gl. (4) voraus oder die

Messung der entsprechenden Admittanzelemente $y_{11} \ldots y_{22}$ in Gl. (6):

$$\begin{bmatrix} v_1 \\ v_2 \end{bmatrix} = \begin{bmatrix} y_{11} & y_{12} \\ y_{12} & y_{22} \end{bmatrix} \begin{bmatrix} -F_1 \\ -F_2 \end{bmatrix} \tag{6}$$

Aus diesen kann man die Impedanzmatrix Gl. (4) berechnen. Diese y-Werte sind ziemlich komlizierte komplexe Funktionen der Frequenz ω. F. Eggers /6/ hat am Cello Betrag und Phase von $w_1 = 1/y_{11}$ und $w_2 = 1/y_{22}$ gemessen. Hier sind an der Violine alle 4 Admittanzelemente gemessen und als digitale Wertepaare (Real- und Imaginärteil) registriert worden, vgl. /5/. Der Meßaufbau ist in Bild 22 skizziert. Die elektrischen Spannungen U_b und U_F aus dem Impedanzkopf 3 entsprechen der Beschleunigung b und der Kraft F an den Punkten des Corpus, auf die der Steg im Betriebsfall mit seinen Füßen aufgesetzt wird. U_b wird zu U_r differenziert und wie U_F in einem Mitlauffilterpaar 7 und 8 sehr schmalbandig gefiltert, ohne daß die Phasenbeziehung zwischen beiden verfälscht wird. Deshalb entspricht der Quotient aus beiden Spannungen, gebildet am Co/Quad-Analysator 1, der gesuchten komplexen Admittanz. Sie wurde am x-y-Schreiber 11 und am Digitaldrucker 12 in Abhängigkeit von der Frequenz registriert. Dabei hat sich gezeigt, daß die Kernadmittanz y_{12}, die ein Maß für die Verkopplung zwischen den beiden Corpus-Eingängen ist, kleiner als y_{11} und y_{22} ist. Außerdem gibt der Steg mit seinen beiden Ausgangsimpedanzen $\left| F_1/v_1 \right|$ und $\left| F_2/v_2 \right|$ nur eine verhältnismäßig geringe Belastung für den Corpus ab. Das heißt andererseits, daß sich der Steg auf dem Instrument dem starr aufgesetzten ähnlicher verhält als dem Steg, dessen Füße frei sind. Mit den gemessenen Werten für $w_1 = 1/y_{11}$ und $w_2 = 1/y_{22}$ wurde aus Gl. (5) ausgerechnet, mit welcher Ausgangsimpedanz F_1/v_1 ein Stegfuß den Corpus belastet und zwar für die Fälle, daß der andere Stegfuß mit w_1, mit w_2 oder mit $w = \infty$ abgeschlossen ist. Die Ergebnisse zeigen die Bilder 20, 21(a) und (b). Man erkennt, daß zwischen allen 3 Fällen kein großer Unterschied besteht. Dagegen ergab die

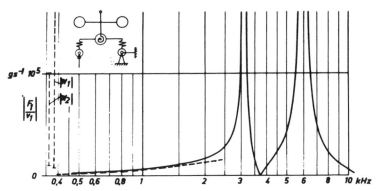

Bild 20: Ausgangsimpedanz des Stegmodells, ein Fuß festgebremst

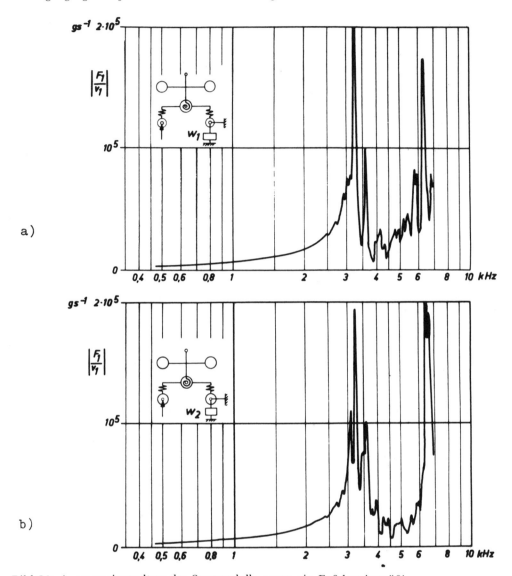

Bild 21: Ausgangsimpedanz des Stegmodells, wenn ein Fuß betriesmäßig
      (a) mit w_1
      (b) mit w_2
  abgeschlossen ist

Berechnung der Impedanz F_1/v_1 einen ganz anderen Verlauf, wenn der zweite Stegfuß frei blieb ($w = 0$).

Im Bild 20 ist links gestrichelt der Bereich angedeutet, in dem die Beträge von w_1 und w_2 schwanken. Mit Ausnahme der Resonanzgebiete um f_{yy} und f_{zz} ist die Stegausgangsimpedanz $|F_1/v_1|$ viel kleiner als die mittleren Werte von $|w_1|$ und

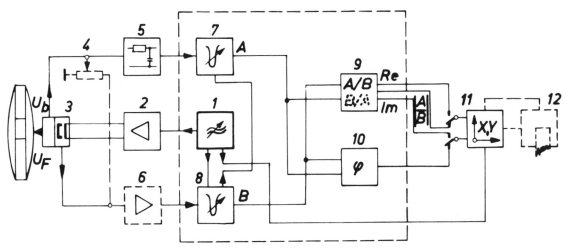

1 Kippschwingungsgenerator, Spectral Dynamics Corp., SD 104
2 Leistungsverstärker, Stange u. Wolfrum
3 Impedanzkopf Endevco 2222 und dynamisches Erregersystem oder Brüel u. Kjaer 8001 und 4810
4 Masse-Kompensationseinheit, Brüel u. Kjaer 5565
5 Integrierglied u. Vorverstärker, Brüel u. Kjaer 1606 und 2112
6 Vorverstärker, Stange u. Wolfrum
7 Dynamischer Analysator, Spectral Dynamics Corp., SD 101 B
8 Dynamischer Analysator, Spectral Dynamics Corp., SD 101 B
9 Co/Quad-Analysator, Spectral Dynamics Corp., SD 109 B
10 Phasenmesser, Spectral Dynamics Corp., SD 110
11 *XY*-Schreiber mit 2-Kanal-Plottereinschub, Hewlett u. Packard 1734A und 17176 A
12 Digitaldrucker, Kienzle D 11-E mit vorgeschaltetem Analog-Digital-Wandler

Bild 22: Vereinfachtes Blockschaltbild des Versuchsaufbaus zur Messung komplexer Corpus-Admittanzen

$|w_2|$. Das hat sich bei F. Eggers für das Cello bereits daran gezeigt, daß die Corpusimpedanzen an den Stegfußpunkten kaum davon abhingen, ob der Steg aufgesetzt oder entfernt war.

Für den Spezialfall, daß eine der mittleren Saiten mit dem Bogen in Richtung y angestrichen wird, findet die Saite am Ende die Impedanz $z_y = F_y/v_y$ vor. Da die Saite den Steg kaum belastet, ist dabei $F_z \approx 0$ zu setzen, wenn man Gl. (5) nach F_y/v_y auflöst. Das Ergebnis dieser Rechnung ist in Bild 23 als dick ausgezogene Kurve gezeigt. Dazu ist der gemessene Verlauf der Betriebs-Eingangsimpedanz als Funktion der Frequenz aufgetragen (punktweise). Diese Meßkurve spaltet sich oberhalb von 4, 7 KHz in zwei unterschiedliche Verläufe, je nachdem, ob der Impedanzkopf über einen kurzen Draht oder Stab an den Stegeingang gekoppelt wurde. In beiden Fällen verfälscht die Eigenimpedanz dieses Koppelgliedes das Meßergebnis oberhalb $f = 4, 7$ KHz. Schließlich ist gestrichelt die Impedanz $\omega\mu m_2$ der Stegoberteilmasse eingezeichnet. Dieser Kurve muß sich die Stegeingangsimpedanz nähern, wenn sich

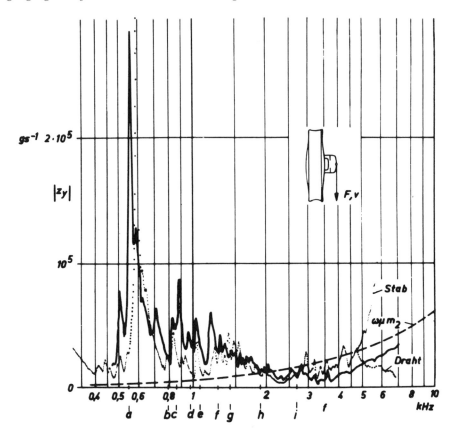

Bild 23: Betrag der Betriebs-Stegimpedanz
— — — —gemessen ————berechnet

die Oberteilmasse bei sehr höhen Frequenzen vom ganzen restlichen System "abkoppelt". Dies geschieht, wie nachgewiesen werden konnte, schon oberhalb mittlerer Frequenzen, wenn die Oberteilmasse durch ein Sordino vergrößert wird.

Werden dagegen die Stegschlitze durch kleine Holzteile verstopft, dann erhält man das Ergebnis nach Bild 24. Man erkennt, daß sich der Betrag von z_y im Mittel vergrößert hat, und zwar auch bei tieferen Frequenzen. Verhielte sich der Steg dabei wie ein *starrer* Hebel, dann hätten die Holzteile keinen Einfluß haben dürfen. Offenbar spielt die Nachgiebigkeit des Steges auch bei tiefen Frequenzen eine Rolle, so daß sich sein Verhalten als federnder Hebel (mit dem schon berechneten Kraftübersetzungsverhältnis, vgl. Bild 12), vor einer vergleichsweise großen (Corpus-) Impedanz beschreiben läßt. Der Verlauf von Stegeingangsimpedanzen ist an anderer Stelle vgl. /6/ und /7/, diskutiert worden, wir brauchen ihn hier nicht zu erörtern.

Abschließend ist die Abklingzeit einer a^1-Saite berechnet worden, die durch die rechnerische Stegimpedanz z_y abgeschlossen ist. Das Ergebnis ist die in Bild 3

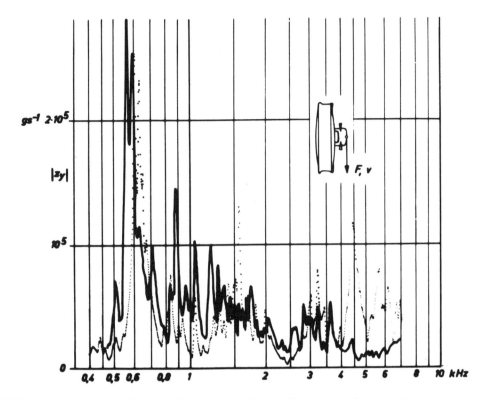

Bild 24: Betrag der Betriebs-Stegimpedanz bei vergrößerten Steifen s_1 und s_2
— — — —gemessen ————berechnet

eingezeichnete, wild schwankende Kurve. Sie weicht nur oberhalb von 4 KHz stark von den Meßwerten (Schraffur) ab. Da bei diesen höhen Frequenzen die Meßwerte für $\tau/2$ kaum davon abhängen, ob die Saite auf starre Stege (glatte Kurve) oder über die Violine (Schraffur) gespannt ist, darf man annehmen, daß dort nicht mehr die Saiten-Abschlußimpedanz z_y maßgebend für $\tau/2$ ist, sondern die inneren Verluste der Saite.

Literaturverzeichnis

1. Cremer, L. "Die Geige aus der Sicht des Physikers," *Nachr. Akad. Wiss. Göttingen: II. Math. Physik. Kl.* **12** (Jan. 1971).
2. Steinkopf, G. "Impedanzen von Stegen bei Streichinstrumenten," Unveröff. Diplomarbeit am Institut f. Techn. Akustik, Technischen Universität, Berlin, 1963.
3. Zimmerman, P. "Theoretische Untersuchungen zur Funktion des Steges bei Streichinstrumenten," *Acustica* **18**, 287– 299 (1967).
4. Stetson, K. A., and R. L. Powell. "Interferometric Vibration Analysis by Wavefront Reconstruction," *J. Opt. Soc. Amer.,* **55**, 1593– 1598 (1965).

5. Reinicke, W. "Die Übertragungseigenschaften des Streichinstrumentensteges," dissertation, Technischen Universität, Berlin, 1972.
6. Eggers, F. "Untersuchung von Corpus-Schwingungen am Violoncello," *Acustica 9*, 453 – 465, (1959).
7. Reinicke, W., and L. Cremer. "Application of Holographic Interferometry to Vibrations of the Bodies of String Instruments," *J. Acoust. Soc. Amer.*, **48**, 988 – 992 (1970).

17

Transfer Properties of String-Instrument Bridges

WALTER REINICKE
ABSTRACT by John C. Schelleng

This paper presents a thorough study of the vibrations of the bridge of a bowed string instrument, with special relation to that of the violin and cello, which differ in important respects. Holography is used to display the motion of a bridge blocked at the feet and energized at the top by a force in the plane of the bridge. The essential action of this important component of the instrument at the higher frequencies in terms of a breakdown of its distributed mass and stiffness into manageable lumped sections is represented in terms of mechanical circuitry and studied mathematically. This is followed by experimental confirmation. As frequency rises to the point where elastic compliance needs to be considered, the motion of the upper section in the violin is one primarily of rotation about a point in its narrow waist, whereas in the cello it is mainly a sidewise translation because of bending in its slender legs, with a less important contrary motion due to rotation in the top. The distortions change in form, however, as frequency rises still higher.

IV

The Soundpost

So critical to the action of instruments of the violin family that the French call it *"l'âme"* or "soul," the soundpost is a major factor in proper tone production. The selection of this cylindrical bit of spruce, its shaping, and its optimum adjustment in each instrument are a challenge to the skill of even the experienced violin maker.

The two papers included here were written 141 years apart. The first reports a now classical series of experiments by Félix Savart (1791–1841) in which he clearly demonstrated the important effects of the soundpost on the tone quality of the violin. The second, by John C. Schelleng, describes the physical mechanism by which the soundpost enhances the radiation of sound from the violin plates. Further studies via hologram interferometry, such as those of Jansson et al. (1), will undoubtedly provide more definitive evidence on this fascinating problem that has eluded researchers for years.

Reference

1. Jansson, E., N.-E. Molin, and H. Sundin. "Resonances of a Violing Body Studied by Hologram Interferometry and Acoustical Methods," *Phys. Scripta,* **2,** 243–256 (1970). (Included in *Musical Acoustics, Part II.)*

Editor's Comments on Papers 18 and 19

18   **Savart:** *The Violin*

19   **Schelleng:** *The Action of the Soundpost*

Félix Savart (1791–1841), a pioneer acoustician, is generally regarded as the grandfather of violin research. Although he did not pursue his early training in medicine, much of his work on hearing and voice production, as well as his experiments with sound waves traveling through air, water, solids, and various elastic media, show the effect of his medical interests. He was known for the ingenious design and expert execution of the mechanical aspects of experiments, attributes beautifully exemplified in the selection given here, which includes his study of the action of the soundpost.

To date the writings of Savart on the violin have been located in the journal *L'Institut*, founded by Napoleon in the early nineteenth century, and in Savart's *Mémoire sur la construction des instruments à cordes et à archet* (Deterville, Paris, 1819). Apparently his technical reports in other fields were all published in the *Annales de chimie et de physique*, a journal founded by Lavoisier and others in 1789. A biographical study of Savart and an extensive bibliography can be found in the monograph "Félix Savart, Physician–Physicist: Early Studies Pertinent to the Understanding of Murmurs," by Victor A. McKusick and Kenneth H. Wiskind, published in *J. Hist. Med. Allied Sci.*, **16**(4) (1959).

The companion volume to this book contains a further report of Savart's writings on the violin from both the *Mémoire* and *L'Institut*.

Information about the life and work of John C. Schelleng can be found in "Editor's Comments on Paper 5."

18

The Violin*

FÉLIX SAVART

This article was translated expressly for this
Benchmark volume by Donald A. Fletcher from
L'Institut, 8, 54–56 (1840)

The finest of all the instruments is the violin; it has been called the king of instruments. It is composed of two plates. The upper is always of pine (spruce) and reinforced by a longitudinal bar. The lower plate, which is always of a different nature than the upper plate, is called the back. They are joined around their contour by strips of wood called the ribs. They are given solidity by means of other little strips called lining strips. In addition, in the angles formed by the different parts of the ribs are little pieces of wood whose purpose is to increase the resistance of the instrument; they are called corner blocks and end blocks. To the box of the violin is fixed, as everyone knows, a neck at the end of which are placed the pegs, around which the strings are wound. The form of the neck and its length have great importance in the construction of violins. Finally, between the two plates and near the piece that supports the strings, which is called the bridge, there is placed a little cylinder of wood that puts the plates into communication, which is called the soundpost. This mobile piece plays a most noteworthy part in a violin, for without this piece its sound is very weak and of poor quality. It is to its influence on the intensity of the sound and on its nature that this part of the instrument owes its name, in French, "the soul." We cannot say that it serves alone to convey movement from one plate to the other, for they are in communication through the ribs and the blocks. The soundpost has different functions, which we shall examine with care.

Maupertuis has given a theory of the violin that is nothing but a grave error, and it must be rejected here. He maintained that all stringed instruments ought to be composed of fibers of different lengths, so that the number of the vibrations of the strings could be reproduced by a fiber which, vibrating in unison with the string, would reinforce the sound; and he then maintained, which still exists as an assumption, that to break a violin and repair it improves it, because that augments

---

*These notes on the violin are an extract from an as yet unpublished memoir of M. Savart—signed A. Masson.

the number of fibers of different lengths, thus obtaining many more wood-strings vibrating at different rates. But it is easy to see how far this explanation of the reinforcement of sound by the box of the violin is false. The two plates enter into vibration through their entire length, as can be confirmed by spreading sand on their surface, and a body that has been broken and then reglued behaves absolutely, before and after the breakage, in the same manner; it produces just the same figures with the sand as can be observed with disks of wood, rock crystal, or metal, a result, indeed, of the laws of the conservation of motion. On both faces a violin shows nodes of vibrations and *ventres* where the sand is seen to be strongly agitated. And so the plates behave like slabs, and not as if composed of fibers vibrating separately.

A violin is composed of a great number of elements all of which have individual and correlated functions that we shall enumerate successively.

Let us examine first the role played by the *soundpost.* If we remove this piece, the sound no longer has any vigor, no bite, it is poor; it regains its intensity and its richness as soon as this important element has been put back in place. It is not to be supposed that the soundpost acts as a conductor of sound, serving only to spread the motion, for this piece can be put not in the violin but on the violin, and its action remains the same; its influence is not changed. If we put on the violin a kind of arch of wood, fastened and glued to the corner blocks, and if we put the soundpost between this arch and the upper table, the effect is the same as in the ordinary circumstance when it is within the box. The arch is formed of two uprights glued to the blocks and supporting a bar that is perpendicular to them. Instead of the soundpost, we can fit to the arch a screw whose head can put greater or less pressure on the top plate. The soundpost effect is produced strongly when the screw is tightened. If we put a similar arch under the violin and make a hole in the back so that a screw, without touching the back, can rest against the top plate, as soon as we tighten this screw against the top plate we get the same results as with a soundpost.

Furthermore, if we simply put a heavy weight on the top plate of a violin without a soundpost, the violin resounds again as though it had the post, provided the weight used is above a certain limit. This suitable point is found by pouring greater or less quantities of mercury into a little *eprouvette.* This soundpost effect relates especially to the top plate, for with a violin without a back plate, but provided only with a transverse bar, the soundpost effect can be produced by means of a screw pressing hard against the single plate of the instrument. The soundpost does not play, as one might suppose, a role analogous to the *chevalet* of the marine trompette, for, in gluing it, the effects do not suffer any alteration. We can get the same results again by squeezing the soundpost against the plates of the instrument by means of the screws and arches mentioned above, and even by pressing it in a clamp, which is accomplished by inserting iron strips into the jaws of the clamp.

The function of the soundpost is to render perpendicular the vibrations of the plates. To prove this, first take an apparatus formed of a horizontal disk on which rests a bridge used to sustain a string. When the string is made to vibrate parallel to the disk, the sound is very weak; on the contrary, it gains great intensity when the vibrations of the string are perpendicular to the disk.

Take a violin whose plates are pierced so that a bow may pass through, and remove the soundpost. If the strings are made to vibrate parallel to the plates the sound is very weak; but if, passing the bow through the violin, the strings are made to vibrate in a plane perpendicular to these plates the sound is considerably reinforced and just as good as with a soundpost. Replace the soundpost and see the effect it will produce. If the vibrations are perpendicular, we get a quality of sound that is the same as for any other direction; nothing is changed in the intensity of the sound in whatever direction the strings are vibrated. Take a violin shaped like a trapezoid; instead of being on the upper base the strings are on the side. This makes a very flat violin having very nearly the same mass of air as in ordinary violins. When the strings are set in motion, they oscillate perpendicularly to the plate; if we insert a soundpost, there is no change in the intensity of the sound and its nature.

The role of the soundpost evidently seems to be to render the vibrations always perpendicular to the plate, and it does not produce an effect either of pulsation or of communication.

We can further prove by a decisive experiment that the role of the soundpost is to render perpendicular the vibrations of the plates. A cylindrical violin was made having an air mass of nearly the same volume as in the ordinary violin. But we know that a cylindrical vase always divides into several vibrating parts, all of which vibrate perpendicularly, and here we have the same conditions; the soundpost is not needed to render the vibrations perpendicular. If a soundpost is put into such an instrument, the sound is altered and the motion is less intense; the post plays the role of a muffler. But how can the soundpost render perpendicular a movement that in appearance ought to be tangential? To explain it let us go back to an experiment already cited. If, taking a rod vibrating longitudinally, we add a new strip at a right angle with it, the latter strip may, under certain circumstances, vibrate tangentially instead of producing perpendicular vibrations, as should be to conform with the laws we have stated relative to the communication of vibratory movement. We have in fact demonstrated that, in a rod that is the seat of longitudinal vibrations, there were contractions, sinuosities, which gave birth to transverse half-oscillations having the same duration as the longitudinal oscillations; consequently, if with another rod we touch the body of the strip that vibrates tangentially, the inflexions ought to communicate themselves to the perpendicular rod. There will be shocks, which at each bending will produce longitudinal vibrations, so that, according to the position of the perpendicular rod, the rod may become the seat of a perpendicular or tangential movement. The same phenomenon is produced in the violin. The transverse oscillations of the plates produce in the soundpost a longitudinal movement, which, reacting on that of the plates, determines there a perpendicular movement. It is an apparent exception, as we have said, to the laws of the communication of movement; the soundpost becomes a sort of bow with respect to the plates. If we were considering only one blow, as in the guitar, the vibration, being instantaneous, would be stifled by the soundpost; but in the violin, the cause engendering the vibrations is unceasing. It is the sum of very small but continuous actions that ends in producing a very intense and pronounced effect; it works like the bow and produces a shock corresponding to each collision produced by this latter. It is to be noted that the constitution of pine makes the transverse bendings easier. Indeed, if we take care, as is indispensable, to put all the fibers of the wood parallel to the greatest length of the instrument, the vibrations that will be perpendicular to them will determine the movements of bending, which will take place more easily than in any other direction, because it is in following this direction that the resistance to bending will be least. Thus it is necessary to put the fibers of the pine parallel to the greatest length of the violin. The soundpost has not only the function we have just attributed to it; but we do not put it inside a violin to support the plates, since it can be put outside, and since moreover it can be reduced to a very weak stick of dry resistant wood. The soundpost plays a very important role, which proves the necessity of giving to the bridge a certain form. It is always put behind the right foot of the bridge; it has in effect to maintain this foot in an almost complete state of immobility, so that the left foot may, as in the *trompette marine*, communicate its movements to the bar of the instrument. In all violins there is, under the plate, a bar destined to give resistance; at the same time it determines throughout the length of the instrument the movement that has been communicated to it. This bar is at the left of the instrument and receives the blows produced by the left foot of the bridge. We shall cite experiments in support of what we are saying.

We take a violin and make a hole in the plate where it should receive the end of the soundpost, so that the right foot of the bridge rests directly on the soundpost, which does not in any way touch the instrument. The sound is then a bit dull, but the effect of the soundpost is produced. In another violin we isolated the left foot, the bar being in the middle; the effect of the soundpost still showed clearly. To do this, a piece of wood was cut out of the upper plate, and without letting it touch this plate, we held it in place by a special device; it supported the left foot of the bridge, which did not in any way touch the plate. The effect of the bar is to produce, throughout the length of the violin, the disturbances communicated to it by the bridge; the bar vibrates in totality. There is no division in its length, nor in that of the plate.

And so, in review, we see that the soundpost has three functions: (1) it communicates the movement from one plate to the other; (2) it renders perpendicular the vibrations of the plates; (3) it renders motionless the right foot of the bridge. The soundpost agitates the plate in its entire extent, and whatever the direction of the agitation, the soundpost renders it always perpendicular in the plates.

Let us now examine the box of the instrument, which is composed of the plate, the back, the ribs, and the blocks. The back is always of beech or maple, and the ribs as well; the belly of pine. Maple is preferable to beech for making the back of a violin.

First we are going to consider a box in its greatest simplicity; we shall make it rectangular and shall examine, in making it, the role of each piece. If we take first a single small strip of wood, it will give a certain sound, F5, for example; at the two ends we place at right angles two little blocks or little shorter strips. If, while making the longest strip vibrate still perpendicularly, we cause it to produce the same manner of division as before, we find a lower note, B4; then adding on the two blocks and, parallel to the first strip, an identical strip, we find, on causing the system to vibrate, G4; the manner of division of the two strips is the same as in the first strip vibrating alone. We have chosen the longest strips in such a way that they are at a perfect unison, that they produce exactly the same manner of division, and that, being joined together, whichever one is agitated, each produces the same system of nodal lines. If now we reduce the thickness of one of the strips, the nodal lines are displaced; they no longer correspond on the two strips. The nodes are nearer together for the thinner strip; but nevertheless the sound is still the same whichever strip is agitated. We can easily conceive that in the thinner strip, since the number of vibrations is to remain the same as in the other, the nodal lines should get closer, since the vibrating parts would be smaller.

We can conceive that the same phenomena could occur in the box of a violin. Isolated, the plates will not give the same number of vibrations; united, they will give the same sound. To prove it, let us work with closed rectangular boxes, pierced at the center of the large surfaces by a circular hole used to receive a wick of horsehair. Furthermore, these boxes have on their lateral parts rectangular openings to give freedom to the vibrations of the contained air. If we set the box in vibration, a nodal line is produced around its perimeter, and the sound is the same whichever face is agitated; if the strips have the same thickness, are identical, the nodal lines are the same, and this is a means of verifying the equality of constitution. If the plates are not of equal thickness, the sound remains the same whichever one is agitated. But the manner of division is different: for the thicker, the nodal line is outside in one experiment, and inside for the thinner.

It is evident, from these experiments that in a violin the plates always vibrate in unison.

We shall examine in a future article the role played by the air enclosed in the box of this instrument.

Reprinted from *Catgut Acoust. Soc. Newsletter,* No. 16, 11 – 15 (1971)

The Action of the Soundpost

JOHN C. SCHELLENG

While no one will deny the importance of the soundpost in the violin family, there seems to be much misunderstanding concerning the reason for its use. In considering this insignificant-looking component, which the French call the soul of the violin, a convenient starting condition is to assume the instrument complete in every respect except that the bassbar as well as the soundpost have as yet not been put in place, an impracticable procedure in assembly that is assumed merely for the sake of discussion. We consider the bridge to be part of the top plate and the effect of strings on body resonance small.

The function of the bridge that is of interest here is that, particularly when playing on the middle strings, it converts forces exerted parallel into forces perpendicular to the top plate. Without bassbar and soundpost the instrument is perfectly symmetrical bilaterally. When the left foot of the bridge pushes down, the right in effect pulls up, and insofar as the outside air is agitated the two motions annul each other except for second-order effects. It is essential to a healthy tone that the instrument behave, in the language of acoustics, like a "simple source"; this is particularly true in the lower octaves where the wavelength in air is large compared with the dimensions of the instrument. The violin must actually undergo a cyclical change in the total volume of air that it displaces. As it vibrates it must swell more in one area than it shrinks in another. This change in volume is of course small at midrange and forte, being much less than a cubic centimeter for a violin. *But it is crucially important.* Obviously, the instrument must not be bilaterally symmetrical.

If we now install the bassbar, we have introduced dissymmetry and tone will perhaps be somewhat increased. However, this is not the reason for the bassbar, its chief acoustical purpose being to keep the upper and lower bouts of the top plate in step with the left foot of the bridge.

In regard to the usually stated function of the soundpost, that it is to transmit motion to the back, it would be more to the point to say that it is to *reduce* locally the canceling motion of the top plate. Properly interpreted, the ingenious experiments of Savart (1) prove that its first acoustical purpose is to introduce assymmetry. It serves in other ways, some difficult to isolate. Finally of course it provides a means of adjustment in the finished instrument.

Everyone agrees that the top plate predominates over the back in producing radiation of sound. It has even been asserted, though few will agree, that the effect of the back is negligible. However, since every decibel that a fiddle loses is costly, it is nearer the truth to say that we could ill afford to dispense with the radiation from the back, that discarding its contribution might lower the market value by hundreds of dollars. Nevertheless, since the instrument that we are now to construct in our minds will never appear on the market and since it does illustrate the essential action of the soundpost, we assume first that the back is rigid and heavy. Later we shall try to be more realistic. The following discussion repeats ideas in my memorandum circulated in 1963 within the then small membership of the CAS and entitled "Vibrations in the Violin Body."

We simplify even further by considering the body to be in the form of a closed cigar box, the first broad face of which is heavy and stiff while the four edges of the second are supported immovably as to position, though the surface within is free to move as to angle: that is, the edges of the top plate are "hinged." With geometry thus simplified the mode shapes become predictable. The broken lines in Fig. 1a represent the locus of points of zero motion; that is, they are nodal lines. The rectangle at the left is for the mode of lowest frequency; the hinged edges are the only nodes. If we were to study the vibrational displacement along horizontal line AA, we would find that it has a sinusoidal shape such as that indicated in Fig. 1b by the parenthesis (1, 1), in which the first 1 means that for this mode there is one sinusoidal lobe in a vertical section, and the second 1 means one lobe in a horizontal section. In the rectangle at the right are indicated the nodal lines for a higher frequency, a vertical nodal line having been added so that along AA there are two sinusoidal lobes, as shown in Fig. 1b. We also assume that dimensions and materials are such that there is no other resonance with frequencies lying between those of these two modes.

If we now insert an immovable soundpost at point B, neither mode can vibrate *alone* because B is not a node for either. However, the soundpost, by nullifying total motion at B, still permits *joint* activity in the two modes provided amplitudes and phases are such that the *sum* of their motions is zero. Whereas (1,1) and (1,2) previously vibrated independently at their own frequencies, they are now constrained so that they can vibrate only as a pair at a new frequency. The important point is this: mode (1,1) is the mode *par excellence* that in vibration changes the displaced volume of air since all over the surface at any moment its motion is in the same direction. It is therefore an effective radiator of sound waves. But alone it cannot be excited by the rocking of the bridge. On the other hand, mode (1,2), though an inefficient radiator, is ideally adapted to be set in motion by the bridge. Working as a team, they satisfy both conditions.

What is the frequency of resonance of the newly formed mode? When we tap the box in the new condition, there is a new resonance frequency, there is no motion at the soundpost, but there is a considerable force at that point. Ignoring the effect

329

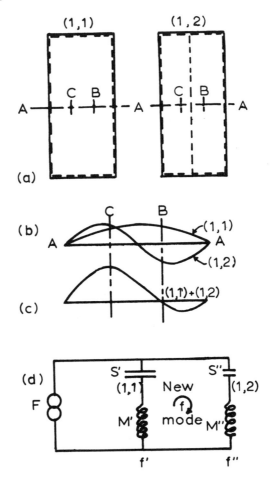

Figure 1. Effect of soundpost in enhancing radiation of sound from a rectangular plate; without it, the best-radiating mode (1,1) is not excited.

of internal friction in the plate, the admittance at B at the new frequency therefore is zero, that is, zero motion divided by finite force. This can happen only if the admittance of mode (1,1) is exactly equal and opposite in sign to the admittance for (1,2), and this can be satisfied only when the new frequency is higher than that of (1,1) and lower than that of (1,2)—in other words when the new frequency lies between the old ones. Circuitwise this is indicated in Fig. 1d; the new frequency is one of *antiresonance* and lies between the old frequencies of *resonance*. When originally there are two modes (as assumed above), it works out that the square of the new frequency is as follows:

$$f^2 = f'f'' \; \frac{f'/f'' + \alpha'/\alpha''}{1 + \alpha'/\alpha'' \cdot f'/f''} \tag{1}$$

330

where the prime means (1,1) and the double prime (1,2), f = frequency, and α' means $(SM)^2$. S is equivalent stiffness of one of the series tuned original modes referred to point B and M the equivalent mass. While this equation has been stated in terms of the cigar box, it applies whenever the soundpost is immovable.

Thus if $\alpha' = \alpha''$, which seems reasonable as to order of magnitude, the new frequency is the geometric mean of the two originals, in musical measure halfway between them. It is also apparent that as α'' decreases with respect to α' the resultant frequency moves toward f'', as one would expect when B moves toward a node of f''.

These considerations apply throughout the frequency range; between consecutive resonance frequencies of the original series there will be points of antiresonance whose positions will depend on the placing of the soundpost. These *anti*resonances will respond as new *resonances* when force is applied at other points, such as the left foot of the bridge. Frequencies of the original series that have a node at the soundpost but not at the left foot will still appear as active resonances after the post is inserted.

Figures 1b and 1c indicate the shape of the new mode along section AA. Though motion at B, the position of the soundpost, is zero, that at C, which is the position of the left foot of the bridge, is at a maximum of motion for the new mode as shown in Fig. 1c. The new condition therefore brings with it an entirely new shape of vibrational pattern depending mostly on modes of adjacent frequencies, but in part on all possible modes except those having a null at the soundpost. Thus one-to-one identification of new modes with the old is not possible. Nevertheless the new shapes can be explained in terms of the old ones provided that we know the "recipe" (to use A. H. Benade's apt expression) for the amplitudes of the new mode in terms of the old.

For those of our readers who have access to a copy of the excellent paper by Jansson, Molin, and Sundin (2) it will be of interest to observe how the shape in Fig. 10a for the principal resonance of the supported top plate for a violin with an immovable soundpost depends on the shapes of Figs. 9a and 9b taken without the soundpost. The displacements in Fig. 9a are evidently all in phase; arbitrarily we take them as positive. The displacements in Fig. 9b will then be regarded as positive on the left side and negative on the right so as to give cancellation at the soundpost. By addition of displacements one arrives at shapes of contour very much like those of Fig. 10a (closer match would be obtained if a vibration of greater amplitude than that in Fig. 9b were assumed). Interestingly enough, the geometric mean between the resonances in Fig. 9 (465 and 600 Hz) is 530 Hz, while the frequency in Fig. 10a is 540 Hz.

We have next to consider for the actual violin how the new frequencies are determined in the light of the fact that the back is actually movable rather than fixed. The condition is easily derived. We can no longer say that the soundpost is immovable as we did in the previous discussion. Though nearly stationary in the lower octaves, it moves significantly in some of the higher resonances. Since the post is fitted reasonably snugly, at least after the strings are tightened, the ends travel with their respective plates. The mass of the post is about 0.7 g and its stiffness, end to end, about 70×10^8 cgs units. This mass is somewhere near that of the bridge and to a first approximation may be considered negligible. Longitudinal stiffness is in the order of 50 times that of the top plate, for practical purposes infinite. The velocities of the two ends up and down are exactly the same in view of the great stiffness, and the forces exerted against the two plates are sensibly equal and opposite because of the lowness of mass. At least for present purposes we ignore the possibility of bending.

The situation when the body is tapped into vibration is therefore as shown in Figure 2. Let F be force and V velocity. A will represent admittance (velocity per unit of force) and subscripts will represent the location under consideration. Thus, F_1 will be the vibratory force applied at the top of the soundpost and V_1 will be the velocity at the same point, both F and V being considered positive if upwardly directed. A_{11} will be velocity at point 1 produced by unit force at point 1, while A_{12} will be velocity at point 1 produced by unit force at point 2. A_{12} and A_{21} therefore represent the action of the ribs in transmitting motion between the plates. Again ignoring losses in the tap tone, the velocity at point 1 will be $F_1 A_{11} + F_2 A_{12}$, and it will equal velocity at point 2, which is given by $F_1 A_{21} + F_2 A_{22}$. Since $F_1 = - F_2$ and since by reciprocity $A_{12} = A_{21}$, it follows that

$$A_{11} + A_{22} - 2A_{12} = 0 \qquad (2)$$

It is interesting that the effect of transfer admittance via the ribs, A_{12}, is doubled. There are reasons for believing that A_{22} is usually the smallest of the three admittances. These admittances, which contain the effect of all modes lumped together, can be measured in the primary condition without the soundpost being in place, and any frequency that satisfies Eq. 2 will be a resonance frequency of a new mode (secondary mode) that will appear when the post is inserted.

Notice that if the back is immovable the equation becomes $A_{11} = 0$, which is the condition of antiresonance noticed in the cigarbox fiddle.

If A_{12} is zero we are dealing with two plates isolated from each other. To accomplish this both plates would be mounted on a heavy jig and the air between pumped out, since in the actual violin the plates are coupled by movable ribs and air. The extent of this coupling is not known quantitatively, but the assumption that all motion in the back is to be ascribed to the soundpost is entirely gratuitous.

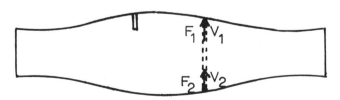

Figure 2. Section of violin at soundpost.

If we would understand what the skilled luthier is doing when he adjusts a soundpost, we must begin with a knowledge of the vibrational pattern displayed by the instrument before its insertion, particularly within a reasonable distance from its normal position. If within this range the amplitude changes but little with position, it seems unlikely that dramatic effects will follow a change in position of the soundpost. Without experimental mapping, however, we are in no position on theoretical grounds to make a prediction of the effect of change, nor to say that nothing is to be gained by such a study. Since the bridge is located according to the f-hole notches, moving the soundpost does two things: it alters its place in the vibrational patterns of the plates existing without the bridge, and it changes the distance from bridge to soundpost. In the present state of knowledge it would be hazardous to guess the relative importance of these two factors. About all that can be said is that such a study should illuminate any critical effect found in adjusting the soundpost and that the matter is worth studying experimentally.

Notes and References

1. Heron-Allen describes these experiments as follows: Savart "removed the soundpost from a violin, and applied it *outside*, and on the top of the belly by means of two uprights on the corner blocks, and a crossbar, between which and the belly the soundpost was set up. Again, fixing this arch to the back of the instrument; by cutting a hole in the back, he set the post up against the belly, *without touching* the back at all. Again, removing the soundpost altogether, he applied a weight to the belly, and in all three experiments the same results were produced, as if the soundpost were there in the normal position." Since measuring means were probably not available, I understand the intent of these statements to have been qualitative rather than quantitative. (Heron-Allen, ed., *Violin Making as It Was and Is*, Ward Lock Co., London, 1885, p. 150.*

2. "Resonances of a Violin Body Studied by Hologram Interferometry and Acoustical Methods," *Phys. Scripta*, 2, 243–256 (1970). Included in *Musical Acoustics, Part II.)*

Editor's Note: See Paper 18 in this volume for a complete discussion of the experiments of Savart.

V

Wood for the Violin Family

It is remarkable that modern technology so far offers the luthier no materials superior, or even equal, to those that became standard for violins in generations long past. Buried in centuries of the empirical development of stringed instruments in Western culture is the craftsman–musician's selection of spruce (*Picea*) for the top plate and flamed, or curly, maple (*Acer*) for the back and sides of instruments of both the violin and viol families (Paper 1). Because of its special characteristics, spruce has also been chosen for the soundboards of guitars, harpsichords, and pianos.

The superiority of spruce when properly selected, cut, and seasoned stems from a high elastic modulus per unit of density and from its low density. These factors are no doubt partly due to the fact that the spruces have one less ring of cellulose fibers in their cell microstructure than other conifers such as pine (*Pinus*) and fir (*Abies*). Two other factors in favor of spruce are its low internal friction (low logarithmic decrement) as compared to other conifers—recognized by its clear ringing sound when tapped—and its relatively high flexibility across the grain. The flexibility across the grain in spruce is nearly ten times that along the grain.

Figure 1 shows a section of a spruce log indicating how the two halves of a violin top would be cut (preferably split out) so that the outside of the tree where the wood grows more slowly comes down the center join of the top. It is important for the two halves to be made from adjacent pieces to assure bilateral symmetry of the grain. Violin makers ideally look for straight-grain spruce where the grain widths progress evenly from narrow in the center of the plate to wider at the edges, preferably with the grain vertical to final plate arching. The strips in the drawing show how variously oriented test strips are cut. An effective and relatively easy method of

335

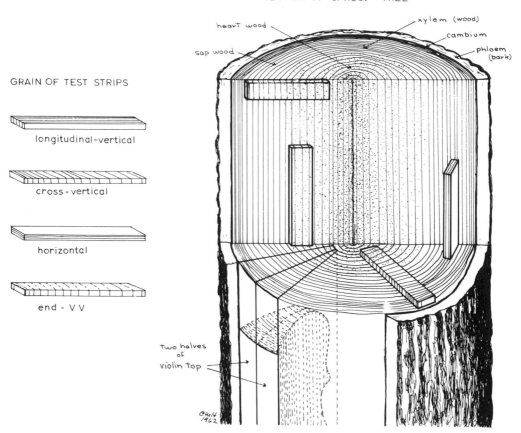

Part V

SECTION OF SPRUCE TREE

heart wood

sap wood

xylem (wood)

cambium

phloem (bark)

GRAIN OF TEST STRIPS

longitudinal-vertical

cross-vertical

horizontal

end - V V

two halves
of
violin top

Figure 1

comparing the acoustical properties of wood in a general way is to make up strips of similar dimensions and grain orientation, being sure that no dimension is a simple multiple of another. The strips are mounted like xylophone bars so that two thin supports each come about one fifth the length of the bar from the ends—at the nodes for the first bending mode. By listening to the pitch and the duration of the "ring" or sound when the bars are struck with a finger or soft hammer, it is possible to compare roughly the frequency and damping characteristics of different woods.

Wood for the back and sides of violins is usually maple (*Acer*) or other woods of nearly similar characteristics, such as sycamore (*Platanus*), pear (*Pyrus*), apple (*Malus*), beech (*Fagus*), and poplar (*Populus*). Important in its selection is strength to resist bending across the grain, since the wood must support considerable downward thrust from the soundpost. Also important are damping characteristics. Our own recent studies (still in progress) indicate that it is desirable to have similar damping in the resonances of both the top and back plates when finished, in spite of the very different physical structure of the two plates. The effect of the "curl," which is an actual bending or curling of the wood fibers usually at right angles to the axis of the grain produced by the annual growth rings, is to provide a more nearly isotropic material in which the stiffness ratios across the grain and along it are about 3 to 5. The phenomenon of "curl" occurs in many species of wood, although curly maple seems to be more plentiful than other species. (This may be a so-called genetic defect, since trees of a given species with curly grain seem to grow in certain localized areas.) Backs are traditionally of one or two pieces cut either on the "quarter" or on the "slab," as the wood warrants. Certainly the beauty of curly maple as well as its desirable physical and acoustical characteristics has encouraged its worldwide use in instruments of both the violin and viol families. Details of other woods used in small amounts in violin construction are described on page 15 in Paper 1.

The eight papers selected for inclusion here represent the best-known works on the acoustical characteristics of wood for violins. From the many woods tested by Barducci and Pasqualini (Paper 22), Schelleng has selected a representative group and plotted them on a graph (Paper 5), indicating the relationship between velocity, density, and damping. This shows spruce (*Picea*) isolated from other species, while there are a number of woods with characteristics nearly like those of maple (*Acer*).

Most of the research to date on the acoustical characteristics of woods has been done in Europe and Japan, where there is industrial production of violin-family instruments. The United States Department of Agriculture, Forest Products Laboratory, Madison, Wisconsin 53705, has some information but indicates that no direct research has been done on wood for musical instruments. Traditionally, European spruce and maple are considered the best woods for violins. Our own work, however, is showing that, when properly handled, certain species of American spruce and maple are equal or even superior to the European ones in the construction of fine-sounding instruments.

Additional Writings on Wood

Hearmon, R. F. S. "The Influence of Shear and Rotary Inertia on the Free Flexural Vibration of Wooden Beams," *Brit. J. Appl. Phys.*, **9**, 381–388 (1958).

Jayne, B. A. "Indices of Quality—Vibrational Properties of Wood," *Forest Prod.*, **9**(11), 413–416 (1959).

Kollman, F., and H. Krech. "Dynamische Messung der elastischen Holzeigenschaften und der Dämpfung" (Dynamic Measurement of Damping Capacity and Elastic Properties of Wood), *Holz Roh- Werkstoff*, **18**(2), 41–54 (1960).

James, W. L. "Vibration, Static Strength, and Elastic Properties of Clear Douglas-Fir at Various Levels of Moisture Content," *Forest Prod.*, **14**(9), 409–413 (1964).

Pearson, F. G. O., and Constance Webster. *Timbers Used in the Musical Instrument Industry*, British Forest Prod. Res. Lab., Princes Risborough, Aylesbury, Bucks, England, n.d. (Describes species of wood for musical instruments and where to obtain them.)

Editor's Comments on Papers 20 and 21

20 **Krüger and Rohloff:** *Über die innere Reibung von Holz*
English translation: *Internal Damping of Wood* (abstract only)

21 **Rohloff:** *Über die innere Reibung und die Strahlungsdämpfung von Geigen*
English translation: *Internal Friction and Damping of Violin Vibrations* (abstract only)

Dr. habil Ernst Rohloff (1911–) studied mathematics and physics at the universities of Berlin and Greifswald. In addition to his technical interests he is active as a violin soloist and art painter. As assistant at the Physikalisch Institut of the University of Greifswald, he investigated violins and *Hochstrombogen,* and continued this work after 1945 in Lübek as an *Oberstudienrat* at the gymnasium. For further research on the violin he has had the support of the Deutschen Forschungsgemeinschaft.

As far as we know, Rohloff is the first to study the damping characteristics of various kinds of wood, including wood for violins—a subject of particular interest to those who make or study musical instruments. Two of his papers on this subject follow. A third one, "Ansprache der Geigenklänge" (The Speaking of Violin Sounds), is included in the section "Psychoacoustics" in *Musical Acoustics, Part II.*

Rohloff's work on the damping characteristics of wood as related to violins has stimulated other researchers to pursue the subject further; see the articles in this volume by Meinel, Barducci and Pasqualini, Itokawa and Kumagai, Fukada, Beldie, and Schelleng. In Rohloff's study the question does not arise as to whether or not the logarithmic decrement contains additions due to shear. He gives the total log decrement from whatever cause. Both Meinel and Schelleng report an increase in the log decrement with rising frequency. This rather fundamental difference may be due to a difference in method. Rohloff started with a long strip of wood and shortened it to get higher frequencies—always measuring the lowest mode of the strip. Others have vibrated a given-length strip at its higher modes. Since there is something to be said for maintaining the length of the strip as more representative of what happens in the violin, this poses a question that needs answering.

Professor Freidrich Krüger (1877–1940) began his teaching career as Professor in the Technical Highschool in Danzig. After 1921 he was Professor of Physics and Director of the Physical, Astronomical, and Mathematical Institutes of the University of Greifswald. His publications number nearly one hundred in different fields of physics.

Selected Publications by Dr. habil Ernst Rohloff

*"Über die innere Reibung von Holz," with F. Krüger, *Z. Physik*, **110**, 58–68 (1938).
*"Über die innere Reibung und die Strahlungsdämpfung von Geigen," *Ann. Physik*, **38**(5), 177–198 (1940).

---

*Included in this volume.

"Über die innere Reibung von Geigenholz," *Z. Physik,* **117,** 64–66 (1940).

"Messung des durch die innere Reibung bedingten In-Dämpfungsdekrements verscheidener Hölzer," with W. Lawrynowicz, *Z. Tech. Physik,* **22,** 110–111 (1941).

"Der Klangcharakter altitalienischer Meistergeigen," *Z. Naturforsch.,* **3a,** 184 (1948).

"Der Klangcharakter altitalienischer Meistergeigen," *Z. Angew. Physik,* **2,** 145–150 (1950).

"Herstellung leicht ansprechender Streichinstrumente," *Die Musikforsch.,* **12,** 86–88 (1959).

†"Ansprache der Geigenklänge," *Z. Angew. Physik,* **17,** 62–63 (1964).

"Konsonantische Anteile der Geigenklänge," *Z. Angew. Physik,* **22,** 174–175 (1967).

"Verbesserung der Ansprache der Geigenklänge," *Z. Angew. Physik,* **18,** 105–107 (1964).

"Geigenteile und Geigendämpfer," *Z. Angew. Physik,* **25,** 349–350 (1968).

"Geigenstege," *Z. Angew. Physik,* **31,** 84–86 (1971).

---

†Included in *Musical Acoustics, Part II.*

20

Reprinted from Z. *Physik*, **110**(1 – 2), 58 – 68 (1938)

Über die innere Reibung von Holz.[*]

Von **F. Krüger** und **E. Rohloff**.

Mit 5 Abbildungen. (Eingegangen am 20. Mai 1938.)

Es wurde das logarithmische Dekrement der Schwingungsdämpfung für verschiedene Holzarten (Fichte, Ahorn, Kiefer, Eiche) einerseits bei Transversalschwingungen und Frequenzen zwischen 10 bis 700 Hz, andererseits bei Longitudinalschwingungen und Frequenzen zwischen 2000 bis 10000 Hz bestimmt. Das auf die Amplitude Null extrapolierte logarithmische Dekrement erwies sich für beide Arten der Schwingungen in den angewandten Frequenzintervallen als frequenzunabhängig. Das steht in Übereinstimmung mit der Theorie für die innere Reibung fester Körper von Boltzmann. Das logarithmische Dekrement ergab sich als am kleinsten für Fichtenholz, in zunehmender Reihe größer für Ahorn-, Kiefern- und Eichenholz. Für Stäbe, die senkrecht zur Faserrichtung geschnitten waren, war das Dekrement 3,5mal größer als für solche parallel zur Faserrichtung geschnittenen. Überziehen der Stäbe mit Ahornlack ergab eine Erhöhung des Dekrements um etwa 40%. Ein mit Wasser während 12 Stunden getränkter Stab ergab ein um 75% erhöhtes Dekrement. In Übereinstimmung mit der für das logarithmische Dekrement hier gefundenen Reihe der Hölzer wird für Resonanzböden in erster Linie Fichten- und Ahornholz als am geeignetsten für Resonanzböden von Musikinstrumenten benutzt, da bei ihnen die Energieverluste durch innere Reibung am kleinsten sind.

Die innere Reibung von Holz ist bisher noch fast gar nicht untersucht worden, jedenfalls nicht in Abhängigkeit von der Art des Holzes. Alle bisherigen Untersuchungen über die innere Reibung fester Körper erstrecken sich fast ausschließlich auf Metalle. Dabei soll hier die innere Reibung gemeint sein, die einer reversiblen Verschiebung der Teile im Innern der Metalle gegeneinander entspricht, nicht die Reibung, wie sie beim Fließen stark gedrückter Metalle auftritt. Die Messungen ergeben dabei zunächst den Wert des logarithmischen Dekrementes infolge der Dämpfung. Die weitere Berechnung des Koeffizienten der inneren Reibung ist in den bisher vorliegenden Theorien noch keineswegs eindeutig geklärt. Während in der Theorie von Voigt[1]) und der daran anschließenden von Honda und Konno[2]) ein Ansatz gemacht wird, der in Analogie zu dem bei der inneren Reibung der Flüssigkeiten steht, geht Boltzmann[3]) bei seiner Theorie von der Voraussetzung der elastischen Nachwirkung aus. Nach der ersten Theorie ergibt sich der Koeffizient der inneren Reibung proportional der Schwingungsdauer der für die Untersuchung angewandten Schwingung, nach der

---

[1]) W. Voigt, Ann., ganze Folge **283**, 671, 1892. — [2]) K. Honda u. S. Konno, Phil. Mag. (6) **42**, 115, 1921. — [3]) L. Boltzmann, Pogg. Ann., Ergänzungsband **7**, 624, 1876.

---

[*]*Editor's Note:* An abstract in English follows this article.

Boltzmannschen Theorie dagegen ist der Koeffizient der inneren Reibung unabhängig von der Frequenz. Wie schon Voigt gezeigt hat, erweist sich bei einigen Materialien seine Theorie, bei anderen dagegen die Theorie von Boltzmann als erfüllt, während bei noch anderen eine Frequenzabhängigkeit gefunden wird, die einem Mittelwert aus den beiden Theorien entspricht.

Unter diesen Umständen wird es geraten sein, sich vorläufig in den Resultaten der Messungen auf die Wiedergabe des logarithmischen Dämpfungsdekrements zu beschränken und die Berechnung des Koeffizienten der inneren Reibung einer späteren exakten Theorie zu überlassen.

Bei der vorliegenden Arbeit erscheint das um so mehr als angebracht, als ihr hauptsächlicher Zweck die Bestimmung der Abhängigkeit des Dämpfungsdekrementes des Holzes von der Verschiedenheit der Holzarten ist. Diese Größe dürfte nämlich von ausschlaggebender Bedeutung sein in all den Fällen, in denen Holz als Resonanzkörper in Musikinstrumenten verwandt wird. Denn es ist klar, daß ein erheblicher Teil der zu der Erregung der Schwingungen des Musikinstrumentes aufgewandten Energie in nutzlose Wärme verwandelt wird, wenn die innere Reibung des benutzten Holzes groß ist. Dieser Teil geht dann für die Ausstrahlung des Instruments in Form von Schallwellen verloren. Schon aus diesem Grunde wird die Wahl des Holzes für die Resonanzböden von Musikinstrumenten in Rücksicht auf die innere Reibung des verwandten Holzes wichtig sein. Man wird also die Eignung des Holzes für diesen Zweck durch die vorausgehende Untersuchung seiner inneren Reibung feststellen können. Um hierfür eine Grundlage zu gewinnen, ist zunächst an einer Anzahl verschiedener Hölzer das logarithmische Dekrement bei Schwingungen verschiedener Frequenz, und zwar für Transversalschwingungen zwischen 10 bis 700 Hz, für Longitudinalschwingungen zwischen 2000 bis 10000 Hz bestimmt worden, also für das Intervall, das hauptsächlich für Musikinstrumente von Bedeutung ist.

Apparatur.

a) Apparatur für transversale Schwingungen (Fig. 1). Das Licht einer Bogenlampe fällt durch einen schmalen Spalt, dann durch eine Linse auf einen kleinen Spiegel, der am Ende des zu untersuchenden Stabes angeklebt ist. Der Stab ist mit dem anderen Ende in einen festen Block eingespannt. Der Lichtstrahl fällt dann von dem Spiegel, durch eine Zylinderlinse punktförmig vereinigt, auf einen Photopapierstreifen, der auf eine Trommel gespannt ist, die sich dreht. Der Abstand Spiegel—Film beträgt etwa 2 m. Wird der Stab angeschlagen oder (bei höheren Frequenzen besser angerissen), so wird auf dem Photopapier die gedämpfte Stabschwingung aufgezeichnet. Durch Ausmessen dieser Aufnahmen erhält man dann das logarithmische Dämpfungsdekrement. Mißt man die Amplitude A und nach zehn Schwingungen die Amplitude B, so ist das logarithmische Dämpfungsdekrement

$$\delta = \frac{\log A - \log B}{10}.$$

Um auch die Frequenz genau bestimmen zu können, fiel ein Teil des durch den Spalt tretenden Lichtes auf einen Spiegel, der an der einen Zinke einer

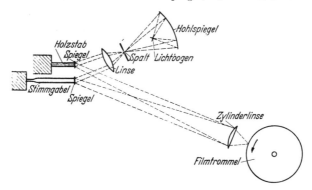

Fig. 1. Anordnung für Transversalschwingungen.

Stimmgabel von 64 Hz angebracht war. Das von ihm reflektierte Licht fiel ebenfalls durch eine Zylinderlinse auf das Photopapier und registrierte die Stimmgabelschwingungen.

Wie aus der Arbeit von Bennewitz und Rötger[1] hervorgeht, ist bei den hier benutzten Frequenzen und Stabdicken die Dämpfung infolge der Luftreibung zu vernachlässigen.

b) Apparatur für longitudinale Schwingungen (Fig. 2). Der zu untersuchende Stab wird in der Mitte zwischen zwei Schneiden festgeklemmt. An den beiden Enden ist ein dünnes Eisenblättchen aufgeleimt. Gegenüber dem einen Ende befindet sich ein Elektromagnet, der von einem Sender gespeist wird. Gegenüber dem anderen Ende befindet sich ebenfalls ein Elektromagnet, der über einen Verstärker mit einem Milliamperemeter verbunden ist. Die durch den Sender hervorgerufenen Stärkeschwankungen des ersten Elektromagneten ziehen das Eisenblättchen des einen Stabendes in demselben Rhythmus an, wodurch der Stab in Schwingungen gerät. Das Eisenblättchen an dem anderen Stabende induziert in demselben

[1] K. Bennewitz u. H. Rötger, Phys. ZS. **37**, 584, 1936.

Rhythmus im zweiten Elektromagneten Ströme, die verstärkt werden und in einem Hitzdrahtmilliamperemeter ein Maß für die Stabamplituden geben. Als Frequenzmesser diente der sogenannte „Frequenzanzeiger" der Allgemeinen Elektricitäts-Gesellschaft, der in mehreren Stufen Frequenzen bis 200, bis 600, bis 2000, bis 6000, bis 20000 und bis 60000 Hz zu messen gestattet. Die Meßgenauigkeit in bezug auf das Dekrement δ nach untenstehender Formel beträgt in den hier benutzten Stufen von

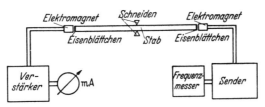

Fig. 2 Anordnung für Longitudinalschwingungen.

2000, 6000 und 20000 Hz etwa 4 bis 5%. Es wurde dann mit dieser Vorrichtung die Resonanzkurve einer Eigenfrequenz der Longitudinalschwingungen des Stabes aufgenommen, und zwar in der Regel der Grundfrequenz, in einzelnen Fällen auch die einer höheren Eigenfrequenz. Trägt man diese Kurve graphisch auf, indem man als Abszisse die Frequenz und als Ordinate das Quadrat des Milliamperemeter-Ausschlags wählt, so erhält man aus der Halbwertsbreite der Kurve in bekannter Weise das Dekrement zu

$$\delta = \pi \frac{\text{Halbwertsbreite}}{\text{Resonanzfrequenz}}.$$

Messungen.

a) Messungen bei transversalen Schwingungen. Es wurden Stäbe von Fichte, Ahorn, Kiefer, Eiche untersucht. Die Stäbe waren etwa 20 mm breit, etwa 4 mm stark in der Schwingungsebene und je nach der Frequenz 50 cm bis 6 cm lang. Bei 50 cm Länge betrug die Frequenz etwa 10 Hz, bei 6 cm Länge etwa 700 Hz, natürlich für die einzelnen Holzarten verschieden. Die Stäbe waren durch etwa halbjähriges Lagern im Zimmer gut getrocknet. Die verschieden langen Stäbe waren durch Verkürzen des ursprünglichen längsten Stabes mit der niedrigsten Frequenz hergestellt.

Es wurden zunächst Stäbe untersucht, deren Längsrichtung mit der Faserrichtung übereinstimmt. Für Ahorn mögen hier die genauen Messungen der Photoaufnahmen als Beispiel folgen. Für die anderen Hölzer wird nur das logarithmische Dämpfungsdekrement angegeben werden.

Tabelle 1. Ahorn.

Frequenz 15,5 Hz. Zeichen in Fig. 3: ●.

Anzahl der Schwingungen	Gemessene doppelte Amplitude in mm	Gemittelte doppelte Amplitude in mm	Natürliches logarithmisches Dämpfungsdekrement
0	27,0		
		23,5	0,0300
10	20,0		
		17,5	288
20	15,0		
		13,15	286
30	11,3		
		10,0	262
40	8,7		
		7,65	276
50	6,6		
		5,8	278
60	5,5		
		4,4	275
70	3,8		

Frequenz 84 Hz. Zeichen in Fig. 3: +.

Anzahl der Schwingungen	Gemessene doppelte Amplitude in mm	Gemittelte doppelte Amplitude in mm	Natürliches logarithmisches Dämpfungsdekrement
0	30,2		
		26,1	0,0317
10	22,0		
		19,1	306
20	16,2		
		14,05	309
30	11,9		
		10,4	313
40	8,7		
		7,55	307
50	6,4		
		5,6	288
60	4,8		
		4,2	288
70	3,6		
		3,15	288
80	2,7		

Frequenz 153 Hz. Zeichen in Fig. 3: ○.

Anzahl der Schwingungen	Gemessene doppelte Amplitude in mm	Gemittelte doppelte Amplitude in mm	Natürliches logarithmisches Dämpfungsdekrement
0	26,2		
		22,7	0,0311
10	19,2		
		16,6	288
20	14,4		
		12,5	307
30	10,6		
		9,3	281
40	8,0		
		7,0	288
50	6,0		
		5,25	288
60	4,5		
		3,95	310
70	3,4		
		2,95	278
80	2,5		
		2,2	274
90	1,9		

Frequenz 356 Hz. Zeichen in Fig. 3: ×.

Anzahl der Schwingungen	Gemessene doppelte Amplitude in mm	Gemittelte doppelte Amplitude in mm	Natürliches logarithmisches Dämpfungsdekrement
0	8,7		
		7,55	0,0307
10	6,4		
		5,55	309
20	4,7		
		4,1	295
30	3,5		
		3,05	297
40	2,6		

Frequenz 552 Hz. Zeichen in Fig. 3: *.

Anzahl der Schwingungen	Gemessene doppelte Amplitude in mm	Gemittelte doppelte Amplitude in mm	Natürliches logarithmisches Dämpfungsdekrement
0	24,5		
		20,25	0,0426
10	16,0		
		13,6	357
20	11,2		
		9,55	349
30	7,9		
		6,8	326
40	5,7		
		4,95	306
50	4,2		

Frequenz 665 Hz. Zeichen in Fig. 3: □.

Anzahl der Schwingungen	Gemessene doppelte Amplitude in mm	Gemittelte doppelte Amplitude in mm	Natürliches logarithmisches Dämpfungsdekrement
0	14,0		
		11,6	0,0420
10	9,2		
		7,7	395
20	6,2		
		5,25	366
30	4,3		
		3,7	327
40	3,1		

Die Ergebnisse dieser Messungen sind in Fig. 3 graphisch dargestellt.

Die Kurven zeigen, daß das Dekrement mit der Amplitude ansteigt. Dazu ist folgendes zu sagen. Belastet man zwei verschieden lange Stäbe gleich, so sind die Ausschläge bekanntlich proportional der dritten Potenz der Stablängen. Sollen also die kurzen Stäbe auf dem Photopapier die gleichen Amplituden zeigen wie die langen Stäbe, so müssen sie umgekehrt proportional der dritten Potenz ihrer Länge stärker angeschlagen werden. In demselben Verhältnis etwa, so zeigen die Kurven, steigt das Dekrement mit der Amplitude bei einem kurzen Stab schneller an als bei einem langen. Dies gilt also, wenn der kurze und der lange Stab denselben Ausschlag auf dem Photopapier zeigen. Reduziert man aber auf gleiche Anschlagsstärke der Stäbe, so steigt das Dekrement mit der Amplitude für alle Stablängen gleich an.

Die punktierten Linien (Fig. 3) zeigen die Extrapolation des logarithmischen Dämpfungsdekrements δ auf die Amplitude Null. Dies ergibt sich aus der Fig. 3 für alle gemessenen Frequenzen zu

$$\delta = 0{,}028 \text{ für Ahorn.}$$

Für die anderen Hölzer beträgt das auf die Amplitude Null extrapolierte logarithmische Dämpfungsdekrement für alle gemessenen Frequenzen zwischen 10 und 700 Hz:

$$\delta = 0{,}024 \text{ für Fichte,}$$
$$\delta = 0{,}029 \text{ für Kiefer,}$$
$$\delta = 0{,}034 \text{ für Eiche.}$$

Die auf die Amplitude Null extrapolierten Dekremente sind also materialabhängig, aber frequenzunabhängig. Fichte, das hauptsächliche Resonanzholz, hat also das kleinste Dekrement. An zweiter Stelle folgt dann Ahorn,

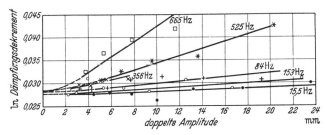

Fig. 3. Meßergebnisse der Transversalschwingungen bei Ahornholz.

das Bodenholz der Geige. Es handelt sich bei diesen Messungen um gewöhnliches Holz. Sicher wird gutes Geigenholz geringere Dekremente aufweisen.

Die Genauigkeit der Einzelmessungen beträgt $\pm 10\%$, wozu noch die Ungenauigkeit durch die Strukturungleichheit der Hölzer hinzukommt. Die Mittelung durch die Kurven verringert den Fehler natürlich. Es traten manchmal Energieabwanderungen in die Einspannvorrichtung auf. Sie konnten durch festeres Einspannen der Hölzer beseitigt werden. Ferner ergaben sich öfter Resonanzen mit Teilen des Experimentiertisches. Die sich dann in der Photoaufnahme zeigenden Schwebungen waren oft so schwach, daß sie nur bei genauem Ausmessen zu erkennen waren. Dies zeigt den Wert der photographischen Registrierung. Bei einer visuellen Beobachtung des Abklingens, etwa in einem Mikroskop, würde man diese Schwebungen nicht erkennen und folglich eine Erhöhung des Dekrements messen. Die Schwebungen konnten durch Beschweren des Tisches mit großen Gewichten beseitigt werden.

Um festzustellen, ob das Dämpfungsdekrement sich mit der Dicke des Stabes ändert, wurde die Dicke ein und desselben Stabes allmählich von 4 mm bis auf 1 mm vermindert. Außerdem wurde noch ein Stab von 10 mm Dicke gemessen. Das Dekrement blieb jedoch ungeändert, was auch zeigt, daß die Strahlungsdämpfung vernachlässigbar war.

Das Dekrement von Stäben, die senkrecht zur Faserrichtung geschnitten waren, war etwa 3,5mal größer.

Stäbe, die mit gewöhnlichem Ahornlack lackiert waren, zeigten eine Erhöhung des Dekrements um 40%. Das Dekrement des Holzes wird also durch Lackierung erhöht, eine Erscheinung, die völlig damit übereinstimmt, daß Geigen durch starke Lackierung oft ihre Klanggüte verlieren.

Ein Stab, eine Nacht in Wasser gelegt, hatte am nächsten Tage ein um 75% erhöhtes Dekrement.

Die Messungen des Dekrements mittels Transversalschwingungen konnten nicht bis zu höheren Frequenzen weitergetrieben werden, da bei den dann sehr kurzen Stäben die Amplituden zu klein wurden.

Die Bestimmung des Dekrements bei höheren Frequenzen wurde daher mittels Longitudinalschwingungen vorgenommen. Hierbei konnte als niedrigste Schwingungszahl allerdings erst die von 2000 Hz gewählt werden, weil die Länge der Stäbe, die dann schon 1 m betrug, sonst zu groß wurde.

b) Messungen bei longitudinalen Schwingungen. Es wurden bei verschiedenen Resonanzfrequenzen die Resonanzkurven aufgenommen.

Beispiel: Fichte.

Tabelle 2.

Milliamp.	(Milliamp.)^2	Frequenz in Hz	Milliamp.	(Milliamp.)^2	Frequenz in Hz
19,0	361	4785	56,0	3136	4854
27,0	729	4812	53,8	2894	4857
34,0	1156	4827	44,4	1971	4863
42,7	1823	4836	30,0	900	4878
53,3	2841	4845	20,4	416,2	4890

Die Messungen der Tabelle 2 sind in Fig. 4 graphisch dargestellt.

Die Amplituden der longitudinalen Schwingungen betragen nur einige μ. Förster und Köster[1]) haben gezeigt, daß für solche kleinen Amplituden das Dekrement von der Amplitude unabhängig ist. Die so erhaltenen Werte entsprechen also den bei den Transversalschwingungen auf die Amplitude Null extrapolierten Werten.

---

[1]) F. Förster, ZS. f. Metallkunde **29**, 109, 1937; F. Förster u. W. Köster, ebenda **29**, 116, 1937.

Die Meßgenauigkeit beträgt etwa ± 10%, wozu dann noch die Ungenauigkeit durch die Strukturungleichheit der Hölzer kommt.

Die Messungen wurden zunächst an denselben Stäben von 1 m Länge vorgenommen, die auch für die Messungen mittels Transversalschwingungen benutzt waren.

Tabelle 3.

Material	Natürliches log. Dekrement für Longitudinalschwingungen	Natürliches log. Dekrement für Transversalschwingungen
Fichte	0,020	0,024
Ahorn	0,026	0,028
Kiefer	0,029	0,029
Eiche	0,037	0,034

Der Vergleich der hier erhaltenen Resultate mit den oben bei Transversalschwingungen erhaltenen zeigt Tabelle 3.

Die für beide Schwingungsarten so erhaltenen logarithmischen Dekremente stimmen also weitgehend, vor allem in der Reihenfolge, überein. Daß dies auch für höhere Frequenzen der Longitudinalschwingungen zutrifft, zeigen die folgenden Messungen.

Die Messungen für höhere Frequenzen wurden an Stäben gemacht, die quadratischen Querschnitt hatten, etwa 1 × 1 cm, und 1 m bis 20 cm lang waren. Die Frequenz betrug bei 1 m Länge etwa 2000 Hz, bei 20 cm Länge etwa 10 000 Hz, natürlich für die einzelnen Holzarten verschieden wegen der Verschiedenheit der Schallgeschwindigkeit in den Hölzern[1]. Die Hölzer waren auch wieder gut getrocknet und parallel zur

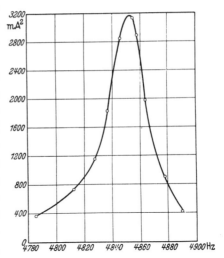

Fig. 4. Resonanzkurve für einen Stab aus Fichtenholz.

Faserrichtung geschnitten. Es wurden gelegentlich zwei bis drei Resonanzfrequenzen dicht nebeneinander gefunden, wie das auch schon Förster und Köster (l. c.) beobachtet haben. Dies führte zu verzerrten Resonanz-

---

[1] Siehe Handb. d. Phys. Bd. VIII, Artikel Lübcke, S. 647. Berlin, Julius Springer, 1927.

Zeitschrift für Physik. Bd. 110. 5

kurven, die nicht verwertbar waren. Der Grund für die Erscheinung dürfte nach Förster und Köster darin bestehen, daß die benutzten Stäbe in verschiedenen Richtungen unterschiedliche Werte des Elastizitätsmoduls besaßen, wie es bei Metallen beim Vorliegen einer Walztextur vorkommt, bei Hölzern natürlich durch die Faserstruktur bedingt sein dürfte. Durch geringfügige Veränderungen der Form oder durch Verkürzung der Stäbe ließ sich die störende Unregelmäßigkeit beheben.

Die Ergebnisse dieser Messungen an longitudinal schwingenden Stäben sind in den folgenden Tabellen wiedergegeben.

Tabelle 4. Fichte.

Resonanz-frequenz	Natürliches logarith-misches Dekrement
2 888	0,0239
2 951	214
3 801	192
4 852	214
5 872	214
6 570	258
7 390	185
8 500	267
10 020	248
10 775	264

Tabelle 5. Ahorn.

Resonanz-frequenz	Natürliches logarith-misches Dekrement
2317	0,0245
2362	260
2378	256
2429	245
3089	336
3148	236
4083	223
4188	240
4840	236
5069	270
6152	286
7528	257
9117	283

Tabelle 6. Kiefer.

Resonanz-frequenz	Natürliches logarith-misches Dekrement
2445	0,0257
2665,5	261
3255	246
3581	298
4393	243
5190	248
5790	245
7032	242
8885	261
9390	317

Tabelle 7. Eiche.

Resonanz-frequenz	Natürliches logarith-misches Dekrement
1921	0,0311
1941	324
2388	355
2531	342
2772	305
3462	327
3519	305
4282	311
5308	355
6610	298
6796	295
8380	327

Die graphische Darstellung dieser Ergebnisse zeigt Fig. 5.

Das Dekrement erweist sich also auch für diese höheren Frequenzen als von der Frequenz unabhängig. Die Mittelwerte für alle Longitudinalschwingungen sind:

Tabelle 8.

Material	Natürliches logarithmisches Dekrement	Material	Natürliches logarithmisches Dekrement
Fichte	0,023	Kiefer	0,026
Ahorn	0,026	Eiche	0,032

Von früheren Messungen der inneren Reibung des Holzes liegt einmal eine solche von Kimball und Lovell[1]) vor, aus deren Torsion-Biegungsschwingungsmessungen bei Frequenzen zwischen etwa 2 bis 200 Hz für Ahornholz sich das logarithmische Dekrement in Übereinstimmung mit den vorliegenden Messungen als frequenzunabhängig zu 0,022 berechnet. Aus den

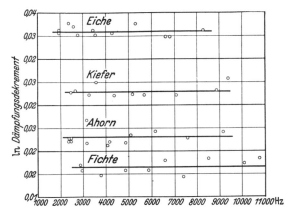

Fig. 5. Meßergebnisse für Longitudinalschwingungen.

Längs- und Biegungsschwingungsmessungen von Schmidt[2]) ergibt sich das logarithmische Dekrement von einem Kiefernholzbalken von 4 m Länge bei Frequenzen zwischen etwa 30 bis 1800 Hz ebenfalls als frequenzunabhängig zu 0,034. Dieser Wert ist zwar etwas höher als der hier für Kiefernholz gefundene, weicht jedoch nicht stark davon ab und mag durch die Verschiedenheit der Dimensionen der benutzten Hölzer bedingt sein.

Diskussion.

Das Resultat, daß das logarithmische Dekrement bei den Hölzern unabhängig von der Frequenz ist, steht im Einklang mit der in der Einleitung erwähnten Theorie von Boltzmann, der die innere Reibung aus der elastischen Nachwirkung ableitet. Die Voigtsche Theorie ist damit nicht vereinbar. Nun ist einerseits von Bennewitz und Rötger[3]), anderer-

---

[1]) A. L. Kimball u. D. E. Lowell, Phys. Rev. (2) **30**, 948, 1927. — [2]) R. Schmidt, Ingineur-Archiv **5**, 352, 1934. — [3]) K. Bennewitz u. H. Rötger, Phys. ZS. **37**, 578, 1936.

5*

seits von Zener[1]) eine Theorie entwickelt worden, nach der die innere Reibung fester Körper nach jenen Autoren in dem Massentransport durch Diffusion oder einer Energiegröße durch Wärmeleitung, nach diesem nur in der letzteren, in beiden Fällen aber in den durch sie bedingten irreversiblen Ausgleichvorgängen zwischen den Kompressions- und Dilatationsstellen der Schwingungen besteht. Bei Zugrundelegung dieser Theorie ergibt sich ein mehr oder weniger breites Maximum für eine bestimmte Frequenz, das bei einer Anzahl von Metallen auch experimentell gefunden wurde. Nach der Formel von Zener würde das Maximum bei Holz für dünne Stäbe wegen der geringen Wärmeleitfähigkeit des Holzes bei Schwingungen von wenigen Hertz liegen, also weit unterhalb der hier angewandten Frequenzen. Dieser Einfluß der irreversiblen Wärmeleitung spielt also für die hier gemessenen Vorgänge der Dämpfung bzw. inneren Reibung des Holzes keine Rolle. Sie muß also anders bedingt sein, nach dem oben Gesagten am ehesten durch die elastische Nachwirkung.

In bezug auf die praktische Bedeutung der inneren Reibung des Holzes für die Verwendung der Hölzer für die Resonanzböden von Musikinstrumenten ist es zunächst gleichgültig, welches ihre eigentliche Ursache ist. Wichtig ist nur die Frequenzunabhängigkeit des logarithmischen Dämpfungsdekrements und die Abhängigkeit desselben von der Natur des Holzes. Aus der Reihenfolge der Hölzer in den Tabellen 2 und 7 ergibt sich, daß in Rücksicht auf die Energieverluste durch innere Reibung bei den Schwingungen für Resonanzböden am besten Fichtenholz geeignet ist, dann Ahornholz, weniger gut Kiefernholz und am schlechtesten Eichenholz. In Übereinstimmung hiermit verwendet man in der Tat als Material für Resonanzböden einerseits Fichten-, andererseits Ahornholz.

Ob eine möglichst geringe innere Reibung des Holzes der Resonanzböden bei gleicher Holzart in der Tat den Unterschied in der Güte von Geigen oder anderen Musikinstrumenten mit bedingt, und wie groß etwa der durch innere Reibung verzehrte Energieanteil bei den Schwingungen ist, der dann für die Schallausstrahlung verloren geht, soll durch weitere Messungen an Musikinstrumenten selbst bestimmt werden.

Für die Überlassung des Frequenzanzeigers danken wir der Deutschen Forschungsgemeinschaft (Notgemeinschaft der Deutschen Wissenschaft).

Greifswald, Physikalisches Institut der Universität, 19. Mai 1938.

---

[1]) C. Zener, Phys. Rev. (2) **52**, 230, 1937; **53**, 90, 1938; C. Zener, W. Otis u. R. Nuckolls, ebenda (2) **53**, 100, 1938; C. Zener, ebenda (2) **53**, 582, 1938.

20

Internal Damping of Wood

F. KRÜGER AND E. ROHLOFF

ABSTRACT by John C. Schelleng

Logarithmic decrements were determined for vibration damping for various kinds of wood—spruce, maple, pine, oak—first, with transverse vibrations between 10 and 700 Hz, and second, with longitudinal vibrations from 2,000 to 10,000 Hz. The log decrement in the frequency ranges tested turns out to be independent of frequency. This is in agreement with Boltzmann's theory of internal friction for solid bodies. Decrement was smallest for spruce and increased in the order maple, pine, oak. For test strips perpendicular to the grain, the decrement was three and one half times greater than in those cut along the grain. Coating the strip with maple varnish (Ahornlack) increased decrement about 40 percent. A test strip soaked in water for 12 hours showed a 75 percent increase in decrement. In accordance with these decrements, spruce and maple are preferred for resonance wood in musical instruments since they have the least loss of energy through internal friction.

Addendum to Abstract

Ernst Rohloff's paper "Uber die Innere Reibung von Geigenholz," Z. *Physik*, **117**, 64–66 (1940), fills the gap in the frequency range of his 1938 paper. The following abstract of this paper is also by John C. Schelleng.

> The log decrement for transversely flexural vibrating wood bars, which were cut from spruce used in violin making and of a thickness as in the violin top, was found to be independent of frequency in the range 10 to 2,000 Hz. In agreement with the 1938 paper (above) the result is that, in the range 10 to 10,000 Hz, which includes the entire range of sounds of stringed instruments, the log decrement of wood of the thickness mentioned is independent of frequency. In this paper the samples were cut along the grain, but the annual rings were not perpendicular to the

surface as were those in the 1938 paper. The log decrement found was 0.029.

Comment: Recent work indicates that the logarithmetic decrement for good violin spruce is around 0.002. Rohloff's findings may be the result of the difference in the orientation of the strips or possibly his testing "spruce such as is used in the cheapest violins made in Mittenwald."

21

Reprinted from *Ann. Physik.*, Folge 5, **38**(3), 177–198 (1940)

Über die innere Reibung und die Strahlungsdämpfung von Geigen[*]

Von E. Rohloff

(Mit 14 Abbildungen)

In einer früheren Arbeit[1]) ist die innere Reibung verschiedener Hölzer untersucht worden. In Anbetracht des Umstandes, daß keine vollständige Theorie der inneren Reibung fester Körper existiert, schien es geraten, sich dabei in den Resultaten der Messungen auf die Wiedergabe des logarithmischen Dämpfungsdekrements zu beschränken. Es wurde gezeigt, daß dieses im Frequenzbereich von 10—700 Hz und 2000—10000 Hz frequenzunabhängig war. Auf Grund noch laufender Messungen kann schon jetzt gesagt werden, daß auch in dem fehlenden Frequenzbereich von 700—2000 Hz keine Änderung des Dekrements auftritt. Das Dekrement der Hölzer ist also bei den in der Arbeit verwendeten Holzdicken von einigen Millimetern (etwa die Dicke der Platten von Geigen) in dem Frequenzintervall, das für Musikinstrumente in Betracht kommt, frequenzunabhängig. Ferner wurde in der eben zitierten Arbeit festgestellt, daß von den untersuchten Hölzern Fichte das geringste Dekrement besaß, dann folgten mit der Reihenfolge entsprechend höheren Dekrementen Ahorn-, Kiefern- und Eichenholz. Es ist bemerkenswert, daß Fichte und Ahorn, die wichtigsten Resonanzhölzer, die geringsten Dekremente besitzen.

In der vorliegenden Arbeit sollen nun Dekrementsmessungen beschrieben werden, die an Geigen selber angestellt wurden. Dabei wurde nach Vorschlag von Prof. Krüger folgendes Verfahren angewandt.

Die nach dem Absetzen des Geigenbogens abklingenden Schwingungen des Geigenkörpers werden photographisch registriert, und zwar einmal in Luft und ein anderes Mal im Vakuum. Aus jeder Aufnahme wird dann das logarithmische Dämpfungsdekrement bestimmt. Das Dekrement in Luft gibt die gesamte Dämpfung, d. h. die Summe der Dekremente aus der inneren Reibung des Geigen-

---

1) F. Krüger u. E. Rohloff, Ztschr. f. Phys. **110**. S. 58. 1938. Im folgenden kurz I genannt.

Editor's Note: An abstract in English follows this article.

körpers und der Strahlungsdämpfung. Das Dekrement im Vakuum gibt nur die innere Reibung des Geigenkörpers. Die Differenz der Dekremente in Luft und im Vakuum gibt das Dekrement der Strahlungsdämpfung. Backhaus[1]) hat akustisch die Gesamtdämpfung von Geigen verschiedener Güte in Luft bestimmt und glaubt gefunden zu haben, daß diese Größe kein Charakteristikum für die Güte von Geigen sei. Hier soll geprüft werden, ob die anderen genannten Größen für verschieden gute Geigen verschiedene Werte annehmen.

Apparatur

Auf einem schweren eisernen Tisch wurde die Geige wie folgt eingespannt (vgl. Abb. 2): An einer kräftigen Stange befand sich eine Schraubzwinge, die den Hals der Geige in einem Gummilager festspannte. Das Einhängeknöpfchen am unteren Teil der Geige wurde dabei fest gegen einen großen Gummistopfen gedrückt. Auf dem Boden der Geige wurde ein dünnes Eisenblättchen von 1×1 cm Fläche mit etwas Klebwachs befestigt. Diesem Blättchen gegenüber befand sich ein Elektromagnet. Die Schwingungen des Geigenkörpers erzeugten durch das Blättchen Wechselströme in dem Elektromagneten. Diese wurden, um Obertöne zu beseitigen, über einen Siemensstromreiniger geschickt, von dort in einen Verstärker und dann in eine Oszillographenschleife (Eigenfrequenz 6500 Hz). Verzerrungen der Schwingungsform, die unter Umständen in dem Elektromagneten auftreten, fallen bei diesen Messungen nicht ins Gewicht, da es ja bei der Bestimmung des Dekrements nur auf das Verhältnis der Schwingungsamplituden zueinander ankommt.

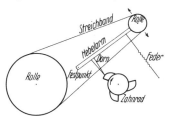

Abb. 1. Schema des periodischen Anstreichens und Absetzens des Streichbandes

Das Anstreichen der Gegensaite geschah mit einem endlosen Streichband, das durch ein Grammophonwerk angetrieben wurde. Das Band bestand aus etwa 5—6 Lagen einfacher Nähseide und wurde gut mit Kolophonium eingerieben. Da ja das Abklingen der Töne nach dem Absetzen des Streichbandes registriert werden und dies auch im Vakuum geschehen sollte, mußte für ein selbsttätiges Absetzen des Streichbandes von der Geigensaite gesorgt werden. Dies wurde so erreicht: Eine der beiden Rollen, über die das Streich-

---

1) H. Backhaus, Ztschr. f. Phys. **18**. S. 98. 1937.

band lief, wurde auf einem freien Hebelarm angebracht. Ein Dorn dieses Hebelarms wurde durch eine Feder gegen ein langsam rotierendes Zahnrad gedrückt. Die Zähne hatten breite Kuppen. Befand sich der Dorn auf einem solchen Zahn, so wurde das Streichband gegen die Geigensaite gedrückt. Bei weiterem Drehen des Zahnrades glitt dann der Dorn plötzlich in die Vertiefung zwischen zwei Zähnen: Das Streichband setzte von der Saite ab. Abb. 1 mag dies veranschaulichen. Durch einen Elektromagneten konnte das Streichband angehalten werden.

Ferner mußte eine Tonänderung im Vakuum möglich gemacht werden. Die dazu konstruierte Apparatur ist auf der Photographie, Abb. 2, zu erkennen. Ein Motor bewegt den „Finger", der mit einem gummibespannten Rädchen versehen ist, das auf der Saite aufliegt, an einer Spindel auf- und abwärts. Eine starke Spiralfeder drückt den „Finger" mit der nötigen Kraft auf die Saite.

Über die in Abb. 2 gezeigte Apparatur wurde für die Vakuumaufnahmen eine große Glasglocke gestülpt. Die Abdichtung zwischen Eisentisch und Glasglocke erfolgte durch einen Gummiring. Die sämtlichen elektrischen Zuleitungen zur Apparatur wurden durch Gummistopfen durch die Eisenplatte gelegt, ebenso die Saugleitung zur Pumpe. Um die Geigen nicht zu gefährden, war eine

Abb. 2. Die Geige auf dem Eisentisch mit Streichband, Apparatur zur Änderung der Tonhöhe und Elektromagnet zum Abnehmen der Schwingungen

Evakuierung nur bis zum Dampfdruck des Wassers bei Zimmertemperatur möglich. Die Vakuumaufnahmen wurden daher alle bei 1,5 cm Quecksilber gemacht.

Messungen

Photographiert wurde ein Teil des stationären Tones mit dem anschließenden Abklingen. Die Frequenz N des Tones wurde aus dem stationären Teil durch Auszählen und Vergleich mit den gleichfalls photographierten Schwingungen einer 50 Hz Stimmgabel gewonnen. Dann wurde die Amplitude A im Moment des Absetzens des Streichbandes gemessen, dann die Amplitude $1/_{10} A$ im Verlauf des Abklingens aufgesucht. Betrug dann der zeitliche Abstand von der Amplitude A bis zur Amplitude $^1/_{10} A$ a Stimmgabelschwingungen, so ergab sich das natürliche logarithmische Dämpfungsdekrement δ zu:

$$\delta = 2{,}3 \, \frac{\log A - \log \frac{1}{10} A}{\frac{N}{50} \cdot a}$$

oder

$$\delta = \frac{2{,}3}{\frac{N}{50} \cdot a} \cdot$$

Zeigte das Abklingen Schwebungen, so wurden die Maxima durch eine Exponentialkurve miteinander verbunden, und der Abfall dieser Exponentialkurve auf $^1/_{10}$ der stationären Amplitude wurde gemessen.

Um zu prüfen, ob verschiedene Stellen des Geigenbodens evtl. verschiedene Dekremente haben, wurde derselbe Ton mehrmals aufgenommen, wobei das Eisenblättchen jedesmal auf eine andere Stelle des Geigenbodens geklebt wurde. Es zeigte sich, daß die Form des Abklingens zwar verschieden war: an manchen Stellen des Geigenbodens traten Schwebungen auf, an anderen nicht. Die aus den verschiedenen Aufnahmen errechneten Dekremente stimmten aber miteinander überein.

Es seien nun zunächst in Abb. 4—7 (S. 185/186) die Ergebnisse graphisch dargestellt, die an vier verschieden guten Geigen erhalten wurden. Von jeder Seite wurden die Dekremente in Luft und die Dekremente im Vakuum bestimmt und zwar über einen Frequenzbereich von etwa einer Oktave (von der bloßen Saite bis etwa zur Oktave). Die einzelnen Meßpunkte wurden möglichst eng gewählt, die Unterschiede betrugen oft nur wenige Hertz. In den Abb. 4—7 ist als Abszisse die jeweils aus den Aufnahmen durch Auszählen bestimmte Frequenz des stationären Teils des Tones im logarithmischen Maßstab aufgetragen, so daß die Tonfolge g, gis usw. linear erscheint. Die Ordinate gibt das ln-Dämpfungsdekrement. Die Kurven mögen kurz *Dekrementskurven* genannt werden. Die Kurven für Luft sind ausgezogen gezeichnet, die für Vakuum gestrichelt. In jeder der Abb. 4—7 sind die Ergebnisse der gleichen Saite der vier verschiedenen Geigen nebeneinander gestellt. Es handelt sich um folgende vier Geigen:

1. Eine größtenteils schlecht klingende Geige ohne jegliches Herstellerzeichen (Wert RM. 20.—).

2. Eine Geige von Joseph Gagliano aus dem Jahre 1763 (Wert RM. 2500.—).

3. Eine Geige von Jo. Bapt. Rogerius Bon aus dem Jahre 1690 (Wert RM. 18000.—).

4. Eine Geige von Antonius Stradivarius aus dem Jahre 1717 (Wert RM. 140000.—).

Zunächst ist an diesen Kurven auffällig, daß die Vakuumkurven (gestrichelt) nicht überall unter den Luftkurven (ausgezogen) liegen. Im Vakuum fehlt die Strahlungsdämpfung, folglich müßten die Vakuumkurven ständig unter den Luftkurven liegen. Nun beschränken sich aber die Abweichungen fast ausschließlich auf die Gebiete, in denen starke Größenschwankungen der Dekremente auftreten, wie z. B. in den Maxima der Kurven. Als besonderes Beispiel sei hier das zweite Maximum der e_2-Saite der schlechten Geige (Abb. 7) genannt. In solchen Stellen liegen aber offenbar nur Verschiebungen der Maxima nach höheren oder tieferen Frequenzen vor. Diese Verschiebungen sind erklärlich, wenn man bedenkt, daß die Maxima, wie wir noch später sehen werden, den Resonanzstellen der Geige entsprechen. Daß die Evakuierung eine leichte Verschiebung dieser Resonanzstellen bewirken kann, ist klar. — Manchmal sind die Maxima im Vakuum höher als in Luft. Dies wird meistens daran liegen, daß in Luft die Resonanzstellen nicht ganz genau gefunden wurden, was erklärlich ist, wenn man bedenkt, daß ein musikalisches Halbtonintervall auf der g-Saite z. B. von g nach gis 6 Hz, auf der e_2-Saite z. B. von e_2 nach f_2 schon etwa 40 Hz beträgt. In einem solchen recht großen Frequenzintervall genau die Resonanzstelle zu finden, ist schwierig, da ja die Halbtöne auf der Geigensaite so dicht aufeinanderfolgen, daß sie unmittelbar nebeneinander gegriffen werden. Und diese „Greifengigkeit" wird auf jeder Saite um so größer, in je höherer Lage man spielt. In den Resonanzstellen und ihrer Umgebung ist also der Unterschied zwischen den Kurven in Luft und den Kurven im Vakuum ungenau durch die öfter auftretende Verschiebung der Resonanzstellen im Vakuum, und dadurch, daß die Resonanzstellen manchmal nicht genau erreicht wurden. Wo aber ein großer Unterschied zwischen der Luft- und der Vakuumkurve in einer Resonanzstelle auftrat, wurden so viele Messungen gemacht, bis der Unterschied hier gewährleistet war. Unerklärt ist dagegen das Ergebnis auf der e_2-Saite der Gaglianogeige (Abb. 7). Die Vakuumkurve liegt fast dauernd über der Luftkurve. Da diese Geige, wie auch die anderen guten Geigen, nur kurze Zeit zur Verfügung stand, war dies nicht zu klären. Vielleicht liegt nur eine größere Verschiebung im Vakuum gegenüber Luft vor. Auf jeden Fall ist von diesem Ergebnis abzusehen, zumal dies der einzige so krasse Fall ist. Berücksichtigt man die hier genannten Verhältnisse — und man muß dies tun, denn die Geige ist ja weit komplizierter als etwa ein schwingender Eisenstab; man wird also bei ihr nicht so einfache Ergebnisse erwarten können —, so geben die Kurven doch ein leidliches Bild des Verlaufs des Dekrements in Luft und im Vakuum.

Nun ist die Frage zu klären, warum überhaupt solche Maxima des Dekrements auftreten. Wie schon gesagt, geschieht dies in den Resonanzstellen der Geige. Die Geige besteht aus verschiedenen Teilen, die miteinander gekoppelt sind: Saite, Steg, Körper usw. Wieweit diese Teile nun alle selber schwingen und wie sie miteinander gekoppelt sind, oder ob manche von ihnen starr miteinander verbunden sind, ist nicht überall klar. An wichtigsten Einzeldekrementen sind folgende zu nennen. Die Saiten haben Dekremente von etwa 0,004 [nach Backhaus[1]]. Der Steg hat ein Dekrement, das je nach der Größe der Amplitude um den Wert 0,05 schwankt. Dies ergibt sich aus den Messungen an Holzstäben in I und aus den Stegamplitudenmessungen von M. Minnaert und C. C. Vlam[2]. Ferner hat gewöhnliches Fichtenholz nach I parallel zur Faserrichtung ein Dekrement von etwa 0,024, senkrecht zur Faserrichtung ein Dekrement von etwa 0,084. Der unterscheidende Faktor zwischen der kleinsten Dämpfung (Saite) und der größten Dämpfung (Holz senkrecht zur Faserrichtung) beträgt also etwa 20! Kommen nun die einzelnen schwingenden Teile in Resonanz miteinander, so treten Verhältnisse auf, die wohl vergleichbar sind mit denen, die M. Wien[3] für zwei schwingende miteinander gekoppelte Körper berechnet hat. Danach treten in den schwingenden Körpern in der Resonanz verschiedene Schwingungen auf, die, unter gewissen Voraussetzungen über die Größe der Kopplung, die auch hier bei der Geige erfüllt sind, ein Dekrement haben, das dem Mittelwert der Einzeldekremente entspricht. Dies gilt übrigens auch von elektrischen Kettenleitern [Vermutung von H. Riegger[4], Bestätigung von H. Backhaus[5]]. Sind wir weit genug von einer Resonanzstelle entfernt, so tritt keine Beeinflussung der einzelnen Dekremente untereinander auf. Vielmehr klingen dann die stark gedämpften Schwingungen rasch ab. Die schwach gedämpften Schwingungen sind dann allein maßgebend für das Dekrement des gesamten Systems. Nach diesen Wienschen Ergebnissen berechnet sich mit

---

1) H. Backhaus, Ztschr. f. techn. Phys. **18**. S. 98. 1937.
2) M. Minnaert u. C. C. Vlam, Physica **4**. S. 361. 1937.
3) M. Wien, Wied. Ann. **297**. S. 151. 1897.
4) H. Riegger, Wiss. Veröff. Siem.-Konz. Bd. I. Nr. 3. S. 127.
5) H. Backhaus, Wiss. Veröff. Siem.-Konz. Bd. IV. Nr. 2. S. 209.

den obigen Einzeldekrementen ein Durchschnittsdekrement für Resonanzstellen der Geige zu etwa 0,04. Dies stimmt in groben Zügen mit der Höhe der Maxima der Kurven (Abb. 4—7) überein. Daß auch kleinere Maxima vorkommen, ist wohl nicht erstaunlich. Dort kann z. B. eine Resonanz des Steges vorliegen ohne Körperresonanz. Darauf, daß die guten Geigen im allgemeinen höhere Maxima zeigen als die schlechten Geigen, wird später eingegangen werden.

Ein Anteil der Höhe eines jeden Maximums wird aber bedingt durch die Vergrößerung der Amplituden in den Resonanzstellen. Die Dekrementsmaxima fallen mit den Körperamplitudenmaxima und auch mit den Schalldruckamplitudenmaxima zusammen (vgl. Abb. 12). Meinel[1] fand, grob gesprochen, etwa eine Verzehnfachung der Körperamplituden in der Resonanzstelle gegenüber einer Nichtresonanzstelle. Da in dieser Arbeit die Dekremente gemessen wurden von Beginn des Abklingens an, wie es ja auch für die Geige das Richtige ist, da es ja beim Spiel immer nur auf die Verhältnisse beim Streichen und kurz nach dem Absetzen ankommt, da es aber andererseits erwünscht war, zu wissen, wieviel die erhöhte Amplitude das Dekrement erhöht, so wurden Messungen bei gleicher Amplitude gemacht. Einige solcher Messungen sind in Abb. 3 dargestellt. Kurve *1* (Abb. 3) zeigt das Dekrement in einer Nichtresonanzstelle, *2* in einer Resonanzstelle bei gleicher Amplitude wie *1*, *2'* ist das Dekrement des gleichen Tones wie *2*, aber bei großer Amplitude (gleich nach dem Absetzen des Streichbandes) gemessen (vgl. auch Unterschrift zu Abb. 3). Man sieht, daß die wesentliche Erhöhung des Dekrements in der Resonanz nicht in der größeren Amplitude begründet ist, sondern eben in den oben genannten Verhältnissen in der Resonanzstelle. *1*, *2* und *2'* sind im Vakuum gemessen, *3* (gleiche Amplitude wie *1* und *2*) und *3'* (große Amplituden) in Luft, aber bei einer anderen Resonanz wie *2* uud *2'*. Auch bei *3* und *3'* ist die wesentliche Erhöhung durch die Resonanzlage, eine unbedeutendere weitere Erhöhung durch die Vergrößerung der Amplitude bedingt. Die Kurven verlaufen alle leidlich parallel zueinander. Dies besagt, daß die Dämpfung der Geige immer von einer Einzeldämpfung (von Körper, Steg, Saite usw.) oder von gleichen Einzeldämpfungen bedingt ist. Dies stimmt mit dem oben genannten Wienschen Ergebnis überein: In der Resonanz gleiche durchschnittliche Dämpfung der Einzelteile, außerhalb der Resonanz völliges Überwiegen der geringen Dämpfungen. Kurz vor

---

1) H. Meinel, Elektr. Nachr. Techn. **14**. S. 119. 1937.

und hinter der Resonanzstelle verlaufen die Kurven jedoch steiler.
Auch dies stimmt mit dem Wienschen Ergebnis überein: Im ersten
Teil des Abklinges herrscht das — allerdings schon infolge der
Nähe der Resonanz verminderte — höchste Dekrement, dann, nach
dem Abklingen dieser Schwingung, jedoch das — jetzt allerdings
auch erhöhte — niedrige Dekrement vor.

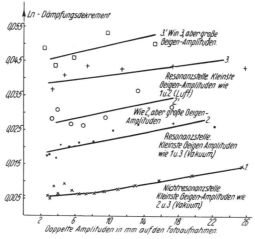

Abb. 3. Abhängigkeit des ln-Dämpfungsdekrements
von der Amplitude der Geigenschwingungen.
(Die Kurven *1, 2, 3* wurden bei gleicher Verstärkung erhalten. Es lagen also
gleiche Amplituden der Geigenschwingungen selbst vor. *2′* und *3′* dagegen
wurden durch geringere Verstärkung als *1, 2, 3* auf die selben Amplituden
auf den Photoaufnahmen gebracht. Die Geigenamplituden waren also bei *2′*
und *3′* wesentlich größer als bei *1, 2, 3*)

Es wurde darauf geachtet, ob die Amplituden im Vakuum
größer waren als in Luft. Dies konnte einwandfrei nur in den
Frequenzgebieten geschehen, wo das Dekrement möglichst unabhängig von der Frequenz war. Es wurden selten Vergrößerungen
der Amplituden im Vakuum gegenüber denen in Luft von etwa $10\,^0/_0$
gemessen. Das bedingt eine Vergrößerung des Dekrements nach
den Kurven der Abb. 3 von nur $2\!-\!3\,^0/_0$. Das liegt unter der
Genauigkeit der Dekrementskurven, die in solchen Stellen etwa mit
$5\,^0/_0$ Genauigkeit zwischen den Meßpunkten durchgelegt werden
können. Die Genauigkeit in den steilen Anstiegen zu den Maxima
sowie in den Maxima selbst ist natürlich aus den oben gegebenen
Gründen weit geringer.

Es kann nun an die Auswertung der Dekrementskurven der
Abb. 4—7 gegangen werden, und zwar im Hinblick auf die Größe
des Unterschieds zwischen der Luftkurve und der Vakuumkurve bei

den einzelnen Instrumenten. Um dies übersichtlicher zu gestalten, sind nach dem Vorschlag von Prof. Krüger in Abb. 8—11 die „Nutzeffektkurven" aus den Abb. 4—7 gezeichnet. Es ist auf-

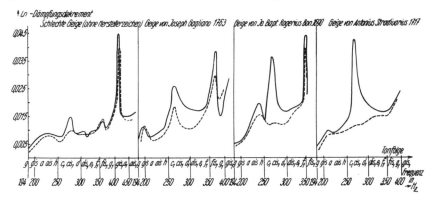

Abb. 4. g-Saite.
——— ln-Dämpfungsdekrement in Luft
– – – – ln-Dämpfungsdekrement im Vakuum

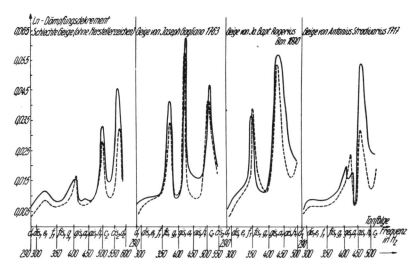

Abb. 5. d_1-Saite.
——— ln-Dämpfungsdekrement in Luft
– – – – ln-Dämpfungsdekrement im Vakuum

getragen die Differenz: ln-Dämpfungsdekrement in Luft minus ln-Dämpfungsdekrement im Vakuum in Prozent des ln-Dämpfungs-dekrements in Luft in Abhängigkeit von der Frequenz, d. h. also der akustische Nutzeffekt in Abhängigkeit von der Frequenz. Hierbei ist zu bemerken, daß in Abb. 11 die große Verschiebung des zweiten Maximums bei der schlechten Geige auf der e_2-Saite korrigiert (Verschieben des Maximums im Vakuum zur Frequenz

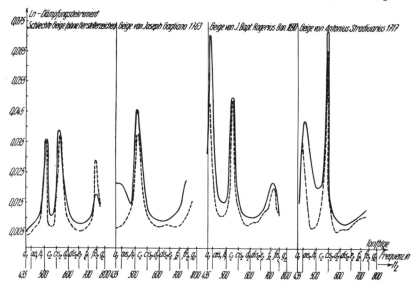

Abb. 6. a_1-Saite.
————— ln-Dämpfungsdekrement in Luft
‒ ‒ ‒ ‒ ln-Dämpfungsdekrement im Vakuum

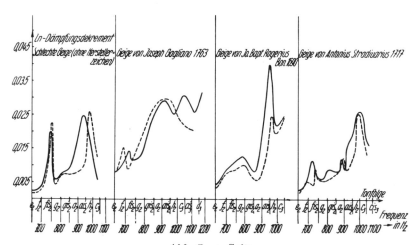

Abb. 7. e_2-Saite.
————— ln-Dämpfungsdekrement in Luft
‒ ‒ ‒ ‒ ln-Dämpfungsdekrement im Vakuum

des Maximums in Luft) ist. Es ergibt sich dadurch der gestrichelte Teil der Kurve. Es empfiehlt sich, beim Lesen dieser Nutzeffektskurven immer zu den Dekrementskurven zurückzuschlagen, da die Bestimmung des Nutzeffekts in den Resonanzstellen, vor allem aber in den steilen Anstiegen zu den Resonanzstellen, nur ungenau zu ermitteln ist.

Der Vergleich der g-Saiten (Abb. 4 und 8) zeigt auffällig Unterschiede zwischen der schlechten Geige und den guten Geigen.

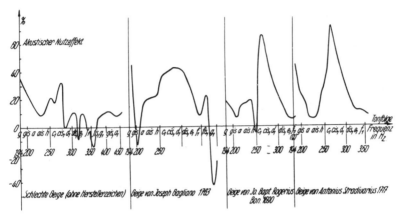

Abb. 8. g-Saite

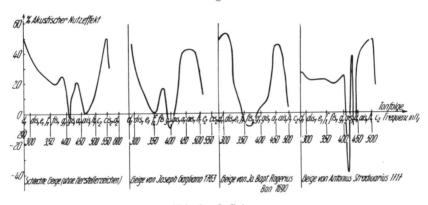

Abb. 9. d_1-Saite

Bei der schlechten Geige ist die Resonanz des Luftraumes bei cis_1 nur angedeutet, während sie bei den guten Geigen auffallend der Güte entsprechend weit größer ist. Ferner schließt sich bei den guten Geigen an diese Luftraumresonanz nach höheren Frequenzen zu ein ausgesprochen günstiges Strahlungsgebiet an, von dem bei der schlechten Geige nichts zu finden ist.

Es sollen hier nicht weiter besonders günstige Nutzeffektsstellen aufgezählt werden. Die Kurven zeigen am deutlichsten Unterschiede in Lage und Höhe von günstigen Nutzeffektsgebieten bei den verschieden guten Geigen. Man kann im allgemeinen ein Überwiegen der günstigen Gebiete bei den guten Geigen feststellen. Die e_2-Saite der Gaglianogeige ist dabei, wie oben schon gesagt, auszuschließen.

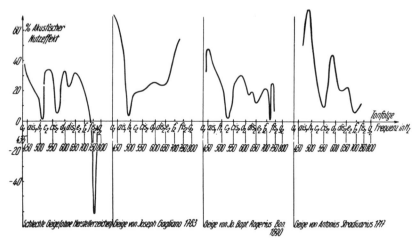

Abb. 10. a_1-Saite

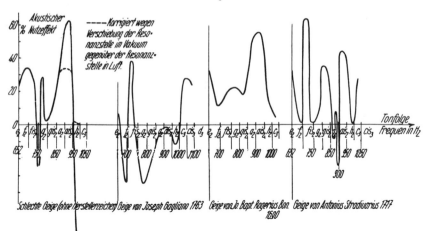

Abb. 11. e_2-Saite

Besonders erfreulich war es, daß die hier untersuchte Stradivarigeige bereits früher von Dr. H. Meinel untersucht worden ist, und zwar auf die Abhängigkeit der Schalldruckamplitude von der Frequenz [sogenannte Frequenzkurve[1])] und auf die Abhängigkeit der Lautstärke von der Frequenz [sogenannte Lautstärkekurve[1])]

---

1) Über Lautstärkekurven vgl. H. Meinel, Ztschr. f. techn. Phys. **19**. S. 302. 1938. — Über Frequenzkurven vgl. H. Meinel, Akust. Ztschr. **2**. S. 22, 62. 1937.

hin. Die Frequenzkurven geben ebenso wie die Dekrementskurven und die Nutzeffektkurven nur die Werte des Grundtones. Die Laut-

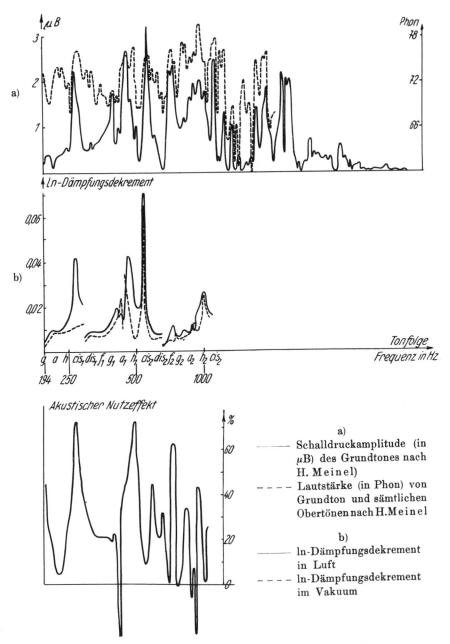

Abb. 12. Geige von Antonius Stradivarius 1717

a)
────── Schalldruckamplitude (in μB) des Grundtones nach H. Meinel)

‒ ‒ ‒ ‒ Lautstärke (in Phon) von Grundton und sämtlichen Obertönen nach H. Meinel

b)
────── ln-Dämpfungsdekrement in Luft

‒ ‒ ‒ ‒ ln-Dämpfungsdekrement im Vakuum

stärkekurve dagegen ist mit einem Verstärker aufgenommen, der
gehörähnlich arbeitet, und umfaßt auch die Obertöne mit. Die
sämtlichen Kurven (d. h. also Frequenz-, Lautstärke-, Dekrement-
und Nutzeffektkurven) sind in Abb. 12 (S. 189) übereinander gezeichnet. Die Kurven sind hier so gezeichnet, daß beim Erreichen
des Tones der nächstfolgenden bloßen Saite auf dieser fortgeschritten
wurde. Es zeigt sich, daß die hier gefundenen günstigen Ab-
strahlungsgebiete ebenfalls mit Lautstärkemaxima zusammenfallen,
so oberhalb der Luftraumresonanz bei cis_1, dann bei h_1, d_2, über
dis_2, zwischen f_2 und fis_2, zwischen g_2 und gis_2, zwischen ais_2 und h_2.
Diese Maxima der Lautstärke fallen zum Teil nicht mit Maxima
der Frequenzkurve zusammen. Sie sind nach H. Meinel[1]) bedingt
durch die größeren Obertonamplituden. Es liegt aber wohl auf der
Hand, daß hier der günstige Nutzeffekt ein große Rolle mit spielt.
Man kann überhaupt annehmen, daß in ähnlichem Maße, wie die
Verteilung der Größe der Schalldruckamplituden über die Frequenzskala für die Güte einer Geige von Bedeutung ist (dies ist von
Meinel in seiner ganzen Bedeutung klargestellt worden), die Verteilung der Größe des Nutzeffekts über die Frequenzskala ebenfalls
für die Güte einer Geige wichtig ist. Hier im einzelnen weitere
Parallelen als die eben bei der Stradivari gezeigten zu verfolgen,
ist jedoch nicht der Zweck dieser Arbeit. Es sei nur noch folgendes
erwähnt. Eine Reihe der guten Nutzeffektstellen findet sich bei
den drei guten Geigen bei derselben Frequenz, z. B. eine solche
Stelle auf der d_1-Saite bei ais_1. Bei der schlechten Geige ist bei
dieser Frequenz jedoch ein ganz geringer Nutzeffekt vorhanden.
Die Übereinstimmung der drei guten Geigen ist bestimmt kein
Zufall, zumal Gagliano ganz, Rogeri zum Teil in Stradivaris
Fußspuren gewandelt sind.

Tabelle 1
Durchschnittliche Nutzeffekte

	Schlechte Geige in %	Gagliano- geige in %	Rogerigeige in %	Stradivari- geige in %
g-Saite	15,2	27,2	21,6	26,5
d_1-Saite	20,9	21,7	21,6	22,4
a_1-Saite	24,1	33,8	23,8	30,0
e_2-Saite	18,0	—	27,0	24,5
Allgem. Durchschnitt für alle 4 Saiten	19,55	—	23,5	25,85
Allgem. Durchschnitt für die 3 tieferen Saiten	20,07	27,9	22,33	26,3

---

1) H. Meinel, Ztschr. f. techn. Phys. **19**. S. 302. 1938.

Hier sollen nun in Tab. 1 die durchschnittlichen Nutzeffekte der einzelnen Geigen angegeben werden. Diese Bestimmung der durchschnittlichen Nutzeffekte war nur möglich unter bestimmten Einschränkungen. Aus den Nutzeffektskurven (Abb. 8—11) wurde für jeden musikalisch gebrauchten Ton, also g, gis, a usw. der Nutzeffekt abgelesen. War dieser infolge der oben erwähnten Unregelmäßigkeiten negativ, so wurde er als klein angenommen und gleich Null gesetzt. Da sich solche Stellen bei allen Geigen finden, so haben diese Fehler nur geringen Einfluß, da es sich ja hier immer um Vergleiche handelt.

Die schlechte Geige hat also auf allen Saiten (mit Ausnahme der a_1-Saite der Rogerigeige, auf der eigenartigerweise auch beim Anhören schwache und unangenehme Töne auffielen, im Gegensatz zu den anderen Saiten dieses Instrumentes) und auch im ganzen einen geringeren durchschnittlichen Nutzeffekt als die guten Geigen. Nutzeffektmessungen (d. h. allerdings nur Werte, die dem Nutzeffekt proportional sind) an guten und schlechten Geigen gibt auch Saunders[1]) an. Saunders schreibt, daß bei einer sehr schlechten und einer sehr guten Geige die Nutzeffekte nur um $10^0/_0$ differierten, während die Preise im Verhältnis $1:600$ standen. Aus den hier angegebenen Werten der Tab. 1 ergibt sich, daß der Nutzeffekt der Stradivarigeige $130^0/_0$ des Nutzeffektes der schlechten Geige beträgt, also um $30^0/_0$ größer ist. Auffällig ist, daß die Gaglianogeige einen noch etwas größeren Nutzeffekt hat als die Stradivarigeige. Dagegen liegt der Nutzeffekt der Rogerigeige, die im Preise zwischen Stradivarigeige und Gaglianogeige steht, unter diesen beiden. Saunders fand, daß bei ein und demselben Instrument der Nutzeffekt bei verschiedenen Tönen nur um den Faktor 2—3 schwankt. Aus den hier gezeigten Nutzeffektskurven (Abb. 8—11) ergeben sich dagegen Schwankungen bis zum Faktor 7. Saunders räumt auf Grund seiner Messungen dem Nutzeffekt keine Bedeutung für die Güte der Geigen ein. Die hier angegebenen Messungen zeigen aber offenbar das Gegenteil, daß nämlich der Nutzeffekt von Bedeutung für die Güte der Geigen sein muß. Bei den Geigen sprechen eben zu viele Dinge mit. Ist der Nutzeffekt bei der einen Geige vielleicht nicht so wesentlich, da vielleicht im allgemeinen stärkere Amplituden vorliegen, so kann er bei einer anderen Geige mit schwächeren Amplituden wesentlich sein. Ganz eindeutige Ergebnisse sind ja überhaupt bei Geigen bisher nur sehr wenige gefunden worden. Dies ist auch bei einem so komplizierten Gebilde kaum anders zu erwarten.

---

1) F. A. Saunders, Journ. of the acoust. Soc. of America **9**. S. 81. 1937.

Man kann hier auf Grund dieser Messungen wohl annehmen, daß der günstigere Nutzeffekt auch ein Grund dafür ist, daß gute italienische Geigen im Freien viel weiter zu hören sind (bis zu 1 km) als andere Geigen. Messungen hierüber hat C. Metzner[1]) ausgeführt.

Nachdem sich so gezeigt hat, daß der Nutzeffekt der Geigen sicher nicht zu vernachlässigen ist, soll nun die Größe der inneren Reibung der Geigen betrachtet werden. Wie schon weiter oben einmal gesagt wurde, sind die Maxima der Dekrementskurven der guten Geigen meist höher als die der schlechten Geigen. Und dies gilt auch ganz allgemein für die ganzen Dekrementskurven. Es sind daher wieder die Durchschnittsdekremente bestimmt worden in ähnlicher Weise, wie oben die Durchschnittsnutzeffekte. Für jeden musikalisch gebrauchten Ton wurde der Wert des Dekrements der inneren Reibung (im Vakuum) an den gestrichelten Kurven der Abb. 4—7 abgelesen. Dann wurde über die einzelnen Saiten, über die einzelnen Geigen gemittelt. Ferner wurden bestimmt die tiefsten Dekremente auf jeder Saite, der Durchschnitt davon für jede Geige und schließlich der absolut tiefste Wert für jede Geige. Die Werte sind in Tabelle 2 und 3 zusammengestellt.

Dabei zeigt sich ganz auffällig, daß die schlechte Geige immer das geringste Dekrement hat. Das höchste Dekrement hat fast immer die Gaglianogeige, seltener die Rogerigeige. Die Werte der Stradivarigeige sind meistens wieder tiefer, erreichen allerdings nur selten die Tiefe der schlechten Geige. Daß die guten Geigen aus Holz gebaut sind, das viel Energie im Innern verschluckt, wird man kaum behaupten wollen. Im Gegenteil, viel näherliegend ist es, anzunehmen, daß das Holz der guten Geigen eine möglichst geringe innere Reibung hat. Es liegt daher nahe, den Grund für die höheren

Tabelle 2
Durchschnittliche ln-Dämpfungsdekremente (Vakuum)

	Schlechte Geige	Gagliano-geige	Rogerigeige	Stradivari-geige
g-Saite	0,0077	0,0111	0,0126	0,0105
d_1-Saite	0,0088	0,0117	0,0174	0,0121
a_1-Saite	0,0098	0,0118	0,0174	0,0136
e_2-Saite	0.0094	0,0207	0,0109	0,0093
Allgem. Durchschnitt	0,0089	0,0138	0,0146	0,0114

1) C. Metzner, Kunst und Wissenschaft im Geigenbau, 1920. Verlag Georg Bratfisch, Frankfurt a. O. Die von C. Metzner gegebene Erklärung für die genannte Erscheinung widerspricht dem Superpositionsprinzip.

Tabelle 3
Tiefste ln-Dämpfungsdekremente (Vakuum)
(Dies sind immer die Werte der bloßen Saiten bis auf die a_1-Saiten
der Rogeri- und Stradivarigeige)

	Schlechte Geige	Gagliano-geige	Rogerigeige	Stradivari-geige
g-Saite	0,0025	0,0055	0.0042	0,0030
d_1-Saite	0,0035	0,0045	0,0032	0,0047
a_1-Saite	0,0038	0,0065	0,0075	0,0055
e_2-Saite	0,0020	0,0080	0,0030	0,0020
Durchschnitt	0,0030	0,0061	0,0045	0,0038
Absolut tiefste Werte	0,0020	0,0045	0,0032	0,0020

Dekremente bei den guten Geigen in irgendeiner Eigenart des Baues
zu suchen, die die Dekremente erhöht. Diese Eigenart wurde tatsächlich gefunden. Einer der besten Geigenbauer des vorigen Jahr-
hunderts, August Riechers, der Freund Joachims, schreibt in
seinem Buch[1]), daß Stradivari die Zargen vom unteren Klotz bis
zum oberen Eckklotz genau auf 30 mm Höhe abgerichtet hat. Vom
oberen Eckklotz bis zum Oberklotz nimmt die Höhe der Zargen
dann aber um $2^1/_2$ mm, also auf $27^1/_2$ mm, ab. Die Decke erhält
dadurch eine Spannung. Ferner: der Baßbalken wurde so geschnitten, daß er, lose auf die Decke gelegt, an jedem Ende 2 mm
von der Decke absteht. Er wird dann unter Druck an die Decke
angeleimt. Auch hierdurch kommt Spannung in Decke und Balken.
Diese Spannungen sind natürlich nicht zu verwechseln mit der
Wölbung, die ja durch Schnitzen erzeugt wird. Schließlich schreibt
Riechers, daß der Ton hell und scharf wird, wenn man den Balken
zu stark spannt; bei geringerer Spannung wird der Ton weich und
voll. An den Angaben Riechers ist nicht zu zweifeln. Er gilt
als einer der erfahrensten Kenner der Stradivarigeigen. Seine
eigenen Instrumente sind sehr gesucht. — Auf Grund dieser Be-
merkungen ist hier das Dekrement von Fichtenholzstäben (wie in I:
4 mm stark, 20 mm breit) unter transversaler Verspannung gemessen
worden. Der Stab wurde zunächst ohne Verspannung an beiden
Enden auf einem großen Metallblock fest eingeklemmt. Das Dekrement, nach der ersten Methode von I gemessen, ergab den Wert
0,024 (auf die Amplitude Null extrapoliert). Der Wert stimmt mit
dem Fichtenholzwert aus I genau überein, was zu erwarten war,
da das Holz aus demselben Bestand war. Dann wurde der Stab

---

1) August Riechers, Die Geige und ihr Bau, 6. Aufl. Berlin 1922 bei
Franz Wunder.

so eingeklemmt, daß dabei die Mitte etwas hochgewölbt, der Stab also transversal verspannt war. Nun war das Dekrement je nach der Stärke der Spannung mehr oder weniger erhöht. Die Abb. 13 zeigt drei so erhaltene Kurven.

Das Ergebnis der Abb. 13 theoretisch zu erklären, ist außerordentlich schwierig. Zur Zeit ist keine einwandfreie Theorie der inneren Reibung fester Körper vorhanden. Noch viel weniger ist dies der Fall bei verspannten Körpern. Immerhin erscheint es vielleicht einleuchtend, daß eine Verspannung eine Unordnung in das Gefüge des Körpers bringt, und daß dadurch eine Erhöhung der inneren Reibung eintritt.

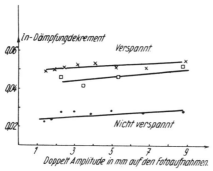

Abb. 13. ln-Dämpfungsdekremente transversal verspannter Fichtenholzstäbe

Das experimentelle Ergebnis, das in Abb. 13 dargestellt ist, zeigt also, daß die oben beschriebenen Spannungen in der Geige das Dekrement erhöhen müssen. Die hier gemessene schlechte Geige hat überall gleich hohe Zargen. Dies war auch bei den anderen schlechten Geigen, die zur Verfügung standen, der Fall. Die Deckenspannung fällt also fort. Ob der Balken gespannt ist, ist unbekannt, aber kaum anzunehmen, denn man wird sich mit billigen Instrumenten wohl kaum so viel Mühe machen. Wie stark die Spannung bei den guten Geigen ist, ist natürlich auch nicht bekannt. Es liegt aber nahe, zu schließen, daß das Dekrement der guten Geigen ohne Spannung tiefer liegen würde, als das der schlechten Geigen, zumal sich die gemessenen Dekremente der besten Geige, der Stradivarigeige, wieder erheblich den Werten der schlechten Geige nähern. Durch gar nicht besonders starke Verspannung wurde das Dekrement des Fichtenstabes verdoppelt (vgl. Abb. 13). Es ist also nicht ausgeschlossen, daß die Stradivarigeige ohne Spannung ein Dekrement zeigen würde, daß etwa halb so groß wäre wie das hier gemessene, und das etwa ein Drittel kleiner wäre als das der schlechten Geige. Danach müßte das Holz, aus dem die Stradivarigeige gebaut ist, Dekremente zeigen, die etwa ein Drittel kleiner sind als die Dekremente, die in I für gewöhnliche Fichtenholzstäbe gefunden wurden. Es sei hier noch einmal, wie schon weiter oben gesagt, daß Gagliano ganz, Rogeri zum Teil in Stradivaris Fußspuren gewandelt sind. Die Spannungsverhältnisse werden also wohl bei ihnen ähnlich sein wie bei Stradivari.

Schon vor vielen Jahren hat der Geigenbauer Prof. Koch[1]) in Dresden die Meinung vertreten, daß die alten Italiener ihre Geigen mit irgendwelchen Substanzen tränkten, um dadurch den Splint (weiche Teile des Jahreszuwachses der Bäume) dem Jahr (harte Teile des Jahreszuwachses der Bäume) anzupassen. Kochs Geigen, die nach diesem Prinzip gebaut wurden, haben sich Weltruhm erworben und sind oft in verdeckten Probespielen neben und sogar über alte italienische Geigen gestellt worden. Nichts liegt näher als anzunehmen, daß die Anpassung des weichen Splints an das härtere Jahr die Dämpfung herabsetzt. Dies wird auch bestätigt durch die Strukturuntersuchungen an Geigen mit Röntgenstrahlen[2]): Gute alte italienische Geigen zeigen im Holz geringere Faserstruktur als neue Geigen. Durch die so erzielte geringere innere Reibung hat der Geigenbauer es in der Hand, der Geige Spannungen zu geben. Diese Spannungen wirken sich auf den Ton entscheidend aus, wie Riechers (a. a. O.) schreibt. Ist dagegen die innere Reibung des Holzes hoch, so kann durch Spannungen die Dämpfung der Geige leicht zu hoch werden, so daß die Geige zu viel Energie verschluckt. Daß die Spannung den Ton entscheidend beeinflußt, ist erklärlich, denn durch bestimmte Spannungen der strahlenden Platten werden sicher bestimmte Obertöne gekräftigt, andere wieder geschwächt. Man wird also durch Spannungen den Ton der Geige genau so beeinflussen können, wie man es durch Veränderungen der Dickenabmessungen der Platten tun kann, welch letzteres Meinel (a. a. O.) mit großer Präzision durchgeführt hat. Diese Beeinflussung des Tones durch die Spannung wird um so verständlicher, als es hervorragende alte italienische Geigen (z. B. von Guarneri d. G.) gibt, die den normalen Dickenverhältnissen widersprechen.

Es ist also nach all diesem klar, daß für den Geigenbau die Verwendung von Holz mit möglichst geringer innerer Reibung von der größten Wichtigkeit ist.

Schließlich sei noch folgendes erwähnt. Ein eigenartiges Ergebnis lieferten die Aufnahmen mit schlechten (*nicht silber*umsponnenen) g-Saiten auf der schlechten Geige. Abb. 14 zeigt dies. Das Dekrement im Vakuum liegt überall über dem Dekrement in Luft! Ähnlich verhielt sich eine andere g-Saite von derselben Sorte auf einer anderen schlechten Geige. Nur in der Luftraumresonanzstelle lag bei letzterer

---

1) F. J. Koch, Ztschr. f. Instrumentenbau 1915. Nr. 32—33. S. 34.
2) K. Lark-Horovitz u. W. I. Caldwell, Naturwiss. 1934. S. 450f.

13*

Geige das Luftdekrement über dem Vakuumdekrement. Der Grund für dieses Verhalten ist nur in der Saite zu suchen, da eine gute silberumsponnene Saite auf derselben Geige dies nicht zeigte (vgl. Abb. 4). Über das eben Gesagte hinaus zeigte sich bei der schlechten g-Saite noch folgendes: Ließ man nach einer Vakuumaufnahme wieder Luft in die Glasglocke, so stieg das Dekrement noch weiter über den Vakuumswert an und fiel erst allmählich in etwa 30 Min. auf seinen normalen Luftwert zurück. Damit parallel laufend wurde folgendes beobachtet: Nach der Evakuierung war der Umwicklungsdraht auf der schlechten Saite so locker, daß man ihn ganz leicht auf der Darmseele verschieben konnte. Erst allmählich wurde er wieder fest. Die Erhöhung der Dämpfung im Vakuum ist also klar: Der Umwicklungsdraht hängt als Ballast an der Darmseele und bedingt vor allem auf Steg und Sattel ein zusätzliches Dekrement, da die Schwingungen der Saite durch das Rutschen des Drahtes auf Steg und Sattel gedämpft werden. Die Erhöhung des Geigendekrements beim Wiedereinlassen der Luft in die Glasglocke ist dann also durch die nun hinzukommende Strahlungsdämpfung der Geige bedingt. Der allmähliche Rückgang des Dekrements zu kleineren Werten ist dann abhängig von der allmählichen Festigung der Wicklung. Die gute Saite zeigte niemals ein Lockerwerden der Wicklung. Der Grund für das Lockerwerden der Wicklung der schlechten Saite ist wohl der, daß die Darmseele durch Luft und Feuchtigkeit quillt. Wurde eine Schale mit Wasser in die Glasglocke gestellt, so war die Lockerung der Drahtumwicklung etwas geringer. Die Darmseele der guten Saite ist also offenbar weit unabhängig von Luft- und Feuchtigkeitseinflüssen. — Es wurden noch statische Durchbiegungsversuche an diesen Saiten gemacht. Die

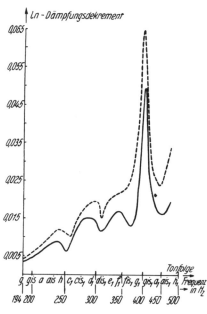

Abb. 14. Schlechte, *nicht silberdraht*umsponnene g-Saite auf schlechter Geige.
——— ln-Dämpfungsdekrement in Luft
– – – – ln-Dämpfungsdekrement im Vakuum

schlechten Saiten (es wurden mehrere dieser Sorte untersucht) zeigten schwankende Effekte. Manchmal wurde die Durchbiegung im Vakuum geringer, manchmal stärker. Der oben beschriebene Dekrementseffekt war aber ständig vorhanden, genau so wie die Lockerung der Wicklung. Die gute Saite zeigte auch Durchbiegungsschwankungen, jedoch geringer als die schlechte Saite. Der Grund für den Dekrementseffekt der schlechten Saiten ist also sicher die Lockerung der Wicklung im Vakuum.

Zusammenfassung

Von Geigentönen wurde das Dämpfungsdekrement des Grundtones in Luft und im Vakuum bestimmt. Untersucht wurden 4 Geigen von verschiedener Klanggüte: eine moderne, schlechtklingende Geige und drei alte italienische, gutklingende Geigen, und zwar eine Gagliano-, Rogeri- und eine Stradivarigeige. Auf jeder Geigensaite wurde ein Tonbereich von etwa einer Oktave untersucht. Der akustische Nutzeffekt, aus der Differenz der Dekremente in Luft und im Vakuum bestimmt, erwies sich bei den guten alten italienischen Geigen im Durchschnitt um etwa 30 % höher, als bei einer schlechten Geige, was sicher für die Fülle des Tones der italienischen Geigen und für seine Reichweite wichtig ist. Ferner zeigten sich ganz bestimmte Gebiete mit gutem Nutzeffekt, die bei den guten alten italienischen Geigen (sie waren alle aus gleicher Schule) auffallende Ähnlichkeit in Ausdehnung und Frequenzlage hatten. Diese Gebiete mit gutem Nutzeffekt stimmen mit Lautstärkemaxima überein. Die Verteilung der günstigen Nutzeffektsgebiete über die Frequenzskala ist also sicher von Bedeutung für die Klangfarbe der Geige. Es wurde ferner gezeigt, daß in den Messungen das Dekrement der guten italienischen Geigen wahrscheinlich nur deshalb im Durchschnitt höher erscheint als das der schlechten Geige, weil die guten italienischen Instrumente Spannungen haben, die von den Erbauern bewußt herbeigeführt worden sind. Messungen an transversal verspannten Stäben bewiesen dies, denn das Dekrement solcher Stäbe war erheblich erhöht gegenüber dem der nicht verspannten Stäbe. Es ist daher klar, daß das Geigenholz an sich eine möglichst geringe innere Reibung haben muß, um solche Verspannungen, die für die Klangfarbe sehr wichtig sind, möglich zu machen, ohne daß das Dekrement der Geige dann zu hoch wird. Die Italiener sollen ihre Instrumente mit irgendeiner Substanz getränkt haben, ein Verfahren, das Prof. Koch in Dresden seit langer Zeit erkannt hat und mit großem Erfolge anwendet. Es ist nun in der Durchtränkung des Holzes wahrscheinlich das Verfahren zu sehen, das die Dämpfung des Holzes vermindert. Es ist also die Schlußfolgerung zu ziehen, daß eine wichtige Voraussetzung für den Bau einer guten Geige in der Verwendung von Holz mit möglichst geringer innerer Reibung liegt.

Schließlich wurde gezeigt, daß schlechte, nicht silberdraht-umsponnene g-Saiten im Vakuum besonderes Verhalten zeigen, das darauf beruht, daß die Umwicklung im Vakuum locker wird.

Meinem hochverehrten Lehrer, Herrn Prof. Dr. F. Krüger †, danke ich für die Anregung zu dieser Arbeit und für die stete Förderung derselben.

Herrn Prof. Dr. O. Reinkober danke ich für Diskussion.

Herrn Dr. H. Meinel, Berlin, danke ich für die Überlassung der Frequenz- und Lautstärkekurven der Stradivarigeige.

Ferner ist zu danken für die leihweise Überlassung von Instrumenten und Apparaten: Frl. Pernice, Greifswald, für die Gaglianogeige, der Firma Emil Herrmann, Berlin, die in außerordentlich entgegenkommender Weise die wertvolle Rogeri- und Stradivarigeige Herrn Prof. Krüger zur Verfügung stellte, der Firma Siemens & Halske für Stromreiniger und Kondensatorleitungen und der Deutschen Forschungsgemeinschaft für einen Oszillographen. Schließlich danke ich dem Feinmechaniker Herrn Kurt Schweden für die Konstruktion der automatischen Anstreichvorrichtung.

Greifswald, Physikalisches Institut der Universität.

(Eingegangen 9. Juni 1940)

21

Internal Friction and Damping of Violin Vibrations

E. ROHLOFF

ABSTRACT by John C. Schelleng

This paper describes an experimental comparison between a bad violin of modern construction and three old instruments of fine tone, with practical deductions for construction.

The purpose of this work was to measure the logarithmic decrement of a violin (that is, of the strings as normally mounted thereon) in such a way as to separate the decrement caused by radiation from that of internal friction in the body of the instrument, thus measuring radiation efficiency. An apparatus was assembled by which a violin could be "fingered" and "bowed," and the "bow" then lifted to allow the vibration to subside—all by remote control in a vacuum. The decrement was calculated from measurements of the decay time to one tenth amplitude. Four violins, as described, were measured in this way through the first octave on each of the four strings. Efficiency of radiation was found to be about 30 percent higher in the good instruments than in the bad one. The author believes that the reason for this difference is that good Italian instruments have intentionally been assembled so that the wood contains stresses. Measurements on transversely stressed rods yielded greater decrements than those not stressed. It follows that violin wood should have an inner friction as low as possible to avoid decrements that are too high when these stresses, which are very important for tone quality, are added. Italian makers apparently soaked their wood with some kind of substance, a procedure successfully exploited by Professor Koch in Dresden. Such treatment probably decreases damping in the wood.

Editor's Comments on Paper 22

22   Barducci and Pasqualini: *Misura dell'attrito interno e delle costanti elastiche del legno*
English translation: *Measurement of the Internal Friction and the Elastic Constants of Wood*

The scientific work of Professor Italo Barducci (1917–   ) has ranged through many different fields in acoustics—telecommunication, molecular acoustics, architectural acoustics, psychoacoustics, and musical acoustics. Since 1943 he has carried on his investigations at the Istituto Nazionale di Elettroacustica, now the Istituto di Acustica "O.M. Corbino" of the Italian National Research Council (Consiglio Nazionale delle Ricerche). His findings are contained in sixty-five papers. For 9 years he taught electroacoustics and is presently conducting the course in Technical Physics for Engineers at the University of Rome.

Professor Gioacchino Pasqualini (1902–   ), a concert violinist from an early age, was also trained in mathematics and acoustics, receiving his diploma from the University of Rome in 1938. He has maintained a lifelong interest in the art and science of violin making as well as in the study of ancient instruments. As sponsor and contributor to many national and international conferences in violin acoustics and violin making, he has done much to engender the revival of violin making in Italy, and was a prime influence in the reestablishment of the Cremona Violin Making School. His research in electroacoustics applied to the resonances of violins and his extensive correspondence on the subject have brought Pasqualini a worldwide reputation. In 1962 Pasqualini presented his extensive collection of stringed instruments, ancient and modern, to the Saint Cecilia Academy of Rome in memory of his son.

The joint paper by Barducci and Pasqualini included here represents the most thorough and extensive study of wood for violins that is known. It is reproduced in the original Italian, with an English translation by Elizabeth B. Abetti. Giving the elastic constants and internal damping characteristics of over eighty species of wood —with both Latin and common names—this work represents the standard reference in the field. Since the intent of the authors was to procure statistical data on a large number of samples rather than to make precise measurements on a small number of wood strips, there were no systematic tests on the effects of humidity—although the work was done under reasonably comparable conditions.

With this wealth of information as a guideline it is hoped that a thorough study of a few of the most promising species will be made in relation to humidity, length of seasoning, dimensional changes, and acoustical properties, using the type of dynamic test developed by Barducci and Pasqualini.

Additional Writings by Gioacchino Pasqualini

The following bibliography is reproduced from Lina Gabrielli, *Gioacchino Pasqualini* (Tipolitográfica Ed., Ascoli P., 1954), pp. 12–13.

«*Proprietà del corpo di risonanza degli strumenti ad arco rivelate con metodi elettroacustici*». (Annuario dell'Accademia di Santa Cecilia - Roma, 1938 - 1939).

«*L'Elettroacustica applicata alla Liuteria*». (Annuario dell'Accademia di Santa Cecilia - Roma, 1938 - 1939).

(N. 20 della Raccolta Scientifica dell'Istituto Nazionale di Elettroacustica - Roma, 1940).

«*Relazionè sulle prove eseguite presso l'Istituto Nazionale di Elettroacustica per addivenire ad una valutazione obbiettiva sulle qualità acustiche di alcuni violini*». («Ricerca Scientifica» del C. N. R. N. 9 , Roma, 9 Settembre 1940, pag. 622).

(N. 22 della Raccolta Scientifica dell'Istituto Nazionale di Elettroacustica - Roma, 1940).

«*Nuovi risultati conseguiti nello studio della cassa armonica dei violini con metodi elettroacustici*». («Ricerca Scientifica» del C.N.R. - N. 2, Roma, Febbraio 1943, pag. 111).

(N. 44 della Raccolta Scientifica dell'Istituto Nazionale di Elettroacustica - Roma, 1943).

In collaborazione con l'Ing. Italo Barducci, Ricercatore presso l'Istituto Nazionale di Ultracustica:

«*Misura dell'attrito interno e delle costanti elastiche del legno*». («Il Nuovo Cimento» - Vol. V, N. 5, pag. 416 - Bologna, 1948).

(N. 87 della Raccolta Scientifica dell'Istituto Nazionale di Elettroacustica - Roma, 1948).

In collaborazione con l'Ing. Hermann Briner dell'Istituto di Fisica dell'Università di Friburgo (Svizzera):

«*Nota circa gli effetti dell'assorbimento ambientale sul suono dei violini antichi e moderni*». («Rivista Musicale Italiana» - Anno LII, fasc. II - Milano, 1950).

Pubblicazioni recenti:

«*La questione del diapason. Necessità di definire la frequenza della nota d'accordo*». (Rivista «Santa Cecilia» dell'Accademia Nazionale di Santa Cecilia - N. 1, Roma, Aprile 1953).

«*L'electro-acoustique appliquée à la lutherie*». (Relazione presentata al I Congresso Internazionale di Liuteria - Milano, 20 Maggio 1953 - pubblicata a cura della Camera Industria e Commercio di Milano).

«*Récents résultats obtenus dans l'étude éléctroacoustique de la caisse harmonique des instruments à archet*». (Rivista International Journal on Acoustics «Acustica» - Anno IV, N. 1, Zurigo, Gennaio 1954).

(«Proceedings of the First Ica Congress Electro-Acoustics». Edito da C. W. Kosten and Kasteleyn. Ed. W. D. Meinema, Ldt. Delft (Olanda), 1954).

22

Reprinted from *Il Nuovo Cimento*, **5**(5), 416–466 (1948)

Misura dell'attrito interno e delle costanti elastiche del legno.*

I. Barducci e G. Pasqualini

Istituto Nazionale di Elettroacustica «O. M. Corbino»

(ricevuto il 13 Luglio 1948)

Riassunto. — Le costanti elastiche e la dissipazione interna di energia nel legno sono state misurate con un metodo elettroacustico basato sulla misura delle frequenze e dei coefficienti di risonanza. Le principali conclusioni cui si è giunti dall'esame dei risultati sperimentali sono: *a)* la velocità di propagazione *c* delle onde elastiche longitudinali ed il coefficiente di risonanza *Q* variano fortemente con la direzione di propagazione; ma il loro rapporto *c/Q* è pressochè indipendente dalla direzione, oltrechè dalla frequenza; *b)* l'angolo fra la direzione di propagazione e quella delle fibre è, in prima approssimazione, sufficiente a determinare l'orientazione della sbarretta in prova, per quello che riguarda la misura di *c* e di *Q*; *c)* fra le proprietà meccaniche delle essenze resinose sono state riscontrate regolarità evidenti: leggi semplici dello stesso tipo non sono state riscontrate per le latifoglie; *d)* le costanti elastiche ed il coefficiente di risonanza dei legni molto antichi sembrano notevolmente modificati da un processo di invecchiamento. Si riportano infine alcune tabelle di valori medi delle principali costanti elastiche e dei coefficienti di risonanza per le 85 specie legnose provate.

1. – Premessa.

La presente ricerca sperimentale sulle proprietà elastiche e sull'attrito interno del legno ha avuto un duplice scopo.

In primo luogo uno degli A., nelle sue ricerche con metodi elettroacustici sulle qualità acustiche dei violini (¹), ha avuto modo di accertare l'utilità di prove tendenti ad analizzare il comportamento di tali strumenti sulle singole parti componenti. Era quindi naturale pensare di completare queste ricerche con prove sulle proprietà meccaniche del legno, cioè del meteriale costituente gli strumenti stessi (²).

(¹) G. Pasqualini: *Annuario Acc. S. Cecilia* (1938-39); *Ricerca Scientifica*, **11**, 622 (1940); **14**, 111 (1943).

(²) Ad analoga conclusione è giunto F. A. Saunders, in un lavoro pubblicato da *J.A.S.A.*, **17**, 169 (1946) quando la presente ricerca era in corso.

Editor's Note: An English translation follows this article.

In secondo luogo è sembrato che la determinazione sistematica dei parametri elastici e dell'attrito interno su varie specie di legno potesse presentare un interesse alquanto più vasto di quello legato alle applicazioni di acustica musicale. Una ricerca di questo genere, infatti, offre possibilità di trarre qualche indicazione più precisa di quelle esistenti su di un materiale, quale è il legno, tipicamente anisotropo ed a struttura fibrosa.

È vero infatti che si hanno da tempo diversi dati sulle costanti elastiche dei legnami e che recentemente se ne sono aggiunti altri sull'attrito interno. Volendo, però, cercare di giungere a qualche conclusione un po' generale, è evidentemente necessario che le ricerche siano eseguite in condizioni, per quanto è possibile, confrontabili e sopra campioni le cui caratteristiche generali (specie, provenienza, direzione delle fibre, ecc.) siano bene conosciute.

Il metodo sperimentale adoperato deriva direttamente da quello, recentemente attuato da P. G. BORDONI [3] fondato sulla misura delle frequenze e dei coefficienti di risonanza (o dei decrementi logaritmici) di sbarrette vibranti, costituite dal materiale in esame.

Alcune modificazioni sono state tuttavia suggerite dalle esigenze particolari della ricerca, e principalmente dalla necessità di operare su un materiale, come il legno, a smorzamento relativamente elevato e sul quale non è possibile lavorare *sotto vuoto* per determinare l'attrito interno senza alterarne le caratteristiche in modo non ben precisabile.

Nei paragrafi seguenti, dopo una breve descrizione del metodo di misura ed un cenno ad alcune conclusioni che si possono trarre dai risultati ottenuti da altri Autori, si discuteranno i nuovi risultati raggiunti, deducendone alcune conclusioni di carattere generale sul comportamento del legno, dalle quali potrebbero anche essere tratti suggerimenti per ulteriori ricerche.

2. – Fondamenti del metodo di misura.

È noto che ponendo in vibrazione forzata per flessione una sbarretta del materiale in esame, libera da ogni vincolo, e determinando una delle sue frequenze di risonanza, si può risalire alla velocità di propagazione c delle onde elastiche longitudinali in una sbarra (infinitamente sottile) dello stesso materiale [4].

Se la sbarretta ha sezione rettangolare vale allora la relazione:

$$(1) \qquad c = a_n \frac{l^2}{h} f_n \quad (\text{cm} \cdot \text{sec}^{-1})$$

essendo: f_n (Hz) la n^{ma} autofrequenza flessionale della sbarretta; h (cm) l'al-

---

[3] P. G. BORDONI: *Nuovo Cimento*, **4**, 177 (1947).
[4] La c non è ovviamente da confondere con la velocità di propagazione delle stesse onde longitudinali in un solido indefinito.

tezza (vedi fig. 4) della sezione rettangolare della sbarretta; l (cm) la lunghezza della sbarretta ed a_n un coefficiente numerico dipendente dall'ordine n dell'autofrequenza.

I valori di a_n per i primi modi sono: $a_0 = 0,975$ per il modo fondamentale (con due sezioni nodali distanti $0,224\,l$ (cm) dagli estremi della sbarretta); $a_1 = 0,353$ per il primo modo superiore (con tre sezioni nodali; una centrale e due poste a $0,132\,l$ dagli estremi); $a_2 = 0,180$ per il secondo modo superiore (con quattro sezioni nodali poste a $0,094\,l$ ed a $0,356\,l$ dagli estremi).

Nota c e la densità ρ del materiale, si calcola immediatamente il modulo di YOUNG E, essendo:

$$(2) \qquad\qquad E = \rho c^2 \,.$$

Per caratterizzare l'attrito interno, si è scelto il coefficiente di risonanza Q delle sbarrette che, per i materiali a smorzamento abbastanza forte, come il legno, è di più facile misura del decremento logaritmico d, al quale d'altronde è legato dalla relazione semplicissima: $Q = \pi/d$.

Se f_n è una delle frequenze di risonanza della sbarretta si ha, come è noto:

$$(3) \qquad\qquad Q_n = \frac{f_n}{(\Delta f)_n}$$

dove: $(\Delta f)_n$ è l'intervallo di frequenza fra i due punti della curva di risonanza (relativa ad f_n) per i quali l'ampiezza di vibrazione è eguale ad $1/\sqrt{2}$ del valore massimo che si ha in corrispondenza ad f_n.

3. – Disposizione sperimentale.

In fig. 1 è rappresentato lo schema elettrico generale dell'apparecchiatura impiegata per la misura delle frequenze di risonanza f_n e degli intervalli di frequenza $(\Delta f)_n$. Come si vede, per eccitare le vibrazioni delle sbarrette ci si è serviti dell'attrazione elettrostatica e per la rivelazione delle vibrazioni ci si serviti della modulazione di frequenza, secondo il metodo con un solo elettrodo ausiliario, senza polarizzazione continua, descritto nel lavoro già citato [3] al quale si rimanda per una descrizione più diffusa.

Ci limitiamo qui ad alcuni cenni sulle modificazioni introdotte nell'apparato sperimentale, per adattarlo alle esigenze particolari della ricerca.

In primo luogo è stato costruito un nuovo sostegno (fig. 2) adatto per sbarrette di sezione rettangolare, vibranti per flessione. La sbarretta S è sospesa orizzontalmente, in corrispondenza dei nodi di vibrazione, mediante due sottili fili conduttori F (rame, diam. $= 0,05$ mm) attaccati mediante mastice di FARADAY ai ponticelli P. Tali ponticelli sono, a loro volta, tenuti al sostegno per mezzo di perni scorrevoli nella scanalatura G, che permette di variarne la posizione e l'orizzontamento relativi, e quindi disporre i fili nella posizione voluta. I fili F servono anche a collegare elettricamente alla massa la superficie della sbarretta. Questa è resa conduttrice da un sottilissimo strato di grafite che non ne altera sensibilmente le caratteristiche meccaniche. La posizione dell'elettrodo E e la sua distanza dalla sbarretta sono regolabili, come è mostrato chiaramente nella figura.

La presenza del sottile strato d'aria, compreso fra sbarretta ed elettrodo, introduce una resistenza di smorzamento addizionale che può abbassare

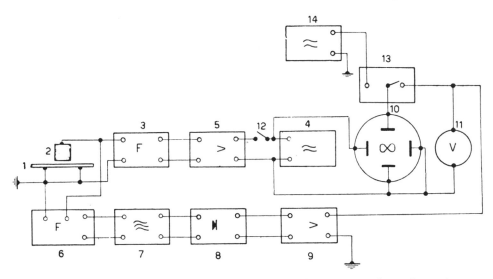

Fig. 1. - Schema elettrico generale: 1, sbarretta vibrante; 2, elettrodo eccitatore; 3, filtro passa basso; 4, oscillatore a battimenti; 5, amplificatore a frequenza acustica; 6, filtro passa alto; 7, oscillatore in alta frequenza; 8, rivelatore per modulazione di frequenza; 9, amplificatore a frequenza acustica; 10, oscillografo catodico; 11, voltmetro; 12, interruttore a pulsante; 13, commutatore; 14, elettrodiapason campione (400 Hz).

notevolmente il coefficiente di risonanza apparente della sbarretta. Tale effetto è trattato per disteso in appendice; teoria ed esperienza mostrano che è necessario tenere l'elettrodo ad una distanza di almeno $0,7 \div 0,8$ mm. Non è invece possibile evitare la difficoltà operando sotto vuoto, perchè, come si è osservato, si altererebbero in modo imprevedibile le proprietà del legno [5].

La distanza relativamente grande fra elettrodo eccitato e sbarra impone che la tensione eccitatrice a frequenza acustica, generata dall'oscillatore a

---

[5] Quanto all'effetto dei fili di sospensione, si potrebbe mostrare, con un metodo analogo a quello adoperato da P. G. BORDONI nel lavoro citato, che esso può essere trascurato anche supponendo che l'errore commesso nel disporre i fili nei nodi sia sensibile, per es. mezzo millimetro.

battimenti, sia $400 \div 500$ V affinchè la sensibilità del metodo risulti sufficiente. Inoltre l'oscillatore a battimenti deve consentire una regolazione assai
fine della frequenza, perchè, sia possibile misurare gli intervalli di frequenza
Δf, e quindi i coefficienti di risonanza, con adeguata precisione. A questo
scopo l'oscillatore adoperato ha, in parallelo col condensatore variabile principale, un secondo condensatore di capacità assai più piccola, il quale con-
sente variazioni totali di frequenza di 100 Hz.

La frequenza di risonanza può essere determinata servendosi delle scale
graduate dei due condensatori, purchè si controlli, prima di ogni misura,

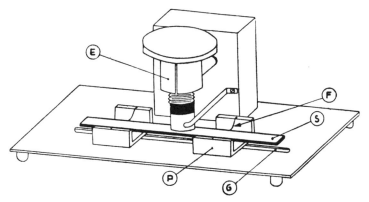

Fig. 2. – Sostegno delle sbarrette: S, sbarretta; E, elettrodo eccitatore; F, fili di sospensione; P, ponticelli spostabili; G, scanalatura per lo spostamento dei ponticelli.

la taratura dell'oscillatore, mediante confronto all'oscillografo, con i 400 Hz
dell'elettrodiapason campione 14 (fig. 1).

Si può ritenere che la precisione di misura del coefficiente di risonanza
sia di circa $\pm 5 \%$; più che sufficiente quindi per lo scopo da raggiungere.
Assai più piccolo è invece l'errore che si commette nella misura delle frequenze di risonanza (minore dell'1 %); il suo effetto nella misura delle costanti elastiche (velocità del suono; modulo di YOUNG) può anzi essere sicu-
ramente trascurato di fronte ad altre cause di errore (vedi § 6).

4. – Caratteristiche del metodo impiegato.

Le principali caratteristiche del metodo adoperato, le quali lo distinguono
da altri analoghi, che pure possono essere usati per ricerche sulle proprietà
elastiche e l'attrito interno dei solidi [6], sono le seguenti:

a) Le proprietà del campione da provare·non vengono sensibilmente
alterate dall'apparato di misura. Questo risultato, che costituisce il princi-

---

[6] Vedere ad esempio: S. SIEGEL: *J.A.S.A.*, **16**, 26 (1944).

pale pregio del metodo, è una conseguenza dell'avere operato su sbarrette sensibilmente libere [7] e dell'aver usato l'eccitazione elettrostatica delle vibrazioni, la quale esige soltanto che si avvicini un elettrodo, senza toccare la sbarretta. Questa caratteristica è fondamentale specialmente per la misura dei coefficienti di risonanza.

b) È possibile lavorare su campioni di dimensioni assai ridotte. Anche questo risultato è facilitato dall'eccitazione elettrostatica e dal rilievo delle vibrazioni mediante la modulazione di frequenza, con un solo elettrodo ausiliario. La maggior parte delle sbarrette adoperate, aveva le dimensioni di $10 \times 1{,}0 \times 0{,}2$ cm³, ma sarebbe possibile adoperarne di dimensioni ancora minori. Ne consegue che si possono analizzare più facilmente le proprietà *locali* del materiale, e le loro variazioni con la posizione e l'orientamento dei campioni, riscontrando disomogeneità che non sarebbero invece rivelabili su sbarrette di dimensioni maggiori [8].

c) Le sollecitazioni applicate alle sbarrette e le ampiezze di vibrazione, sono piccolissime. I loro ordini di grandezza sono rispettivamente di 10^{-4} Nw per la forza applicata e di $1\,\mu$ circa per l'ampiezza di vibrazione [9]. In altri termini il materiale è sollecitato solamente in un piccolissimo tratto iniziale della sua caratteristica elastica.

d) L'esecuzione delle misure è semplice, specialmente se si confronta il metodo da noi prescelto con altri adoperati in ricerche analoghe per la misura del decremento logaritmico. È stato possibile in tal modo eseguire le misure su un grande numero di campioni.

5. – Richiami su alcuni risultati sperimentali di altri autori.

Prima di esporre i risultati della presente ricerca fermiamo l'attenzione su alcuni dei risultati ottenuti da altri ricercatori.

La velocità di propagazione nel legno varia considerevolmente da essenza ad essenza fra valori di oltre 5000 m · sec⁻¹ (abete), cioè maggiori di quelli più elevati dei metalli, a valori assai più bassi (intorno a 3000 m · sec⁻¹). La velocità di propagazione (così come il modulo di YOUNG) varia inoltre notevolmente con la direzione. I valori precedenti si riferiscono alla direzione parallela alle fibre; nelle direzioni perpendicolari si hanno invece valori da 3 a 4 volte minori.

---

[7] Si potrebbe pensare di lavorare, invece, su sbarrette incastrate ad un estremo, come effettivamente è stato fatto da altri sperimentatori (vedere i lavori citati nella nota 10).

[8] In particolare è stato possibile misurare le differenze di caratteristiche tra la *zona primaverile* e la *zona tardiva* di una stessa specie legnosa (vedi § 7).

[9] Essi possono essere calcolati con un metodo analogo a quello usato nelle appendici A e D del lavoro citato di P. G. BORDONI.

Quanto alle notizie, relativamente scarse. sulla dissipazione interna [10], interessa qui specialmente la conclusione cui è giunto ROHLOFF, secondo il quale il decremento delle sbarre di legno è sensibilmente indipendente dalla frequenza, tanto per vibrazioni longitudinali quanto per vibrazioni flessionali; almeno nel campo di frequenza nel quale sono state eseguite le misure, cioè, fra 10 Hz e 10 000 Hz. Inoltre il decremento ha lo stesso valore per i due tipi di vibrazioni. Questi risultati sono assai importanti, perchè mostrano che, nel caso del legno, la scelta del tipo di vibrazioni non è essenziale, a differenza di quanto avviene, ad esempio, per i metalli [11].

Siccome si è visto (§ 3) che è necessario porre l'elettrodo ausiliario ad una distanza abbastanza grande dalla sbarra perchè l'effetto dell'aria possa essere trascurato, se ne deduce che conviene lavorare con vibrazioni flessionali, per le quali l'ampiezza di vibrazione e la sensibilità del metodo sono relativamente assai maggiori.

Il fatto che, per sbarre di legno, il decremento (e quindi anche il coefficiente di risonanza) è in prima approssimazione indipendente dalla frequenza e dalle dimensioni, mostra inoltre che esso può essere considerato come un vero parametro caratteristico del materiale, come ad esempio la densità e la velocità del suono. Risulta anche dai lavori di ROHLOFF che il decremento è notevolmente più elevato per le direzioni normali alle fibre (circa 3,5 volte maggiore) che per la direzione parallela alle fibre.

Tutte le caratteristiche elastiche ed il decremento dipendono, infine, in modo non indifferente, per una stessa essenza, da numerosi parametri (provenienza, stagionatura, umidità, etc.).

6. – Misure eseguite.

I campioni di legno provati sono oltre 400, appartenenti ad 85 specie diverse. Quasi tutte le sbarrette avevano le dimensioni di $10 \times 10 \times 0,2$ cm^3 circa; solamente per alcune verifiche sperimentali di cui si parlerà più avanti, si sono adoperati campioni di dimensioni differenti.

Sulla scelta delle dimensioni delle sbarrette influiscono diversi fattori. Le sbarrette più piccole si prestano meglio per studiare come variano localmente le proprietà fisiche di una stessa specie legnosa; però al diminuire dello spessore h aumentano le difficoltà di una lavorazione precisa. Inoltre l'effetto smorzante dello strato d'aria compreso frà sbarretta ed elettrodo è, percentualmente, maggiore per sbarrette più piccole, a parità di frequenza.

Si è trovato che non è conveniente scendere con lo spessore al di sotto

[10] F. KRÜGER e E. ROHLOFF: *Zeits. f. Phys.*, **110**, 58 (1938); E. ROHLOFF: *Zeits. f. Physik*, **117**, 64 (1941); R. B. ABBOTT e G. H. PURCELL: *J.A.S.A.*, **13**, 54 (1941).
[11] C. ZENER: *Phys. Rev.*, **52**, 230 (1937).

di 2 mm circa; una volta scelto h, la lunghezza della sbarretta va fissata tenendo conto della frequenza prescelta. Con i mezzi sperimentali a disposizione, i valori più convenienti per la frequenza vanno da alcune centinaia di Hz fino a un migliaio di Hz, o poco più. Lavorando sull'autofrequenza più bassa della sbarretta, si ricava dalla (1) che il valore più conveniente di l si aggira intorno ai dieci centimetri. La larghezza b della sbarretta può essere scelta con il solo criterio che sia notevolmente minore della lunghezza l, affinchè siano soddisfatte le ipotesi che permettono di scrivere la (1).

Su ogni sbarretta sono state misurate direttamente 6 grandezze: le tre dimensioni geometriche l, b, h, il peso P, la frequenza di risonanza f del modo fondamentale (o in qualche caso del 1º modo superiore) e l'intervallo di frequenza Δf (vedi § 2) [12].

Da queste grandezze si sono potuti calcolare i valori di 6 parametri fisici caratteristici del materiale, essi sono: la densità ρ, la velocità di propagazione c, il modulo di YOUNG E, l'impedenza caratteristica $\sqrt{\rho(1/E)} = \rho c$, il coefficiente di risonanza Q ed il prodotto del decremento logaritmico d per ρc, cioè $k = d\rho c$. L'importanza di questo ultimo parametro risulterà nelle considerazioni del paragrafo seguente.

Nella determinazione dello spessore h della sbarretta si è fatto la media di 5 valori relativi a 5 sezioni egualmente distanziate; si tenga presente che su piccoli campioni, è sempre la misura di h, piuttosto che quella della frequenza, che limita la precisione con cui può essere determinata la velocità c (alcune unità %).

Come si è accennato, sono state eseguite alcune verifiche sperimentali aventi lo scopo di confermare la più importante delle conclusioni dei lavori di ROHLOFF, e cioè che il coefficiente di risonanza, è, in prima approssimazione, indipendente, sia dalle dimensioni della sbarretta, sia dal tipo di fenomeno vibratorio (vibrazioni flessionali o longitudinali), sia dalla frequenza.

Nella tab. I, sono riportati i risultati di una di tali prove, che vale a confermare l'indipendenza di Q dalle dimensioni, almeno nei limiti della prova eseguita.

7. – Discussione dei risultati.

Per orientarsi sull'esame dei risultati sperimentali conviene fare riferimento ad un caso tipico, riportato nella tab. II. Si tratta di 10 sbarrette diversamente orientate, ma ricavate tutte dallo stesso tronco di abete rosso

---

[12] Poichè lo scopo principale del lavoro era quello di procurarsi i dati per una statistica su un grande numero di campioni, piuttosto che eseguire misure di grande precisione su un numero limitato di sbarrette, non si è ritenuto necessario eseguire prove sistematiche sull'effetto dell'umidità, ma ci si è limitati ad operare in condizioni sufficientemente paragonabili.

(*Picea excelsa*). L'orientazione di ciascuna sbarretta si identifica facilmente con l'aiuto della fig. 3. Come si vede, l'orientazione della sbarretta è rappresentata mediante una o due lettere che indicano il piano di giacitura e mediante l'angolo rispetto ad un asse di riferimento scelto in questo piano,

TABELLA I.

Picea excelsa (Abete rosso (^{o}); Tirolo - 1895).

Dimensioni $l \times b \times h$ (cm)	f (Hz)	Q (^{oo})	ρ (g $\cdot$ cm^{-3})	c (km $\cdot$ sec^{-1})
$9,98 \times 0,995 \times 0,215$	1180	120	0,450	5,35
$12,00 \times 0,960 \times 0,300$	1150	120	0,435	5,40
$14,00 \times 0,995 \times 0,400$	1090	120	0,435	5,20

(^{o}) Sbarrette tagliate l'una sopra l'altra, in direzione R-0° (vedi fig. 4).
(^{oo}) Errore di misura: $\pm 3\%$ circa.

TABELLA II.

Picea excelsa (abete rosso (^{o}); Foresta del Cansiglio - 1935).

Orientazione	ρ (g $\cdot$ cm^{-3})	c (km $\cdot$ sec^{-1})	Q	$d\rho c$ (g $\cdot$ cm^{-2} $\cdot$ sec^{-1})
R-0°	0,415	5,60	135	$5,45 \cdot 10^3$
T-0°	0,405	5,80	115	$6,30 \cdot 10^9$
RT-0°	0,340	5,20	115	$4,75 \cdot 10^3$
TR-0°	0,355	5,35	120	$5,05 \cdot 10^3$
R-90°	0,400	1,10	33	$4,10 \cdot 10^3$
N-90°	0,410	1,35	31	$5,50 \cdot 10^3$
T-90°	0,420	1,—	31	$4,20 \cdot 10^3$
N-0°	0,425	1,10	30	$4,85 \cdot 10^3$
RT-90°	0,400	0,75	26	$3,50 \cdot 10^3$
N-45°	0,400	0,90	25	$4,65 \cdot 10^3$

(^{o}) Sbarrette tratte da uno stesso tronco.

e precisamente : una retta parallela all'asse del tronco per i piani *radiale* (R) e *tangenziale* (T) ed una tangente agli anelli annuali di accrescimento per il piano *trasversale* (N). Inoltre è stato indicato con (RT) uno dei piani bisettori del diedro formato da un piano radiale ed uno tangenziale; con (TR) si è indicato l'altro piano bisettore, normale al primo.

Nella tabella le sbarrette sono state raggruppate in cinque coppie, mettendo sempre in una stessa coppia due sbarrette ricavate con i lati maggiori l paralleli, e le larghezze b giacenti in piani fra loro perpendicolari.

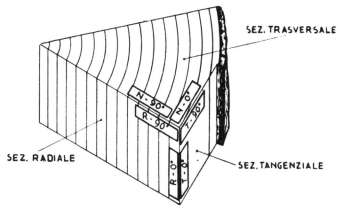

Fig. 3. – Direzione di taglio delle sbarrette: R-0°, sbarretta giacente in un piano radiale e con il lato maggiore parallelo all'asse del tronco; R-90°, sbarretta radiale con il lato maggiore normale all'asse; T-90°, sbarretta tangenziale con il lato maggiore normale all'asse; N-0°, sbarretta trasversale con il lato maggiore tangente agli anelli annuali; N-90°, sbarretta trasversale con il lato maggiore normale agli anelli annuali.

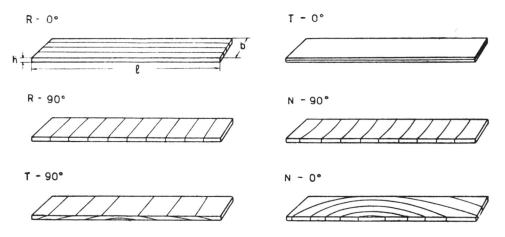

Fig. 4. – Disposizione degli anelli annuali nelle sbarrette. Le sbarrette sono contrassegnate come nella precedente fig. 3.

I valori della densità ρ, della velocità c e del coefficiente di risonanza Q sono relativamente poco diversi per le due sbarrette di una stessa coppia, tanto che le differenze riscontrate sono paragonabili a quelle che si hanno fra diverse sbarrette egualmente orientate, ma ricavate, per es., da tronchi diversi, od anche da parti differenti dello stesso tronco (confrontare la tab. II con la tab. III).

TABELLA III.

Caratteristiche di singole sbarrette (Direzione di taglio: R-$0°$).

Essenza	Dati caratteristici	ρ (g·cm^{-3})	c (km·sec^{-1})	Q	$\dfrac{dc}{\rho c}$ (g·cm^{-2}·sec^{-1})
Picea excelsa	Tirolo - 1880	0,475	5,40	120	6,75 · 10^3
	Tirolo - 1880	0,450	5,25	115	6,40 · 10^3
	Tirolo - 1895	0,465	5,35	120	6,50 · 10^3
	Tirolo - 1895	0,440	5,40	130	5,75 · 10^3
	Foresta del Cansiglio - 1935	0,415	5,60	135	5,45 · 10^3
	Russia - 1938	0,550	5,05	130	6,70 · 10^3
Acer platanoides	Francia - 1930	0,785	4,15	80	12,5 · 10^3
	Abruzzi - 1930	0,565	3,70	75	8,5 · 10^3
	Ungheria - 1934	0,630	2,60	55	9,0 · 10^3
	Lombardia - 1934	0,665	4,30	105	8,5 · 10^3
	Abruzzi - 1938 (°)	0,720	4,05	80	12,0 · 10^3
	Abruzzi - 1938 (°)	0,705	3,30	60	12,5 · 10^3

(°) Sbarrette tratte dallo stesso tronco.

Con una approssimazione un po' minore le sbarrette della tab. II possono essere ancora classificate in due gruppi. Un primo gruppo comprende 4 sbarrette (R-$0°$; T-$0°$; RT-$0°$; TR-$0°$) aventi tutte il lato maggiore *parallelo* all'asse del tronco; un secondo gruppo è formato da tutte le rimanenti 6 sbarrette, aventi invece il lato l *normale* all'asse (ma giacenti in piani diversamente orientati).

Le sbarrette del primo gruppo hanno velocità di propagazione e coefficiente di risonanza assai più elevati di quelle del secondo.

Se ne deduce una prima conclusione importante, cioè che è lecito parlare in prima approssimazione di valori dei parametri relativi alla orientazione parallela all'asse del tronco (sbarrette con il lato maggiore parallelo all'asse) e di valori relativi alla orientazione ortogonale (sbarrette con il lato maggiore ortogonale all'asse del tronco).

Questo risultato era sostanzialmente noto, ma mediante la tab. II è possibile rendersi conto dell'entità dell'approssimazione che si introduce schematizzando le cose in questo modo.

Da un esame più particolare si vede ancora che fra le due coppie di sbarrette con l'asse parallelo a quello del tronco esistono differenze di valore (almeno per la velocità c) alquanto più grandi di quelle che si hanno fra sbarrette della stessa coppia. La causa di questo fatto, tuttavia, potrebbe anche essere ricercata nelle differenze di caratteristiche fra parti diverse di uno stesso tronco; ciò sembrerebbe confermato dal valor medio della densità, che è nettamente minore per le sbarretta della seconda coppia, rispetto a quelle della prima. Inoltre la disposizione degli anelli di accrescimento annuale nelle sbarrette della seconda coppia (RT-$0°$; TR-$0°$) non è simmetrica rispetto al piano di sollecitazione e ciò può, forse, in qualche modo infirmare l'applicabilità della teoria semplice svolta nel § 2.

Risultati analoghi sono stati trovati per le altre specie legnose esaminate.

Da quanto si è detto risulta che *conviene assumere come valori della velocità di propagazione c e del coefficiente di risonanza Q in direzione parallela all'asse, la media dei valori che gli stessi parametri hanno per le orientazioni: R-0° e T-0°.*

Considerazioni analoghe valgono per le altre tre coppie di sbarrette, tagliate in direzioni ortogonali all'asse del tronco. Si nota che le due coppie di sbarrette (R-$90°$; N-$90°$) e (T-$90°$; N-$0°$) dànno per c e Q valori un po' maggiori di quelli dell'ultima coppia (RT-$90°$; N-$45°$); *come valori relativi all' orientazione ortogonale conviene scegliere la media di quelli relativi alle prime quattro sbarrette (R-90°; N-90°; T-90°; N-0°)* ([13]).

a) Valori dei parametri nella direzione assiale. – I valori dei parametri caratteristici (c, E, Q, $k = d\rho_0$) relativi alla propagazione in direzione parallela all'asse del tronco per 85 specie legnose diverse, sono raccolti nella tab. IV. In essa, accanto ai valori medi dei 6 parametri fisici determinati, è indicato, per ogni specie, il numero di campioni provati; esso può servire come *indice di attendibilità* dei dati relativi ad ogni specie.

Quanto alle differenze di caratteristiche fisiche fra campioni appartenenti ad una stessa specie, il caso dell'abete rosso, di cui un esempio è stato riportato nella tab. III, è probabilmente il più favorevole. Per quasi tutte le altre specie per le quali si è potuto provare un discreto numero di campioni, si sono trovate differenze, dall'una all'altra sbarretta, notevolmente maggiori che nell'abete rosso, sebbene per quest'ultimo si avesse a disposizione un numero di campioni (diciassette) assai maggiore di quello di ognuna delle altre specie (sette al massimo). La spiegazione sta, evidentemente, nella grandissima regolarità della struttura fibrosa dell'abete rosso la quale, come è noto, è quella che lo rende particolarmente adatto alla costruzione di strumenti musicali.

([13]) Non sembra che si possa parlare, almeno in generale, di differenze *sicure e ben definite* fra le sbarrette del gruppo (R-$90°$) ed (N-$90°$) e quelle del gruppo (T-$90°$) ed (N- $0°$), come talvolta è stato fatto, introducendo, ad es., una velocità del suono relativa alla direzione radiale ed una relativa alla direzione tangenziale.

TABELLA IV.

Caratteristiche medie delle varie specie (in direzione parallela alle fibre).

ESSENZE	N. campioni	ρ g·cm^{-3}	c km·sec^{-1}	ρc g·cm^{-2}sec^{-1}	E dine cm^{-2}	Q	$d\rho c$ g·cm^{-2}sec^{-1}
Abies concolor . . . . (abete concolore)	4	0,59	4,65	$2,70 \cdot 10^5$	$12,5 \cdot 10^{10}$	110	$7,7 \cdot 10^3$
Abies pectinata . . . (abete bianco)	5	0,62	4,85	$3 \cdot 10^5$	$14,5 \cdot 10^{10}$	115	$8,1 \cdot 10^3$
Acacia melanoxylon . (ebano violetto)	1	1,03	4,40	$4,50 \cdot 10^5$	$20 \cdot 10^{10}$	135	$10,5 \cdot 10^3$
Acer campestris . . . (loppio)	1	0,62	3,30	$2,05 \cdot 10^5$	$6,7 \cdot 10^{10}$	65	$9,8 \cdot 10^3$
Acer platanoides . . . (acero)	7	0,68	3,80	$2,60 \cdot 10^5$	$9,9 \cdot 10^{10}$	80	$9,9 \cdot 10^3$
Acer pseudoplatanus . (acero riccio)	4	0,63	3,40	$2,15 \cdot 10^5$	$7,3 \cdot 10^{10}$	75	$9 \cdot 10^3$
Aesculus ippocastanum (ippocastano)	1	0,65	3,95	$2,55 \cdot 10^5$	$10 \cdot 10^{10}$	100	$8 \cdot 10^3$
Ailantus glandulosa . (ailanto)	3	0,57	4,00	$2,30 \cdot 10^5$	$9,2 \cdot 10^{10}$	95	$7,6 \cdot 10^3$
Alnus glutinosa . . . (ontano)	1	0,62	3,45	$2,15 \cdot 10^5$	$7,4 \cdot 10^{10}$	65	$10,5 \cdot 10$
Aspidosperma quebracho (quebracio)	1	0,91	3,65	$3,30 \cdot 10^5$	$10,5 \cdot 10^{10}$	85	$12,5 \cdot 10^3$
Aucoumea klaineana . (okoumè)	2	0,48	4,65	$2,25 \cdot 10^5$	$10,5 \cdot 10^{10}$	95	$7,2 \cdot 10^3$
Bambusa aurea . . . (bambù)	1	0,71	4,35	$3,10 \cdot 10^5$	$13,5 \cdot 10^{10}$	105	$9,3 \cdot 10^3$
Betula alba . . . . . (betulla)	1	0,66	4,60	$3,05 \cdot 10^5$	$14 \cdot 10^{10}$	100	$9,3 \cdot 10^3$
Buxus sempervirens . (bosso)	2	0,94	3,40	$3,20 \cdot 10^5$	$11 \cdot 10^{10}$	60	$16 \cdot 10^3$
Caesalpinia brasiliensis (verzino)	3	0,95	4,40	$4,20 \cdot 10^5$	$18,5 \cdot 10^{10}$	210	$6,3 \cdot 10^3$
Caesalpinia echinata . (pernambuco marrone)	1	1,25	3,90	$4,90 \cdot 10^5$	$19,5 \cdot 10^{10}$	100	$16 \cdot 10^3$
Caesalpinia pulcherrima (frijol)	2	0,53	5,40	$2,85 \cdot 10^5$	$15,5 \cdot 10^{10}$	160	$5,6 \cdot 10^3$
Calamus communis . (canna)	1	0,49	3,70	$1,80 \cdot 10^5$	$6,5 \cdot 10^{10}$	50	$11,5 \cdot 10^3$
Calamus rotang . . . (canna d'India)	2	0,55	2,30	$1,30 \cdot 10^5$	$3,1 \cdot 10^{10}$	40	$9,5 \cdot 10^3$
Castanea vesca . . . . (castagno)	6	0,50	4,45	$2,25 \cdot 10^5$	$10 \cdot 10^{10}$	125	$5,7 \cdot 10^3$
Carpinus betulus . . . (carpino bianco)	1	0,63	3,45	$2,15 \cdot 10^5$	$7,4 \cdot 10^{10}$	55	$12 \cdot 10^3$

Segue TABELLA IV.

ESSENZE	N. campioni	$\rho \cdot cm^{-3}$ g·cm⁻³	c km·sec⁻¹	ρc g·cm⁻²sec⁻¹	E dine cm⁻²	Q	$d\rho c$ g·cm⁻²sec⁻¹
Cedrela odorata . . . (cedrela)	2	0,49	4,95	$2,40 \cdot 10^5$	$12 \cdot 10^{10}$	115	$6,7 \cdot 10^3$
Citrus aurantium . . (arancio)	1	0,83	4,05	$3,35 \cdot 10^5$	$14 \cdot 10^{10}$	80	$13,1 \cdot 10^3$
Cupressus sempervirens (cipresso)	3	0,63	4,65	$2,95 \cdot 10^5$	$14 \cdot 10^{10}$	130	$7,1 \cdot 10^3$
Cupressus torulosa . . (cipresso dell'Imalaja)	2	0,60	4,15	$2,50 \cdot 10^5$	$10 \cdot 10^{10}$	125	$6,2 \cdot 10^3$
Dahlbergia nigra . . . (palissandro)	3	0,74	3,90	$2,90 \cdot 10^5$	$11 \cdot 10^{10}$	135	$6,7 \cdot 10^3$
Diospyros sp. . . . . (ebano)	1	1,25	4,25	$5,30 \cdot 10^5$	$22,5 \cdot 10^{10}$	135	$12,3 \cdot 10^3$
Erica scoparia . . . . (erica)	2	0,66	4,05	$2,65 \cdot 10^5$	$11 \cdot 10^{10}$	80	$10,5 \cdot 10^3$
Eucalyptus sp. . . . . (eucalipto)	2	0,80	3,25	$2,60 \cdot 10^5$	$8,5 \cdot 10^{10}$	90	$9,1 \cdot 10^3$
Euphorbia abyssinica . (euforbia)	2	0,24	4,60	$1,10 \cdot 10^5$	$5,1 \cdot 10^{10}$	80	$4,3 \cdot 10^3$
Fagus silvatica . . . . (faggio)	1	0,58	4,55	$2,65 \cdot 10^5$	$12 \cdot 10^{10}$	95	$8,7 \cdot 10^3$
Fraxinus excelsa . . . (frassino)	4	0,75	4,40	$3,25 \cdot 10^5$	$14,5 \cdot 10^{10}$	105	$9,8 \cdot 10^3$
Glycine chinensis . . . (glicine)	2	0,41	2,15	$0,88 \cdot 10^5$	$1,85 \cdot 10^{10}$	40	$7 \cdot 10^3$
Guajacum sanctum . . (guaia o)	1	0,74	3,80	$2,80 \cdot 10^5$	$10,5 \cdot 10^{10}$	90	$9,8 \cdot 10^3$
Hedera helix . . . . . (edera)	2	0,52	2,50	$1,40 \cdot 10^5$	$3,2 \cdot 10^{10}$	40	$10 \cdot 10^3$
Ixora ferrea . . . . . (legno ferro)	1	1,05	4,45	$4,70 \cdot 10^5$	$21 \cdot 10^{10}$	210	$7 \cdot 10^3$
Juglans nigra . . . . (noce d'America)	4	0,66	4,30	$2,85 \cdot 10^5$	$12 \cdot 10^{10}$	85	$10,5 \cdot 10^3$
Juglans regia . . . . (noce)	5	0,63	4,05	$2,55 \cdot 10^5$	$10,5 \cdot 10^{10}$	100	$8 \cdot 10^3$
Larix europaea . . . (larice)	4	0,61	4,35	$2,65 \cdot 10^5$	$11,5 \cdot 10^{10}$	105	$7,9 \cdot 10^3$
Laurus camphora . . (canforo)	2	0,63	3,85	$2,45 \cdot 10^5$	$9,2 \cdot 10^{10}$	90	$8,5 \cdot 10^3$
Laurus nobilis . . . . (alloro)	2	0,70	4,30	$3 \cdot 10^5$	$13 \cdot 10^{10}$	115	$8,2 \cdot 10^3$
Liquidambar styraciflua (noce satiné)	3	0,56	4,20	$2,35 \cdot 10^5$	$10 \cdot 10^{10}$	85	$8,7 \cdot 10^3$
Ligustrum Japonica (ligustro)	2	0,73	4,10	$3 \cdot 10^5$	$12 \cdot 10^{10}$	80	$12 \cdot 10^3$

Segue TABELLA IV.

ESSENZE	N. campioni	ρ g·cm⁻³	c km·sec⁻¹	ρc g·cm⁻²sec⁻¹	E dine cm⁻²	Q	$d\rho c$ g·cm⁻²sec⁻¹
Mimosa sp. . . . . . (mimosa)	2	0,59	3,30	$1,95 \cdot 10^5$	$6,4 \cdot 10^{10}$	60	$10 \cdot 10^3$
Morus nigra . . . . . (gelso)	2	0,51	3,25	$1,65 \cdot 10^5$	$5,4 \cdot 10^{10}$	90	$5,8 \cdot 10^3$
Ochroma layopis . . . (balsa)	1	0,30	5,50	$1,70 \cdot 10^5$	$9,2 \cdot 10^{10}$	90	$5,8 \cdot 10^3$
Olea europaea . . . . (olivo)	3	0,79	3,30	$2,65 \cdot 10^5$	$8,7 \cdot 10^{10}$	70	$12 \cdot 10^3$
Physocalymna scaberrimum (legno rosa)	1	0,81	4,50	$3,65 \cdot 10^5$	$16,5 \cdot 10^{10}$	175	$6,5 \cdot 10^3$
Picea excelsa . . . . (abete rosso)	17	0,46	5,35	$2,50 \cdot 10^5$	$13,5 \cdot 10^{10}$	125	$6,3 \cdot 10^3$
Pinus sp. . . . . . (pitch-pine)	4	0,65	4,85	$3,20 \cdot 10^5$	$15,5 \cdot 10^{10}$	125	$7,9 \cdot 10^3$
Pinus cembra . . . . (cirmolo)	3	0,40	4,90	$1,95 \cdot 10^5$	$9,6 \cdot 10^{10}$	105	$5,8 \cdot 10^3$
Pinus nigra . . . . . (pino nero)	2	0,55	4,35	$2,40 \cdot 10^5$	$10,5 \cdot 10^{10}$	90	$8,4 \cdot 10^3$
Pinus pinaster . . . . (pino marittimo)	2	0,57	4,60	$2,60 \cdot 10^5$	$12 \cdot 10^{10}$	100	$8,2 \cdot 10^3$
Pinus silvestsis . . . (pino silvestre)	6	0,53	4.25	$2,25 \cdot 10^5$	$9,5 \cdot 10^{10}$	90	$7,8 \cdot 10^3$
Pirus communis . . . (pero)	4	0,80	3,85	$3,10 \cdot 10^5$	$12 \cdot 10^{10}$	80	$12 \cdot 10^3$
Pittosporus lobira . . (pittosporo)	2	0,69	3,35	$2,30 \cdot 10^5$	$8 \cdot 10^{10}$	65	$11 \cdot 10^3$
Platanus orientalis . . (platano)	6	0,64	3,40	$2,15 \cdot 10^5$	$7,4 \cdot 10^{10}$	85	$8 \cdot 10^3$
Populus alba . . . . (alberello)	2	0,51	5,70	$2,85 \cdot 10^5$	$17 \cdot 10^{10}$	105	$8,5 \cdot 10^3$
Populus canadensis . . (pioppo del Canadà)	5	0,44	5,15	$2,25 \cdot 10^5$	$11,5 \cdot 10^{10}$	100	$7 \cdot 10^3$
Populus nigra . . . . (pioppo)	1	0,43	5,—	$2,15 \cdot 10^5$	$11 \cdot 10^{10}$	90	$7,2 \cdot 10^3$
Prunus armeniaca . . (albicocco)	1	0,85	3,95	$3,35 \cdot 10^5$	$13,5 \cdot 10^{10}$	135	$7,8 \cdot 10^3$
Prunus avium . . . . (ciliegio)	4	0,61	4,35	$2,65 \cdot 10^5$	$11,5 \cdot 10^{10}$	125	$6,7 \cdot 10^3$
Pseudotsuga taxifolia . (abete americano)	2	0,56	5 —	$2,80 \cdot 10^5$	$14 \cdot 10^{10}$	140	$6,3 \cdot 10^3$
Pterocarpus indicus . . (paduca)	1	0,67	5,70	$3,80 \cdot 10^5$	$21,5 \cdot 10^{10}$	200	$6 \cdot 10^3$
Quercus cerris . . . . (cerro)	1	0,97	3,60	$3,50 \cdot 10^5$	$12,5 \cdot 10^{10}$	70	$15,5 \cdot 10^3$

Segue TABELLA IV.

ESSENZE	N. campioni	$\rho \cdot sec^{-3}$ g ·	c km · sec^{-1}	ρc g · cm^{-2}sec^{-1}	E dine cm^{-2}	Q	$d\rho c$ g · cm^{-2}sec^{-1}
Quercus ilex . . . . . (leccio)	4	0,84	4,05	$3,40 \cdot 10^5$	$13,5 \cdot 10^{10}$	85	$12,5 \cdot 10^3$
Quercus pedunculata . (farnia)	2	0,79	3,85	$3,05 \cdot 10^5$	$11,5 \cdot 10^{10}$	75	$12,5 \cdot 10^3$
Quercus robus . . . . (rovere)	3	0,68	4,20	$2,85 \cdot 10^5$	$12 \cdot 10^{10}$	105	$8,5 \cdot 10^3$
Quercus rubra . . . . (quercia rossa)	2	0,71	4,40	$3,10 \cdot 10^5$	$13,5 \cdot 10^{10}$	90	$11 \cdot 10^3$
Quercus suber ⁒. . . . (quercia sughera)	2	0,73	3,50	$2,55 \cdot 10^5$	$8,9 \cdot 10^{10}$	65	$12,5 \cdot 10^3$
Robinia pseudoacacia . (robinia)	4	0,85	4,15	$3,55 \cdot 10^5$	$14,5 \cdot 10^{10}$	130	$8,6 \cdot 10^3$
Rosmarinus officinalis (rosmarino)	1	0,78	3,30	$2,60 \cdot 10^5$	$8,4 \cdot 10^{10}$	65	$12,5 \cdot 10^3$
Sambucus nigra . . . (sambuco)	2	0,56	3,70	$2,05 \cdot 10^5$	$7,7 \cdot 10^{10}$	70	$9,3 \cdot 10^3$
Sophora japonica . . . (sofora)	2	0,71	3,90	$2,75 \cdot 10^5$	$11 \cdot 10^{10}$	100	$8,7 \cdot 10^3$
Sorbus aucuparis (sorbo degli uccellatori)	1	0,86	4,55	$3,90 \cdot 10^5$	$17,5 \cdot 10^{10}$	100	$12,5 \cdot 10^3$
Swietenia macrophylla . (mogano chiaro) ·	1	0,57	4,35	$2,50 \cdot 10^5$	$11 \cdot 10^{10}$	105	$7,5 \cdot 10^3$
Swietenia Mahagoni . (mogano)	4	0,65	3,95	$2,60 \cdot 10^5$	$10 \cdot 10^{10}$	85	$9,5 \cdot 10^3$
Taxus brevifolia (tasso a foglie piccole)	4	0,70	3,75	$2,60 \cdot 10^5$	$9,7 \cdot 10^{10}$	105	$7,8 \cdot 10^3$
Tecoma pentaphylla . (zapatero)	2	0,80	4,05	$3,25 \cdot 10^5$	$13 \cdot 10^{10}$	90	$11,5 \cdot 10^3$
Tectonia grandis . . . (teak)	2	0,80	5 —	$4 \cdot 10^5$	$10 \cdot 10^{10}$	175	$7,1 \cdot 10^3$
Thuja plicata . . . . (tuja)	2	0,43	5,50	$2,35 \cdot 10^5$	$13 \cdot 10^{10}$	160	$4,6 \cdot 10^3$
Tilia europaea . . . . (tiglio)	5	0,45	4,55	$2.05 \cdot 10^5$	$9,4 \cdot 10^{10}$	105	$6,1 \cdot 10^3$
Ulmus campestris . . (olmo)	1	0,76	4,05	$3,10 \cdot 10^5$	$12,5 \cdot 10^{10}$	105	$9,2 \cdot 10^3$
Ulmus chinensis . . . (olmo cinese)	2	0,93	3,60	$3,35 \cdot 10^5$	$12 \cdot 10^{10}$	75	$14 \cdot 10^3$
Vitis vinifera . . . . (vite)	1	0,56	2,05	$1,15 \cdot 10^5$	$2,3 \cdot 10^{10}$	15	$21 \cdot 10^3$

A riprova di ciò si riporta nella stessa tab. III un esempio che può essere considerato opposto a quello dell'abete rosso. Nella specïe ora considerata (*Acer platanoides*) le notevoli diversità riscontrate fra l'uno e l'altro campione, si spiegano appunto con la struttura irregolare, a fibre intrecciate, propria dell'acero.

Per brevità non si è ritenuto opportuno riportare nella tab. IV i limiti di variabilità misurati per ogni parametro; i due esempi riportati dell'abete rosso e dell'acero (tab. III) possono però servire di orientamento tenendo conto che, per la maggior parte delle altre specie, la situazione è intermedia fra questi due casi estremi.

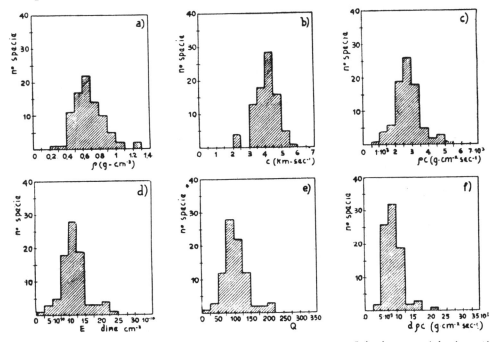

Fig. 5. – Distribuzione statistica delle varie essenze per ognuno dei sei parametri misurati·

Il comportamento assai diverso delle varie essenze, rispetto alla omogeneità di caratteristiche, ha come conseguenza che, noti i valori medi dei vari parametri fisici per una data specie, le caratteristiche di un campione appartenente a quella specie possono essere previste con una approssimazione notevolmente diversa da una specie all'altra; ciò resterebbe vero anche se i valori medi fossero ricavati su un numero uguale ed abbastanza grande di campioni, e perciò ugualmente attendibili nel senso già detto.

I grafici della fig. 5 mostrano la distribuzione ·statistica dei valori di ognuno dei sei parametri determinati (ρ; c; ρc; E; Q; $d\rho c$) fra le 85 specie esaminate. Si vede abbastanza bene che l'addensamento relativo è minimo per la densità, mentre è più elevato per ·altri parametri specialmente per la velocità del suono e per il prodotto $d\rho c = k$.

Nella tab. V sono riportati i valori massimi e minimi riscontrati per ognuno dei 6 parametri fra tutte le 85 specie esaminate.

Il rapporto fra valor massimo e valor minimo di uno stesso parametro varia fra 2,8 per la velocità di propagazione e 14 per il coefficiente di risonanza. Si vede anche che i valori massimi e minimi di parametri diversi appartengono, spesso, alle stesse specie, le quali rappresentano perciò casi eccezionali. Ad esempio la specie *Diospyros* (ebano) compare tre volte nella tabella, perchè ad essa competono i valori massimi di ρ; ρc; E. La specie *Vitis vinifera* compare egualmente tre volte (valori minimi di c e di Q; valore massimo di k).

I massimi di ρc; E; Q appartengono tutti ad essenze molto pesanti. I minimi di c; ρc; E; Q ed il massimo di k appartengono a due arbusti (*vite* e *glicine*).

Un esame più accurato, fatto con l'ausilio di tabelle (non riportate) nelle quali le essenze siano ordinate secondo i valori crescenti (o decrescenti) di una delle loro costanti fisiche, mostrerebbe che alcuni dei casi riportati nella tab. V possono essere considerati come assolutamente eccezionali. Ad esempio per la velocità c i quattro valori minori (compresi fra 2,50 e 2,05 km · sec^{-1}) appartengono a quattro arbusti (edera, canna d'India, glicine, vite); per tutte le altre 81 specie il valore minimo di c è 3,25 (gelso, eucalipto). Agli stessi 4 arbusti appartengono i valori minimi del modulo di YOUNG (fra $2 \cdot 10^{10}$ e $3 \cdot 10^{10}$) e quelli del coefficiente di risonanza (fra 15 e 40), ed essi ancora, insieme ad una quinta specie (euforbia), dànno i minimi di ρc.

Non tenendo conto delle 4 specie suddette, i campi di variabilità di quasi tutti i parametri considerati si riducono notevolmente ed il rapporto fra valor massimo e minimo diviene: 1,75 per c; 4,8 per ρc; 4,5 per E; 4,2 per Q; 3,5 per k.

In questo stesso ordine di idee si può ancora osservare che le velocità di propagazione più elevate (maggior di 5 km · sec^{-1}) appartengono, per lo più, ad essenze leggere (Thuia, balsa, abete rosso, pioppo, etc.), alle quali competono valori intermedi del coefficiente di risonanza ($Q = 90 \div 160$). I massimi Q (cioè il minimo attrito interno) si hanno, invece, più spesso in alcuni legni pesanti (legno ferro, verzino, legno rosa). Due fra le specie provate sembrano però unire valori elevati della velocità e del coefficiente di risonanza (*Pterocarpus indicus*; *Tectonia grandis*).

b) Valori dei parametri relativi alla orientazione ortogonale all'asse. — L'esame delle caratteristiche per l'orientazione ortogonale all'asse del tronco è stato fatto su un numero di specie (11) molto minore di quello per cui furono rilevate le caratteristiche in direzione parallela all'asse. Soltanto per una essenza (abete rosso) sono state provate più serie di sbarrette ricavate da tronchi diversi. La ragione di ciò sta soprattutto nella notevole difficoltà di

Tabella V.

Limiti di variabilità dei vari parametri.

Para-metro	Unità	VALORE MASSIMO		VALORE MINIMO		Rapporto max/min
		max	Specie	min	Specie	
ρ	$g \cdot cm^{-3}$	1,25	Caesalpinia echinata; Diospyros sp.	0,24	Euphorbia abyssinica · · · · ·	5,2
c	$km \cdot sec^{-1}$	5,70	Pterocarpus indicus; Populus alba	2,05	Vitis vinifera · · · · · ·	2,8
ρc	$g \cdot cm^{-2} \cdot sec^{-1}$	$5,30 \cdot 10^{5}$	Diospyros sp. · · · · · ·	$0,88 \cdot 10^{5}$	Glycine chinensis · · · · ·	6,0
E	$dine \cdot cm^{-2}$	$22,5 \cdot 10^{10}$	Diospyros sp. · · · · · ·	$2,0 \cdot 10^{10}$	Glycine chinensis · · · · ·	11,3
Q	—	210	Ixora ferrea; Caesalpinia brasiliensis	15	Vitis vinifera · · · · · ·	14
$k = \delta\rho c$	$g \cdot cm^{-2} \cdot sec^{-1}$	$21 \cdot 10^{3}$	Vitis vinifera · · · · · ·	$4,5 \cdot 10^{3}$	Euphorbia abyssinica; Thuja plicata	4,7

tagliare, con la precisione necessaria e senza danneggiarle, sottili sbarrette normalmente orientate alle fibre, specialmente per i legni meno compatti.

Le specie esaminate hanno valori di ρ, di c, e di Q non troppo diversi fra loro ed intermedi fra gli estremi possibili per tali parametri, riportati nella citata tab. V.

Le caratteristiche *ortogonali* (nel senso già chiarito), riassunte nella tab. VI, non rappresentano perciò le proprietà medie delle varie specie, ma possono, piuttosto, dare un'idea di quanto accade in casi singoli, scelti tuttavia fra le specie più significative.

In accordo con quanto era già noto (§ 5) sia per la velocità di propagazione c_n; sia per il coefficiente di risonanza Q_n nelle direzioni normali alle fibre sono stati misurati valori assai minori di quelli corrispondenti c_l, Q_l, relativi alle direzioni parallele alle fibre.

Per le specie esaminate c_n varia fra circa 1 e 1,70 km · sec^{-1}, mentre Q_n varia fra 30 e 50 circa.

A differenza della densità, la velocità c, il modulo di YOUNG, ed il coefficiente di risonanza dipendono perciò fortemente dall'orientazione e sono assai minori nelle direzioni normali.

Se si considerano, inoltre, i rapporti (c_l/c_n) e (Q_l/Q_n) fra valori relativi alle orientazioni parallela ed ortogonale (colonne 6a e 7a della tab. VI), si osserva che i due rapporti hanno, in generale, valori non troppo diversi per una stessa specie; anzi, spesso sono quasi eguali. Il fatto che la velocità c ed il coefficiente di risonanza variano quasi nello stesso rapporto passando dall'orientazione parallela a quella ortogonale, significa che il rapporto fra velocità del suono e coefficiente di risonanza, è, in prima grossolana approssimazione, indipendente dall'orientazione; cioè $c_l/Q_l \cong c_n/Q_n$.

I valori di c Q potrebbero essere assunti come parametro caratteristico accanto, ad esempio, alla densità. Per ragioni che saranno chiarite fra breve, sembra più significativo il prodotto $k = d\rho c = \pi \rho c/Q$ che, evidentemente, dipende anche esso assai poco dalla direzione, come si vede nelle ultime due colonne della tab. VI. Nel caso più sfavorevole (*Picea excelsa*; anno 1935) il prodotto k varia solo del 23 % fra la direzione parallela e quella ortogonale, mentre la velocità del suono varia di quasi il 500 % ed il coefficiente di risonanza varia di oltre il 400 %.

I rapporti c_l/c_n non sono troppo diversi fra l'uno e l'altro caso, e così quelli Q_l/Q_n; non si può essere sicuri, però, che questo fatto rispecchi una proprietà generale, perchè si è già detto che quasi tutte le 11 specie esaminate hanno proprietà relativamente poco diverse fra loro.

c) *Invecchiamento del legno*. – Nella tab. VI si osserva che il terzo esemplare di abete rosso ha velocità di propagazione e coefficiente di risonanza assai minori di quelli dei primi due esemplari ed in contrasto con le caratteristiche medie della specie, riportate nella precedente tab. IV.

TABELLA VI.

Variazione delle caratteristiche fisiche con la direzione delle fibre.

ESSENZA	ρ g·cm⁻³	C_l km·sec⁻¹	C_n km·sec⁻¹	Q_l	Q_n	C_l/C_n	Q_l/Q_n	$(d\rho c)_l$ g·cm⁻²sec⁻¹	$(d\rho c)_n$ g·cm⁻²sec⁻¹
RESINÓSE									
Pinus cembra (anno 1938)	0,395	4,90	—	110	—	—	—	5 500	—
» (» 1560)	0,415	3,95	1,30	90	31	3,05	2,90	5 700	5 200
Picea excelsa (» 1935)	0,410	5,70	1,15	125	31	4,95	4,05	5 900	4 800
» (» 1895)	0,455	5,35	1,50	125	35	3,55	3,65	6 000	6 100
» (» 1560)	0,450	4,20	0,95	95	27	4,45	3,45	6 400	5 000
Larix europaea	0,680	4,65	1,40	120	31	3,30	3,90	8 200	9 600
Pinus nigra	0,570	4,35	1,40	90	37	3,10	2,45	8 600	6 800
LATIFOGLIE									
Prunus avium	0,545	4,65	1,70	130	51	2,75	2,55	5 800	6 000
Ailantus glandulosa . . .	0,550	4,00	1,45	95	38	2,75	2,50	7 700	6 400
Juglans regia	0,610	4,20	1,65	105	42	2,55	2,50	7 700	7 500
Liquidambar styraciflua .	0,560	4,05	1,25	85	33	3,25	2,60	8 300	6 800
Swietenia Mahagoni . . .	0,715	4,35	1,40	90	31	3,10	2,95	10 500	10 000
Fraxinus excelsa	0,750	4,45	1,55	100	37	2,85	2,75	10 500	9 800
Quercus ilex	0,760	4,30	1,50	105	30	2,85	3,45	10 500	11 000

Essendo questo campione assai antico, si ritiene che la differenza di caratteristiche sia dovuta ad un processo di invecchiamento; l'ipotesi è avvalorata da un secondo campione, appartenente ad un'altra essenza (cirmolo) che risale pure alla stessa epoca (XVI secolo).

Questo risultato è assai importante per le applicazioni alla liuteria, perchè può forse spiegare la nota diversità di caratteristiche degli strumenti ad arco antichi rispetto a quelli moderni.

È notevole che alla stessa conclusione, che cioè le proprietà degli strumenti antichi possono essere spiegate con un processo di invecchiamento del legno, è giunto anche F. SAUNDERS ([2]) per una via del tutto diversa.

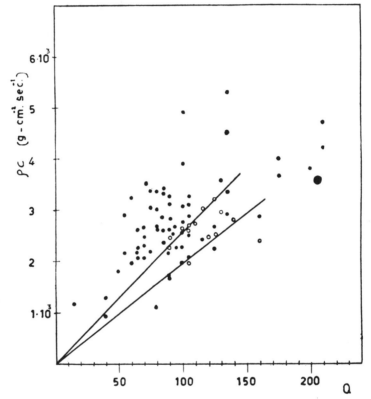

Fig. 6. – Valori medi di ρc in funzione di Q per le varie essenze (i punti si riferiscono ad essenze latifoglie; i circoletti ad essenze resinose).

d) *Caratteristiche delle essenze resinose.* – Finora non si è fatta distinzione fra essenze resinose e latifoglie. Era facile però prevedere che le diverse essenze resinose dovessero presentare notevoli analogie fra le loro proprietà, data la grande regolarità della loro struttura fibrosa.

Nella fig. 6 sono riportati in un grafico i valori di c in funzione del coefficiente di risonanza Q per le 85 specie le cui caratteristiche sono riportate in tab. IV. Ad ognuna delle specie corrisponde, nel grafico, un punto (latifoglie) o un circoletto (resinose) che ne rappresenta le proprietà medie.

Si vede subito che i punti rappresentativi delle essenze resinose si dispon-

TABELLA VII.

Caratteristiche delle resinose in direzione parallela alle fibre (valori medi).

ESSENZA	N. campioni	$d\rho c$ $g \cdot cm^{-2}sec^{-1}$	ρ $g \cdot cm^{-3}$	c $km \cdot sec^{-1}$	ρc $g \cdot cm^{-2}sec^{-1}$	E dine cm^{-2}	Q
Thuja plicata . . . . (tuia)	2	$4,6 \cdot 10^3$	0,43	5,50	$2,4 \cdot 10^5$	$13 : 10^{10}$	160
Pinus cembra . . . . (cirmolo)	3	$5,8 \cdot 10^3$	0,40	4,90	$2,0 \cdot 10^5$	$9,5 \cdot 10^{10}$	105
Picea excelsa . . . . (abete rosso)	17	$6,3 \cdot 10^3$	0,46	5,35	$2,5 \cdot 10^5$	$13 \cdot 10^{10}$	125
Pseudotsuga taxifolia . (abete americano)	2	$6,3 \cdot 10^3$	0,56	5,00	$2,8 \cdot 10^5$	$14 \cdot 10^{10}$	140
Cupressus torulosa . . (cipresso dell'Imalaja)	2	$6,3 \cdot 10^3$	0,60	4,15	$2,5 \cdot 10^5$	$10 \cdot 10^{10}$	125
Cupressus sempervirens (cipresso)	3	$7,1 \cdot 10^3$	0,63	4,65	$3,0 \cdot 10^5$	$14 \cdot 10^{10}$	130
Abies concolor . . . . (abete concolore)	4	$7,7 \cdot 10^3$	0,59	4,65	$2,7 \cdot 10^5$	$12 \cdot 10^{10}$	110
Pinus silvestris . . . (pino silvestre)	6	$7,8 \cdot 10^3$	0,53	4,25	$2,3 \cdot 10^5$	$9 \cdot 10^{10}$	90
Taxus brevifolia (tasso a foglie piccole)	4	$7,8 \cdot 10^3$	0,70	3,75	$2,6 \cdot 10^5$	$9 \cdot 10^{10}$	105
Pinus sp. . . . . . . (pitch-pine)	4	$7,9 \cdot 10^3$	0,65	4,85	$3,2 \cdot 10^5$	$15 \cdot 10^{10}$	125
Larix europaea. . . . (larice)	4	$7,9 \cdot 10^3$	0,61	4,35	$2,7 \cdot 10^5$	$11 \cdot 10^{10}$	105
Abies pectinata . . . (abete bianco)	5	$8,1 \cdot 10^3$	0,62	4,85	$3,0 \cdot 10^5$	$14 \cdot 10^{10}$	115
Pinus pinaster . . . . (pino marittimo)	2	$8,2 \cdot 10^3$	0,57	4,60	$2,6 \cdot 10^5$	$12 \cdot 10^{10}$	100
Pinus nigra . . . . . (pino nero)	2	$8,4 \cdot 10^3$	0,55	4,35	$2,4 \cdot 10^5$	$10 \cdot 10^{10}$	90

gono quasi tutti intorno a due rette passanti per l'origine. Ciò significa che quasi tutte le resinose esaminate si dipongono in due gruppi, per ciascuno dei quali il rapporto c/Q è pressochè costante.

Essendo $\pi\rho c/Q = d\rho c = k$, la precedente constatazione vale assai bene a confermare l'importanza del parametro k, già mostrata dall'essere, esso, indipendente dalla direzione.

Tutto ciò risulta ancor meglio dalla tab. VII, nella quale le 14 essenze resinose esaminate sono state ordinate secondo i valori (medi) crescenti del prodotto k. Si notano anche, nella tabella, due grandi gruppi; formati il primo di 8 specie (k compreso fra $7,7 \cdot 10^3$ e $8,4 \cdot 10^3$) e l'altro da 4 specie ($k = 5,8 \cdot 10^3 \div 6,3 \cdot 10^3$). Specialmente nel primo gruppo, più numeroso, l'accostamento non è casuale, perchè ben 7 delle 8 specie appartengono ad una stessa famiglia di resinose (pinacee).

Regolarità così evidenti non si notano per le resinose se si ordinano le specie secondo i valori crescenti di ρ, o di c, Q, etc.; si osserva in questi casi piuttosto una distribuzione statistica ([14]).

Una prova che è valsa a chiarire le probabili ragioni di tale regolarità è quella i cui risultati sono riportati nella tab. VIII. In essa sono riportate

TABELLA VIII.

Abies concolor (Abete concolore; California - 1933).

Caratteristica delle due zone (primaverile e tardiva) in una stessa specie.

CARATTERISTICHE	ρ g $\cdot$ cm^{-3}	c km $\cdot$ sec^{-1}	E dine $\cdot$ cm^{-2}	Q	$k = d\rho c$ (g $\cdot$ cm^{-2} $\cdot$ sec^{-1})
Zona primaverile: T-0°	0,335	4,55	$16,0 \cdot 10^{10}$	110	$4,5 \cdot 10^3$
Zona tardiva: T-0°	0,775	4,75	$7,2 \cdot 10^{10}$	105	$11,0 \cdot 10^3$
Caratteristiche medie (R-0°; T-0°)	0,590	4,65	$12,5 \cdot 10^{10}$	110	$7,7 \cdot 10^3$

le caratteristiche di due sbarrette di *abete concolore* ricavate la prima interamente nella zona primaverile, l'altra interamente in quella tardiva di uno stesso anello di accrescimento annuale.

Tale prova è stata possibile per la notevole estensione radiale degli anelli annuali nell'esemplare scelto e per lo spessore assai ridotto (2 mm) che è sufficiente dare alle sbarrette con il metodo di misura adoperato.

([14]) Una ragione di questo genere è quella che ha fatto preferire il prodotto k al rapporto c/Q come parametro caratteristico; ciò si vedrebbe riportando le proprietà delle resinose in un diagramma che dia c in funzione di Q costruito analogamente a quello della fig. 6.

Passando dalla zona primaverile a quella tardiva la velocità di propagazione resta pressochè la stessa, pur crescendo la densità in un rapporto maggiore di 1 : 2. Questo risultato, che può apparire inaspettato, significa soltanto che anche il modulo di YOUNG cresce, all'incirca, nello stesso rapporto della densità, passando dall'una all'altra zona. Il coefficiente di risonanza resta, invece, pressochè invariato, come la velocità c.

Supponendo in prima approssimazione che la densità ρ ed il prodotto k della zona primaverile e quelli della zona tardiva di una data specie (per esempio *abete concolore*) siano costanti, i valori degli stessi parametri che si misurano su un dato campione tagliato in posizione generica (ma sempre orientato con l'asse parallelo a quello del tronco) dipendono allora soltanto dalle proporzioni di zona primaverile e zona tardiva comprese nel campione.

L'ipotesi più semplice per spiegare l'uguaglianza del valor medio di k per parecchie specie resinose sarebbe allora che il k di ciascuna delle due zone di cui si compone ogni anello annuale, vari poco dall'una all'altra essenza e che le proporzioni fra le due zone siano mediamente le stesse. Tuttavia risulta da quanto si sa sulle proprietà delle resinose che almeno la seconda affermazione contenuta nell'ipotesi precedente non è, in generale, accettabile. Bisogna ammettere perciò che passando da una ad un'altra delle resinose che hanno lo stesso k medio, si abbiano variazioni sia nelle proporzioni delle due zone in uno stesso anello, sia nel k di ciascuna zona; ma gli effetti di queste variazioni debbono tendere a compensarsi.

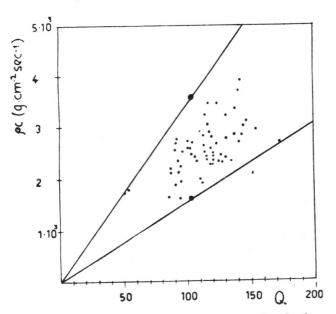

Fig. 7. – Valori di ρc in funzione di Q per singole sbarrette di essenze resinose. (I due punti rappresentati con simbolo diverso — punto chiuso in un circoletto — si riferiscono alle due zone primaverile e tardiva nell'abete concolore).

Quanto precede è chiarito anche meglio dalla fig. 7 nella quale sono rappresentati i valori di ρc in funzione di Q per tutte le singole sbarrette appartenenti alle specie resinose, anzichè i valori medi di ogni specie, come nella precedente fig. 6.

In figura sono rappresentati anche (con simbolo diverso) i due punti corrispondenti alle due sbarrette di *abete concolore* cui si riferisce la tab. VIII.

Ogni retta passante per l'origine rappresenta nel diagramma un dato valore di k.

Si vede che quasi tutti i punti rappresentativi cadono nel settore compreso fra le rette che passano per quelli corrispondenti alla zona primaverile ed a quella tardiva dell'abete concolore ([15]).

I punti rappresentativi sono distribuiti in tutto il settore e non si addensano intorno a pochi valori privilegiati di k, come avviene invece per le caratteristiche medie delle varie specie. Ciò si spiega, ovviamente, con la presenza di variazioni casuali da esemplare a esemplare, sia nelle caratteristiche delle due zone, sia nelle loro proporzioni. Se infatti si rappresentassero in modo diverso i punti corrispondenti a specie differenti (ciò che per chiarezza non è stato fatto nella figura) si vedrebbe che per l'abete rosso, la cui struttura è assai regolare, l'addensamento intorno ad una retta media è assai più netto che per tutte le altre specie, sebbene il numero di campioni provati sia di gran lunga il maggiore.

e) Cause dell'anisotropia. – Dai risultati sperimentali fin qui riassunti si deducono facilmente le probabili cause che fanno variare con la direzione la velocità del suono e il coefficiente di risonanza. Si è visto infatti che tutte le sbarrette con l'asse parallelo a quello del tronco si comportano in modo abbastanza simile, qualunque sia l'orientazione del loro piano rispetto agli anelli annuali. Ed altrettanto può dirsi per tutte le sbarrette tagliate con il loro asse normale a quello del tronco.

Ciò si spiega, nel modo più naturale, supponendo che la causa più importante della anisotropia del legno stia nella sua minuta struttura fibrosa, piuttosto che nelle diversità macroscopiche legate alla presenza degli anelli annuali. Non si comprenderebbe, altrimenti, perchè due sbarrette come la (R-90°) — nella quale gli anelli annuali sono orientati normalmente al suo asse (v. fig. 4) — e la (N-0°) — con gli anelli annuali approssimativamente longitudinali — diano quasi gli stessi valori di c e di Q. Quello che le due sbarrette hanno di comune è soltanto che il loro lato maggiore è normale alle fibre.

Con l'ipotesi fatta, i valori di c e di Q in direzione parallela alle fibre sono determinati, principalmente, dalle proprietà delle singole fibre; sui corrispondenti valori in direzione ortogonale alle fibre ha, invece, importanza assai maggior la connessione tra fibre diverse. Naturalmente le piccole differenze che esistono fra sbarrette tutte dell'uno o tutte dell'altro tipo, possono benissimo essere dovute alla presenza degli anelli di accrescimento.

([15]) Il punto che fa eccezione appartiene a una specie a bassissima densità (Thuja) nella quale, quindi, le caratteristiche delle due zone sono certamente diverse da quelle misurate sull'abete concolore. La stessa prova è stata compiuta con identico risultato, anche su un campione di larice ad anelli annuali larghissimi.

Questa interpretazione è avvalorata dal fatto che la zona primaverile e quella tardiva hanno, come si è visto, valori quasi uguali della velocità di propagazione e del coefficiente di risonanza, almeno nelle resinose [16].

Una verifica diretta di tutto ciò si è avuta con la prova eseguita su 4 sbarrette di larice e riportata nella tab. IX.

TABELLA IX.

Larix europaea (Larice: Sila - 1946) (anelli larghi).

CARATTERISTICHE	c km $\cdot$ sec^{-1}	ρ g $\cdot$ cm^{-3}	E dine $\cdot$ cm^{-2}
R-0°	3,20	0,485	$5,0 \cdot 10^{10}$
T-0° (zona tardiva)	3,35	0,670	$7,5 \cdot 10^{10}$
T-0° (zona primaverile) . .	3,05	0,365	$3,4 \cdot 10^{10}$
T-30° (zona primaverile) . .	2,35	0,390	$2,2 \cdot 10^{10}$

Le prime tre sbarrette esaminate sono tutte con l'asse parallelo alle fibre, ma una giace in un piano radiale ed è composta in parte di zona primaverile ed in parte di zona tardiva; la seconda e la terza, invece, sono ambedue giacenti in un piano tangenziale, ma composte esclusivamente dall'una o dall'altra zona. La velocità c nelle tre sbarrette è quasi uguale, in accordo con quanto si è visto anche sull'abete concolore.

Nella quarta sbarretta, che differisce dalla terza soltanto perchè il suo asse fa un angolo di circa 30° rispetto alle fibre, la velocità di propagazione è notevolmente minore.

Ciò prova in modo inequivocabile che la orientazione rispetto alle fibre è la causa dell'anisotropia.

Supponendo in prima approssimazione che la legge di variazione di c sia ellittica, si ricava $c_n \cong 1,5 \cdot$ km^{-1}, in accordo con i dati sperimentali.

f) Confronto con le proprietà di altri solidi. – È interessante confrontare le proprietà del legno con quelle di altri solidi, in particolare dei metalli. Si è visto che il prodotto k è, per il legno, un parametro assai significativo; lo si prenderà quindi come termine di confronto.

Per tutte le 85 essenze esaminate k varia fra $4,5 \cdot 10^3$ e $21 \cdot 10^3$ (g $\cdot$ cm^{-2} sec^{-1}); se però si escludono poche specie eccezionali (sette) il campo di variabilità si riduce di molto e va da $5,5 \cdot 10^3$ a $13 \cdot 10^3$.

---

[16] Nessuna prova analoga è ancora stata fatta su essenze latifoglie.

Nei metalli k assume valori da 10 a 100 volte minori, quando si prenda per il d il decremento misurato per vibrazioni longitudinali, decremento che ha spesso valori piccolissimi (qualche unità per 10^{-5}).

Se si prende, però, per d il valore *massimo* che esso può assumere (ad una certa frequenza) per vibrazioni flessionali, il valore di k massimo per i metalli così calcolato, ha lo stesso ordine di grandezza che lo stesso prodotto k ha nel legno, su tutto il campo delle frequenze acustiche ([17]); in altre parole i valori di k che sono normali per il legno, costituiscono un limite superiore per i metalli.

I coefficienti si risonanza minimi nei metalli variano fra 250 e 1500 circa, cioè da un valore poco più grande del massimo misurato nel legno fino a un valore circa 6 volte maggiore.

8. - Conclusione.

Le conclusioni che si traggono dalla presente ricerca sperimentale sono le seguenti:

a) La velocità di propagazione delle onde elastiche longitudinali ed il coefficiente di risonanza di sbarrette di legno variano fortemente con l'angolo fra la direzione considerata e quella delle fibre. In prima approssimazione si può ritenere che questo angolo sia sufficiente a determinare l'orientazione del campione in prova e conviene riferirsi quindi ai valori della velocità c_l e del coefficiente di risonanza Q_l in direzione parallela alle fibre ed ai valori c_n, Q_n nelle direzioni ortogonali. Inoltre il coefficiente di risonanza, come aveva mostrato Rohloff, è sensibilmente indipendente dalla frequenza e dal tipo di vibrazioni.

b) Il rapporto c/Q, è invece, pressochè indipendente dalla direzione, oltrechè dalla frequenza; lo stesso vale anche per il prodotto $k = d\rho c$.

c) Nelle essenze resinose, il legno della *zona primaverile* e quello della *zona tardiva* hanno velocità di propagazione e coefficiente di risonanza quasi eguali, pur variando la densità ed il modulo di Young in un rapporto circa $1:2$ dall'una all'altra zona. Le proprietà di un campione qualunque sono intermedie fra quelle delle due zone.

d) Fra le proprietà delle essenze resinose si riscontrano delle evidenti regolarità; in particolare 12 fra le 14 resinose provate si dispongono in due gruppi, in ciascuno dei quali k varia pochissimo. Una probabile spiegazione di questo fatto sperimentale è suggerita da uno dei risultati già ricordati (vedi più sopra in c).

e) Leggi semplici di questo tipo non sono state riscontrate per le latifoglie. Ciò non meraviglia perchè la struttura delle resinose è molto più regolare.

([17]) Ciò si vedrebbe facilmente servendosi dei dati contenuti nei lavori di C. Zener. Vedere, ad esempio, quello citato ne.la precedente nota 10.

Si nota tuttavia che i valori di k per tutte le 85 specie legnose esaminate variano entro limiti non troppo vasti, in relazione al campo di variabilità di ciascuno dei tre fattori. Se poi si escludono pochissime specie eccezionali, per tutte le altre k varia fra $5,5 \cdot 10^3$ e $13 \cdot 10^3$ g · cm^{-2} sec^{-1}; cioè in un rapporto $1 : 2,35$ soltanto.

Dello stesso ordine di grandezza sono pure i massimi valori che k può assumere per moltissimi altri solidi, fra cui i metalli ; solamente però a certe frequenze e per vibrazioni flessionali. Per tali solidi infatti, k, come il decremento, varia moltissimo con la frequenza nel caso di vibrazioni flessionali, tanto che i suoi valori minimi misurabili sono fino a qualche centinaio di volte minori. Sembra lecito dedurre da tutto ciò che il prodotto del decremento per la impedenza caratteristica è un parametro assai significativo come termine di confronto del comportamento dei solidi nei fenomeni vibratori.

f) Le costanti elastiche ed il coefficiente di risonanza di alcuni legni molto antichi (XVI secolo) sembrano notevolmente modificati da un processo di invecchiamento.

Gli AA. sentono il dovere di ringraziare anzitutto il prof. A. GIACOMINI, direttore dell'Istituto Nazionale di Elettroacustica « O. M. Corbino », per il benevolo interessamento dimostrato a questo lavoro. Ringraziano inoltre: il prof. G. GIORDANO, della Cattedra di Tecnologia del Legno e Utilizzazioni Forestali presso l'Università di Firenze, per la sua collaborazione alla interpretazione fisica dei risultati sperimentali, oltrechè per le numerose listerelle fornite ; il prof. R. CORMIO, direttore della « Civica Siloteca Cormio » di Milano; ed il liutaio G. TOMASSUCCI, che hanno fornito gran parte dei campioni, e l'apprendista liutaio E. MASSENZIO che con grande abilità ha preparato le numerose listerelle.

APPENDICE

Effetto dell'aria sulle vibrazioni flessionali della sbarretta.

Lo stra erello d'aria compreso fra sbarretta ed elettrodo aggiunge alla impedenza meccanica della sbarretta una rigidezza ed una resistenza; inoltre la radiazione introduce una reattanza inerziale ed una resistenza. L'effetto più importante, ed il solo che meriti di essere considerato nel caso presente, è però quello che ha la resistenza dello straterello d'aria, compreso fra sbarretta ed elettrodo, sul coefficiente di risonanza della sbarretta. Questa resistenza è data da [18]:

$$\mathcal{R}' = \frac{3}{2} \frac{\eta}{\pi d^3} S^2 \quad (g \cdot sec^{-1})$$

[18] P. G. BORDONI: *Ricerca Scient.*, **15**, 415 (1945).

dove: $\eta = 1,8 \cdot 10^{-4}$ (g $\cdot$ cm^{-1} sec^{-1}) = viscosità dinamica dell'aria; d (cm) = = distanza fra elettrodo e sbarretta; S (cm^2) = superficie utile dell'elettrodo. Con l'elettrodo impiegato è $S \cong 1$ cm^2. Si ha allora:

$$\mathscr{R}' = 86 \cdot 10^{-6}; \quad \frac{1}{d^3} \text{ (g $\cdot$ sec^{-1})}.$$

Come si vede, la resistenza aggiuntiva $\mathscr{R}'$ cresce assai rapidamente al diminuire della distanza d.

Mediante la formula precedente e ricordando l'espressione del coefficiente di risonanza in funzione dei parametri equivalenti della sbarretta ([19]) si cal-

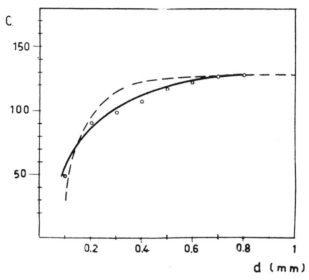

Fig. 8. – Coefficiente di risonanza apparente di una sbarretta in funzione della sua distanza dall'elettrodo eccitatore. (La linea continua rappresenta la curva sperimentale, la linea a tratti rappresenta la curva teorica).

cola facilmente quale valore deve avere la distanza di lavoro d perchè l'effetto di $\mathscr{R}'$ sul coefficiente di risonanza di una sbarretta sia contenuto entro limiti tollerabili.

Se si indica con $Q' < Q$ il coefficiente di risonanza apparente della sbarretta misurato in presenza dell'elettrodo, si trova che la variazione relativa del coefficiente di risonanza dovuta alla resistenza addizionale $\mathscr{R}'$ è:

$$\frac{\Delta Q}{Q} = \frac{Q - Q'}{Q} = \frac{\mathscr{R}'}{\mathscr{R}_0 + \mathscr{R}'},$$

dove $\mathscr{R}_0$ è la resistenza meccanica equivalente della sbarretta; od anche,

---

([19]) Cioè le costanti del sistema ad un grado di libertà equivalente alla sbarretta, nel modo di vibrazione considerato.

approssimativamente, se si tien conto che in condizioni di lavoro deve essere
sempre $\mathscr{R}' \ll \mathscr{R}_0$:

$$\frac{\Delta Q}{Q} \cong \frac{\mathscr{R}'}{\mathscr{R}_0} = \frac{86 \cdot 10^{-6}}{\mathscr{R}_0 d^3}$$

Si può allora calcolare il valore minimo da dare alla distanza d perchè, con
una sbarretta avente certi Q ed $\mathscr{R}$, la variazione relativa $\Delta Q/Q$ non superi
un valore assegnato.

Considerando, come esempio, un caso tipico (sbarretta di abete rosso,
avente una massa $m = 0,81$ g, un coefficiente di risonanza $Q = 130$ ed una
frequenza fondamentale di risonanza $f = 1035$ Hz) si ha:

$$\mathscr{R}_0 = \frac{2\pi f \mathfrak{M}_0}{Q} \cong 27 \text{ g} \cdot \sec^{-1}$$

$\mathfrak{M}_0 =$ massa equivalente della sbarretta, eguale a 0,677 m per il modo fondamentale ed a 0,5 m per i modi superiori 3°-5°, etc. ([20])). Se si vuole, ad
esempio, che sia $100\,\Delta Q/Q \leq 2$ % (valore accettabile data l'approssimazione
che si può sperare nelle misure di Q) si trova che d deve essere compreso fra
5 o 6 decimi di millimetro. Se la distanza scendesse invece, per esempio,
a 4/10 si avrebbe già un errore del 5 % su Q ed un errore maggiore del
10 % si avrebbe per $d = 3/10$.

Nella fig. 8 è rappresentato il coefficiente di risonanza apparente in funzione della distanza, per la sbarretta cui si riferiscono i calcoli precedenti.
Come si vede, l'andamento sperimentale del coefficiente di risonanza si ac-
corda abbastanza bene con quello previsto dalla teoria, ma è tale da consi-
gliare una distanza di lavoro un po' maggiore (0,7 ÷ 0,8 mm).

([20]) Tali valori si calcolano con un metodo analogo a quello seguito, per es., in un
lavoro di I. BARDUCCI: *Atti Accad. Naz. Lincei*, **1**, 206 (1946).

SUMMARY

Elastic parameters and internal friction have been determined on strips of wood
by a dynamical method, involving the measurement of the resonance frequency and
the Q of a vibrating reed.

A discussion of the experimental results leads to the following conclusions: a) the
sound velocity c and the Q are strongly dependent on the direction of the sample, while,
as a first approximation, their ratio c/Q is not a directional parameter. Moreover c/Q
is independent on frequency; b) as a first approximation, for that which concerns
the c and Q measurements, the angle between elastic waves direction and direction
of fibers is sufficient to determine the direction of the sample; c) the conifers show
interesting regularities in their elastic and dissipative properties; d) the elastic parameters and the Q of very old woods seem to ben highly modified by an ageing process.

Tables are given containing the mean values of some elastic and dissipative parameters for 85 different woods.

22

Measurement of the Internal
Friction and the Elastic Constants
of Wood

I. BARDUCCI and G. PASQUALINI

*This article was translated expressly for this
Benchmark volume by Elizabeth B. Abetti from*
Il Nuovo Cimento, **5**(5), *416–466 (1948).*

1. Foreword

The present experiment on the elastic properties and the internal friction of wood has a twofold purpose. First, one of the authors, in his work on the acoustical quality of violins done with electroacoustical methods (1), has ascertained the usefulness of experiments that analyze the separate component parts of the instruments. We have thought it proper, therefore, to complete these experiments by analyzing the mechanical properties of wood, of which these musical instruments are made (2). Second, we have thought that the systematic determination of elastic parameters and internal friction for various kinds of wood might have a wider application than just to musical acoustics. With research of this kind we can draw very precise conclusions about anisotropic and fibrous materials such as wood.

It is true that we have had data for some time on the elastic constants of wood and additional data on internal friction. If we wish to be more general, we must make our experiments, as far as possible, under controlled conditions and with specimens whose general characteristics (species, locality, direction of fibers, etc.) are well known.

The experimental method used derives from one recently employed by P. G. Bordoni (3), which is basically the measurement of frequencies and coefficients of resonance (or logarithmic decrements) on vibrating strips of the material in question.

Editor's Note: See end of Italian text for English summary. Figures appear in the preceding original article.

Some modifications have been warranted by the peculiar nature of the experiment, largely due to the necessity of working on a highly damped material like wood, for which it is not possible in a vacuum to determine the internal friction without altering the characteristics uncontrollably.

In the following paragraphs, after a brief description of the method of measurement and a glimpse at some conclusions drawn from the results of other authors, we shall discuss our new experiments and our general conclusions on the behavior of wood, hoping that this will lead to even further research.

2. Fundamentals of the Method of Measuring

It is well known that when we subject a free strip of wood of a certain material to forced vibration by bending, and when we have determined one of its resonant frequencies, we can attain the velocity of propagation c of the longitudinal elastic waves in a strip (infinitely thin) of the same material (4).

If the strip has a rectangular cross section, then

(1)
$$c = a_n - \frac{l_2}{h} f_n \quad \text{cm} \cdot \text{s}^{-1}$$

f_n (Hz) = nth natural frequency of bending of the strip
h (cm) = height of the rectangular section
l (cm) = length of the strip
$\quad a_n$ = numerical coefficient depending on the order n of the natural frequency.

The values of a_n for the first overtones are $a_0 = 0.975$ for the fundamental [with two nodal sections distant $0.224\ l$ (cm) from the ends of the strip]; $a_1 = 0.353$ for first overtone (with three nodal sections; one central and two placed at $0.132\ l$ from the ends); $a_2 = 0.180$ for the second overtone (with four nodal sections placed at $0.094\ l$ and at $0.356\ l$ from the ends).

Knowing c and the density ρ of the material one can calculate Young's modulus:

(2)
$$E = \rho c^2$$

To characterize the internal friction of the strips, we have chosen the coefficient of resonance Q, which, for heavily damped materials like wood, is more easily found by the logarithmic decrement d. We have the simple relation $Q = \pi/d$.

If f_n is one of the resonant frequencies of the strip, we have

(3)
$$Q_n = \frac{f_n}{(\Delta f)_n}$$

411

where $(\Delta f)_n$ is the frequency interval between the two points of the resonance curve (relative to f_n) whose amplitude of vibration is equal to ½ of the maximum value that we have corresponding to f_n.

3. Experimental Setup

Figure 1 represents the general electrical layout of the apparatus used for measuring frequencies of resonance f_n and of intervals of frequency $(\Delta f)_n$. To induce vibrations in the strips of wood, we have made use of electrostatic attraction. For detecting the vibrations we have used the frequency-modulation method, employing only one auxiliary electrode without continuous polarization, as described in another paper (3).

Here we limit ourselves to some modifications introduced in the experimental apparatus to adapt it to our particular problems. In the first place we have constructed a new mounting (Figure 2) adapted for strips of rectangular cross section, vibrating due to bending. The strip of wood S is suspended horizontally at the nodes of vibration by two thin wires F (copper, diameter 0.05 mm) attached to the bridges P by Faraday's gum. These bridges in turn are supported on sliding pivots in the groove G, which allow the relative position and alignment to be varied; thus the wires can be adjusted as desired. The wires F also serve to join electrically the surface of the strip of wood to mass. The strip is made a conductor by a very thin layer of graphite, which does not alter essentially the mechanical characteristics. The position of the electrode E and its distance from the strip of wood are adjustable, as is clearly shown in the diagram.

The presence of a thin layer of air between the wood and the electrode introduces a resistance of additional damping, which can lower considerably the apparent coefficient of resonance of the strip of wood. Such an effect is treated more at length in the appendix. Theory and experiment show that it is necessary to keep the electrode at a distance of at least 0.7–0.8 mm. Nor is it possible to avoid the difficulty by working in a vacuum, because, as has been shown, the properties of the wood would be necessarily altered (5). The rather large distance between the charged electrode and the strip of wood necessitates a charging voltage at acoustical frequency generated by the beat oscillator of 400–500 V in order for the accuracy of the method to be sufficient. Besides, the best oscillator must yield a rather fine regulation of the frequency so that we can measure the frequency intervals Δf, and thus the coefficients of resonance, with adequate precision. To this end the oscillator has a second condenser of rather low capacity in parallel with the main variable condenser, which yields total variations of frequency of 100 Hz.

The resonant frequency can be determined by using graduated scales on the two condensers, provided that we control the vibrations of the oscillator before every measurement by comparing the oscillograph with the 400 Hz of the reference tuning fork 14 (Fig. 1). We can ascertain our precision in measuring the coefficient of resonance to be about ±5 percent, which is more than sufficient for our purpose.

The error that we make in measuring resonant frequency is even smaller (less than 1 percent). Its effect on the measurement of elastic constants (velocity of sound; Young's modulus) can be overlooked in view of other causes of error. (See Section 6.)

4. Characteristics of the Method Used

The principal characteristics of this method that distinguish it from other methods used in research on the elastic properties and internal friction of solids (6) are the following:

a. The properties of the sample to be tested are not appreciably altered by the measuring apparatus. This fact, which is the main advantage of the method, results from working on a bar that is essentially free (7) and from using the electrostatic charge of the vibrations, which necessitates the approach of only one electrode, without touching the bar.

b. It is possible to work on rather small samples. This also is facilitated by use of an electrostatic charge and by the reinforcement of the vibrations through frequency modulation with only one auxiliary electrode. Most of the bars tested had dimensions of 10 by 1.0 by 0.2 cm³, but it would be possible to use bars of even smaller dimensions. It follows that we can analyze more easily the "local" properties of the material and the variations according to the position and alignment of the samples, finding irregularities that might not be noticeable in bars of larger size (8).

c. The forces applied to the bars are very small, as are the amplitudes of vibration. The orders of magnitude are, respectively, 10^{-4} Nw for the applied force and 1 μ for the amplitude of vibration (9). In other words, the material is excited only during a very small initial phase of its elastic characteristic.

d. The taking of the measurements is simple, especially when we compare our method with other similar research on the measurement of the logarithmic decrement. It has thus been possible to take measurements on a great number of samples.

5. Claims from Experiments Made by Other Authors

Before demonstrating results from the present research, we shall focus our attention on some results obtained by other workers. The velocity of propagation in wood varies considerably among species from values beyond 5,000 m·s⁻¹ (*abete*), that is, higher than the highest values of metals, to values much lower (about 3,000 m·s⁻¹). The velocity of propagation (just like Young's modulus) also varies considerably with direction. The preceding values refer to the direction parallel to the grain. For perpendicular directions we have values three to four times less.

As to the relatively scarce information on internal dissipation (10), we are especially interested in Rohloff's conclusions according to which the decrement of the wooden bars is essentially independent of the frequency for longitudinal vibrations as well as for transverse ones, at least for the range of frequency in which the experiments were made, that is, between 10 and 10,000 Hz. Besides, the decrement has the same value for the two types of vibration. These results are rather important because they show that the type of vibration is not important in the case of wood, as it is for example in metals (11). Since we have seen that the auxiliary electrode must be placed at a great enough distance from the bar so that the effect from the air is negligible, we can deduce that it is better to work with transverse vibrations; both the amplitude of vibration and the sensitivity of the method are greater.

The fact is that for bars of wood the decrement (and thus the coefficient of resonance) is independent of frequency and dimension (at first approximation). This shows moreover that the decrement can be considered as a true characteristic parameter of the material, like for example the density and velocity of sound. Rohloff has also found that the decrement is much higher for directions normal to the grain (about three and one half times greater) than for a direction parallel to the grain.

In conclusion, all the elastic characteristics and the decrement in one sample of wood depend on numerous parameters (location, seasoning, humidity, etc.).

6. Measurements Made

The samples of wood tested number over 400 and belong to eighty-five different species. Almost all the bars had dimensions of about 10 by 1.0 by 0.2 cm^3. Only for experimental verification, of which we shall speak later, did we use samples of different dimensions. In the choice of dimensions for the bars we were influenced by various factors. The smaller bars lend themselves better to the study of how the physical properties of the same species may vary locally. However, by diminishing the thickness h we increase the difficulty of a precise experiment. Besides, the damping effect of the layer of air between the bar and the electrode is proportionally greater for smaller bars, the frequency being equal.

We have found it is not feasible to reduce the thickness below 2 mm; once the thickness h has been chosen, the length of the bar is determined according to the frequency desired. With the experimental means at our disposal the most convenient frequency values range from some hundreds to some thousands of hertz or little more. Working on the lowest natural frequency of the bar, we find from formula (1) that the most convenient value of l is around 10 cm. The width b of the bar can be chosen with the sole stipulation that it be much less than the length l in order that the conditions of formula (1) be fulfilled.

On every bar we make six direct measurements: the three geometric measurements, l, b, h, the weight P, the resonant frequency f of the fundamental (or in some cases the first overtone), and the frequency interval of Δf (see Section 2) (12).

From these measurements we can calculate the values of six physical parameters characteristic of the material, that is, the density ρ, the velocity of propagation c, Young's modulus E, the characteristic impedence $\rho(1/E) = \rho c$, the coefficient of resonance Q, and the product of the logarithmic decrement d and ρc, that is $K = d\rho c$. The importance of this last parameter will be apparent in section 7. In the determination of the thickness h we have taken the mean value of h for five equally divided sections. We must remember that for small samples it is always the value of h, not the frequency, which limits the precision with which we can determine the velocity c (by some percent).

As we have mentioned before, we have made some experimental tests trying to prove the most important of Rohloff's conclusions: that the coefficient of resonance is independent (in a first approximation) of the dimensions of the bar, of the type of vibratory phenomenon (transversal or longitudinal), as well as of the frequency. In Table I are listed the results of one such test, which shows Q as independent of dimensions (at least within the limits of the experiment).

7. Discussion of Results

To examine the experimental results we must refer to a typical case, found in Table II. We are using ten bars, all oriented differently, but all taken from the same trunk of *abete rosso* (*Picea excelsa*). The orientation of each bar can be easily identified with the help of Figure 3. As we can see, the orientation of the bar is represented by one or two letters that indicate the plane of position and by the angle to a reference axis in this plane: more precisely, a straight line parallel to the axis of the trunk for the radial planes (R) and the tangential planes (T), and a tangent to the annual growth rings for the transversal plane (N). We have also indicated one of the bisecting planes (RT) of the angle formed by a radial and a tangential plane, with the other bisecting plane (TR) normal to the first.

In the table the bars are grouped in five pairs; within the same pair there are always two bars with the longer sides l parallel and the widths b lying in perpendicular planes.

The values of density ρ, of velocity c, and of the coefficient of resonance Q are nearly the same for two bars in the same pair. What differences there are are similar to those found in different bars equally oriented but taken (for example) from different trunks, or even from different parts of the same trunk. With a little less approximation we can classify the bars of Table II into two groups. The first group comprises four bars (R-$0°$; T-$0°$; RT-$0°$, TR-$0°$) all having their longer side parallel to the axis of the trunk. The second group comprises the remaining six bars, all having their side normal to the axis (but lying in planes oriented differently).

The bars of the first group have a velocity of propagation and coefficient of resonance somewhat higher than those of the second. We deduce one important conclusion. In a first approximation we may speak of parameter values according to a parallel orientation (bars with the longer side parallel to the axis) and of values

415

according to a perpendicular orientation (bars with the longer side perpendicular to the axis of the trunk). This result was already known before, but with the help of Table II we can see the entirety of the approximation schematically.

On closer examination we can see that there are differences of value (at least for the velocity *c*) between the two pairs of bars with their axes parallel to that of the trunk, and these differences are quite a bit larger than those existing between bars of the same pair. This could be caused by the difference in characteristics among various parts of the same trunk. The mean density is clearly less for the second pair than for the first. Besides, the disposition of the annual growth rings in the bars of the second pair (*RT*-0°; *TR*-0°) is not symmetrical with respect to the plane of excitation, and that perhaps could make less valid the simple theory shown in Section 2. Comparable results have been found for the other specimens of wood. From what we have said, *we must assume, for values of the velocity of propagation and of the coefficient of resonance Q in a direction parallel to the axis, the mean value which the same parameters have for that orientation R-0° and T-0°.*

Similar considerations apply to the other three pairs of bars, cut in directions normal to the axis of the trunk. We note that the two pairs of bars (*R*-90°; *N*-90°) and *T*-90°; *N*-0°) give values for *c* and *Q* somewhat greater than those of the last pair (*RT*-90°; *N*-45°); For values relating to the right-angle orientation we must choose the mean value corresponding to the first four bars (*R*/90°; *N*-90°; *T*-90°; *N*-0°) (13).

a. Values of Parameters in the Axial Direction

The values of the characteristic parameters (c, E, Q, $k = d\rho c$) as a function of the propagation in a direction parallel to the axis of the trunk have been listed in Table IV for eighty-five different kinds of wood. Next to the mean values of the six physical parameters, we have indicated the number of samples tested for each species. This can serve as an index of confidence of the relative data for each species.

As to the difference in physical characteristics among samples of the same species, the case for the *abete rosso* (Table III) is the most favorable. For almost all the other species for which we could find a suitable number of specimens, we have found differences from one bar to another much greater than for the *abete rosso*, although for the *abete rosso* we tested seventeen samples, as against seven samples as the highest number for any other species. The explanation lies evidently in the very great regularity of the fibrous structure of *abete rosso*. This is what makes the wood so particularly adapted to the making of musical instruments.

In contrast we notice in Table III an example that can be considered the opposite of *abete rosso*. In this species (*Acer platanoides*) the notable discrepancies found among the samples are explained by the irregular structure with twisted fibers of maple.

For brevity we have not shown in Table IV the limits of variation for each parameter. The two examples cited, fir and maple (Table III), can serve as a guide,

416

if we remember that for most other species the situation is somewhere between these two extreme cases.

The different behavior of the various species with respect to homogeneity of characteristics has this consequence. The characteristics of a sample belonging to a given species can be predicted with an approximation, very different from one species to another, if we note the mean values of the various physical parameters for the given species. This will remain true even if the mean values have been found on an equal number, and on a large enough number, of samples. They will thus be all equally useful in the aforementioned sense.

The graphs in Figure 5 show the statistical distribution of the values of each of the six parameters determined (ρ, c, ρc, E, Q, $d\rho c$) from the eighty-five species of wood examined. We can see that the relative grouping is least for density and closer for the other parameters, especially for the velocity of sound and for $k = d\rho c$.

In Table V are listed the maximum and minimum values found for each of the six parameters from among all the eighty-five species examined.

The difference between maximum and minimum value of a given parameter varies between 2.8 for the velocity of propagation and 14 for the coefficient of resonance. We see also that the maximum and minimum values of different parameters belong often to the same species (but those species represent exceptional cases, however). For example the species *Diospyros* (ebony) appears three times in the table, because it represents the three highest values of ρ, ρc, and E. The species *Vitis vinifera* represents three extreme values, minimum values of c and Q and the maximum value of k.

The maximums of ρc, E, and Q all belong to very heavy woods. The minimums of c, ρc, E, and Q and the maximum of k belong to two vines (grape and wisteria).

A more accurate examination made with the help of tables (not shown), in which the specimens are listed in order of increasing (or decreasing) values of one of their physical constants, would show that some of the cases listed in Table V are absolutely exceptional. For example, for the velocity c the four smallest values (from 2.50 to 2.05 km $\cdot$ s^{-1}) belong to four vines (ivy, bamboo, wisteria, and grape). For all the other eighty-one species the minimum value of c is 3.25 (mulberry, eucalyptus). To the same four vines belong the minimum values of Young's modulus (between 2.10^{10} and 3.10^{10}) and of the coefficient of resonance (from 15 and 40); they also, along with a fifth species (*Euphorbia*) give minimums of ρc.

Not counting the four species mentioned above, the fields of variability of almost all the parameters considered are reduced considerably and the range between maximum and minimum value becomes 1.75 for c, 4.8 for ρc, 4.5 for E, 4.2 for Q, and 3.2 for k.

Along this same line we can also observe that the highest velocity of propagation (greater than 5 km $\cdot$ sec) belongs for the most part to light woods (thuja, balsa, *Picea* and poplar) which all have intermediate values of coefficients of resonance ($Q = 90$–160). The largest Q (that is, the least internal friction) is found instead in some heavy woods. Two of the tested woods seem to combine high values of velocity and of the coefficient of resonance, *Pterocarpus* and teak.

417

b. Values of Parameters as a Result of Orientation Normal to the Axis

An examination of the characteristics due to an orientation normal to the axis of the trunk has been made on a number of species (11), although much less than for a parallel axis. Only for one kind of wood (*abete rosso*) have we tried several bars cut from different trees. The reason for this has been the difficulty in cutting thin strips, perpendicular to the grain, with the necessary precision and without damaging the strips, especially for the less dense woods.

The specimens examined have values of ρ, c, and Q not too different one from another, and midway between the possible extremes for such parameters, all cited in Table V.

The orthogonal characteristics (orthogonal in the sense already discussed) gathered together in Table VI do not represent the average properties of the various species, but rather they give an idea of what happens in single cases, chosen nevertheless from among the more significant species.

In accordance with what we know (Section 5) about the velocity of propagation c_n and the coefficient of resonance Q_n in directions normal to the grain, we have measured corresponding values c_1 and Q_1, somewhat smaller and parallel to the grain.

For the woods under experiment, c_n varies between 1 and 1.70 km $\cdot$ s^{-1}, while Q_n varies between 30 and 50 (approximately).

In contrast to the density, the velocity c, Young's modulus, and the coefficient of resonance are strongly affected by the orientation, and are somewhat less when in a perpendicular direction.

When we consider the ratios (c_1/c_n) and (Q_1/Q_n) among values resulting from both parallel and orthogonal orientations, we observe that the two ratios have in general values not too different for a given species. In fact, they are often equal. The fact that velocity c and the coefficient of resonance vary almost in the same ratio passing from a parallel to an orthogonal orientation signifies that the ratio between the velocity of sound and coefficient of resonance is largely independent of orientation; that is, $c_1/Q_1 \cong c_n/Q_n$.

The values of c/Q could be assumed as a characteristic parameter, for example, next to density. For reasons that we shall explain shortly, the product $k = d\rho c = \pi\rho c/Q$ seems more significant, and it also depends but little on direction as we see in the last two columns of Table VI. In the most unfavorable case (*Picea excelsa*, year 1935) the product k varies only 23 percent between parallel and orthogonal direction, while the velocity of sound varies almost 500 percent and the coefficient of resonance varies beyond 400 percent.

The ratios c_1/c_n are not too different for one and the other case, as also are Q_1/Q_n. We cannot be sure, however, that this fact induces a general property, because we have already said that almost all our eleven species had very similar properties.

c. Aging of the Wood

In Table VI we observe that the third sample of *abete rosso* has a velocity of propagation and coefficient of resonance quite a bit lower than that of the first two samples, and that they are opposed to the mean characteristics of the species as listed in Table IV.

Since this sample was rather old, we assume that the difference in characteristics is due to the process of aging. The hypothesis is confirmed by a second sample, belonging to another species (*cirmolo*) which dates from the same period (sixteenth century).

This result is rather important for its application to violin making, because it perhaps can explain the marked difference between ancient and modern bowed instruments.

It is worth noting that F. Saunders (2) has arrived by an entirely different method at the same conclusion; that is, the properties of ancient instruments can be explained by a process of aging within the wood.

d. Characteristics of Resinous Woods

Until now we have made no distinction between resinous (conifers) and broad-leaf (deciduous) woods. It is easy, however, to see that the various resinous woods must present very similar properties, given the great regularity of their fibrous (grain) structure.

Figure 6 is a graph in which the values of c are functions of the coefficient of resonance Q for the eighty-five species whose characteristics are listed in Table IV. To each of the species there corresponds a point (broadleaf) or a small circle (resinous) that represents the mean properties.

We see immediately that the points corresponding to the resinous woods are almost all aligned along two lines passing through the origin. This signifies that almost all the resinous woods fall into one of two groups for each one of which the ratio c/Q is constant.

Given $\pi \rho c/Q = d\rho c = k$, the preceding statement confirms the importance of the parameter k, already which has been shown to be independent of direction.

All this is even clearer in Table VII where fourteen resinous woods are grouped according to increasing (mean) values of k. We also notice two large groups in the table, the first formed of eight species (k between 7.7×10^3 and 8.4×10^3) and the other by four species ($k = 5.8 \times 10^3$ to 6.3×10^3). In the first group especially, the grouping is not accidental, because seven out of the eight species belong to the same resinous family (pine).

We do not find such marked regularity for the resinous woods if we arrange the species according to increasing values of ρ, c, Q, and so on. In such cases there is only a statistical distribution (14).

419

The results of one valid test that will explain the probable reasons of such regularity are tabulated in Table VIII. Here we list the characteristics of two strips of *abete concolore,* one cut in the spring zone, the other cut in the fall zone of the same annual growth ring.

Such an experiment has been possible because of the marked radial extension of the annual rings (in the specimen chosen) and because of the reduced thickness (2 mm) of the bars, which is, however, sufficient for our method of measurement.

Passing from the spring zone to the fall zone, the velocity of propagation stays about the same, although the density increases by more than 1 to 2. This result, which might seem surprising, just shows that Young's modulus increases proportionately with the density, passing from one zone to another. The coefficient of resonance remains about the same, however, as does the velocity c.

Let us assume that the density ρ and the product k of the spring zone are the same for the fall for a given species (for example, *abete concolore*). The values of these parameters measured on a given specimen cut arbitrarily (but always with the axis parallel to the trunk) depend only on the proportion of spring and fall zones included in the specimen.

The simplest explanation of the equal values of k for several resinous kinds of wood is that k, in each of the two zones of the annual growth ring, varies little from one variety of wood to another and that the proportions of the two zones are about the same. However, this second hypothesis is not generally acceptable from what we know of resinous woods. We must admit that, passing from one resinous wood to another that has the same k, we have variations in the proportions of the two zones in the same growth ring and in k of each zone. But the effects of these variations must tend to compensate each other.

Figure 7 explains the preceding more clearly. The values ρc are shown as a function of Q for all the individual bars belonging to the resinous woods, instead of the mean values of each species as in Fig. 6.

Figure 7 also shows the two points corresponding to the two bars of *abete concolore* to which Table VIII refers. Each straight line passing through the origin in the diagram represents a given value of k.

We see that almost all the points lie in the sector bound by the straight lines which pass through the points corresponding to the spring zone and the fall zone (of *abete concolore*) (15).

The points are distributed over the entire sector and are not grouped together around a few special values of k, as happens instead for the mean characteristics of the various species. This can be easily explained by the presence of random variations from one example to another, both in the characteristics of the two zones and in their proportions. If the points corresponding to different species were represented in a different way (which for clarity has not been done in the figure), we would see for *abete rosso,* whose structure is rather regular, that the grouping together about a mean line is much clearer than for all the other species, even though the number of tested samples is far greater.

e. Cause of Anisotropy

From all these experimental results we can easily deduce the probable factors that make the velocity of sound and the coefficient of resonance vary with the direction. We have seen in fact that all the bars with axes parallel to that of the tree trunk act more or less alike, regardless of their orientation with respect to the growth rings. And we can say the same for all bars cut with their axes normal to that of the tree trunk (cross grain).

We can explain this most naturally by supposing that the most important cause of anistropy in wood lies in its minute fibrous structure rather than in the larger variations due to the presence of the annual rings. Otherwise, we would not understand why two bars like (R-90°), in which the annual rings are normal to the axis, and (N-0°), in which the growth rings are longitudinal, give the same values for c and Q. The only thing the two bars have in common is that their longer side is normal to the grain.

Given our hypothesis, the values of c and Q in a direction parallel to the grain are determined mainly by the properties of the single fibers. For c and Q in a direction normal to the grain, the connection between the different fibers is of greater importance. Naturally, the all small differences of one type or another that exist among bars can very well be due to the presence of the growth rings.

This interpretation is made more valid by the fact that the spring and the fall zones have almost equal values for the velocity of propagation and the coefficient of resonance, at least in resinous woods (16).

We have a direct proof of this from an experiment made on four larch bars, as reported in Table IX.

The first three bars all have the axis parallel to the grain, but one lies in a radial plane and is composed partly of the spring zone and partly of the fall zone. The second and third are both lying in a tangential plane, but composed solely of either one zone or the other. The velocity c in the three bars is nearly equal, in agreement with what we have seen for *abete concolore*. In the fourth bar, which differs from the third only in that its axis makes an angle of 30 degrees to the grain, the velocity of propagation is considerably less. This proves decisively that the direction of the grain is the cause of anisotropy.

Let us suppose for a first approximation that the law of variation for c is elliptical. Then we have $c_n \cong 1.5 \times$ km^{-1}, in agreement with our experimental data.

f. Comparison of Properties of Other Solids

It is interesting to compare wood with other solids, particularly metals. We have seen for wood that the product k is a rather significant parameter. We shall use it as a basis for comparison. For all eighty-five specimens tested k varies from 4.5×10^3 to 21×10^3 g· cm^{-2}/s^{-1}. If we omit a few (seven) exceptional specimens, the field of variation is reduced considerably from 5.5×10^3 to 13×10^3.

For metals, k assumes values from ten to one hundred times less, when we take for d, the decrement measured for longitudinal vibrations, a decrement that often has very small values (sometimes 1×10^{-5}).

If we take for d the largest value that we can assume (at a given frequency) for transverse vibrations, the value of k for metals thus calculated has the same order of magnitude as k for wood in the whole field of acoustical frequencies (17). In other words, the values of k that are normal for wood constitute an upper limit for metals.

The minimum coefficients of resonance for metals range from 250 to 1,500, that is, from a value a little larger than the largest measured for wood to a value six times greater.

8. Conclusion

The conclusions derived from the foregoing paper are the following:

a. The velocity of propagation of elastic longitudinal waves and the coefficient of resonance in bars of wood vary greatly with the angle between the direction considered and the direction of the grain. For a first approximation we can say that this angle is sufficient to determine the orientation of the sample under test. It is convenient to refer to values of the velocity c_1 and the coefficient of resonance Q_1 for directions parallel to the grain, and to values c_n and Q_n for directions normal to the grain. Besides, the coefficient of resonance, as Rohloff has shown, is essentially independent of the frequency and of the type of vibration.

b. The ratio c/Q is independent both of direction and frequency. The same is true for the product $k = d\rho c$.

c. In resinous woods (conifers) the woods from the spring and the fall zones (of the annual growth rings) have nearly equal velocities of propagation and coefficients of resonance. The density and Young's modulus vary in a ratio of 1:2 from one zone to another. The properties of any sample whatsoever stand somewhere between the properties of the two zones.

d. The properties of resinous woods show marked regularity. In particular, twelve out of fourteen of the tested woods lie in two groups, and in none of them does k vary very much. A probable explanation of this experimental fact is suggested by one result already noted (see conclusion c).

e. Simple laws like these have not been found for deciduous trees. This is not astonishing since the structure of conifers is much more regular. In any case, we find that k does not vary tremendously for any of the eighty-five woods examined, in relation to the variation of k's three factors. If we exclude a few exceptional species, k varies only from 5.5×10^3 to 13×10^3 g·cm^{-2} s^{-1}, that is, in a ratio of 1:2.35.

The maximum k values for many other solids, for example, metals are of the same order of magnitude but only at certain frequencies and for transverse vibrations. In fact, for such solids (metals) k, as a decrement, varies considerably with the frequency in the case of transverse vibrations, and the minimum measurable values are some hundreds of times less. It seems safe to assume from all this that the product of decrement and characteristic impedance is a rather significant parameter for comparison of vibratory phenomena in solids.

f. The elastic constants and the coefficient of resonance of some aged woods seem highly modified by the process of aging.

Notes and References

1. N.T. (Not translated; see original.)
2. F. A. Saunders arrived at a similar conclusion, in a work published in *J. Acoust. Soc. Amer.*, **17**, 169 (1946), while the present paper was in process.
3. N.T.
4. We must not confuse c with the velocity of propagation of the same longitudinal waves in an undefined solid.
5. As for the effect of suspension wires, we can show, by a method similar to the one used by P. G. Bordoni already cited, that it can be ignored, even if the error made by setting the wires at the nodes is noticeable, e. g., ½ mm.
6. N.T.
7. One could work instead with strips fixed at one end, as has been done effectively by other experimenters. See note 10.
8. In particular it is possible to measure the characteristic differences between the spring and the fall zones of the wood.
9. They can be calculated by a method as in Appendixes A and D and as cited by P. G. Bordoni.
10. N.T.
11. N.T.
12. Since the principal scope of this work was to procure statistical data from a large number of samples rather than to make precise measurements on a small number of wood strips, we have not made systematic tests on the effect of humidity, although we have worked under sufficiently comparable conditions.
13. It does not seem necessary to speak (at least in general) of precise and well-defined differences between the strips of group R-90° and N-90° and those of group T-90° and N-90°, as is sometimes done, introducing, for example, a velocity of sound relative to the radial and one relative to the tangential direction.
14. One reason for this is when the product k is preferred to the ratio c/Q as a characteristic parameter. This becomes apparent when the properties of resinous woods are entered in a diagram that makes c a function of Q as in Figure 6.
15. One exception is a low density wood (*Thuja*), where the characteristics of the two zones are quite different from those of *abete concolore*. The same test has been made on larch wood with very wide annual rings with identical results.
16. No similar test has been made on deciduous specimens.
17. This is easily seen by looking at the data of C. Zener, Note 10.

Editor's Comments on Papers A and 23B

23A Fukada: *The Vibrational Properties of Wood I*

23B Fukada: *The Vibrational Properties of Wood II*

Dr. Eiichi Fukada (1926–) received his undergraduate training in physics at the University of Tokyo, graduating in 1944, when he joined the Kobayashi Institute of Physical Research, where he became Chief Research Member in 1959. He obtained his D. Sci. degree from the University of Tokyo in 1960, and later moved to the Institute of Physical and Chemical Research, Tokyo and Saitma, where he is at present Principal Scientist and Chief of the Biopolymer Physics Laboratory. Since 1961 he has also been Professor of Physics at Gakushuin University, Tokyo. In 1956–1958 he was a British Council Scholar at Imperial College, London, and in 1965–1966 Visiting Professor at New York University. He was appointed Fellow of the International Academy of Wood Science in 1966, and received the Yamaji Natural Science Award in 1970.

His first paper presented here gives findings on the relations among Young's modulus, logarithmic decrement, and frequency in twenty-eight different species of wood, showing an interesting variation in log decrement with frequency between broadleaf (deciduous) and needle-leaf trees (conifers). (Spruce is a conifer and maple a deciduous tree.)

The second paper explores the temperature and moisture-content dependence of Young's modulus and log decrement with frequency for one conifer and one deciduous type. It is hoped that someday these findings can be used to help explain the variations in tone qualities of violins that take place with changes in temperature and relative humidity.

Reprinted from *J. Phys. Soc. Japan*, **5**, 321 – 327 (1950)

The Vibrational Properties of Wood I.

By Eiichi FUKADA.

Kobayashi Institute of Physical Research, Kokubunji, Tokyo.

(Read May 1, 1949; Received January 26, 1950)

§ 1. Introduction.

The vibrational properties of wood are very interesting as it is used popularly in many musical instruments. They are very different according to its original location in the tree, the direction of fibers, the moisture content, the temperature, and the previous treatment. Therefore it is desirable to repeat the measurement so many times that the reliable results could be obtained. We have measured the logarithmic decrement and the Young's modulus of wood by means of a new method which seems to be comparatively simpler than the previously reported ones[1][2][3]. In the present paper we will report the data on the various woods under the condition of air-dried state.

§ 2. Measuring Method.

Fig. 1. gives the schematic diagram of measuring apparatus. The dimensions of test piece M are about 30 cm long, 2 cm wide, and 0.5 cm thick. We suspend it with two fine threads F at the nodal lines corresponding to each mode of lateral vibration. The thin iron piece P is attached on both ends of specimen and a pole of the electromagnetic coil L_1, L_2, is respectively set with a small air gap toward it. When the alternating current of the valve oscillator is sent through coil L_1, the specimen begins the lateral forced vibration and the induced electromotive force arises in coil L_2. When the frequency of alternating current in exciting coil L_1 is adjusted to coincide with the natural frequency of wood specimen, the induced electromotive force in receiving coil L_2 has a maximum value. This maximum is determined by the microammeter A. The Young's modulus of wood is calculated from this frequency f according to the following formula,

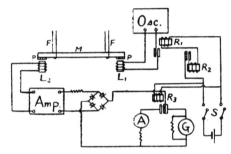

Fig. 1. The schematic diagram of measuring apparatus.

$$E = \frac{48\pi^2}{m^4} \cdot \frac{l^2}{a^4} \cdot \rho \cdot f^2, \qquad (1)$$

where l is the length, a the thickness, ρ the density of specimen, and m the numerical constant determined by the mode of vibration.

The method of measuring the logarithmic decrement is as follows; on the condition of resonance we switch off the exciting current and rectify the damping current in receiving coil L_2 caused by the damping vibration of specimen and measure its total charge by a ballistic galvanometer G. When a switch S is closed and the contact of relay R_2 is touched with hand, then relay R_1 operates and the alternating current flows through the exciting coil L_1, which makes the specimen vibrate. The alternating current induced in the receiving coil L_2 is amplified and full-wave rectified. When the switch S is open, the contact of relay R_1 separates, and the exciting current disappears, and the specimen ceases to vibrate. If the switch S is closed again, as the contact of relay R_2 separates, the exciting

current does not flow and the specimen does not vibrate. When the switch S is opened, the contact of relay R_3 touches on the side of ballistic galvanometer G, by which the total charge of rectified damping current caused by the damping oscillation of specimen is measured. Owing to the relay R_2 we can also measure the charge in the arbitrary time interval while the switch S is open, and reduce the error due to the hum of amplifier by opening the switch S only for the necessary small interval. It is already known that the damping of vibration of wood is just exponential at small amplitude[a]. Therefore the amplitude of alternating current induced in the receiving coil L_2 decreases as $\exp(-\alpha t)$ with the time t, where α is the damping constant. Denoting the initial amplitude of current I_0 and its angular frequency ω, the total charge obtained is

$$Q = \int_0^\infty I_0 \exp(-\alpha t)|\cos \omega t|dt$$

$$= I_0 \left(\frac{\alpha}{x^2 + \omega^2} + \frac{\omega}{x^2 + \omega^2} \operatorname{cosech} \frac{\pi\alpha}{2\omega} \right). \qquad (2)$$

As $\qquad \omega^2 \gg \alpha^2$,

$$Q = I_0 \cdot \frac{2}{\pi\alpha}. \qquad (3)$$

When the specimen is vibrating steadily, the current I in the microammeter is

$$I = \frac{\omega}{2\pi} \int_0^{\frac{2\pi}{\omega}} I_0 |\cos \omega t| dt = I_0 \frac{2}{\pi}. \qquad (4)$$

From (3) and (4), we have

$$Q = I/\alpha. \qquad (5)$$

This is the same as the representation of the total charge of the direct current the amplitude of which is initialy I and attenuates as $\exp(-\alpha t)$. The logarithmic decrement λ is defined as

$$\lambda = \log_e{}^{\alpha T} = \alpha T = \alpha/f. \qquad (6)$$

where T is the period of vibration. From (5)

$$\lambda = I/fQ. \qquad (7)$$

Using this relation the logarithmic decrement λ can be calculated.

But here is an indispensable correction. For after opening the switch S and separating the contact of relay R_1, a finite time interval $\varDelta t$

elapses before the contact of relay R_3 touches on the side of ballistic galvanometer. Accordingly the electric charge do not flow in this interval through the galvanometer and it causes the systematic error. Now if the direct current I decreases as $\exp(-\alpha t)$, its total charge Q is

$$Q = \int_0^{\Delta t} Ie^{-\alpha t}dt + \int_{\Delta t}^\infty Ie^{-\alpha t}dt$$
$$= \Delta Q + Q_m, \qquad (8)$$

where Q_m is the measured charge and ΔQ the unmeasured charge. We indicate by λ_m the logarithmic decrement calculated from Q_m using the relation (7) and by λ the true logarithmic decrement, then

$$\frac{\lambda}{\lambda_m} = \frac{Q_m}{Q} = \frac{Q - \Delta Q}{Q} = 1 - (1 - e^{-\alpha \Delta t})$$
$$= e^{-\alpha \Delta t} = 1 - \alpha \Delta t + \frac{\alpha^2 \Delta t^2}{2}. \qquad (9)$$

Therefore we must add the correction term $\Delta \lambda$ to the measured value λ_m, that is,

$$\lambda = \lambda_m - \Delta \lambda, \qquad (10)$$

where

$$\Delta \lambda = \lambda_m^2 \cdot f \cdot \Delta t \,(1 - \lambda_m \cdot f \cdot \Delta t/2). \qquad (11)$$

To determine this time interval Δt we charged a condenser C to which a resistance R was connected in parallel and discharged the charge on this condenser through the relays used in the apparatus and measured the difference between the discharged charge when the parallel resistance is connected and that when not connected. From this amount and the value of R and C, Δt was obtained after a simple calculation. In our apparatus Δt was about 2×10^{-3} seconds. Further by appling the voltage of exciting coil L_1 and that of ballistic galvanometer G simultaneously to the oscillograph, we photographed the change of the exciting and damping current. This photograph showed that Δt was of the above-mentioned order of magnitude. The correction containing Δt is considerably large when the frequency of vibration or the logarithmic decrement is large, so that this correction must not be neglected.

In the range of about 100—5000 cycles per second we used the resonance at each mode of vibration and the highest mode of vibration used

was about sixth overtone. According to each mode of vibration each nodal line was suspended by thread. Though the suspending position deviated a few millimeters from the true nodal line its effect was negligibly small[3]. The amplitude of vibration was less than the order of one tenth millimeters and the logarithmic decrement was indifferent to the amplitude.

§ 3. Experimental Results.

Fig. 2 and 3 give respectively the experimen-

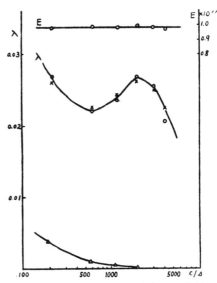

Fig. 2.   The Young's modulus and the logarithmic decrement of air-dried Tabu.

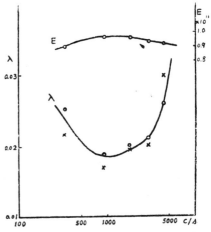

Fig. 3.   The Young's modulus and the logarithmic decrement of air-dried Momi.

tal results (the mark ○) for Tabu (a broad-leaved tree) and Momi (a needle-leaf tree). To confirm the precision of the new method the values measured by the resonance method are shown with the mark ×. When we denote by Δt the frequency interval at which the amplitude of receiving current takes the half value of the maximum at resonance and f_r the resonant frequency, the logarithmic decrement λ is given as follows;

$$\lambda = \frac{\pi}{\sqrt{3}} \frac{\Delta f}{f_r}. \tag{12}$$

It is seen that the two methods give the same results. The value obtained from the photograph of damping current taken by the oscillograph also nearly coincided with the above results. The measured values for the same specimen, the length of which was cut half or one third of original one, could be also plotted on the same curve within the limit of error. Therefore the variation of λ due to frequency is not originated from the mode of vibration but from the frequency itself.

The lower curve in Fig. 2. shows the attenuation due to air, that is, the difference of the logarithmic decrement measured in the atmospheric pressure from that measured in about 10 mm Hg vaccum. The logarithmic decrement of wood itself is, therefore, the measured decrement minus this decrement due to air.

In the measurable frequency range (about 100−5000 c.p.s.) the Young's modulus shows hardly any variation but the decrement varies remarkably with frequency and its behavior differs according to the kind of wood. After investigation of about 30 kinds of wood, we have found that the frequency characters of the logarithmic decrement could be divided into two main types. One is such that the decrement increases gradually with the frequency (A), to which the needle-leaf trees belong. The other is such that the decrement decreases with a maximum value on the way (B), in some cases this maximum disappears (B'). The broad- leaved trees belong to this type. Fig. 4 gives the measured results for Sugi. Hinoki, and Ezomatsu as the examples of the first type, these are all the needle-leaf trees. Fig. 5 gives the data for Kaede, Harunire, and

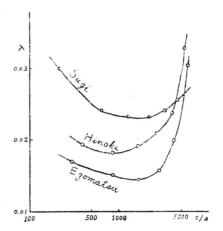

Fig. 4. The logarithmic decrement *vs.* frequency; A type.

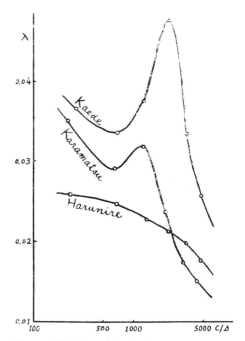

Fig. 5. The logarithmic decrement *vs.* frequency; Band B' type.

Karamatsu as the examples of the second type, where the former two are the broad-leaved trees but the last is the needle-leaf tree. The eighteen kinds of needle-leaf tree used in the experiment showed the first type (A) with only one exception of Karamatsu and the ten kinds of broad-leaved tree all showed the second type (B) or (B'). The distinction between B and B' is almost meaningless, for both change to each other and the height of maximum in the B state also varies according to the moisture content. These precise relations are still unknown.

Table 1 shows the kind of wood, density, Young's modulus, and the logarithmic decrement at the frequency of fundamental mode (in the neighbourhood of 100—200 c.p.s.), and the frequency character of the logarithmic decrement. The meaning of the notation A, B, B' was above described. The woods denoted by No. 1 to No. 18 are the needle-leaf trees and those from No. 19 to No. 20 are the broad-leaved trees. It is seen that the frequency character of the logarithmic decrement is distinctly discriminated according to the kind of tree. These specimens were all in the state of so-called air-dried, namely, dried by being left in the atmosphere for a long time and their moisture content would be about 15 percents by weight. The temerature of measurement was the room temperature in winter namely about 10°C. It must be noticed that the different results would be obtained under the different conditions of moisture content or temperature.

All the specimens in the above table are such that their fibers are orientated in the direction of the length of specimens. The different results might be expected when the fibers are orientated perpendicular to the direction of the length of specimen. Fig. 6 and Fig. 7 show the comparison of two cases for Bishūhinoki. It is seen that the Young's modulus is about one tenth

Table I.

The Experimental Results of the Various Woods.

No.	Local Name.	Botanical Name.	Density	Young's modulus	Logarithmic decrement	Type
1	Sawara	Chamaecyparis pisifera Endl.	0.278	0.58×10^{11}	2.45×10^{-2}	A
2	Sugi (Japan cedar)	Cryptomeria japonica D. Don.	0.358	0.52	3.10	A
3	Sirabe	Abies Veitchii Lindl.	0.390	0.90	1.74	A
4	Hinoki (Japan cypress)	Chamaecyparis obtusa Endl.	0.397	0.95	1.95	A
5	Bishūhinoki	ditto (produced at Bishū).	0.356	0.84	2.31	A
6	Ōtakihinoki	ditto (produced at Ōtaki).	0.377	0.80	1.80	A
7	Taiwanhinoki	ditto (produced at Taiwan).	0.424	1.00	1.73	A
8	Miurehinoki	ditto (produced at Miure).	0.453	1.12	2.00	A
9	Kisohinoki	ditto (produced at Kiso).	0.472	0.98	2.45	A
10	Kōchihinoki	ditto (produced at Kōchi).	0.487	0.66	2.18	A
11	Ezomatsu (Spruce)	Picea jezoensis Carr.	0.399	1.11	1.70	A
12	Todomatsu	Abies sachalinensis.	0.407	0.91	1.85	A
13	Momi (Fir)	Abies firma Sieb. et Zucc.	0.413	0.90	2.38	A
14	Tōhi	Picea hondoensis Mayr. Monog.	0.420	1.15	1.90	A
15	Asunaro	Thujoposis dolabrata Sieb. et Zucc.	0.487	0.90	1.85	A
16	Akamatsu (Red pine)	Pinus densiflora Sieb. et Zucc.	0.516	0.89	2.43	A
17	Tsuga	Tsuga Sieboldii Carr.	0.550	1.28	1.91	A
18	Karamatsu (Larch)	Larix leptolepis Murray.	0.590	0.87	3.50	B
19	Shina (Bass wood)	Tilia japonica Simk.	0.389	0.58	3.10	B'
20	Tochi (Horse-chestnut)	Aesculus turbinata Bl.	0.493	0.79	2.91	B'
21	Harunire	Ulmus davidiana Planch.	0.526	0.94	2.58	B'
22	Nire (Elm)	Ulmus parvifolia Jacq.	0.546	1.14	2.77	B'
23	Kaede (Maple)	Acer palmatum Thunb.	0.635	1.15	3.65	B
24	Buna (Beech)	Fagus crenata Bl.	0.649	1.15	3.90	B
25	Keyaki	Zelkova serrata Makino.	0.669	0.85	2.01	B'
26	Tabu	Machilus Thunbergii Sieb. et Zucc.	0.682	0.98	2.65	B
27	Nara	Quercus serrata Thunb.	0.708	1.24	3.03	B'
28	Kaba (Birch)	Betula Tauschii Koidz.	0.793	1.72	2.46	B

and the logarithmic decrement is about thrice in the case that the direction of fibers are perpendicular to that of the length as compared with the case of parallel. The Young's modulus decreases at the high frequency in the specimen which the fibers are orientated perpendicular to the length. In order to examine whether this is due to the effect of the iron pieces, we attached further a thin iron piece on the previously attached one (about 0.1 grams) in both ends, but the measured results was the same. Therefore it seems that the decrease of Young's modulus at higher frequency is not the effect by the iron pieces.

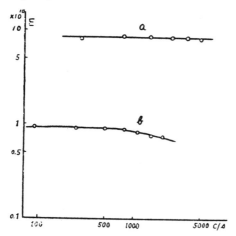

Fig. 6. The Young's modulus of Bishūhinoki.
  a: The direction of fibers is parallel to the length of specimen.
  b: The direction of fibers is perpendicular to the length of specimen.

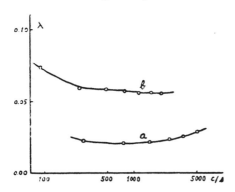

Fig. 7. The logarithmic decrement of Bishūhinoki.
  a: The direction of fibers is parallel to the length of specimen.
  b: The direction of fibers is perpendicular to the length of specimen.

§ 4. Remarks.

We have measured the Young's modulus of various wood from the resonant frequency of lateral vibration and the logarithmic decrement from the electric charge which summed up the rectified damping current after stopping the excitation. In the main audiofrequency range the Young's modulus does not change while the logarithmic decrement shows an interesting variation and its behaviour is characteristic for the needle-leaf trees and the broad-leaved trees respectively. These two kinds of wood are plainly distinguished with only one exception. But these results were obtained under the condition of the air-dried state and in room temperature of winter. It is to be noticed that this character changes rather sensitively with the moisture content and temperature. It was surposed that even in the needle-leaf trees the logarithmic decrement would decrease through a maximum value at the higher frequency. And it was actualy confirmed. The measurement by the resonance method in the longitudinal vibration of the specimen of the needle-leaf tree showed that the logarithmic decrement had a maximum value between 5000 c.p.s. and 10000 c.p.s. The difference of the above described two types A and B is, therefore, only due to the difference of the relaxation time which causes the maximum in the decrement vs. frequency curve. This relaxation time is considered to be given as the reciprocal of the frequency at maximum and may be related to some internal structure of the wood. The further investigation for these behaviours, especially the relations with the moisture content and the temperature are very interesting and their results will be reported in the next paper.

In conclusion the author wishes to express his sincerest gratitude to Dr. Heiji Kawai. Without his original idea and his kindest guidance this work would not have been performed. The author also wishes to thank Mr. Akira Akiyama who has kindly transfered his precious specimens. This study was helped by the research grant from the Ministry of Education.

References.

(1) F. Krüger and E. Rohloff: Zeits. f. Physik. **110**, (1938), 58.

(2) R. B. Abbott and G. H. Purcell: J. Acous. Soc. Am. **13**, (1941), 54.

(3) A. Akiyama: Rep. Inst. Sci. Techn. Univ. Tokyo. **1**, (1947), 38. (in Japanese.)

23B

Reprinted from *J. Phys. Soc. Japan*, **6**(6), 417–421 (1951)

The Vibrational Properties of Wood. II.

By Eiichi FUKADA.

Kobayashi Institute of Physical Research, Kokubunji, Tokyo.

(Received October 20, 1950)

When the moisture content is reduced from the air-dried state, the internal friction of wood increases and the peak in tan δ-frequency curve shifts to the lower frequency. The Young's modulus and tan δ decreases and peak of tan δ shifts to the higher frequency with the increase of temperature at the absolutely dried state. The experimental activation energy was found to be about 1000 cal/mol. It is larger for the broad-leaved tree than for the needle-leaf tree. The frequency character of tan δ can be explained by superposing both Eyring's and Newton's viscosity.

§1. Introduction.

In the previous report[1] it has been shown that the frequency character of the logarithmic decrement of wood is specified to A type for the needle-leaf trees, B and B′ type for the broad-leaved trees. The decrement of A type increases gradually with frequency but the decrement of B type shows the maximum in the audiofrequency range in the air-dried state. In this report there will be shown the change of this behaviour due to the decrease of moisture content by drying as well as the temperature effect in the dried state. As a measure of internal friction, henceforward, we shall take tan δ rather than the logarithmic decrement λ, where δ is the delayed phase angle between the stress and the strain, in view of the theoretical considerations. The relation between these two quantities are as follows,

$$\tan \delta = \lambda/\pi \qquad (1)$$

§2. Variation due to the Moisture Content.

Fig. 1 shows the measuring results of tan δ for Bishūhinoki, one of the needle-leaf trees which belong to A type, in the thoroughly dried state (mark $\triangle$) compared with the air-dried state (mark $\bigcirc$). This specimen is the same with that used in Fig. 6 and Fig. 7 of the previous paper[1]. The figures of percentage show the moisture content in weight and the notation // or $\perp$ means that the axis of fiber is parallel or perpendicular to the direction of the length of the specimen. In both cases tan δ increases with drying and the maximum peak appears like the B type. This behaviour is the same in many other needle-leaf trees. As stated in the remark of the previous paper the needle-leaf trees show also

the peak in the frequency above 5000 c.p.s. in the air-dried state. Therefore the drying shifts the peak to the lower frequency and increases the absolute values of tan δ. The mark $\otimes$ in Fig. 1 shows the measuring results after being left in the atmosphere untill the moisture content becomes nearly the same with the original case. The measured value approaches to the original one but is not the same. The same moisture content and the same temperature of measurement do not necessarily reproduce the same measuring results. This may be due to the hysteresis of the absorption of moisture or some permanent change in the structure by drying. Generally the properties of wood are very sensitive to the moisture and temperature so that it was so difficult to reproduce the same results at will under the various conditions.

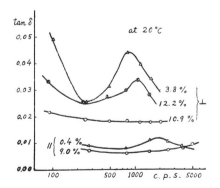

Fig. 1.　Variation of tan δ for Bishūhinoki (needle-leaf tree) due to the moisture content. % shows the moisture content, // or $\perp$ represents that the axis of fibers is parallel or perpendicular to the direction of the length of the specimen respectively.

Fig. 2 shows the variation of $\tan \delta$ for Tabu, one of the broad-leaved trees. The peak shifts to the lower frequency by drying. Young's modulus E hardly changes with the moisture content if we make the correction for the expansion of volume by the absorption of moisture. E_s in Fig. 2 shows the static Young's modulus obtained by measuring the static deflection under the known load. These behaviours were also demonstrated for Kuroezomatsu (Spruce) in another report.[2] In short, when the moisture content increases from zero, the Young's modulus does not change, the peak frequency of $\tan \delta$ increases and the absolute value of $\tan \delta$ decreases. When the moisture content exceeds the so-called fiber saturation point $\tan \delta$ increases again as reported by the other authors.[3]

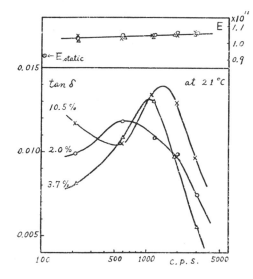

Fig. 2. Variation of $\tan \delta$ for Tabu (broad-leaved tree) due to the moisture content. % shows the moisture content.

§3. Variation due to the Temperature.

Since the moisture in wood evaporates as heated above room temperature we could not measure the variation due to the temperature under the constant moisture content. So we have measured the temperature dependence of Young's modulus E and internal friction $\tan \delta$ at the absolutely dried state, where the absolutely dried state means the state dried for long hours in the thermostat at about $100°C$ until the further decrease of weight of specimen can not be found. The results for Tabu

are shown in Fig. 3. As the temperature rises the absolute value of $\tan \delta$ decreases and the frequency at peak shifts to higher frequency. Young's modulus decreases with the temperature rise.

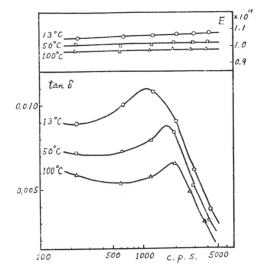

Fig. 3. Temperature dependence of E and $\tan \delta$ for Tabu at absolutely dried state.

This behaviour was also reported[2] for Kuroezomatsu (Spruce), a needle-leaf tree. In figures in that report, Fig. 3 and Fig. 6 in this report part I, the measuring values of Young's modulus were observed to decrease at high frequencies. This apparent decrease was caused by neglecting the correction for the shearing force and the rotatory inertia[4] in the lateral vibration.* When this correction is taken the Young's modulus is nearly constant or slightly increases in the whole measurable frequencies as shown in Fig. 2 and Fig. 3.

Now $\tan \delta$ of wood is divided into two part. One is decreasing with frequency and the other has a maximum peak. In order to investigate the temperature dependence of former part of $\tan \delta$ we made the measurement for Tabu at dried state at about 230 c.p.s. between $-30°C$ and $110°C$. The results are shown in Fig. 4. and Fig. 5. Both Young's modulus and internal friction decreases as the temperature rises. Below $0°C$ there is no particular change, but the temperature coefficient of both E and $\tan \delta$

---

* The author thanks to Dr. O. Nomoto for his remark on this correction.

is a little different above and below 0°C. The mark ○ is the data at cooling and the mark ⊗ is at heating.

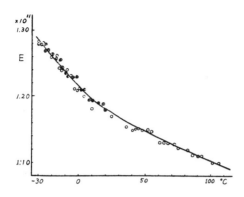

Fig. 4.   Young's modulus vs. temperature for Tabu at about 230 c.p.s.

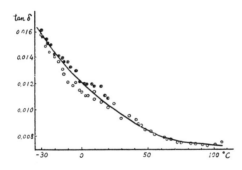

Fig. 5.   Internal friction vs. temperature for Tabu at about 230 c.p.s.

§ 4.   Activation Energy.

The peak frequency in $\tan \delta$ vs. frequency curve shifts to high frequency as the moisture content or measuring temperature increases. It is supposed that this may be due to some relaxation phenomena in the structure of wood. If we plot the reciprocal of frequency at maximum against the reciprocal of the absolute temperature $1/T$, we can obtain a straight line, so that the following relation is valid.

$$1/2\pi f = \tau = \tau_0 \exp (A/RT) \qquad (2)$$

where τ is the relaxation time, τ_0 a constant, R the gas constant and A the activation energy. It was found to be $A = 1200$ cal/mol for Tabu and $A = 890$ cal/mol for Kuroezomatsu.[2] The

activation energy is larger for the broad-leaved tree than for the needle-leaf tree in accord with the experimental results that the peak frequency is lower for the former at the air dried state.

The activation energy of basic part of $\tan \delta$ is obtained from the data in Fig. 5. When we define the complex Young's modulus $E^* = E + i\omega\eta$ as usual, the viscosity coefficient η is calculated from the following relation,

$$\eta = E \cdot \tan \delta / 2\pi f \qquad (3)$$

Fig. 6 shows $\log_{10}\eta$ plotted against $1/T$. From this linear relation we put

$$\eta = \eta_0 \exp (B/RT) \qquad (4)$$

where η_0 is a constant and B the activation energy. It was found to be $B = 1300$ cal/mol for Tabu and $B = 650$ cal/mol for Kuroezomatsu. The activation energy of basic part of $\tan \delta$ is nearly the same with that of the peak part of $\tan \delta$. Therefore the both parts will be caused by the same kind of the dissipative mechanism.

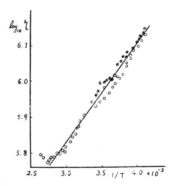

Fig. 6. $\log_{10}\eta$ vs. $1/T$ for Tabu at about 230 c.p.s.

Since the peak frequency of $\tan \delta$ increases and the absolute value decreases as the moisture content increases at the same temperature, the activation energies A and B are considered to diminish with the increase of moisture content. The water seems to play the roll of plasticizer in the viscosity of wood. The experiment varying the moisture content is so difficult that the reliable relation between the activation energy and the moisture content has not yet been obtained.

§ 5. Frequency Dependence of the Internal Friction.

Previously the author has reported[5] that the frequency dependence of the internal friction of polymethylmetacrylate can be derived from the constants in the creep experiment. · The tensile creep of woods by Minami[6] shows the following relation between strain ε and time t;

$$\varepsilon = \varepsilon_1 + k \log_{10} t \qquad (5)$$

where ε_1 is the strain at one second after applying the stress and k is a constant proportional to the magnitude of stress. Then we can derive the above mentioned basic part of the internal friction from the constant $C = 0.434\, k/\varepsilon_1$ and the suitably assumed constant θ as the same manner as polymethyl metacrylate. In curve (1) of Fig. 7, C is estimated to be 0.006 from the Minami's data at small stress and θ is taken as $10^{-3}/2\pi$. The Young's modulus calculated from these values shows hardly any change with frequency, which is in accord with the experimental results.

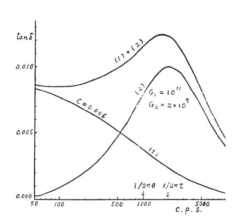

Fig. 7. Two parts of the internal friction.

The linear relation between ε and $\log t$ can be derived using the so-called three element model having the Eyring's viscosity in its dashpot. The rate of flow is described as $A_2 \sinh B_2 \sigma_2$, where σ_2 is applied stress, A_2 and B_2 are the constants.[7] At the very small stress the viscosity will be the usual Newton's type and the rate of flow is represented as σ_2/η, where η is the viscosity coefficient.

Now we suppose that the internal friction due to the Newton's viscosity is superposed

over that due to the Eyring's viscosity. If a spring, whose Young's modulus is G_1, is connected parallel with a dashpot and another spring, whose Young's modulus is G_2, we can get the wellknown dispersion formula.[8] Then the total change of Young's modulus Δ_G is expressed by G_2/G_1 and the relaxation time $\tau = \eta/G_2$ and $(\tan \delta)_{\max} \sim \Delta_G/2$. Experimentally we find no appreciable value of Δ_G in these frequencies and $(\tan \delta)_{\max}$ is of the order of 0.01 or less. G_1 may be the actually measured one, about 10^{11}. If we assume G_2 to be in the order of 10^9, then Δ_G is less than a few percent, that is, within the experimental error and $(\tan \delta)_{\max}$ is fit to the measured value. For instance we take $G_2 = 2 \times 10^9$, then $\Delta_G = 0.02$ and $(\tan \delta)_{\max} = 0.01$. τ is determined from the frequency at maximum of $\tan \delta$ according to the relation $\tau = 1/2\pi f$. If we take $f = 2000$ c.p.s., $\tau = 0.8 \times 10^{-4}$ is obtained.

Curve (2) in Fig. 7 is calculated by taking $G_1 = 10^{11}$, $G_2 = 2 \times 10^9$ and $1/2\pi\tau = 2000$. The superposed curve (1)+(2) represents well the

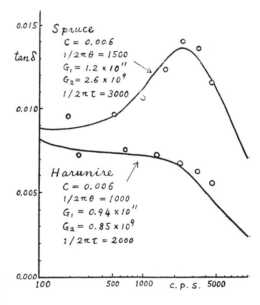

Fig. 8. Comparison between theoretical and experimental results.

actual experimental results. The necessary quantities to calculate this curve are C and θ for (1) and G_1, G_2 and τ for (2). C is obtained from the creep experiment and θ is determined to fit the calculated curve to the measured values. G_1 and τ are obtained from the vibraion experiment and only G_2 is assumed in the order of 10^9 so that the calculated height of peak becomes consistent with the observed value. Various types of $\tan\delta$ vs. frequency curve A, B and B' as reported in part I can be represented by choosing the suitable G_2. The examples are shown in Fig. 8 for Kuroezomatsu and Harunire. The solid line is the calculated curve using the constants written in the figure. The mark $\bigcirc$ is the experimental results.

§ 6. Coexistence of Eyring's and Newton's Viscosity.

We can explain the frequency dependence of internal friction by considering both Eyring's and Newton's viscosity. The former arises at large stress and the latter at small stress. Therefore the large and small stresses are acting at the same time in the specimen. Perhaps this may be due to the magnification and reduction of inner stress by the many cracks or air gaps in the structure of wood.[9] The distribution of stress in the specimen is also caused by the lateral vibration itself. The maximum magnitude of stress used in the ordinary measurement was estimated about 0.5 $\sim$ 5 kg/cm² at the vicinity of the loop of lateral vibration. The magnification ratio of stress due to the presence of the narrow crack would be of the order of 10^2. Therefore the actual inner stress would reach to the experimental range of stress used in the tensile creep by Minami.

When we consider the peak of $\tan\delta$ is caused by the Eyring's viscosity at small stress, we can express the frequency at maximum of $\tan\delta$ as

$$f_m = \frac{kT}{2\pi h} e^{\Delta S/R} e^{-\Delta H/RT} \qquad (6)$$

where ΔS is the entropy of activation and ΔH is the heat of activation. By plotting $\log f_m/T$

against $1/T$ we could obtain $\Delta H = 490$ cal/mol, $\Delta S = -39$ cal/mol. deg. for Tabu and $\Delta H = 220$ cal/mol, $\Delta S = -38$ cal/mol. deg. for Kuroezomatsu.

Our consideration explains well the frequency dependence of internal friction and the temperature dependence of the frequency at maximum of $\tan\delta$, but it fails to elucidate the decrease of the absolute value of $\tan\delta$ with the increase of temperature. Further, in the experiment of creep $B_2\sigma_2$ is nearly constant, so that we can not equate $\sinh B_2\sigma_2 = B_2\sigma_2$ when we approach to the case of small stress. The constants A_2 and B_2 at large stress depend on the magnitude of stress and are different from the constants A_2 and B_2 at very small stress, which relate to the Newton's viscosity.

The author wishes to express his sincerest gratitude to Dr. Heiji Kawai for his helpful guidance and discussions throughout this work. This research is indebted to the grant in aid for the Ministry of Education.

References.

1) E. Fukada: J. Phys. Soc. Japan. **5** (1950), 321.

2) E. Fukada: Nature. **166** (1950), 772.

3) Y. Tani: Ohyobutsuri. **9** (1940), 372. (in Japanese.)

4) S. Timoshenko: Vibration Problems in Engineering. (1937), 337.

5) E. Fukada: J. Phys. Soc. Japan. **6** (1951), 254.

6) Y. Minami: J. Aeron. Res. Inst. Tokyo Imp. Univ. No. 174, (1939), 23. (in Japanese)

7) A. V. Tobolsky and H. Eyring: J. Chem. Phys. **11** (1943), 125.

8) C. Zener: Elasticity and Anelasticity of Metals. (1948), 46.

9) A. A. Griffith: Phil. Trans. Roy. Soc. Lond. **TA221** (1921), 163.

Editor's Comments on Papers 24 and 25

24 Beldie: *Die Bestimmung der Schubmoduln des Fichtenholzes*

25 Ghelmeziu and Beldie: *On the Characteristics of Resonance Spruce Wood*

Mr. Ion Paul Beldie (1935–), a native of Romania, acquired his M.S. in physics at the University of Bucharest in 1957. After some years in industry he developed an interest in the acoustics of stringed instruments at the Research Institute for the Wood Industry in Bucharest between 1962 and 1970. He is now in musical acoustics research under Professor Cremer at the Institute for Technical Acoustics of the Technical University of Berlin.

Two of his papers are included in this volume. The first gives information on the experimentally determined shear moduli and six formulas for the relation between the moduli of shear of the parenchyma and the rays in samples of spruce wood. The second relates some of the acoustical qualities of spruce to the criteria necessary in musical instruments. A third paper, "Chladnische Figuren und Eigentone der Geigenplatten," *Instr. Z.*, **23** (2) (Feb. 1969), appears in *Musical Acoustics, Part II.*

Professor Dr. eng. Nicolae Ghelmeziu (1910–) is presently Scientific Director and Chief of the Research Department of the Institute for Research and Technological Design for the Wood Industry, Bucharest. A native of Rumania, he obtained the diploma of Forest Engineer from the Polytechnical School, Bucharest, in 1933 and that of Doctor Engineer in 1938 from the Technische Hochschule, Berlin. His research and teaching are concerned with the structural, physical, mechanical, and chemical properties of wood and methods to improve them as applied to sawed timber and its utilization in the packaging and building industries.

Selected Publications By I. P. Beldie

Beldie, I. P., I. Stan, A. Toth, T. Penciuc, and E. Bataga. "Chladnische Figuren isotroper und anisotroper Platten," *Studii Cercetari Fiz.*, **18**(9), 983–991 (1966). In Rumanian.

Beldie, I. P. "Die Bestimmung der Schubmoduln des Fichtenholzes," *Holz Roh- Werkstoff,* **26**(7), 261–266 (1968).

Beldie, I. P. "Chladnische Figuren und Eigentöne der Geigenplatten," *Instr. Z.*, **23**(2), 168, 170, 172, 174 (1969).

Ghelmeziu, N., and I. P. Beldie. "On the Characteristics of Resonance Spruce Wood," *Catgut Acoust. Soc. Newsletter,* No. 17, May 1972, pp. 10–16.

24

Reprinted from *Holz Roh- Werkstoff,* **26**(7), 261 – 266 (1968)

Die Bestimmung der Schubmoduln des Fichtenholzes

Von **Ion Paul Beldie**

Zusammenfassung

Die dynamische Messung des Schubmoduls in der yz-Ebene auf Grund von Biegeschwingungen an Fichtenstäben mit longitudinalem und radialem Faserverlauf zeigt ungleiche G_{yz}- und G_{zy}-Werte. Dies kann dadurch erklärt werden, daß man die periodische Inhomogenität des Holzes in der radial gerichteten z-Achse in Betracht zieht und die hierfür experimentell gemessenen Werte der Schubmoduln als gemittelte Werte ansieht. Zur theoretischen Bestimmung dieser Werte wird ein geometrisches Modell für die Fichtenholzstruktur benützt. Man erhält auf diesem Wege sechs Formeln, welche die Beziehung zwischen den Schubmoduln des Parenchyms und der Markstrahlen einerseits und den experimentell bestimmbaren Schubmoduln andererseits angeben. Der theoretische Vergleich zwischen G_{yz} und G_{zy} zeigt ebenfalls, daß G_{yz} größer als G_{zy} ist. Ausgehend von den für G_{yz}, G_{yx}, G_{zx} und G_{zy} gemessenen Werten werden mit Hilfe der entsprechenden Formeln Schätzwerte für die Schubmoduln des Parenchyms und der Markstrahlen errechnet und die Größenordnungen für G_{xy} und G_{xz} abgeschätzt.

Summary *

The dynamic measurement of the modulus of shear in the yz-plane on the basis of bending vibrations taken on spruce samples in the longitudinal and radial direction of grain shows different values for G_{yz} and G_{zy}. This may be explained by the fact that the periodic inhomogeneity of the wood in the radially running z-axis exists and the experimentally measured values of the modulus of shear are considered as mean values. For the theoretical detrmintion of these values a geometric model for the structure of spruce is used. By this method six formulae are obtained which indicate the relation between the moduli of shear of the parenchyma and the rays on the one hand and the experimentally determined moduli of shear on the other. The theoretical comparison between G_{yz} and G_{zy} also shows that G_{yz} is larger than G_{zy}. Based on the values measured for G_{yz}, G_{yx}, G_{zx} and G_{zy} and by the respective formulae estimate-values are calculated for the moduli of shear of the parenchyma and the rays, and the magnitudes for G_{xy} and G_{xz} are evaluated.

Einleitung

Im allgemeinen wird davon ausgegangen, daß das Holzgefüge eine rhombische Symmetrie aufweist. Die im Holz gegebene Anordnung von Früh- und Spätholzzellen, die eine periodische Inhomogenität darstellt, kann man durch Bildung eines Mittelwertes über eine hinreichend große Anzahl von Jahrringen ausgleichen [Keylwerth 1951]. Damit ergibt sich die Möglichkeit, Holz als homogenes, anisotropes Material durch neun elastische Konstanten zu beschreiben. Dabei bleibt aber eine der wichtigsten Eigenschaften des Holzes, nämlich die Struktur des einzelnen Jahrringes, unberücksichtigt.

Die statischen Verfahren zur Bestimmung der drei Schubmoduln des Holzes haben alle den Nachteil, daß es nur in wenigen Fällen möglich ist, eine ausreichend reine Schubverformung zu erzeugen. [Kollmann 1951]. Ein Beispiel hierfür ist das Zwei-Würfelverfahren von Schlüter, bei dem neben der Schubverformung zusätzlich eine Biegeverformung auftritt [Schlüter 1932]. Die Biegeverformung macht sich dabei z. T. deutlich bemerkbar, und die gemessenen Schubmodulwerte fallen deshalb kleiner aus als die tatsächlichen Werte. Tabelle 1 gibt die hier interessierenden Schubmodul- und Elastizitätsmoduln wieder, wie sie von Schlüter [1932] in zwei Meßreihen an Bukowina-Fichte festgestellt worden sind. Nur die Werte für G_{zz} und G_{zz} sind etwa gleich; dagegen sind die Werte für G_{yx} und G_{yz} bemerkenswert verschieden von den Werten für G_{xy} und G_{zy}. Der erste Index gibt die Achse an, die auf der Ebene, in welcher die Schubkraft wirkt, senkrecht steht, der zweite Index die Achse, mit der die Schubkraft parallel läuft.

Das dynamische Verfahren, welches auf der Differentialgleichung der freien Biegeschwingung von Stäben nach Timoschenko beruht, wurde von Kollmann und Krech [1960] verwendet. Dieses Verfahren gibt die Möglichkeit, den Schubmodul in der Biegeebene als Nebenprodukt bei der Bestimmung des Elastizitätsmoduls in Längsrichtung eines Stabes zu bestimmen. Hierbei bleibt der Wert des Schubmoduls frei vom verfälschenden Biegeeinfluß, was beim Zwei-Würfelverfahren nicht der Fall ist. Kollmann und Krech benützten prismatische Probestäbe (Bild 1) mit longitudinaler Orientierung. Deren Längsachse lag parallel zur y-Achse und die Querschnittseiten parallel zur x- bzw. z-Achse. Für jeden Probestab wurden die Eigenschwingungsfrequenzen in den Biegeebenen yx und yz gemessen und durch Heranziehen von Masse und Stababmessungen jeweils

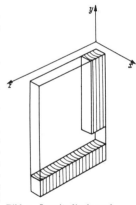

Bild 1. Longitudinale und radiale Probestäbe; x, y, z, Hauptachsen der Holzstruktur

Tabelle 1. Elastizitätskonstanten nach verschiedenen Autoren, in dyn/cm² (der Multiplikand 10^{10} wurde überall weggelassen)

Feuchtigkeitsgehalt %	Longitudinale Probestäbe			Radiale Probestäbe			G_{xy}		G_{xz}		Bemerkungen
	E_y	G_{yx}	G_{yz}	E_z	G_{zx}	G_{zy}	(−3%)	(+3%)	(−3%)	(+3%)	
9,0	15,10 (2,00)	0,621 (0,10)	0,573 (0,10)	1,13 (0,05)	0,031 (0,01)	0,250 (0,10)	0,598...0,564...0,558 (berechnet)		0,066...0,060...0,015		Eigene Messungen und Berechnungen
14,3	18,18	0,612	0,753	—	—	—	—		—		Kollmann und Krech [1960]
7,5	8,13	0,615	0,616	0,904	0,023	0,435	0,404		0,025		Material No. 1 für l/a
6,75	11,30	0,673	0,661	0,974−1,395	0,041	0,516	0,510		0,040		Material No. 70 $= 0,05$ Schlüter [1932]
9,8	16,56	0,796	0,642	0,714	0,041	—	—		—		Keylwerth [1951]

Editor's Note: See the "Revised Summary" at the end of this article.

der E-Modul in Längsrichtung E_y und die Schubmoduln G_{yx} und G_{yz} berechnet. Die von Kollmann und Krech für Fichtenholz ermittelten Werte sind ebenfalls in Tabelle 1 angegeben.

Aus einem regelmäßig gewachsenen und genügend dicken Fichtenstamm lassen sich selbstverständlich ebenso radiale, prismatische Probestäbe ausformen, deren längste Abmessung in Richtung der z-Achse liegt. Die leichte Krümmung der Jahrringe in den Seitenflächen des Prismas ist dabei so gering, daß sie vernachlässigt werden kann (Bild 1). An solchen Stäben lassen sich der radiale E-Modul E_z und die Schubmoduln G_{zz} und G_{zy} bestimmen. Könnte man nun tatsächlich die durch die Jahrringe bedingte Inhomogenität vernachlässigen, was praktisch bei der Aufstellung der Differentialgleichung angenommen wird, und wäre das Näherungsverfahren für die Schubmoduln genügend genau, so wäre eine Gleichheit für G_{yz} und G_{zy} zu erwarten.

Die im Rahmen der hier vorliegenden Arbeit mit dem dynamischen Verfahren durchgeführten Messungen für G_{yz} und G_{zy} zeigen aber, daß sie verschieden sind. Nach-

folgend wird die Durchführung der Messungen beschrieben und eine Erklärung der Ergebnisse versucht.

Messungen

Die für die Messungen verwendeten Probestäbe wurden aus einem 9 mm dicken Fichtenbrett nach dem in Bild 2 gezeigten Schema entnommen, insgesamt also 6 radial und 10 longitudinal orientierte Proben.

Die Versuchseinrichtung und die Durchführung der Messungen entsprachen dem von Kollmann und Krech [1960] beschriebenen Verfahren, mit dem Unterschied, daß ein Präzisionsgenerator benutzt wurde, der die elektronische Zähleinrichtung überflüssig machte. Es ergaben sich bei den Messungen keine besonderen Schwierigkeiten, ausgenommen bei der Bestimmung der Eigenfrequenzen der Biegeschwingungen in der yz-Ebene der radialen Probestäbe. Hier konnten nur die ersten 2 bis 4 Eigenfrequenzen, die im Vergleich zu den Biege-Eigenfrequenzen in der xz-Ebene sehr unregelmäßig auf der Frequenzskala verteilt lagen, gemessen werden. Da die Vermutung nahe lag, daß die radialen Probestäbe im Verhältnis zu ihrem Querschnitt zu kurz waren, wurde ihr Maß in y-Richtung von 12 mm auf 7 mm verringert, anschließend je zwei Stäbe an den Enden zusammengeleimt und schließlich jeder der drei auf diese Weise entstandenen Stäbe (1 + 4; 2 + 3; 5 + 6;) um 10 mm gekürzt. Trotzdem blieb, nach erneuter Messung der (bis zuletzt 11) Eigenfrequenzen, die unregelmäßige Verteilung bestehen. Dies hatte eine verhältnismäßig hohe Streuung der Meßpunkte zur Folge. Bild 3 zeigt die Meßwerte mit den Ausgleichsgeraden für den Probestab (2 + 3). Weitere Meßergebnisse enthält die Tabelle 1, Zeile 1, wobei in Klammern die größte Abweichung von dem darüber stehenden Mittelwert angegeben ist.

Diskussion der Ungleichheit von G_{yz} und G_{zy}

Tabelle 1 zeigt die Ungleichheit der im dynamischen Verfahren ermittelten Werte für G_{yz} und G_{zy}, die theoretisch hätten gleich sein müssen. Eine überlagerte Biegeverformung kann hierfür nicht mehr die Ursache sein. Eine Erklärung bietet sich aber, wenn man die bisher außer acht gelassene, durch die Jahrringe bedingte Inhomogenität berücksichtigt. Man nimmt gewöhnlich an, daß ein Jahrring aus einer Frühholzschicht und einer Spätholzschicht besteht und daß diese beiden Schichten unterschiedliche Elastizitätskonstanten aufweisen. Man nimmt weiter an, daß das Frühholzgefüge in sich homogen ist und folglich seine Schubverformungen mit den drei dafür zur Verfügung stehenden Schubmoduln beschrieben werden kann. Dasselbe gilt sinngemäß für das Spätholzgefüge. Die sechs möglichen einfachen Schubverformungen eines Holzwürfels [Schlüter 1932], der aus mehreren zur z-Achse senkrecht stehenden Früh- und Spätholzschichten besteht, können aber nun nicht mehr durch nur drei für die jeweiligen Gefüge gemittelte Schubmoduln beschrieben werden, da die Schubverformungen ja außerdem von den Ebenen, in denen die Schubkräfte wirksam sind, abhängen.

Die experimentell ermittelten Schubmodulwerte können als über den ganzen Probekörper gemittelte Werte angesehen werden. Man kann diese Werte aber auch theoretisch bestimmen. Für die Berechnung legt man ein geometrisches Modell zugrunde und betrachtet die Schubmoduln der verschiedenen Gefügearten als bekannt. Mit derartigen Gleichungen können dann die Schubmoduln der jeweils in Betracht stehenden Gefügeteile abgeschätzt werden. Für das nachfolgend entwickelte geometrische Modell werden folgende Vorannahmen gemacht (Bild 8):

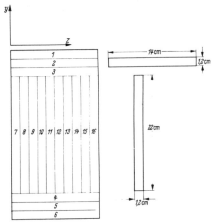

Bild 2. Entnahmeschema für die Probestäbe aus einer 0,9 cm dicken Fichtenholzplatte. 1 bis 6 radiale Probestäbe; 7 bis 16 longitudinale Probestäbe.

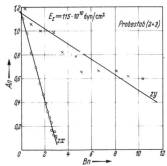

Bild 3. Ausgleichgeraden des radialen Probestabes (2 + 3). zy Schubkräfte parallel zur y-Achse; zx Schubkräfte parallel zur x-Achse. A_n, B_n Koeffizienten der Gleichung $E = A_n + B_n \dfrac{sE}{G}$ mit n Ordnung der Eigenschwingung (nach Kollmann und Krech [1960]). Da es sich hier nur um Größenordnungen handelt, wurde einfachheitshalber $s = 1$ angenommen (statt $s = 1{,}06$). E_z = Elastizitätsmodul in Richtung der z-Achse.

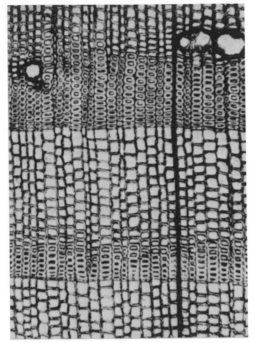

Bild 4. Mikroschnitt in der xz-Ebene; Fichtenholz. Vergr.: 125:1 [nach Ghelmeziu und Suciu 1959].

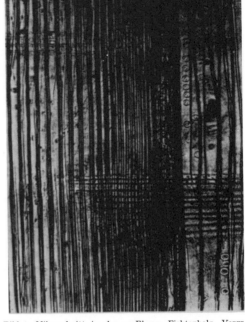

Bild 5. Mikroschnitt in der zy-Ebene. Fichtenholz. Vergr.: 125:1 [nach Ghelmeziu und Suciu 1959].

1. Das Fichtenholz besteht nur aus Markstrahl- und Parenchymschichten, ähnlich wie dies Schniewind [1959] für Eichenholz angenommen hat. Diese Schichten verteilen sich, in wechselnder Folge, senkrecht zur x-Achse.

2. Auf Grund der Bilder 4 und 5 [nach Ghelmeziu und Suciu 1959], die xz- und zy-Mikroschnitte darstellen, wird weiter angenommen, daß die Schubmoduln des Parenchymgewebes innerhalb jedes Jahrringes lineare Funktionen des Abstandes von der Jahrringgrenze sind. Somit ist

$$G(z) = \frac{G_h - G_0}{h}\, z + G_0 = \frac{A}{h}\, z + B \qquad (1)$$

wobei G_h der Schubmodul an der äußeren Spätholzgrenze (Ende des Jahrringes),

G_0 der Schubmodul an der äußeren Frühholzgrenze (Anfang des Jahrringes),

und h die Jahrringbreite ist.

An den jeweiligen Grenzen der Jahrringe ändern sich die elastischen Konstanten sprunghaft, so daß insgesamt die Schubmoduln entlang der z-Achse eine sägezahnförmige, periodische Änderung erfahren. Diese Feststellung entspricht den densiometrischen Rohdichtemessungen, die Polge [1967] an radiographischen Aufnahmen von Fichtenholz durchgeführt hat, und die eine mehr oder weniger andauernde Änderung der Rohdichte innerhalb eines Jahrringes zeigten.

Eine unendlich dünne Schicht dz des Parenchymgewebes, die senkrecht zur z-Achse liegt, kann man als homogen ansehen; sie besitzt eine tetragonale Symmetrie. Die zwei entsprechenden Schubmoduln lauten

$$G_{zx} = G_{xz} = P_0(z)$$
$$G_{xy} = G_{zy} \approx G_{yx} = G_{yz} = P(z)\,. \qquad (2)$$

3. Auf Grund von Bild 6 [nach Ghelmeziu und Suciu 1967], das einen in der xy-Ebene geführten Mikroschnitt

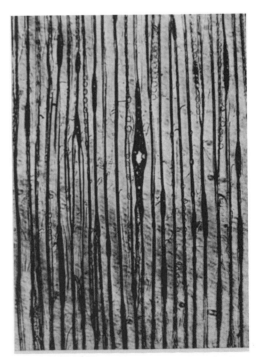

Bild 6. Mikroschnitt in der xy-Ebene. Fichtenholz. Vergr.: 125:1 [nach Ghelmeziu und Suciu 1959].

zeigt, wurde nach dem in Bild 7 angegebenen Verfahren eine statistische Berechnung durchgeführt, aus welcher folgt, daß die Summe aller Breiten der Markstrahlen, die von einer beliebig gelegten, zur x-Achse parallelen Linie geschnitten werden, praktisch konstant ist. Der Anteil der Markstrahlbreite in Richtung der x-Achse beträgt im Mittel $\alpha = 0{,}05$. Es kann ferner angenommen werden, daß sich alle Markstrahlen innerhalb von gleich breiten Schichten befinden, die zur x-Achse senkrecht liegen und sich in der ganzen yz-Ebene erstrecken.

Das Markstrahlgewebe ist homogen und besitzt ebenfalls tetragonale Symmetrie. Die entsprechenden Schubmoduln lauten:

$$G_{xy} = G_{yx} = M_0$$
$$G_{yz} = G_{xz} \approx G_{zy} = G_{zx} = M \, . \qquad (3)$$

4. Weiterhin wird angenommen, daß bei genügend kleinen Probekörpern, wie sie von Schlüter und ebenso in der hier vorliegenden Untersuchung verwendet wurden, alle Jahrringe gleiche Breite und gleiche elastische Konstanten besitzen. Diese Annahme ergibt die Möglichkeit, die Berechnungen verhältnismäßig leicht durchführen zu können und die Ergebnisse als für den ganzen Probekörper zutreffend zu bezeichnen. Tatsächlich wird

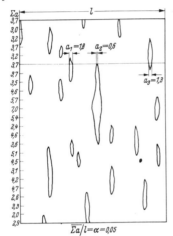

Bild 7. Verfahren zur Berechnung der anteiligen mittleren Breite der Markstrahlen in Richtung der x-Achse, auf Grund des Mikroschnittes in Bild 6. a Breite einer Markstrahlgruppe (in mm); Σa Summe der Breiten aller Markstrahlgruppen auf einer beliebigen Geraden, die parallel zur x-Achse liegt. $\overline{\Sigma a}$ Mittelwert; l Breite der zur Berechnung benutzten Fläche; α mittlerer anteiliger Wert der Markstrahlbreite.

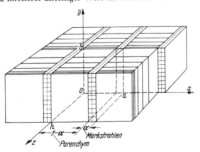

Bild 8. Geometrisches Modell der Fichtenholzstruktur. a gemeinsame Breite einer Parenchym- und einer Markstrahlschicht; h Jahrringbreite; b beliebige Höhe des in den Rechnungen benutzten Prismas (Kantenlängen a, h, und b).

zur Bestimmung der 6 Schubmoduln ein kleines Prisma verwendet (Bild 8), das in der z-Achse einen Jahrring von der Dicke h umfaßt, in der x-Achse so ausgewählt ist, daß es nur eine Markstrahlschicht und nur eine Parenchymschicht enthält von der Dicke a, während die 3. Abmessung mit der Dicke b beliebig ist.

Die wichtigsten Rechenoperationen seien nachfolgend kurz angegeben.

1. Bei der Schubverformung von schwingenden Probestäben mit longitudinaler Faserorientierung liegt die Fläche, in welcher die Schubkräfte wirken, senkrecht zur y-Achse. Liegen die Schubkräfte in Richtung der x-Achse, so folgt aus dem verallgemeinerten Hookeschen Gesetz für eine dünne Schicht dz in der xy-Ebene und für eine unendlich kleine Verformung (Bild 9):

$$\frac{dF(z)}{a \cdot dz} = G_{yx}(z) \, \frac{\Delta x}{b} \, . \qquad (4)$$

Integriert man über die ganze Fläche, in welcher die Schubkräfte wirken, und multipliziert mit $1/h$, so erhält man

$$\frac{1}{h\,a} \int dF(z) = \left[\frac{1}{h} \int\limits_0^h G_{yx}(z) \, dz \right] \frac{\Delta x}{b} \, . \qquad (5)$$

Der Schubmodul dieser Verformung des ganzen Prismas ist somit

$$G_{yx} = \frac{1}{h} \int\limits_0^h G_{yx}(z) \, dz \, . \qquad (6)$$

Um diesen Schubmodul zu berechnen, wird das verallgemeinerte Hookesche Gesetz für jede einzelne Gewebeart angewendet.

Für das Markstrahlgewebe (m): $\dfrac{df_m}{\alpha\, a\, dz} = M_0 \, \dfrac{\Delta x}{b}$.

$$(7)$$

Für das Parenchymgewebe (p): $\dfrac{df_p}{(1-\alpha)\,a\,dz} = P(z)\,\dfrac{\Delta x}{b}$.

Da $df_m + df_p = dF(z)$, folgt aus Gl. (4) und Gl. (7)

$$G_{yx}(z) = \alpha M_0 + (1-\alpha)\, P(z) = \alpha M_0 + (1-\alpha) \left(\frac{A}{h} z + B \right) \qquad (8)$$

Führt man Gl. (8) in Gl. (6) ein und integriert über die ganze Jahrringbreite, so erhält man schließlich

$$G_{yx} = \alpha M_0 + (1-\alpha) \left(\frac{A}{2} + B \right) . \qquad (9)$$

Wirken die Schubkräfte parallel zur z-Achse, so erhält man auf ähnliche Weise

$$G_{yz} = \alpha M + (1-\alpha) \left(\frac{A}{2} + B \right) . \qquad (10)$$

2. Bei den schwingenden Probestäben mit radialer Faserrichtung steht die Fläche, in der die Schubkräfte wirken, senkrecht zur z-Achse. Wirken die Schubkräfte parallel zur y-Achse, so erhält man ähnlich wie oben (Bild 10):

$$\frac{F}{ab} = G_{zy}(z) \, \frac{d(\Delta x)}{dz} \, . \qquad (11)$$

Die Integrierung ergibt, nach Multiplikation mit $1/h$,

$$\frac{F}{ab} \left[\frac{1}{h} \int\limits_0^h \frac{dz}{G_{zy}(z)} \right] = \frac{1}{h} \int d(\Delta_y) \, , \qquad (12)$$

und der darausfolgende Schubmodul G_{zy} ist gegeben durch

$$\frac{1}{G_{zy}} = \frac{1}{h} \int\limits_0^h \frac{dz}{G_{zy}(z)} \, . \qquad (13)$$

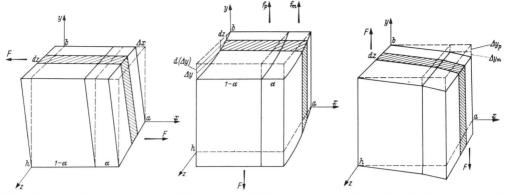

Bild 9. Einfache Schubverformung des Prismas in der xy-Ebene. Die Schubkräfte F wirken in der Ebene, die senkrecht zur y-Achse liegt und haben die Richtung der x-Achse. dz Dicke einer unendlich dünnen Schicht; Δx absolute Verformung des Prismas.

Bild 10. Einfache Schubverformung des Prismas in der yz-Ebene. Die Schubkräfte F wirken in der Ebene, die senkrecht zur z-Achse liegt und haben die Richtung der y-Achse. f_p Teilkraft, die der Parenchymschicht entspricht; f_m Teilkraft, die der Markstrahlschicht entspricht; dz Dicke einer unendlich dünnen Schicht; Δy absolute Verformung des Prismas; $d(\Delta y)$ unendlich kleiner Bruchteil der absoluten Verformung.

Bild 11. Einfache Schubverformung des Prismas in der xy-Ebene. Die Schubkräfte F wirken in der Fläche, die senkrecht zur x-Achse liegt und haben die Richtung der y-Achse. dz Dicke einer unendlich dünnen Schicht; Δy_p absolute Verformung der Parenchymschicht; Δy_m absolute Verformung der Markstrahlschicht.

Um $G_{zy}(z)$ zu ermitteln, schreibt man

$$\frac{f_m}{\alpha\,ab} = M\,\frac{d(\Delta y)}{dz} \qquad \frac{f_p}{(1-\alpha)\,ab} = P(z)\,\frac{d(\Delta y)}{dz}, \qquad (14)$$

und da $f_m + f_p = F$, folgt aus Gl. (11) und Gl. (14)

$$G_{zy}(z) = \alpha M + (1-\alpha)\,P(z) = \alpha M + (1-\alpha)\left(\frac{A}{h}z + B\right). \qquad (15)$$

Führt man Gl. (15) in Gl. (13) ein und integriert, so erhält man

$$\frac{1}{G_{zy}} = \frac{1}{(1-\alpha)\,A}\ln\left[\frac{(1-\alpha)\,A}{(1-\alpha)\,B + \alpha M} + 1\right]. \qquad (16)$$

Wirken die Schubkräfte in Richtung der x-Achse, so ist der Schubmodul G_{zx} des Prismas gegeben durch

$$\frac{1}{G_{zx}} = \frac{1}{(1-\alpha)\,A_0}\ln\left[\frac{(1-\alpha)\,A_0}{(1-\alpha)\,B_0 + \alpha M} + 1\right]. \qquad (17)$$

3. Für die experimentelle Bestimmung der letzten zwei Schubmoduln und zugleich des dritten Elastizitätsmoduls E_x wären tangentiale Probestäbe, deren Längsachse parallel zur x-Achse verläuft, notwendig, die technisch nicht mehr so leicht herstellbar sind. Die Fläche, in der die Schubkräfte wirken würden, läge senkrecht zur x-Achse und falls die Schubkräfte die Richtung der y-Achse hätten, könnte man folgendermaßen schreiben (Bild 11):

$$\frac{dF(z)}{b\,dz} = G_{xy}(z)\,\frac{\Delta y}{a}. \qquad (18)$$

Integriert man und multipliziert mit $1/h$, so folgt,

$$\frac{1}{hb}\int dF(z) = \left[\frac{1}{h}\int_0^h G_{xy}(z)\,dz\right]\frac{\Delta y}{a} \qquad (19)$$

und der Schubmodul des ganzen Prismas ist dann

$$G_{xy} = \frac{1}{h}\int_0^h G_{xy}(z)\,dz. \qquad (20)$$

Für jedes Gewebe getrennt gilt:

$$-\frac{dF(z)}{b\,dz} = M_0\,\frac{\Delta y_m}{\alpha a} \qquad \frac{dF(z)}{b\,dz} = P(z)\,\frac{\Delta y_p}{(1-\alpha)a}. \qquad (21)$$

Da $\Delta y_m + \Delta y_p = \Delta y$, so erhält man durch Einsetzen, aus Gl. (18) und Gl. (21):

$$\frac{1}{G_{xy}(z)} = \frac{\alpha}{M_0} + \frac{1-\alpha}{P(z)} = \frac{\alpha}{M_0} + \frac{1-\alpha}{\dfrac{A}{h}z + B}. \qquad (22)$$

Führt man Gl. (22) in Gl. (20) ein und integriert, so erhält man

$$G_{xy} = \frac{M_0}{\alpha}\left[1 - \frac{1-\alpha}{\alpha}\cdot\frac{M_0}{A}\ln\left(\frac{\alpha A}{\alpha B + (1-\alpha)M_0} + 1\right)\right]. \qquad (23)$$

Verlaufen die Schubkräfte parallel zur z-Achse, so erhält man:

$$G_{xz} = \frac{M}{\alpha}\left[1 - \frac{1-\alpha}{\alpha}\cdot\frac{M}{A_0}\ln\left(\frac{\alpha A_0}{\alpha B_0 + (1-\alpha)M} + 1\right)\right]. \qquad (24)$$

Die sechs Gleichungen (9), (10), (16), (17), (23) und (24) ergeben somit die über einen Jahrring gemittelten Werte der Schubmoduln unter Berücksichtigung der sechs einfachen Schubverformungsarten, denen ein endliches Prisma mit einer bestimmten idealisierten inhomogenen und anisotropen Struktur unterworfen werden kann. Es wurde erwähnt, daß diese Formeln als für den ganzen Probekörper gültig angenommen werden können. Aus den obigen Gleichungen folgt, daß zur Beschreibung der Schubverformungen eines anisotropen und in einer Richtung periodisch inhomogenen Materials sechs verschiedene Schubmoduln nötig sind.

Die Ungleichheit der Schubmoduln G_{zy} und G_{yz} folgt aus dem Vergleich der Gln. (10) und (16). Entwickelt man Gl. (16) nach der Beziehung

$$\ln x = 2\left[\frac{x-1}{x+1} + \frac{1}{3}\left(\frac{x-1}{x+1}\right)^3 + \cdots\right],$$

so erhält man

$$\frac{1}{G_{zy}} = \frac{1}{(1-\alpha)\dfrac{A}{2} + (1-\alpha)\,B + \alpha M} + \qquad (25)$$

$$+ \frac{1}{3(1-\alpha)A}\left[\frac{(1-\alpha)A}{(1-\alpha)\frac{A}{2}+(1-\alpha)B+\alpha M}\right]^3 + \cdots$$

$$= \frac{1}{G_{yz}} + \left\{\cdots\right\} > \frac{1}{G_{yz}},$$

da der zweite Summand stets positiv ist. Es folgt

$$G_{zy} < G_{yz} \tag{26}$$

wie es die dynamischen Messungen gezeigt haben (Tabelle 1).

Die Schätzung von G_{xy} und G_{yx}

Der Messung dieser zwei letzten Schubmoduln stehen wegen der Ausführung der tangentialen Probestäbe wesentlich größere Schwierigkeiten im Wege. Die longitudinalen oder die radialen Probestäbe müssen hierfür in kleine Prismen zerschnitten und in der x-Richtung wieder zusammengeklebt werden.

Eine eigene Untersuchung muß den Einfluß der Klebstoffschichten feststellen und erst dann können G_{xy} und G_{yx} bestimmt werden.

Bis zur Durchführung dieser Messungen wäre es jedoch interessant, eine Schätzung der Größenordnung dieser beiden Schubmoduln vorzunehmen, was mit Hilfe der sechs Formeln, die in Tabelle 2 zusammengefaßt sind, möglich ist.

Tabelle 2. Gleichungen für die Schätzung der Schubmoduln.

$$G_{yx} = \alpha M_0 + (1-\alpha)\left(\frac{A}{2} + B\right) \tag{9}$$

$$G_{yz} = \alpha M + (1-\alpha)\left(\frac{A}{2} + B\right) \tag{10}$$

$$\frac{1}{G_{zy}} = \frac{1}{(1-\alpha)A}\ln\left[\frac{(1-\alpha)A}{(1-\alpha)B+\alpha M}+1\right] \tag{16}$$

$$\frac{1}{G_{zz}} = \frac{1}{(1-\alpha)A_0}\ln\left[\frac{(1-\alpha)A_0}{(1-\alpha)B_0+\alpha M}+1\right] \tag{17}$$

$$G_{xy} = \frac{M_0}{\alpha}\left[1 - \frac{1-\alpha}{\alpha}\cdot\frac{M_0}{A}\ln\left(\frac{\alpha A}{\alpha B+(1-\alpha)M_0}+1\right)\right] \tag{23}$$

$$G_{xz} = \frac{M}{\alpha}\left[1 - \frac{1-\alpha}{\alpha}\cdot\frac{M}{A_0}\ln\left(\frac{\alpha A_0}{\alpha B_0+(1-\alpha)M}+1\right)\right] \tag{24}$$

Betrachtet man $(1-\alpha)B+\alpha M$ als eine einzige Veränderliche, so sind Gl. (10) und Gl. (16) zwei Gleichungen mit zwei Unbekannten, deren Lösung leicht graphisch gefunden werden kann. Die Lösung hängt von G_{yz}, G_{zy} und α ab:

$$(1-\alpha)A = f_1(G_{yz}, G_{zy}, \alpha) \tag{27}$$

$$(1-\alpha)B + \alpha M = f_2(G_{yz}, G_{zy}, \alpha). \tag{28}$$

Der Vergleich der Gl. (16) und (17) zeigt, daß man Gl. (16) aus Gl. (17) erhält, wenn man die letztere Gleichung mit

$$\frac{1}{\gamma} = \frac{G_{zz}}{G_{zy}} \tag{29}$$

multipliziert. Man erhält dann folgende Gleichungen

$$\gamma A_0 = A$$

$$\gamma[(1-\alpha)B_0 + \alpha M] = (1-\alpha)B + \alpha M$$

und mit Gl. (27) und Gl. (28),

$$A_0 = \frac{A}{\gamma} = \frac{f_1}{(1-\alpha)\gamma}, \tag{30}$$

$$(1-\alpha)B_0 + \alpha M = \frac{1}{\gamma}[(1-\alpha)B + \alpha M] = \frac{f_2}{\gamma}. \tag{31}$$

Die letzte Gleichung liefert die Grenzen, zwischen denen sich die Werte von M befinden

$$(1-\alpha)B_0 = \frac{f_2}{\gamma} - \alpha M > 0$$

$$\frac{f_2}{\alpha\gamma} > M > 0, \tag{32}$$

so wie die Grenzen für B_0

$$0 < B_0 < \frac{f_2}{(1-\alpha)\gamma}. \tag{33}$$

Mit Hilfe von Gl. (28) findet man die Grenzen für B

$$(1-\alpha)B = f_2 - \alpha M$$

$$\frac{f_2}{1-\alpha}\left(1 - \frac{1}{\gamma}\right) < B < \frac{f_2}{1-\alpha}. \tag{34}$$

Schließlich ergibt die Subtraktion von Gl. (9) und Gl. (10), $\Delta G = G_{yx} - G_{yz}$, die letzte Unbekannte:

$$M_0 = M + \frac{\Delta G}{\alpha}. \tag{35}$$

Mit den gemessenen Werten aus Tabelle 1 (erste Reihe) und $\alpha = 0,05$ erhält man

$$f_1 = 1,1204 \qquad \gamma = 8,06$$
$$f_2 = 0,0128 \qquad \Delta G = 0,048$$

Die anderen Größen haben die Werte

$$A = 1,180; \quad 0,0118 < B < 0,0135; \quad 0,032 > M > 0$$
$$A_0 = 0,147; \qquad 0 < B_0 < 0,00168; \quad 0,992 > M_0 > 0,960.$$

Die außerordentlich kleinen Werte der Schubmoduln des Frühholzes sind verständlich, wenn man an die ersten, großen, dünnwandigen Zellen denkt. Aber der kleine Wert des Markstrahlschubmoduls M kann schwerlich als der Schubmodul des eigentlichen Markstrahlgewebes angesehen werden. Daran könnte schon eher die schwache Verbindung zwischen den zwei Gewebearten Schuld sein, was nur bei bestimmten Beanspruchungen zur Geltung kommt.

Mit den Gl. (23) und (24) kann man jetzt die beiden letzten Schubmoduln berechnen. Die drei erhaltenen Werte jeweils für G_{xy} und G_{xz}, die in Tabelle 1 eingetragen sind, entsprechen jeweils den drei verwendeten Werten, für B, B_0, M und M_0: Mittelwert aus den oberen und unteren Grenzwerten und die etwa 3% von jeder der Grenzen entfernten Werte. Der Vergleich dieser Werte mit jenen von Schlüter zeigt gute Übereinstimmung in den Größenordnungen.

Schrifttum

Kollmann, F.: Technologie des Holzes und der Holzwerkstoffe. Bd. 1. Berlin/Göttingen/Heidelberg 1951: Springer.
—, u. H. Krech: Dynamische Messung der elastischen Holzeigenschaften und der Dämpfung. Holz als Roh- und Werkstoff Bd. 18 (1960) H. 2, S. 41/54.
Schlüter, R.: Elastische Messungen an Fichtenholz. Braunschweig 1932: Vieweg & Sohn.
Keylwerth, R.: Die anisotrope Elastizität des Holzes und der Lagenhölzer. VDI-Forschungsheft Nr. 430, Ausgabe B, Bd. 17 (1951).
Ghelmeziu, N., u. P. Suciu: Identificarea lemnului. Bucureşti, 1959: Ed. Tehnică.
Schniewind, A.: Transverse Anisotropy of Wood: A Function of Gross Anatomic Structure. For. Prod. J. Vol. IX (1959) No. 10, S. 350/359.
Polge, H.: Etablissement des courbes de variation de la densité du bois par exploration densitométrique de radiographies d'échantillons prélévés à la tarière sur des arbres vivants. Annales Sci. For. Bd. 23 (1967) S. 1/81.

Revised Summary by I. P. Beldie

The dynamic measurement of the modulus of elasticity E on the basis of bending vibration [see Kollman and Krech, *Holz Roh- Werkstoff,* **18**(2) 41–56, (1960), or Ghelmeziu and Beldie, *Catgut Acoust. Soc. Newsletter,* No. 17, 1972, pp. 10–16] on spruce samples in the longitudinal and radial direction of grain has a second result the values of the modulus of shear G in the bending plane, G_{yz}, G_{yx}, G_{zx}, and G_{zy}. Theoretically, G_{yz} and G_{zy} must be equal. The measurements give $G_{yz} = G_{zy}$. This may be explained by the fact that the periodic inhomogeneity of the wood in the radially running z-axis exists and the experimentally measured values of the modulus of shear are considered as mean values. For the theoretical determination of these values a geometric model for the structure of spruce is used. By this method six formulas are obtained which indicate the relation between the moduli of shear of the parenchyma and the rays on the one hand and the experimentally determined moduli of shear on the other. The theoretical comparison between G_{yz} and G_{zy} also shows that G_{yz} is larger than G_{zy}. Based on the values measured for G_{yz}, G_{yx}, G_{zx}, and G_{zy} and by the respective formulas, estimate values are calculated for the moduli of shear of the parenchyma and the rays, and the magnitudes for G_{xy} and G_{xz} are evaluated.

Reprinted from *Catgut Acoust. Soc. Newsletter*, No. 17, 10–16 (1972)

On the Characteristics of Resonance Spruce Wood

NICOLAE GHELMEZIU and I. P. BELDIE

Introduction

Violin makers pay particular attention to the choice of the spruce wood they use for the manufacture of stringed musical instruments. In view of the specific requirements of this wood, several empirical criteria have been defined for the macrostructural characteristics of "resonance spruce wood," namely:

1. Complete lack of any defect.

2. The width of the annual growth has to be 0.8–2.5 mm for the violin and viola, the upper limit increasing to 3.5 mm for the cello and to 5.0 mm for the double bass and guitar.

3. The irregularity of annual growth width shall be as small as possible.

4. The highest permitted proportion of latewood in the annual growth shall be 25 percent.

5. The grain has to be straight.

The fulfillment of these criteria is considered absolutely necessary to ensure a minimum acoustical quality of the finished instrument. Obviously, however, the construction of the instrument is at least as important.

On the other hand, for the characterization of the resonance spruce wood from the acoustical point of view, the following quantities have been defined:

1. The acoustic radiation constant K, given by the expression

$$K = \sqrt{\frac{E_y}{\rho^3}} \tag{1}$$

where E_y = logitudinal modulus of elasticity (Figure 1, y direction) and ρ = apparent wood density.

This expression was first deduced by W. Hahnemann and H. Hecht in 1917 in order to characterize the radiation damping of a circular plate vibrating with its lowest eigenfrequency (quoted in ref. 1). In 1937 N. N. Andreew again found this

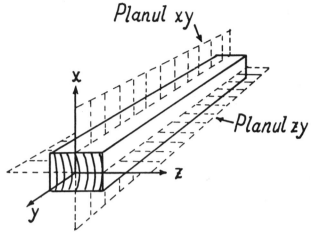

Figure 1. Form and orientation of the wood samples.

relationship starting from the fundamental vibration of an elliptical wooden plate (quoted in ref. 2.). In the Soviet Union (2) and Poland (3) it is admitted that the greater this constant K the more suitable the wood for manufacturing of the top plate of stringed musical instruments. Its lowest permitted limit is 10 m⁴/kg·s for 10 percent moisture content of the wood.

The importance of the constant K has also been pointed out by J. Schelleng (4) based on his electroacoustic analogy of the violin. It follows from his theory that if a violin is to be built identically with a given one, resonance wood with the same K as the wood of the model must be used.

2. The loss factor η is defined by the expression of the complex modulus of elasticity

$$E_y^* = E_y(1 + j\eta) \tag{2}$$

This second acoustical characteristic is related to the loss of vibrational energy through internal friction, and one may suppose that a lower value of η favors the acoustic radiation of the musical instrument.

The purpose of the present study was to measure the macrostructural and acoustic characteristics of the Rumanian resonance spruce wood sorted on the basis of the five empirical criteria mentioned above, and to establish the correlations between these characteristics.

Experimental

The specimens were cut from twelve logs selected as resonance spruce logs— Picea excelsa (L) Karsten—felled in woods from different regions of Rumania, woods known as resonant wood producers.

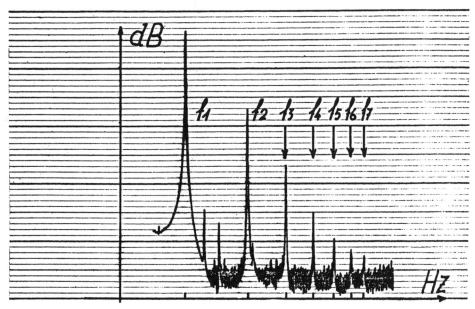

Figure 2. Frequency response curve of the bending vibration of the wooden bar ($f_1 \cdots f_7$ = eigenfrequencies).

Each sample was cut so as to fulfill exactly the empirical criteria 1 and 5 (above), to respect as much as possible criterion 3, and to fulfill a supplementary condition, namely an optimum regularity of the proportion of latewood. The only remaining variables which differ from one sample to another are the mean annual growth width and the mean proportion of latewood (the means are taken over each sample).

The samples had prismatic form (Figure 1) with the axis parallel to the longitudinal direction of the wood, y, with rectangular cross section and the following dimensions: 40 by 2 by 2 cm (thirty-nine samples) and 40 by 2 by 1.6 cm (thirty-three samples). The relative moisture content of the samples was set at 11.3 $\pm$ 1 percent.

The resonance curves embracing the first five to eight bending eigenfrequencies in the plane (xy) of each sample were obtained with the sample in the usual two-point suspension. The frequencies (see Figure 2) and dimensions of the sample were used to calculate the quantities A_n and B_n in the equation

$$E_y = A_n + B_n \, (s/G_{xy})E_y \tag{3}$$

according to the method of Kollmann and Krech (ref. 5). When the values of A_n for the various frequencies are plotted against B_n in Cartesian coordinates, they yield according to this equation a straight line of which the A intercept, E_y, and the slope, $-sE_y/G_{xy}$, are obtained by least-squares methods. Young's moduls and the shear moduls are thus obtained simultaneously. The term s is a constant depending on the distribution of stresses in the cross section of the sample.

447

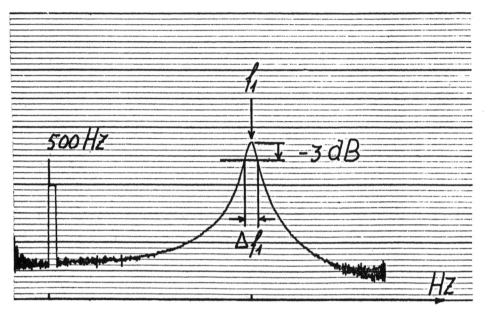

Figure 3. Frequency response curve of the first eigenfrequency.

The loss factor was obtained from the expression

$$\eta = \frac{\Delta f_1}{f_1} \tag{4}$$

where Δf_1 is the width at -3 dB from the peak of the resonance curve of the first eigenvibration, and f_1 is the frequency of the first eigenvibration (Figure 3).

The resonance curve of the bending vibrations in the plane (zy) are similarly traced and the whole calculation is repeated. The value of the longitudinal modulus of elasticity E_y is obtained together with a new shear modulus G_{zy} in the plane (zy). Within the limits of the measurement errors, the two values found for E_y are equal. Also, no essential difference was found between the values of E_y for the samples with rectangular cross section and those for the samples with a square one.

The density p and macrostructural characteristics—mean annual growth width b_m, mean proportion of latewood p_{ltz}, and irregularity of annual growth widths r—were determined for each sample.

Results and Discussion

The results of the measurements and calculations are given in Table 1. Since the logs used for cutting the samples were obtained only from woods known as

448

Table 1. Elastic, acoustic, and macrostructural characteristics of Rumanian resonance spruce wood.

Quantity	Units	Range of Values	Mean Value	Standard Deviation
E_y	kgf/cm²*	84,200 – 173,200	133,000	19,600
G_{xy}/s	kgf/cm²	4,690 – 6,730	5,760	176
G_{zy}/s	kgf/cm²	5,200 – 7,230	5,930	157
η(600–700 Hz)	1	0.0058 – 0.0077	0.0068†	0.0004
ρ	g/cm³	0.353 – 0.492	0.421	0.032
b_m	mm	0.73 – 4.33	2.01	0.84
p_{ltz}	%	0.136 – 0.305	0.206	0.034
r	%	3.0 – 131	33.8	—
$K = \sqrt{E_y/\rho^3}$	$m^4/\text{kg}\cdot\text{s}$	11.67 – 15.20	13.32	0.72

*kilograms force/cm² = 9.81 N/cm²
†Fifty-four included instead of seventy-two used for other characteristics.

resonance log producers, the values given in Table 1 may be considered representative for wood selected on the basis of the five empirical criteria presented above.

In order to establish to what extent the macrostructural characteristics are decisive for the selection of spruce wood of resonance quality, these quantities were correlated with the density, the acoustical characteristics K and η, the longitudinal modulus E_y, and the shear moduli G_{xy} and G_{zy}. The linear correlation coefficients R were calculated. The corresponding regression equations were calculated only if R exceeded the limit of uncertainty ($|R| > 0.31$ for seventy-two samples if a threshold of 1 percent is considered). The results were:

1. The correlation between apparent density and mean width of annual growths, which is not too close ($R = -0.58$) is given in Figure 4. The equation of the

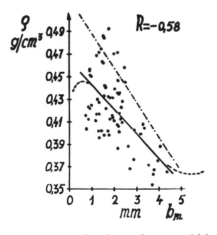

Figure 4. Correlation between apparent density and mean width of annual growth.

449

Table 2. Correlation between apparent density (g/cm³) and mean width of annual growths (mm).

Number of Samples	Relative Moisture Content (%)	Equation of Regression Line	Correlation Coefficients R	Reference
250–450	10	$\rho = (0.519 \pm 0.046) - 0.031 b_m$	−0.58	Mironow, Kulikow (see ref. 2)
520	8	—	−0.56	Ref. 8
72	11.3 ± 1	$\rho = (0.465 \pm 0.026) - 0.022 b_m$	−0.58	Present work

regression line is given in Table 2 and has the form previously found by N. I. Mironow and N. P. Kulikow (quoted in ref. 2),

$$\rho = M_1 - N_1 b_m \tag{5}$$

where M_1 and N_1 are constants, and is valid in the range $1 < b_m < 4$ mm.

Investigations by others, quoted in references 2 and 6, show that at an annual growth width under 1 mm the density reaches a maximum value, and at 5 mm or more, a minimum. This type of relationship between density and annual growth width seems to be an important feature of spruce wood. It is interesting to point out that the measured values of the mean widths of the annual growths lie only between these two extremes.

2. The well-known relationship of A. Ylinen,

$$\rho = M_2 + N_2 p_{ltz} \tag{6}$$

between density and latewood was not borne out in the present data, probably since the range of latewood was very narrow for the selected samples. Combining equations (5) and (6), the following correlation should exist:

$$P_{ltz} = M_3 + N_3 b_m \tag{7}$$

with M_3 and N_3 as constants. However, $R = -0.26$, which is lower than the limit of uncertainty.

3. The correlation between longitudinal modulus of elasticity and density is very close and is represented in Figure 5. The equation of the regression line (Table 3) is of the form

$$E_y - A\rho - B \tag{8}$$

with A and B as constants, as has been noted by others (1, 2, 3, 5, 7, 8). See also Figure 6.

450

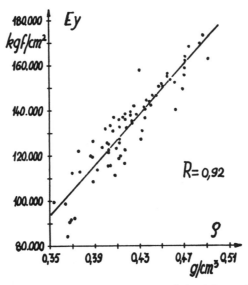

Figure 5. Correlation between longitudinal modulus of elasticity and apparent density.

The high correlation coefficient $R = 0.92$ allows us to consider E_y as a function of ρ. Then, substituting (8) in (1), the acoustic constant K is obtained as a function of density,

$$K = \frac{a}{\rho} + \frac{b}{\rho} \, 2 \qquad (9)$$

with a and b as constants. Thus the value of K characteristic for the acoustic quality of the spruce wood increases with the decrease of the density. The range of measured densities being comparatively small (Table 1), the corresponding part of

Table 3. Correlation between the longitudinal moduls and apparent density.

Number of Samples	Relative Moisture Content (%)	Equation of Regression Line	Correlation Coefficients	Reference
250, 450	10	$E_y = 378,000 - (45,000 \pm 10,300)$	0.74/0.90	Mironow, Kulikow (see ref. 2)
129	11–12	$E_y = 471,000\rho - 59,700$	0.95	Ref. 7
122	10	$E_y = 333,000\rho - 53,000$	—	Ref. 3
65	14.3	$E_y = 466,000\rho - (55,600 \pm 4,100)$	0.85	Ref. 5
520	8	$E_y = 600,800\rho - 120,080$	0.70	Ref. 8
86	8	$E_y = 459,200\rho - (62,310 \pm 11,990\ \ )$	—	Ref. 1
72	11.3	$E_y = 554,000\rho - (100,600 \pm 7,800)$	0.92	Present work

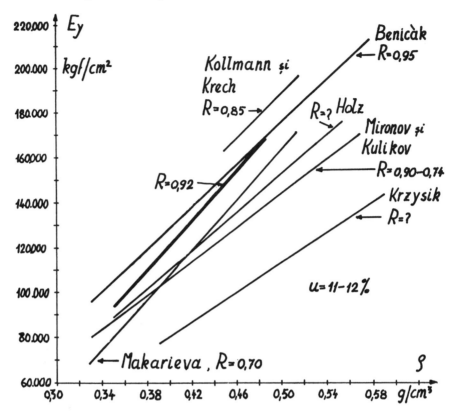

Figure 6. Regression lines of the correlation between longitudinal modulus of elasticity and apparent density obtained by different authors (corrected for a relative moisture content: 11–12 percent).

the curve can be represented as a straight line with negative slope, as found by others (2, 3, 8). In this case the equation of this line may be obtained directly by evaluating the correlation between acoustic radiation constant and density. The following relation is obtained:

$$K = (21.10 \pm 0.48) - 18.58\rho \tag{10}$$

with a correlation coefficient $R = -0.81$. As expected, this coefficient is smaller (absolute value) than R for the correlation (8), since a part of the curve (9) was assimilated to a straight line (10).

It follows that the value of the acoustic radiation constant is determined by the value of the apparent density, which in turn is determined by both mean width of annual growth and mean latewood proportion. Using the same example, one may say that a large width of annual growth and a small proportion of latewood will correspond very likely to a high acoustic radiation constant.

The question arises: Why do violin makers not use, for the top plates of their violins, wood with the maximum allowable width of annual growth of 5 mm, but limit themselves to 2.5 mm?

There are at least two ways of answering this question. The first principal answer is that the acoustic radiation constant is not sufficient for the acoustic characterization of the resonant spruce wood, imposing at the most a minimum allowable value. This may be understood if one notes that the expression of the acoustics radiation constant contains only the first of the two principal moduli of elasticity E_y and E_z (E_z is the modulus in the radial direction z), and they both play a part in the determination of the bending eigenvibrations of the violin's top plate. The second answer has a more practical aspect, namely, that plates of resonance spruce wood sufficiently large for making top plates having both a large width of annual growth and a minimum irregularity (criterion 3) can hardly be found.

4. The irregularity of annual growth width has been determined for each sample using the expression

$$r = \frac{b_{\max} - b_m}{b_m} \cdot 100 \qquad (11)$$

where $b_{\max}$ is the maximum width measured on the sample. This variable did not correlate with any of the other factors, most likely a consequence of the sample selection. However, if one takes into account the correlations expressed by (9) and (5), it follows that a variation of the annual growth width over the top plate of a musical instrument will lead to similar variations of the modulus of elasticity and implicitly of the acoustic radiation constant. If these variations are too large, some difficulties in realizing correct thicknesses of the top plate may arise.

5. Attempts to correlate the shear moduli, G_{xy} and G_{zy}, with macrostructural and elastic characteristics yielded correlation coefficients near zero, except for the correlation between G_{xy} and apparent density, where $R = 0.40$ (only slightly higher than the limit of uncertainty, $R = 0.31$).

6. Similarly, correlation between the loss factor and all macrostructural features was found to be zero, except with density ($R = -0.29$). Although the latter is below the limit of certainty, it does suggest a possible increase of the loss factor with decrease in density.

Conclusions

1. The data given in Table 1 may be considered as characteristic for the Rumanian resonance spruce wood selected on the basis of the empirical criteria described.

2. The dependence between density and (mean) width of annual growth seems to be an important feature of spruce wood and deserves special attention.

3. As a working assumption, the value of the acoustic radiation constant is determined by the magnitude of the apparent density, which in turn is determined simultaneously by the magnitudes of the latewood proportion and the width of annual growth.

4. The studies represent the first attempt to correlate the structure (macro) of Rumanian resonance spruce wood with its acoustic properties. The poor degree of correlation observed in some instances in these studies suggests the desirability of further work in which broader ranges of sample parameters are included.

References

1. Holz, D. "Untersuchungen an Resonanzholz," *Holztechnologie, 8* (4), 221–224 (1967).
2. Rimskii-Korsakow, A. V., and N. A. Diakonow. *Muzikalinie Instrumenti*, Rosgizmestprom, Moscow, 1952.
3. Krzysik, F. "Untersuchungen über den Einfluss der Rohdichte auf die Verwandungs-möglichkeit von Fichtenklangholz" International Symposium, Eberswalde, May 10–12, 1966, pp. 71–81.
4. Schelleng, J. C. "The Violin as a Circuit" *J. Acoust. Soc. Amer.,* 35 (3), 326–338 (1963).
5. Kollmann, F., and H. Krech. "Dynamische Messung der elastischen Holzeigenschaften und der Dämpfung," *Holz Roh-Werkstoff, 18* (3), 41–54 (1960).
6. Kollmann, F. *Technologie des Holzes und der Holzwerkstoffe,* 2nd ed., Vol. I, Springer-Verlag, Berlin, 1951.
7. Benicak, I. "Nondestructive Testing of Wood Using Dynamic Methods" (in Czechoslovakian), *Drevarsky Vyskum,* no. 3, 1962, pp. 261–278.
8. Makariewa, T. A. "Vlianie razlicinich faktorow na znatchenie acustitcheskoi konstanti rezonansowoi drewesini," *Derevoobrabativaiuschiaia promislenosti 17,* (10) 14–15 (Oct. 1968).

Editor's Comments on Paper 26

26 Fryxell: *The Hazards of Weather on the Violin*

Dr. Robert E. Fryxell (1923–   ) is a chemist and an amateur cellist. During the 1950s he participated in discussions of violin acoustics with Professor Saunders and Mrs. Hutchins. These three, together with John Schelleng, who later joined in the discussions, formed the nucleus of the Catgut Acoustical Society, Inc., which was formally organized in 1963. In the capacity of Editor of the Society's *Newsletter* since its inception in 1964, Fryxell has provided encouragement for the publication of a wide variety of articles related to technical and acoustical problems of the violin that might not come within the scope of other journals. Several of these are included in this volume (Papers 13, 17, 19, and 25).

Fryxell's own research on violins has been limited to a study of the elastic contents of a wide variety of woods—a study originally begun as a service for Dr. Saunders. The paper that follows summarizes results demonstrating the magnitude of wood–environment interactions, a factor which merits further attention.

Reprinted from *Amer. String Teacher*, 26–27, (Fall 1965)

The Hazards of Weather On The Violin

R. E. Fryxell

All players of stringed instruments know that an arid climate — or dry indoor heat in the winter — is hard on their instruments. Glued joints may open and cracks may develop. I have even heard of some who keep their fiddles in the bathroom in an attempt to prevent such drying out! At the other extreme, very humid weather leads to other problems such as sticky resin, poor response, and the change in arch which frequently necessitates a change in the bridge. This latter problem is primarily the result of swelling of the wood, particularly in the cross-grain direction, caused by absorption of moisture from the atmosphere.

In addition to these rather obvious effects, moisture "breathing" by the wood may also lead to more subtle changes in the instrument which would manifest themselves in throwing the instrument out of balance in the sense that weak (or strong) notes might develop as the weather changes.

In a recent article elsewhere[1] I discussed the large effect that such weather changes can have on the resonance frequency of a violin top or back. For example, absorption of moisture from the air causes maple to exhibit a ¾% drop in resonance frequency for each percent increase in weight. For spruce, the resonance change is somewhat less, and to the extent that these two materials react *differently*, the instrument may be thrown out of balance.[†] Carleen Hutchins[2] has defined the optimum resonance frequencies of top and back, and this is indeed a delicate tuning requirement (which apparently some of the old makers were aware of on a rule-of-thumb basis). Changes in weather can well upset this tuning enough to be apparent in the behavior of the instrument. In addition, other changes probably occur. The amplitude of the resonance peak may be depressed and damping may be increased. These two phenomena are collectively known to the violin maker as "ring" and he is very aware how sensitive the proper "ring" is to plate graduations. Although unequivocal tests have not been made, it is reasonable to suppose that a change in weight—by moisture breathing—is detrimental in the same sense as a poorly graduated plate.

The data which I mentioned above were all obtained with small rectangular bars of wood since these permitted rapid and reliable checks of resonance fre-

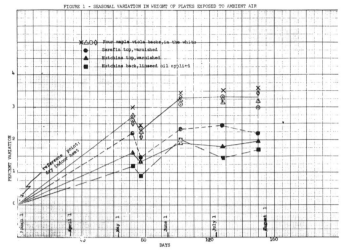

FIGURE 1 - SEASONAL VARIATION IN WEIGHT OF PLATES EXPOSED TO AMBIENT AIR

X△�‍◇○ Four maple viola backs, in the white
● Serafin top, varnished
▲ Hutchins top, varnished
■ Hutchins back, linseed oil applied

quency. The purpose of this note is to describe some further results obtained with actual tops and backs. With these, however, measurement of resonance frequency (tap tone[2]) requires more elaborate equipment than I had available, so I had to be content with measuring weights under various conditions of humidity. It is reasonable to suppose that the corresponding shifts in tap tone (percentagewise) would be the same as obtained with the rectangular specimens.

The experimental conditions were as follows: For the driest condition, used as the reference point for data shown in the graphs, I used weights corresponding to dry indoor heat (February) in a room in which no humidifier was operating. Based on separate studies, these weights were shown to be essentially the same as could be obtained by artificial drying techniques, namely a chemist's desiccator containing calcium sulfate. For the wettest condition, plates were suspended in a sealed container of about five gallons capacity in which a tray of water was placed. After being closed for some days, this of course corresponds to 100% relative humidity. Thus the results represent the extremes which may be anticipated for actual weather variations, the temperature being nominally in the range 70 to 80 degrees. The plates tested are identified as follows:

(1) Four viola backs, maple, in the white, 4 to 20 years old

(2) Two violin backs, maple, in the white, over 70 years old

(3) Four violin tops, spruce, in the white, two of them over 100 years old.

(4) Santus Seraphin violin top, original varnish, exact date unknown, but early 18th century.

(5) Carleen Hutchins viola top, spruce, varnished 1948 but no filler, wood from Germany about 1935

(6) Carleen Hutchins viola back, maple, linseed oil applied 1953 but not varnished, wood from Czechoslovakia about 1930

The results are summarized in three graphs and reveal several interesting features. In Figure 1, weight changes caused by ambient weather changes are given for individual plates. No attempt was made to record the relative humidity, yet it is clear that large changes can occur between dry indoor heat and fairly humid (Cincinnati) summer weather. The four unvarnished maple plates gave very consistent results, as did other groups of similar plates, so in the other graphs, only averages of groups are plotted. It is also clear that the finished plates show appreciably greater stability than those in the white, which also implies greater stability with respect to resonance frequency changes and any other changes which may occur. The Seraphin, however, is certainly no better than the other coated examples, suggesting that age in itself has not "improved" matters. Information of this kind is very scarce in the literature. In one paper[3] Ulrich Arns reports that a violin varied nearly 12 grams in weight during a two year period of observation. This he colorfully describes as equivalent to half a schnapps glass of liquid. If one discounts the heavy neck and scroll as not absorbing moisture (completely enclosed by varnish), twelve grams corresponds to a 6% variation in weight of the top and back plates which together weigh in the neighborhood of 200 grams. Twelve grams is nearly half an ounce, or one tablespoon of water.

[†]These measurements were in the with-grain direction. The total effect is undoubtedly greater than these data indicate inasmuch as in the cross-grain direction, dimensional (and probably frequency) changes are more severe.

In Figure 2, more results are shown which demonstrate the difference between bare and varnished plates when maintained in an environment of 100% relative humidity. Note especially that the water is absorbed very slowly over a period literally of months, yet after removal from such an environment, the original weights are recovered very quickly. In some cases, this occurred within four hours. It would be instructive to relate this difference in rates to the change in behavior of an instrument when moved from a humid to dry environment or vice versa. It is also evident in Figure 2 that the capacity for moisture breathing is roughly the same for maple and spruce. However, plate de-tuning may nevertheless occur, since these materials show rather different sensitivities to a shift in resonance frequencies.

In Figure 3, some additional (averaged) data are shown for the three coated plates. They were put through two cycles between 100% relative humidity and ambient air, the difference being that the starting point was dry indoor heat for one and humid summer weather for the other. Note therefore after the second cycle that the weights returned only part way to the original (dry) weights. Again, there is a striking difference between the rates of weight gain and loss.

To summarize, water pickup and loss is great enough to cause significant changes*in environmental humidity. The difference between pickup and loss rates, I believe, have previously not been reported. Varnishing lessens the magnitude of these changes; future research on varnishes may well include this as a goal.

Further research also might profitably be carried out to find other materials, or other species of wood, which are less vulnerable to moisture breathing. In the course of my tests, I obtained some tentative data which suggest that cherry and sycamore (frequently used for backs) have a lower capacity for absorbing moisture than maple.

I would like to acknowledge the frequent help of Carleen Hutchins not only for encouragement and fruitful discussions, but equally for supplying all the plates. The four viola backs, in addition to the two coated plates, were the product of her craft.

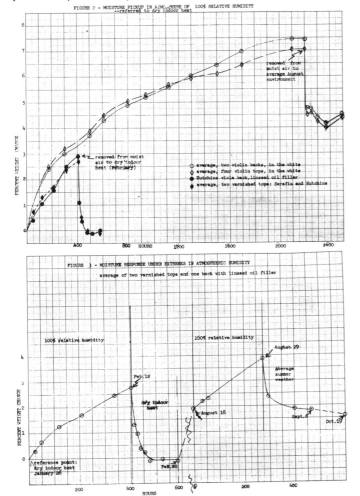

FIGURE 2 - MOISTURE PICKUP IN AIR WHERE OF 100% RELATIVE HUMIDITY
--referred to dry indoor heat

FIGURE 3 - MOISTURE RESPONSE UNDER EXTREMES IN ATMOSPHERIC HUMIDITY
average of two varnished tops and one back with linseed oil filler

REFERENCES

1. Robert E. Fryxell, "Some Thoughts on the Requirements of a Good Varnish," THE STRAD, February, 1965, pages 361-365.

2. Carleen Maley Hutchins, "The Physics of Violins," Scientific American, November, 1962, pages 78-93.

3. Ulrich Arns, "Eine neue Art objektiver Qualitatsfeststellung von Geigen," Gravesaner Blätter, 11, 7/8, 1957, pages 92-116.

* in the sound of a stringed instrument for nominal variations

VI
Varnish

The literature on violin varnish is voluminous but is, unfortunately, limited primarily to the esthetic question of how to duplicate the appearance of instruments made by the old masters. Occasionally there are comments concerning the alleged beneficial effect of varnish on tone, or, conversely, that too heavy a layer can be deleterious. Since musical performance of an instrument rather than its appearance is the subject of this volume, only nine references are included in the bibliography following this commentary.

The one paper included here, "Acoustical Effects of Violin Varnish," by John C. Schelleng, indicates that as far as the acoustical importance of varnish is concerned, the less varnish the better. Schelleng shows the effects of violin varnish on mass, stiffness, and internal friction over a 2-year period. There is no consideration of the effects of various fillers that might be applied to the violin before varnishing.

Information about the life and work of John C. Schelleng can be found in "Editor's Comments on Paper 5."

Additional Writings on Varnish

Reade, Charlen. *Cremona Violins and Varnish* (four letters that appeared in the *Pall Mall Gazette*, 1872), reprinted by Alexander Brodie, Inc., New York, n.d.

Mailand, Eugène. *Découverte des anciens vernis italiens employés pour les instruments à cordes et à archets*, E. Lahure & Co., Paris, 1859; 2nd ed., 1874; 170 pp.

Fry, George. *The Varnishes of the Italian Violin Makers of the Sixteenth, Seventeenth and Eighteenth Centuries and Their Influence on Tone*, Stevens & Sons Ltd., London, 1904, 170 pp.

Part VI

Christ-Iselin, Wilhelm. *The Mystery of Cremona Varnish,* Breitkopf and Härtel, Leipzig, 1923, 43 pp.
Greilsamer, Lucien. "Vernis de Crémone," in *L'Anatomie et la physiologie du violon de l'alto et du violoncelle,* Libraire Delagrave, Paris, 1924, pp. 103–238.
Leipp, Émile. *Essai sur la lutherie, le vernis de crémone,* Chez l'Auteur, Paris, 1946, 90 pp.
Michelman, Joseph. *Violin Varnish,* Joseph Michelman, Cincinnati, Ohio, 1946, 185 pp.
Corbara, Lamberto. *Violin Varnish Old and New,* Cesena, 1963, 97 pp. In Italian.
Arakelian, Sourene. "Mon Vernis à Base de Myrrhe," *Das Musikinstrument,* Frankfurt, 1964, 15 pp.

27

Reprinted from *J. Acoust. Soc. Amer.*, **44**(5), 1175–1183 (1968)

Acoustical Effects of Violin Varnish

JOHN C. SCHELLENG

301 Bendermere Avenue, Asbury Park, New Jersey 07712

Acoustical effects of violin varnish—those of mass, stiffness, and internal friction—are examined qualitatively and quantitatively. Dynamic measurements of Young's modulus, real and imaginary, are made for films of typical oil varnishes on a substrate of cross-grain spruce through aging periods of 1 yr or more and expressed in the practically useful terms of modulus over density. After 1½ yr, the hard varnish had a real modulus per unit density of 2.5×10^{10} cm²/sec² and Q about 23; after 1 yr, the values for the soft varnish were 1.6×10^{10} and 8. A high-grade floor varnish had intermediate values. The imaginary component of the soft one was substantially independent of frequency from 400 to 5600 cps. Exact calculations of added loss and tuning shift, made for a "pseudofiddle" comprising rectangular top and back of orthotropic materials simulating spruce and maple, are believed roughly representative of the violin. Added loss is more apt to be damaging than is detuning of plates. The loss comprises two components: (1) a reduction in sound caused by the reactive parameters of varnish, i.e. mass and stiffness, and effective irrespective of resonance; and (2) one caused by internal friction and greatest at a resonance. With an intermediate varnish, a coat of 0.0127 gm/cm² (thickness about 0.005 in.) causes "flat" loss of about 1 dB and resonance loss about 3 dB; this is excessive. Varnish loss in the top plate is about three times that in the back; unless the spruce is overly resonant, its varnish should be thin. Top is more subject to detuning than back. Loss due to varnish is best controlled by weighing, with final adjustment of wood thickness after varnishing.

INTRODUCTION

OVER the years, there has been endless discussion on the virtues and vices of violin varnish. Much has been well or poorly said about chemical nature and visual appearance and much has been affirmed or implied as to musical importance. R. E. Fryxell[1] has forcefully commented, however, that very little has been done to determine exactly what it is that varnish does beside protecting the outside surface of a fiddle and giving it an attractive appearance: we should also know, "How much?" It is the present purpose to examine the several acoustical effects of varnish quantitatively. While it is important to understand the action of deeply penetrating fillers, they are best treated in connection with the study of violin woods.

In their excellent *Antonio Stradivari, His Life and Work*, the Hill brothers[2] argue at some length to show that of the three categories (1) material (wood), (2) dimensions and construction, and (3) varnish, the last is the most important and the first least. Today, there seems to be general agreement among those who think in terms of physics rather than art that the acoustical virtues of varnish have been greatly exaggerated. Major stress is now placed on dimensions and their adjustment to the materials so as to achieve necessary vibrational patterns, frequencies and impedances, while varnishing has become the final protective operation that is covered by the injunction: not too soft, not too hard, not too much.

Even today there seems to be no agreement in answer to the query: Is varnish helpful or harmful acoustically? We probably should not expect the same answer in all cases. Many years ago, H. Meinel published curves[3] showing how for a certain violin response over the frequency range was affected by application of (alcohol) ground, color, and cover varnishes. His first conclusion was that the over-all effects are small compared with those induced by reductions in wood thickness made in the course of optimizing response, contrary to the somewhat ambiguous statements of the Hills. The maximum amplitudes of the three lowest wood resonances decreased about 2 dB, while the corresponding minima were scarcely affected. Subjectively, the violin seemed

[1] R. E. Fryxell, *The Strad* (Lowe and Brydone Ltd., London, 1965).
[2] W. H. Hill, A. F. Hill, and A. E. Hill, *Antonio Stradivari, His Life and Work* (Dover Publications, Inc., New York, 1963), pp. 159–160, 178–179.
[3] H. Meinel, Akust. Z. **2**, 27–28 (1937).

to "speak" better, to have a somewhat pleasanter sound with a decrease in the previously noted slight roughness. *There was no claim for anything striking.* Incidentally, there was no clue as to how much varnish was used.

The same investigator[4] seems to be the only one to have made quantitative measurements of the acoustical effects of violin varnish. Curves showed an increase in the decrement of his wood strips (grain ran lengthwise) as frequency rises from 100 to 5000 cps, and he considered this to be an important contribution by the varnish toward the suppression of vibration at extremely high frequencies. Data given are insufficient to calculate the magnitude of the effect.

A well-known German work on violin making[5] emphasizes the damage done by varnish. "How many newly-made instruments after completion sound soft and full only to forfeit their tonal beauty after the complete drying of the varnish." This loss depends on the resins, not on the solvent. Hard substances like amber and copal when aged hold the instrument "as in clamps of iron." A bad instrument cannot be ennobled by the best varnish. As a general precaution one avoids heavy layers. "The best-sounding of all violins, the Cremonese, had the thinnest varnish." This same view is held by the large number of makers who hold that a fiddle never sounds so well after varnishing as it did "in the white."

Reversing the varnishing process, Louis Condax has subjectively observed[6] a remarkable improvement as a violin is freed by removal of an old and improper coat of varnish.

The effect of varnish need not be large to be important. According to W. Lottermoser and J. Meyer,[7] 5 dB measure the difference, averaged over peaks and valleys, between an instrument of extraordinary power and one that is too weak. Clearly, the master maker must conserve every decibel with miserly care.

R. K. Myers[8] and associates in a broad investigation aimed at paints generally have studied the rheology of the drying process employing internal reflection of high-frequency shear waves from a quartz surface that has been painted. Such a study of violin varnish might be of interest with reference to the upper octaves of the spectrum where shear in the wood seems to be of importance.

Rheologists have accumulated a large body of information on the viscoelastic properties of polymers. Violin varnish falls in a class of "amorphous polymers of high molecular weight below their glass-transition temperature" and measurements herein are consistent

in orders of magnitude with those encountered in this class.[9]

Like other components in a vibrating system, varnish produces effects corresponding to mass, stiffness, and internal friction. The mass added by varnish to the light spruce plate is not negligible. If we assume a varnish thickness[10] of 0.005 in., the mass of a violin top plate would be increased about 7 gm or about 10%. This by itself would cause a lowering in tap-tone frequencies of 5%, which is the larger part of a semitone and not to be casually ignored.

This lowering would be the same over the frequency range: a vibration controlled by stiffness along the grain would be lowered the same number of cents as one across the grain. In contrast with this are the effects that follow from stiffness added by varnish. Across the grain, the stiffness of spruce is less than that along it by a factor of about one-fifteenth; it is therefore markedly more susceptible to interference from the varnish. Indeed, a stiffness great enough to be important along the grain would be all but catastrophic across it. Our greatest concern therefore is with stiffness effects across the grain.

Whereas the application of varnish leads to an immediate decrease in frequency of the plate because of mass, stiffness effects develop gradually. When strips of spruce are varnished, the effect following the initial drop in frequency is a rise as volatile components evaporate. This continues beyond the time when tackiness disappears and may not be complete after a long period of aging. The change could be calculated in terms of Young's modulus of varnish if measurements thereof were available. In a fiddle, the frequencies of different resonances will be unequally affected because of differences in the relative bending in the two orientations.

While these effects on frequency are to some extent self-canceling, the result of stiffness eventually predominates over mass effect and might ultimately be a cause for serious concern. Even though stiffness nullifies the effect of mass on frequency, a damage remains concealed in an increase of impedance. Broadly speaking, the magnitude of impedance is determined by the square root of stiffness times mass $(SM)^{\frac{1}{2}}$, both factors of which have been increased. If, for example, M increases 10% and S, 14%, then impedance increases 12% and response drops 1 dB.

Finally, we have to evaluate the effect of internal friction in the varnish. Paralleling the effect of the real component of Young's modulus in altering frequency,

[4] H. Meinel, J. Acoust. Soc. Amer. **29**, 831 (1957).
[5] O. Mockel and F. Winckel, *Die Kunst des Geigenbaues* (Voigt, Berlin, 1954), p. 213.
[6] L. Condax (personal communication) (16 May 1963).
[7] Instrumentenbau Z. **12**, 42–45 (1957).
[8] Raymond R. Myers, "The Rheology of the Drying Process," Official Digest, Fed. Soc. Paint Technol. (August, 1961).

[9] J. D. Ferry, *Viscoelastic Properties of Polymers* (John Wiley & Sons, New York, 1961), 1st ed. Curves labeled IV in Figs. 2—6, 2—7, and 2—8, and Figure 14—2, all for polymethyl methacrylate, are of interest.
[10] This is approximately the thickness measured by Joseph Michelman and quoted in his *Violin Varnish*, published by the author (Cincinnati, Ohio, 1946), p. 110. From the context, the inference is that thickness in this order is typical among violin makers.

there is a loss of power because of the imaginary component. This, too, is especially important across the grain. Whereas the impedance effect reduces vibration in "valleys" as well as on "hills," internal friction produces little change in the lows, or "nulls." However, it can be very important at peaks of resonance and is usually the most noticeable result of the use of varnish.

I. EXPERIMENTAL METHOD AND RESULTS

Throughout these studies, the effects of varnish have been measured by observing the degree of change in the resonance bandwidth of a strip of substrate vibrating in its lower-frequency modes. The first series of tests employed glass ($16 \times 1.5 \times 0.2$ cm), but all later ones used quarter-cut cross-grain spruce ($13 \times 1.6 \times 0.38$ cm). Glass has the advantage of stability in the presence of changes in environment. For present purposes, however, its disadvantage is that its stiffness is so great as to obscure the stiffness of a thin coat of varnish whose modulus may be less than 1% of its own. In going to spruce, we avoid this difficulty because its modulus is in the same order as that of dried varnish, but now we court a different trouble, namely, instability of mass and elasticity in the presence of moisture, not only from summer to winter but from day to day as well. A laboratory with controlled atmosphere was not available. Keeping specimens in a small moisture-controlled space failed because changes in frequency set in as soon as the wood was brought out for test.

The method used instead has shortcomings that admittedly limit accuracy, but it does permit a first step in the problem. The procedure is to use a "control" strip of spruce, which suffers the same vagaries as the wood in the varnished strip and can by subtraction eliminate variations introduced by changing ambient. The saving grace is that there are long periods of weather in which changes are not so abrupt as to require rapid change of correction, that, in fact, the "control" strip does work to a useful degree and that in the present state of our knowledge a high degree of accuracy is not necessary.

With the glass strip, the first five tones had frequencies from 400 to 5600 cps. The lowest six tones of the wood ranged from 330 to 5000 cps; most of the measurements, however, were made with the second and third having frequencies of about 900 and 1750 cps. Decrement measurement had an accuracy in the order of 10% with internal consistency somewhat better. A Hewlett–Packard Model 200 AB audio oscillator provided with a simple vernier device served well for excitation. The 60-cps power supply, checked frequently by comparison with a tuning fork, proved amply accurate as a standard of frequency.

Test strips were energized by a piece of 2-mil (0.002 in.) iron at the end, having length equal to the width of the strip and width of a couple of millimeters,

acting under the influence of a Brüel & Kjær magnetic transducer.

The strip to be varnished and the control strip were first measured as to frequency of resonance and Q. The same comparison was made many times after the one was varnished. Changes in Q and frequency were interpreted in terms of real and imaginary Young's modulus of varnish.

In a simple vibrating system, where a stiffness S and mass M can be identified and angular frequency at resonance is therefore $(S/M)^{\frac{1}{2}}$, the incremental change in stiffness occasioned by the thin coat of varnish is

$$\Delta s/s = \Delta M/M + 2\Delta f/f. \tag{1}$$

Actually, the effective mass M of a mode of vibration is not the simple mass obtained by weighing, but it is proportional thereto and undergoes the same percentage change as a result of varnishing. In the dimensionless terms of Eq. 1, it is therefore permissible to treat M and ΔM as results of weighing.

It is readily shown that the real component of Young's modulus for the film of varnish is

$$E_v' = \tfrac{1}{3} E_w' \cdot (H/\Delta H) \cdot (\Delta s/s), \tag{2}$$

where E_w' is the real modulus for the wood substrate, H the thickness of the substrate, and ΔH that of varnish. (Varnish, being on the surface, is in position to increase stiffness more than a similar thickness within the wood; the factor one-third is the result of integrating this advantage over the thickness of wood.) To the extent that varnish penetrates the wood (and it is sometimes not a small effect), ΔH cannot be measured simply. If we interpret ΔH as the thickness that the known weight would have at the normal density ρ_v, Eq. 2 becomes

$$E_v'/\rho = \tfrac{1}{3} E_w' \cdot (V/\Delta M) \cdot (\Delta s/s), \tag{3}$$

in which V is volume of the wood strip. All factors in Eq. 3 are measurable. This establishes the modulus of varnish in terms of that of wood, which is calculated readily by use of standard formulas for the transverse vibrational frequencies of strips.

A little manipulation of Eq. 1 and 3 shows that at the stage of drying (with strips) when f has recovered its original frequency $(\Delta f = 0)$, $E_v'/E_w' = 3\rho_w/\rho_v$. Thus, when $\rho_w = 0.35$ gm/cm^3 (as in some spruce that I have used) and $\rho_v = 1.05$ gm/cm^3 (which seems typical), frequency is recovered when the modulus of varnish is passing through the same value as that of wood. Oil varnish on cross-grain spruce strips may pass through this value after drying for a day, more or less. Along the grain, frequency is never recovered.

The Q of the varnish may be calculated as follows:

$$Q_v = \frac{\pi(\Delta s/s)}{(\delta + \Delta\delta)(1 + \Delta s/s) - \delta}, \tag{4}$$

where δ is the original logarithmic decrement and $\Delta\delta$ is

TABLE I. One set of data for one varnish, with results according to the foregoing equations.

M	$=3.42$ gm	δ	$=0.059$
ΔM	$=0.27$ gm	V	$=7.85$ cm^3
$\Delta M/M$	$=0.079$	$E_w{}'$	$=0.89\times10^{10}$ dyn/cm^2
$\Delta f/f$	$=0.041$	Q	$=7.6$
$\Delta s/s$	$=0.161$	$(E'/\rho)_v$	$=1.4\times10^{10}$ cm^2/sec^2
$\Delta\delta$	$=0.050$	$(E''/\rho)_v$	$=0.185\times10^{10}$ cm^2/sec^2

its increment induced by varnish. Equation 3 gives the real component of modulus over density. Q being the ratio of real to imaginary modulus, the imaginary component is obtained from Eqs. 3 and 4 by division:

$$E_v{}''/\rho = Q_v{}^{-1}\cdot(E_v{}'/\rho_v). \qquad (5)$$

$(E/\rho)_v{}^{\frac{1}{2}}$ has the dimensions of velocity.

Table I indicates magnitude of various quantities involved in a single measurement. Much detail, including data for the control strip, has been omitted.

During drying, the aging of the outer surface is in advance of that within, and the moduli can only be regarded as "effective." It has even been speculated that thick coats of some materials may never attain real homogeneity, though it is not clear how such a condition could be a stable one.

Figure 1 represents the behavior of a moderately hard varnish on cross-grained spruce in the early stages of drying. Based on actual data, it is typical of all the varnishes measured except in matters of detail. Deviations occur depending on manner of application, and the varnishing procedure used for this particular test was different from any that could be reasonably employed in violin making, since the purpose was to produce a heavy layer quickly so as to avoid drying between coats as much as possible. Four coats were applied at 20-min intervals after which measuring was immediately begun. Data for two tones—the second and third near 700 and 1350 cps—were combined. Time is measured from application of the last coat. Stable weather conditions and the brevity of the experiment permitted omission of control strips and gave better continuity in data than was possible in long runs.

Curve I shows the portion of the total decrement that was induced by varnish, the decrement of the wood having been subtracted from the gross measured decrement.

Curve II represents the change in frequency that the varnish produced divided by frequency without varnish. In terms of Eq. 1, it is $\Delta f/f$.

In Curve III is given $\Delta m/M$, the fractional increment of mass added by the coating. Actually, one-half of it is shown, plotted negatively so that the distance between Curves II and III measures half the relative change in stiffness of the strip, $\frac{1}{2}\Delta s/s$ (see Curves and Eq. 1). The hook in the curve near zero time is the result of loss of volatile components.

If the wet coat merely added mass without stiffness, Curves II and III would diverge from the same point

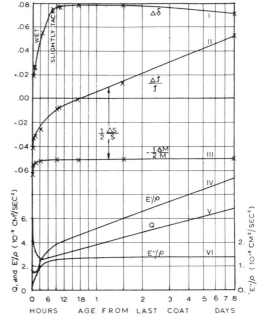

FIG. 1. Early stages in aging of a moderately hard oil varnish: Curves I, II, and III give experimental data for a varnished strip of wood; Curves IV, V, and VI, deduced therefrom, show real and imaginary Young's modulus over density and quality factor Q for varnish alone. Δ indicates changes induced by varnish.

at zero time. Some drying, of course, occurred prior to the last coat. These curves continue to diverge out to eight days, the length of the test. (The time scale is a hybrid, linear up to one day and logarithmic thereafter, being proportioned so as to avoid discontinuity at the point of division.)

From the experimental data of Curves I–III and other information (e.g., dimensions and mass of strip), Curves IV–VI are plotted representing the elastic constants of the varnish without reference to the substrate. The ordinate of Curve IV is $E_v{}'/\rho_v$, the "ordinary" or "real" Young's modulus divided by density: $E_v{}'$ indicates the degree in which the varnish affects the *frequencies* of resonance of a fiddle. Curve VI plots $E_v{}''/\rho_v$, corresponding to the imaginary modulus that governs loss of amplitude at resonance. Curve V gives their ratio, $E_v{}'/E_v{}''$, that is, Q_v.

Returning to Curve I, the first point, which was measured substantially at zero time, shows that immediately upon application, the varnish added somewhat to the losses. Within 9 h, the decrement of the varnished strip had climbed to its highest value. (Slow-drying oil varnishes take more time, alcohol varnishes less.) At this point, the varnish was only slightly tacky. A somewhat similar test with a very slow-drying varnish indicated that while the varnish was still wet to the touch the decrement began to climb, but not until the tacky condition had set in did the stiffness show significant increase. It is the volatile components that account for the moderately high Q at zero, and their relatively prompt disappearance that causes its quick drop, a consequence not at all surprising since the purpose of these components is to reduce viscosity. The increase in internal friction precedes the increase in stiffness, whose long climb takes place largely after the internal friction has ceased to increase, and the surface is no longer tacky. This striking tendency toward constancy or subsequent decrease is present in all the materials measured. A horizontal Curve VI representing decrement of varnish alone does not, in general, lead to a horizontal Curve I for the varnished strip, which displays a gradual drop caused by the increment of stiffness in the varnish. For very thin coats, however, Curve I has the same shape as Curve VI.

Figure 2 displays data obtained with three varnishes over a period beginning in July 1966 and ending in February 1968. The growth of curves such as these depends not only on the varnish but also on the amount in which it is applied; as one expects, the elastic properties of thin coats mature more quickly than those of a thick one. In these tests, however, thickness did not differ radically from sample to sample. The heavy line is for a moderately hard violin varnish, the thin line for a reputedly soft one, and the broken line for a high-class commercial floor varnish. The latter (a tung-oil product) was chosen with the surmise that a material very good for floors might be very bad for violins; it is a penetrating varnish for heavy duty service.

In Fig. 3, the two components of the modulus are plotted against each other instead of against time. In this way, we avoid much of the effect of aging-rate differences that separate the curves of Fig. 2 and obtain a more compact representation. Time remains concealed in the passage from the wet state at the left, along the shaded area and toward the right; the rate of progression along the path differs with different varnishes and different thicknesses.

Data such as these do not, of course, tell the whole story. For example, a product whose consistency does not permit the application of a thin, even coat must be judged bad regardless of its excellent showing in terms of elastic constants.

Various conclusions concerning the elastic constants of varnish follow:

● Since density of varnish is near unity, the real component E' for dry violin varnish films is in the order of 1.0 to 2.5×10^{10} dyn/cm². This is about the same as the modulus for quarter-cut spruce across the grain, but much less than that along the grain.

● Characteristically, the real component of Young's modulus increases with aging, and a decrease, if it occurs, indicates change in ambient. After $1\frac{1}{2}$ yr, the

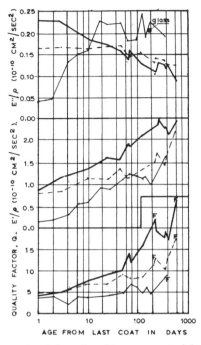

FIG. 2. Aging of three oil varnishes over an extended period: the middle curves are for the real components of Young's modulus over density, the top curves are for imaginary component and the bottom curves for their ratio Q. Heavy lines for a hard varnish, light lines for a soft varnish, broken lines for a tung-oil floor varnish. "F" means February.

hard varnish attained an E'/ρ of about 2.5×10^{10}; after 1 yr, the soft one reached 1.6×10 cm²/sec².

● The imaginary component of Young's modulus characteristically rises early in the drying cycle, levels off, and then undergoes a slow decline. During the decline, its fractional change is less than the accompanying rise in the real component. After $1\frac{1}{2}$ yr, the E''/ρ of the hard varnish was down to about 0.11 cm²/sec²; after 1 yr, the soft one approximated 0.19×10^{10} cm²/sec².

● The Q's of varnishes studied, excluding shellac, ranged up to about 25 (logarithmic decrement down to 0.12). With fixed ambient conditions, the Q always increased with aging. After $1\frac{1}{2}$ yr, the hard ones had a Q of about 23; after 1 yr, the soft one stood at about 8.5 but may not yet have become stable.

● Twelve hours after the varnishing, the E''/ρ of shellac was the highest observed in these tests. Seven months later, E''/ρ was the lowest observed and E'/ρ the highest, and the Q approached that in the substrate, in the order of 50. This extreme condition occurred in the dry winter months; the following summer, it behaved much like the other varnishes. Since temperature

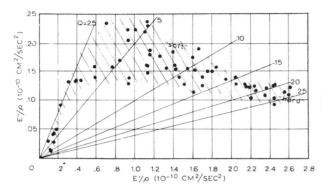

FIG. 3. Mass plot for four varnishes, including those of Fig. 2. All followed the same shaded path from wetness near origin to some terminal condition at the right.

in the laboratory differed little in the two seasons, the implication is that water content in the film follows that in the ambient to some extent. The same accelerated drying in winter was found for the other varnishes, but the recovery in summer was less pronounced. Whatever the reason for these changes, the conditions in the laboratory were throughout the year representative of the conditions in which violins are normally made and used.

● It is not apparent in what way, if any, the tung-oil floor varnish is inferior to the violin varnishes, since its elastic moduli fall between those of the hard and soft types. It may be undesirable in some nonacoustical way.

● Using methods basically the same as those employed here, Meinel[4] found that varnish changed decrement of his wood strips only slightly at a few hundred cps but increased it considerably at 5000 cps, the increase being greater with a soft varnish than with a hard one. I made one test with the soft varnish on glass; this avoided capriciousness in the substrate. After the varnish had aged for 6 mo, the decrement of the strip was measured at five frequencies from 400 to 5600 cps; the varnish was then removed, and readings were immediately repeated. Changes in decrement of the varnish were found to be independent of frequency within experimental error, the largest difference from the average being 7%. The loss modulus over density is indicated on Fig. 2 by the point labeled "glass." Other measurements on wood with harder varnishes, though not as clear-cut, indicate the same conclusion. Differences in varnish may be enough to reconcile the difference in results at the high frequencies, but it is difficult to understand the negligible decrements observed by Meinel at low frequencies.[3]

● While the differences in elastic constants of the varnishes tested are sufficient to produce significant differences when applied to an instrument, they are not as great as lore would lead one to expect. Perhaps not enough types were tested. However, if varnish degrades an instrument, one cannot prove the material at fault unless it is known, preferably by weighing, that the amount used was not excessive.

II. ESTIMATION OF ACOUSTICAL EFFECTS

Estimating the effect of varnish on a violin is beset with many uncertainties owing to the complicated nature of the vibrations and the mathematically bizarre shape of the instrument. It is possible, however, to contrive a simple substitute for which a calculation can be made with considerable confidence. Such a "pseudofiddle" may consist of rectangular plates of spruce and maple assumed perfect in uniformity, each plate having the area and approximate thickness of the violin plate and a ratio of length to width of 2.3, a presumed "average" ratio in the violin. Grain runs lengthwise and the edges are "supported," not "clamped." The deviations from the real instrument—absence of sound post and bass bar, etc.,—are obvious. It is important to recall, however, that the elastic properties of wood and varnish are independent of frequency in the lower octaves. The lowering of frequency of resonance caused by absence of arch and soundpost will therefore have little effect on the dimensionless quantities—decrement and semitones shift—provided that the thicknesses of the plates have been chosen in relation to the thickness of varnish.

Attention is confined to what might be called the principal resonance of the box, namely, that in which the two plates vibrate in the (1,1) mode of a "supported" rectangle, thickness having been chosen to make their resonance frequencies identical. The shapes of vibration of the plates being the same in both dimensions (sinusoidal), the assumption that their edges are immobile, which is implicit in the assumption that they are "supported," is satisfied if their velocities are inversely proportional to their masses and are opposed in direction so that the resultant forces on the ribs cancel. In regard to radiation, it approximates a simple source or what the German writers call a "Nullstrahler."

The method, mathematical details of which are given in Appendix A, consists in (1) extending the standard expression for the vibration of a supported isotropic rectangle to apply to an orthotropic one in which the principal axes of elasticity are parallel to the edges and in which the rigidity coefficient D_{xy} equals $(D_x D_y)^{\frac{1}{2}}$, where D_x is the rigidity coefficient in the x direction and D_y is that in the y direction[11]; (2) recognition of the fact that Young's modulus as it appears in the expression for frequency has a small imaginary component leading to damping; and (3) deriving therefrom the frequency and decrement following standard procedures. Finally, (4) varnishing the wood may be regarded as, in effect, altering the elastic properties of the substrate, and leading to the changes in resonance frequency and damping in which we are interested.

If it were not for penetration into the wood, the measurement of thickness of varnish on test strips would be a simple matter. The equivalent penetration differs with varnish, that of our interest being in the order of a thousandth of an inch in the absence of a sealer. This difficulty is not important, however, since measurement of change in weight is at least as significant and far more practical in the violin itself. For that reason, amount of varnish is here expressed in terms of grams per square centimeter.

Of the assumptions made, the one most likely to be questioned is that a ratio of length to width of 2.3 simulates the violin. The smaller we make this ratio, the more relatively does the Q of the plate depend on the modulus along the grain and the higher it becomes. However, the measured Q of violins corresponds with my assumptions reasonably well. The thickness chosen —3 mm—is midway between thickness commonly found near edge and that in middle of the top plate.

Results of such a computation are shown in Fig. 4. Figure 4(a) is derived from Fig. 3 by drawing a median line through the crosshatched region. Point A is believed to represent a medium hard varnish; B, a very hard varnish; and C, a soft one.

Decrements of the spruce and maple pseudofiddle plates are plotted in Fig. 4(b). It is purely accidental that when unvarnished, they show exactly the same decrement. (Constants used are substantially those for two excellent woods, a spruce from West Germany and a maple from Czechoslovakia.)

Figure 4(c) shows the over-all effect of the varnish in terms having direct significance to the luthier. The dotted line shows the reduction in output of sound produced by increased weight and stiffness, that is, increased impedance: it is the "flat loss" independent of frequency mentioned earlier. To it is added the increase caused by the internal friction, most in evidence at resonances but of little effect between them. The combined effect appears in the curve labeled "total at peaks." In the calculation, radiation has been taken

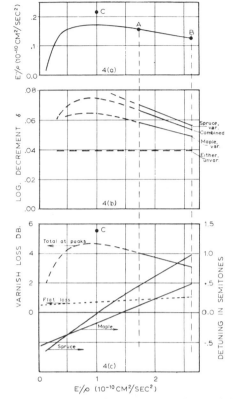

FIG. 4. Calculated results for the pseudofiddle: (a) varnish data used; (b) logarithmic decrements for plates separately and assembled with and without varnish; (c) detuning of separate plates by varnish and varnish losses in the assembly.

into account, and this curve for total loss, therefore, tells what a sound-meter would show. The 5.5-dB loss for soft Varnish C obviously makes the weight of varnish assumed excessive. Averaged over valleys as well as peaks, the loss would of course be much less. Hard Varnish B has about half this loss. In respect to detuning, however, the curves for the separate plates show the soft one to be entirely in the clear, since increased stiffness has canceled the effect of increased weight. The hard varnish, however, may be in trouble because of detuning, since the resonance for spruce has been raised a semitone. In a violin, this might be mitigated to some extent by the fact that the maple follows part way.[12]

---

[11] Manfred A. Heckl, *Compendium of Impedance Formulas* (Armed Serv. Tech. Inform. Agency, Arlington, Virginia, 1961), pp. 17 and 19. The statement is made that for plates made of nonisotropic material, etc., it is true in most practical cases that this equality is approximated. Failure to hold in wood seems unlikely to alter seriously the present estimates.

[12] The thickness of maple back required to tune to the same frequency as the spruce top turned out to be nearly the same as that of the spruce. The reason is that in spite of lower Young's modulus along the grain and greater density, both tending to lower frequency, the maple has a modulus across the grain that is 2.5 times that of spruce. Along with greater stiffness undoubtedly goes greater ultimate strength, making maple better suited than a coniferous wood in withstanding stress from string tension and giving a firm footing for the sound post.

In making the foregoing calculations, the relative motions of the plates had to be taken into account. From the previous assumptions, it follows that in the (1,1) mode, their momenta are equal and opposite and their energies inversely as their masses. Numerically, this ratio is 1.65 for the unvarnished plates. For the violin, data reported by Meinel[13] for the principal resonance indicate a velocity ratio near 1.6 or in the same order, although closeness of agreement is undoubtedly accidental.

Conclusions concerning acoustical effects may be summarized as follows. Unless otherwise stated, numerical results refer to Varnish A weighing 0.0127 gm/cm². Qualitatively, I believe them true of both pseudofiddle and violin; quantitatively, they seem in keeping with experience.

● Neither the observations by Meinel nor the present measurements confirm the popular view of violin varnish as a major contributor to acoustical excellence. Small benefits can occur if the wood used is really overly resonant or if damping by varnish mounts dramatically above 4000 or 5000 cps, but of neither of these suppositions has the generality been established.

● The impedance that the body of a fiddle offers to forces exerted by the strings approximates about a given resonance more or less that of a simple mechanically resonant circuit of stiffness S, mass M, quality factor Q, and numerical reactance X, which passes through zero at resonance; this impedance may be written $(SM)^{\frac{1}{2}} \cdot (1/Q + jX)$. Increase in $(SM)^{\frac{1}{2}}$ decreases motion in the radiating plates more or less uniformly throughout the spectrum, while decrease in Q superposes additional reduction at resonance. For the pseudofiddle, the flat loss caused by impedance increase following application of this varnish is about 1.0 dB. *There can be no doubt that in a violin this effect is undesirable.* Such a coat would increase the mass of the violin about 18 gm.

● Added to this, at resonance peaks, is loss from internal friction. Assuming the same radiation decrement as in a violin and taking as reasonable for an outstanding instrument in the white a radiation efficiency of 35%, we have the decrements shown in Table II for the pseudofiddle at resonance. Corresponding to these decrements, therefore, the loss produced by varnish at resonance is 3.0 dB in addition to the 1.0 dB of flat loss. By way of comparison, the average of 10 decrements for body resonances in fine old violins as measured by F. A. Saunders[14] is 0.099; for 18 fine new violins, 0.090. In the total decrement arrived at here, which is in the same order, an abnormally high varnish loss is probably balanced by a wood loss lower than usual. However, we would not be justified in expecting close agreement.

[13] H. Meinel, Elek. Nachr.–Tech. 4, 119–134 (1937).
[14] F. A. Saunders, J. Acoust. Soc. Amer. 17, 169–186 (1945). Table IX.

TABLE II. Decrements for the pseudofiddle at resonance.

Wood, calculated	0.040
Varnish, calculated	0.026
Radiation	0.022
Total	0.088

● In the calculation, varnish increases the decrement of the top by 0.031, of the back by 0.018. Energy in the first being greater by a factor of about 1.65, *the energy loss per cycle in the top exceeds that in the back by a factor of 0.65×0.031/0.018 or nearly three times regardless of the description of the varnish coat. If one wishes to avoid varnish effects, he naturally reduces varnish on the top to the point of austerity, and is sparing on the back.*

● The opposing effects of added mass and stiffness produce a negligible change in the frequency in the maple (an increase of about 0.5%). In the spruce, it is about a quarter tone. Heavier coats of a stiffer varnish could bring this up to serious detuning of top plate.

● With test strips of the spruce assumed here cut, respectively, along and across the grain, the addition of varnish would change the ratio of stiffness from 12.5 to 10.0.

● Answers to these questions may be unnecessary if quantity of varnish is kept to the minimum needed for protection. For example, *if it is feasible to reduce varnish on the top to one third that on the back, losses in top and back will near equality, total loss will be halved, frequency difference between plates caused by varnish will be substantially eliminated and change in "velocity ratio" reduced. The varnished violin will sound nearly like the violin in the white.*

● A good method of controlling varnish is by weighing before and after varnishing. A coat such as that assumed would weigh about 18 gm, a light coat one quarter to one half as much. (A violin including nonvibrating parts weighs around 400 gm.)

● Final tuning of the plates could be postponed to advantage until just before the final finishing coat is applied. This involves new techniques in varnishing and possible revision of rules for tuning.

● Since, with existing procedures, frequency of the top is raised by varnish with respect to the back, some protection may be obtained by tuning the top slightly flatter while in the white than would be best with unvarnished instruments.

● Other instruments of the family, e.g., the cello, tolerate heavier coats than the violin in proportion to the greater thickness of their plates.

468

III. CONCLUSION

The damping effects of varnish and the proper thickness of the coat are not matters unrelated to properties of the wood. The practice, if not the philosophy, of varnishing seems often to be to obtain wood of such low loss that heavy coats of varnish may be applied with impunity for cosmetic purposes without concern for musical damage. Up to a point, logarithmic decrement is necessary, but the skilled luthier obtains it as far as possible as radiation decrement, a gain in tonal volume,

rather than a loss of energy in varnish. The "secret of varnish," therefore, may well be a secret of varnishing techniques for providing adequate but not excessive protection and acceptable appearance with the least possible material.

ACKNOWLEDGMENT

It is a pleasure to acknowledge my indebtedness to the following persons for help and advice in various ways: R. E. Fryxell, C. M. Hutchins, Melvin Mooney, Mrs. F. A. Saunders, and C. E. Schelleng.

Appendix A: Varnished Orthotropic Plate

Subject to the assumption made above,[11] frequencies and decrements may be calculated following standard methods. Instead of flexural rigidity being proportional to $E/(1-\mu^2)$ as in isotropic material, the corresponding factors in violin wood are $E_x/(1-\mu_{LR}\mu_{RL})$ and $E_y/(1-\mu_{RL}\mu_{LR})$, where L and R refer to longitudinal and radial coordinates of the tree trunk. From measurements for kiln-dried Sitka spruce,[A1] the product of these two Poisson ratios is approximately 0.01 and negligible in the present context. Subject to these conditions, Huber's equation gives angular frequency, as follows, for a supported rectangular plate with edges parallel to the x- and y axes, which lie, respectively, along and perpendicular to the grain,

$$\omega = K(U_x + U_y), \qquad (A1)$$

where

$$K = \pi^2 H/(2[3\rho]^{\frac{1}{2}}),$$
$$U_x = E_x^{\frac{1}{2}}(m/a)^2,$$

and

$$U_y = E_y^{\frac{1}{2}}(n/b)^2,$$

and m and n are positive integers. Taking $E = E' + jE''$, as usual, substituting in Eq. A1, taking into account the fact that in good violin wood the Q is high in both orientations, and writing $E'/E'' = Q$, one obtains

$$\omega = \omega' + j\omega''$$
$$= K[(U_x' + U_y') + \tfrac{1}{2}j(U_x'/Q_x + U_y'/Q_y)], \quad (A2)$$

[A1] J. T. Drow and R. S. McBurney, *Elastic Properties of Wood (Sitka Spruce)*, Forest Products Lab., U. S. Dept. Agriculture, No. 1528-A, (Nov. 1954) (reaffirmed June 1959), Table 5.

and

$$Q_{\text{plate}} = \frac{\tfrac{1}{2}\omega'}{\omega''} = \frac{(U_x' + U_y')}{(U_x'/Q_x + U_y'/Q_y)}.$$

The equations given do not relate directly to a plate that is a composite of varnish and wood, but they may be used by replacing the plate with one of the same thickness, whose density is increased in the same proportion as the total mass, whose moduli E' are increased in proportion to the corresponding total stiffness along or across the grain, as the case may be, and whose Q's are correspondingly altered.

In this way, the new parameters become

$$|\rho| = \rho + \sigma H, \quad \sigma = \Delta M/A, \qquad (A3)$$

where ΔM is the mass of the varnish and A the varnished area.

$$|E'|_w = E_w' + 3[E'/\rho]_v \cdot \sigma/H, \qquad (A4)$$

where w is x or y depending on orientation. Equations A3 and A4 may be substituted in Eq. A1 to obtain the new frequency.

The new longitudinal and transverse Q's needed in Eq. 12 are

$$|Q_w| = \frac{Q_w[1 + \Delta s/s]}{[1 + (\Delta s/s)(Q_w/Q_v)]}, \qquad (A5)$$

where $\Delta s/s = 3[E'/\rho]_v \cdot \sigma/HE_w'$, in which the wood parameters are taken in suitable orientation.

469

Author Citation Index

Subject Index